Polymer Films for Photovoltaic Applications

Polymer Films for Photovoltaic Applications

Editor

Bożena Jarzabek

MDPI • Basel • Beijing • Wuhan • Barcelona • Belgrade • Manchester • Tokyo • Cluj • Tianjin

Editor
Bożena Jarzabek
Laboratory of Functional
Materials Engineering
Centre of Polymer and
Carbon Materials
Polish Academy of Sciences
Zabrze
Poland

Editorial Office
MDPI
St. Alban-Anlage 66
4052 Basel, Switzerland

This is a reprint of articles from the Special Issue published online in the open access journal *Polymers* (ISSN 2073-4360) (available at: www.mdpi.com/journal/polymers/special_issues/poly_films_photovol_appli).

For citation purposes, cite each article independently as indicated on the article page online and as indicated below:

LastName, A.A.; LastName, B.B.; LastName, C.C. Article Title. *Journal Name* **Year**, *Volume Number*, Page Range.

ISBN 978-3-0365-4676-6 (Hbk)
ISBN 978-3-0365-4675-9 (PDF)

© 2022 by the authors. Articles in this book are Open Access and distributed under the Creative Commons Attribution (CC BY) license, which allows users to download, copy and build upon published articles, as long as the author and publisher are properly credited, which ensures maximum dissemination and a wider impact of our publications.

The book as a whole is distributed by MDPI under the terms and conditions of the Creative Commons license CC BY-NC-ND.

Contents

About the Editor vii

Preface to "Polymer Films for Photovoltaic Applications" ix

Paweł Gnida, Muhammad Faisal Amin, Agnieszka Katarzyna Pajak and Bożena Jarzabek
Polymers in High-Efficiency Solar Cells: The Latest Reports
Reprinted from: *Polymers* **2022**, *14*, 1946, doi:10.3390/polym14101946 1

Luke Jonathan, Lina Jaya Diguna, Omnia Samy, Muqoyyanah Muqoyyanah, Suriani Abu Bakar and Muhammad Danang Birowosuto et al.
Hybrid Organic–Inorganic Perovskite Halide Materials for Photovoltaics towards Their Commercialization
Reprinted from: *Polymers* **2022**, *14*, 1059, doi:10.3390/polym14051059 41

Leland Weiss and Tyler Sonsalla
Investigations of Fused Deposition Modeling for Perovskite Active Solar Cells
Reprinted from: *Polymers* **2022**, *14*, 317, doi:10.3390/polym14020317 69

Yiping Guo, Zeyang Li, Mengzhen Sha, Ping Deng, Xinyu Lin and Jun Li et al.
Synthesis of a Low-Cost Thiophene-Indoloquinoxaline Polymer Donor and Its Application to Polymer Solar Cells
Reprinted from: *Polymers* **2022**, *14*, 1554, doi:10.3390/polym14081554 83

Ahmed N. M. Alahmadi
Design of an Efficient PTB7:PC70BM-Based Polymer Solar Cell for 8% Efficiency
Reprinted from: *Polymers* **2022**, *14*, 889, doi:10.3390/polym14050889 93

Bożena Jarzabek, Paweł Nitschke, Marcin Godzierz, Marcin Palewicz, Tomasz Piasecki and Teodor Paweł Gotszalk
Thermo-Optical and Structural Studies of Iodine-Doped Polymer: Fullerene Blend Films, Used in Photovoltaic Structures
Reprinted from: *Polymers* **2022**, *14*, 858, doi:10.3390/polym14050858 105

Gabriela Lewińska, Piotr Jeleń, Jarosław Kanak, Łukasz Walczak, Robert Socha and Maciej Sitarz et al.
Investigation of Dye Dopant Influence on Electrooptical and Morphology Properties of Polymeric Acceptor Matrix Dedicated for Ternary Organic Solar Cells
Reprinted from: *Polymers* **2021**, *13*, 4099, doi:10.3390/polym13234099 123

Ravindra Kumar Gupta, Hamid Shaikh, Ahamad Imran, Idriss Bedja, Abrar Fahad Ajaj and Abdullah Saleh Aldwayyan
Electrical Transport, Structural, Optical and Thermal Properties of [$(1-x)$Succinonitrile: xPEO]-LiTFSI-Co(bpy)$_3$(TFSI)$_2$- Co(bpy)$_3$(TFSI)$_3$ Solid Redox Mediators
Reprinted from: *Polymers* **2022**, *14*, 1870, doi:10.3390/polym14091870 141

Gernot M. Wallner, Baloji Adothu, Robert Pugstaller, Francis R. Costa and Sudhanshu Mallick
Comparison of Crosslinking Kinetics of UV-Transparent Ethylene-Vinyl Acetate Copolymer and Polyolefin Elastomer Encapsulants
Reprinted from: *Polymers* **2022**, *14*, 1441, doi:10.3390/polym14071441 157

Marilena Baiamonte, Claudio Colletti, Antonino Ragonesi, Cosimo Gerardi and Nadka Tz. Dintcheva
Durability and Performance of Encapsulant Films for Bifacial Heterojunction Photovoltaic Modules
Reprinted from: *Polymers* **2022**, *14*, 1052, doi:10.3390/polym14051052 171

João P. Cachaneski-Lopes and Augusto Batagin-Neto
Effects of Mechanical Deformation on the Opto-Electronic Responses, Reactivity, and Performance of Conjugated Polymers: A DFT Study
Reprinted from: *Polymers* **2022**, *14*, 1354, doi:10.3390/polym14071354 185

Manuel Hohgardt, Franka Elisabeth Gädeke, Lucas Wegener and Peter Jomo Walla
A Refined Prediction Parameter for Molecular Alignability in Stretched Polymers and a New Light-Harvesting Material for AlGaAs Photovoltaics
Reprinted from: *Polymers* **2022**, *14*, 532, doi:10.3390/polym14030532 209

Fahad Mateen, Namcheol Lee, Sae Youn Lee, Syed Taj Ud Din, Woochul Yang and Asif Shahzad et al.
Thin-Film Luminescent Solar Concentrator Based on Intramolecular Charge Transfer Fluorophore and Effect of Polymer Matrix on Device Efficiency
Reprinted from: *Polymers* **2021**, *13*, 3770, doi:10.3390/polym13213770 227

Arief Suriadi Budiman, Rahul Sahay, Komal Agarwal, Gregoria Illya, Ryo Geoffrey Widjaja and Avinash Baji et al.
Impact-Resistant and Tough 3D Helicoidally Architected Polymer Composites Enabling Next-Generation Lightweight Silicon Photovoltaics Module Design and Technology
Reprinted from: *Polymers* **2021**, *13*, 3315, doi:10.3390/polym13193315 237

Saif M. H. Qaid, Hamid M. Ghaithan, Khulod K. AlHarbi, Bandar Ali Al-Asbahi and Abdullah S. Aldwayyan
Enhancement of Light Amplification of $CsPbBr_3$ Perovskite Quantum Dot Films via Surface Encapsulation by PMMA Polymer
Reprinted from: *Polymers* **2021**, *13*, 2574, doi:10.3390/polym13152574 253

Paweł Nitschke, Bożena Jarząbek, Marharyta Vasylieva, Marcin Godzierz, Henryk Janeczek and Marta Musioł et al.
The Effect of Alkyl Substitution of Novel Imines on Their Supramolecular Organization, towards Photovoltaic Applications
Reprinted from: *Polymers* **2021**, *13*, 1043, doi:10.3390/polym13071043 269

Ahmed. N. M. Alahmadi and Khasan S. Karimov
A Novel Poly-N-Epoxy Propyl Carbazole Based Memory Device
Reprinted from: *Polymers* **2021**, *13*, 1594, doi:10.3390/polym13101594 287

About the Editor

Bożena Jarzabek

Prof. dr Bożena Jarzabek is an Institute Professor at the Centre of Polymer and Carbon Materials Polish Academy of Sciences in Zabrze (Poland). Main interests include:

Organic materials and conjugated polymers; thin films and nanotechnology; materials characterization; optical spectroscopy; absorption edge parameters; electronic transitions; thermo-optical properties; doping; and polymer films in optoelectronic structures.

Preface to "Polymer Films for Photovoltaic Applications"

We present a reprint of the Special Issue entitled "Polymer films for photovoltaic applications", which focuses on polymer thin films and polymer blend films in photovoltaic (PV) structures. Organic materials are widely used in optoelectronic devices (e.g., organic solar cells (OSCs)), mainly due to the low cost of production (thin films may be deposited at low temperature onto a large surface, flexible substrates, etc.). However, polymer compounds in OSC systems also suffer from certain limitations, such as low efficiency and the short lifetime of devices, resulting from an insufficient thermal and time stability. Polymers, which may be utilized in PV systems, should exhibit the appropriate optical properties (e.g., a wide range of absorption and low energy gap), good durability and stability (not undergoing any phase transitions or degradation in the temperature range in which the system is working), and relevant electronic structure (good alignment of molecular orbitals of donor and acceptor compounds in bulk heterojunction (BHJ) organic solar cells).

Bożena Jarzabek
Editor

Review

Polymers in High-Efficiency Solar Cells: The Latest Reports

Paweł Gnida [1,*], Muhammad Faisal Amin [1], Agnieszka Katarzyna Pająk [2] and Bożena Jarząbek [1,*]

[1] Centre of Polymer and Carbon Materials, Polish Academy of Sciences, 34 M. Curie-Sklodowska Str., 41-819 Zabrze, Poland; mfaisal-amin@cmpw-pan.edu.pl
[2] Institute of Chemistry, University of Silesia, Szkolna 9, 40-006 Katowice, Poland; agpajak@us.edu.pl
* Correspondence: pgnida@cmpw-pan.edu.pl (P.G.); bjarzabek@cmpw-pan.edu.pl (B.J.)

Abstract: Third-generation solar cells, including dye-sensitized solar cells, bulk-heterojunction solar cells, and perovskite solar cells, are being intensively researched to obtain high efficiencies in converting solar energy into electricity. However, it is also important to note their stability over time and the devices' thermal or operating temperature range. Today's widely used polymeric materials are also used at various stages of the preparation of the complete device—it is worth mentioning that in dye-sensitized solar cells, suitable polymers can be used as flexible substrates counter-electrodes, gel electrolytes, and even dyes. In the case of bulk-heterojunction solar cells, they are used primarily as donor materials; however, there are reports in the literature of their use as acceptors. In perovskite devices, they are used as additives to improve the morphology of the perovskite, mainly as hole transport materials and also as additives to electron transport layers. Polymers, thanks to their numerous advantages, such as the possibility of practically any modification of their chemical structure and thus their physical and chemical properties, are increasingly used in devices that convert solar radiation into electrical energy, which is presented in this paper.

Keywords: photovoltaics; dye-sensitized solar cells; bulk-heterojunction solar cells; perovskite solar cells; polymers; thin layers

1. Introduction

Given the ever-increasing demand for electricity and environmental pollution, new, efficient, and, very importantly, environmentally friendly sources of renewable energy are being sought. However, it is worth remembering that a very large amount of electricity is still obtained from fossil fuels; thus, to find alternatives to non-renewable fuels, very efficient and relatively cheap sources of green energy are needed. One of the fastest-growing branches of renewable energy sources is solar energy, specifically photovoltaics. It is worth remembering that solar energy can be used in two ways, not only by photovoltaic cells but also by solar collectors for, among other things, heating [1]. Solar cells are divided into three generations. The first generation is made up of crystalline silicon cells, which are currently the most commercially used [2]. Thin-film devices based on the CdTe, CIGS (Copper Indium Gallium Selenide), GaAs, and a-Si, among others, represent the second generation [3]. Currently, the third generation of solar cells, which includes dye-sensitized solar cells (DSSC), perovskite solar cells (PSC), and bulk-heterojunction solar cells (BHJ), among others, is the most widely researched and rapidly developed [4–6]. Of course, as it is well known, currently, the highest solar-to-electricity conversion efficiencies are demonstrated by multi-junction solar cells overcoming the 47% threshold. However, these devices are very expensive to prepare, and their processes are extremely difficult and complicated, which significantly limits the possibility of their commercial application [7]. The use of organic compounds in photovoltaic devices offers great opportunities through a wide range of possibilities to modify the chemical structure of these compounds and, consequently, change their physical and chemical properties. In addition, it is also worth mentioning their significantly lower production costs, less energy consumption, and simpler preparation methods [8–11]. Research in recent years has confirmed a significant increase in the energy conversion efficiency of third-generation cells. According to the literature reports, perovskite solar cells achieve efficiencies of over 25% [12], DSSCs over 14% [13,14] and BHJs around 18% [15]. Furthermore, a great advantage of organic compounds is that they can be applied to various substrates using various methods. Numerous attempts have been made with often positive results to prepare flexible photovoltaic cells with polymer substrates. Both BHJ [16–18], PSC [19–21] and DSSC [22–24] structured devices are widely used for the preparation of flexible solar cells when new methods of preparing and applying materials to polymer substrates are sought.

In recent years, huge interest in using new polymeric materials in organic photovoltaics (OPV) has emerged. In each of these three types of the third generation of solar cells, polymeric materials find a variety of very important applications. Considering each of the components of solar cells, one can multiply the associated application of polymer materials.

The layers of polymeric materials play different roles starting with dye-sensitized solar cells, which are characterised by their layered structure. The first layer of a DSSC is the substrate. The glass or polymer substrate is covered with a transparent conductive oxide (TCO); very often, fluorine-doped tin oxide (FTO) or indium tin oxide (ITO) are used. However, research on flexible photovoltaic cells is increasingly reported in the literature. The most common flexible substrates used in photovoltaics are made of polymers such as polyethylene naphthalate (PEN) or polyethylene terephthalate (PET) [22,23,25–29]. Subsequently, polymers are used as materials responsible for forming the porous structure of a semiconducting oxide layer, e.g., TiO_2. For this purpose, polymers or copolymers such as polystyrene (PS), polyvinylpyrrolidone (PVP), P123 Pluronic (PEO_{20}–PPO_{70}–PEO_{20}) is a triblock copolymer, copolymer of poly(vinyl chloride) (PVC) and poly(oxyethylene methacrylate) (POEM) [30–33]. Polymers are also widely used in the electrolyte layer. Currently, DSSCs containing a liquid electrolyte containing redox pair are very widely used; however, due to quite significant limitations resulting from the use of a liquid electrolyte, quasi-solid state DSSCs (qs-DSSCs) and solid-state DSSCs (ss-DSSC) are being developed. The polymers used include polyacrylonitrile (PAN), polyethylene oxide (PEO) or poly(ethylene glycol) (PEG) [34–36]. The last layer forming the DSSC is the counter-electrode, which is now often made of polymeric materials or composites thereof. These materials significantly reduce the cost compared to a platinum electrode. Common polymers used are polyaniline (PANI), polypyrrole (PPy), polythiophene (PTh) and its derivatives [37–41].

Polymeric compounds are also widely used in the BHJ solar cells, where the active layer is a mixture of donor and acceptor (D-A) materials. π-conjugated polymers are the most commonly used as electron-donor materials in the active layer, among others [42–46]. However, they are present less frequently as an acceptor among others [47–49]. Furthermore, polymers are also used as buffer layers to form a barrier between the active layer and the electrode. This barrier impedes the fast charge transfer, which leads to serious charge accumulation at the contacts. The charge accumulation increases the probability of recombination and deteriorates the performance of the device. It is, therefore, necessary to speed up the charge separation process and increase the transfer efficiency.

Polymers are also being investigated for use in perovskite solar cells due to their diverse characteristics. To improve the nucleation and crystallization processes in the perovskite layer(s), polymers are added as additives [50–53]. Because of their proper charge mobility and energy level organization, polymers can also be utilized as electron and hole-transporting materials [54–60], as well as an interface layer, to prevent recombination and improve carrier separation efficiency [61–64].

Herein, we present the latest reports on polymeric materials used in photovoltaic solar cells. In our paper, three types of solar cells: dye-sensitized, bulk heterojunction and perovskite solar cells, are presented in three successive chapters, where the role of polymers and polymers thin films are described and discussed. Based on the latest literature reports, the photovoltaic parameters, such as open circuit voltage (V_{oc}), short-circuit current (J_{sc}), fill factor (FF) and power conversion efficiency (PCE) are gathered and compared for these types of solar cells.

2. Polymers in Dye-Sensitized Solar Cells (DSSCs)

DSSCs, increasingly studied worldwide, show an ever-increasing radiation conversion efficiency. A distinction is made between solar cells containing a liquid or gel electrolyte of a similar design and a solid electrolyte. Research on DSSCs began with the use of a liquid electrolyte, and it is this type of cell that is most commonly reported in the literature. This is mainly due to the fact that the preparation of a device with this structure is the least complicated and time-consuming and is ideal when testing new dyes. Additionally, the use of liquid electrolyte, although it improves the photovoltaic parameters of the device, has many disadvantages, among which we can specify the limitation of the temperature range of the device, problems with proper sealing of the solar cell and causing corrosion of materials. Hence, the idea of replacing the liquid electrolyte with a solid or gel electrolyte emerged. When using these two types of electrolytes, it is difficult to achieve the same high parameters as with a liquid electrolyte. Still, numerous reports in the literature emerge to suggest this is being achieved. Replacing the liquid electrolyte, especially with a gel electrolyte, significantly increases the stability of the device over time and the charge mobility is definitely higher than with solid electrolytes. As also mentioned in the introduction, polymers are increasingly used in solar cells with DSSC structures. The following subsections describe the roles they play in them, including polymeric dyes, which are very difficult to find in other review papers. The structures of these devices are shown in Figure 1a, while the chemical structures of polymers used in this type of solar cell are presented in Figure 1b.

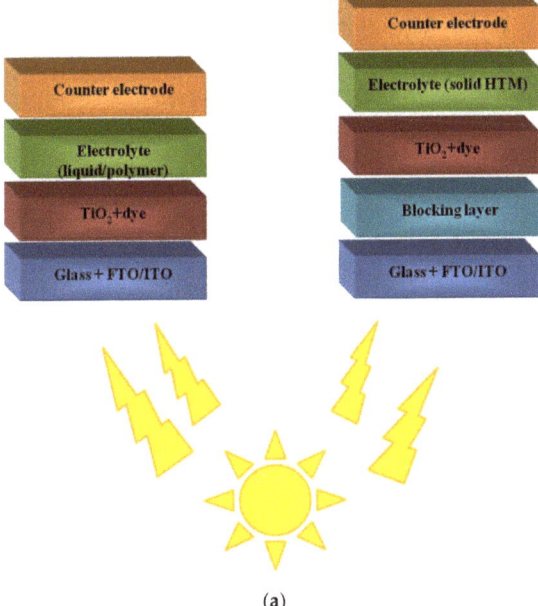

(a)

Figure 1. *Cont.*

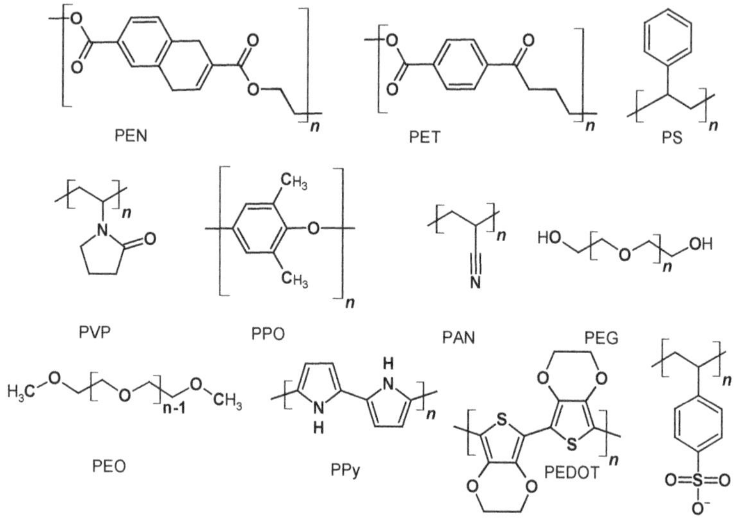

Figure 1. The structures of DSSCs with liquid and solid electrolyte (**a**) and polymer structures often used in this type of solar cell (**b**).

2.1. Polymers as a Flexible Substrates

The commercial application of DSSCs is currently being intensively developed, and they are often used in glass façades that can additionally generate electricity. However, flexible substrates are increasingly being used to be able to make the whole device more flexible and thus greatly expand the application possibilities of this type of solar cell. In laboratory work, fluorine-doped tin oxide coated glass is used, which is being replaced by polymeric materials such as polyethylene terephthalate or polyethylene naphthalate deposited on ITO [22,65,66]. The previously mentioned ITO-embedded polymers are the most commonly used due to their numerous advantages, such as high transparency, low preparation costs, ability to form the required shapes, low weight, flexibility and low resistance [67]. When using flexible substrates, it is very important to prepare a conductive oxide layer such as TiO_2 in a low-temperature manner in the region of 120–150 °C [66]. The resulting parameters for flexible DSSC, such as open-circuit voltage, short-circuit current, fill factor and power conversion efficiency over the last few years, are summarised below, in Table 1. The commercial dye di-tetrabutylammoniumcis-bis(isothiocyanato)bis(2,2′-bipyridyl-4,4′-dicarboxylato) ruthenium(II), denoted as N719, was used to prepare the devices.

Table 1. Photovoltaic parameters of flexible dye-sensitized solar cells, containing dye N719.

Substrate	V_{oc} (mV)	J_{sc} (mA/cm^2)	FF (-)	PCE (%)	Ref.
ITO/PEN	660	8.97	0.45	2.65	
ITO/PEN + G LSL	680	10.62	0.48	3.47	
ITO/PEN + T LSL	690	14.65	0.43	4.33	[26]
ITO/PEN + TG LSL	680	14.32	0.53	5.18	
ITO/PEN (100 mW/cm^2)	400	8.70	0.46	2.60	
ITO/PEN (18 mW/cm^2)	380	1.8	0.53	3.30	[68]
ITO/PET (Pt CE)	685	5.53	0.78	2.95	
ITO/PET (PEDOT:PSS CE)	695	6.09	0.52	2.18	[69]

T—titanium dioxide, G—grapheme, TG—titanium dioxide-graphene, LSL—light-scattering layer.

In [26], the preparation method for flexible photoanodes using the polyethylene naphthalate deposited on ITO was reported. The preparation of titanium dioxide-graphene quantum dot (TG) was used as a light-scattering layer (LSL) using facile electrodeposition and drop-casting. The photoanode denoted as ITO/PEN + T LSL was prepared by the facile electrodeposition from an aqueous solution of 0.1 M titanium tetraisopropoxide in 0.1 M LiClO$_4$. The photoanodes containing ITO/PEN + G LSL and ITO/PEN + TG LSL were prepared by the drop-casting 5 µL solution containing 1 mg/mL GQD onto ITO/PEN and photoanode + T LSL. The highest PCE shown the photoanode + TG LSL (5.18%). The paper [68] presents the possibility of preparing flexible cells with a PEN substrate, which at standard illumination (100 mW/cm^2) reached efficiencies of 2.6%, while at a lower intensity (18 mW/cm^2) the efficiency of the devices increased to a value of 3.3%. Fu et al. [69] prepared flexible solar cells and tested the influence of platinum and polymer counter-electrode. The substrate on which the dye was anchored was ITO on PET covered with a TiO$_2$ layer. The use of a polymer electrode resulted in an increase in V$_{oc}$ and J$_{sc}$ but higher resistances due to a decrease in FF. Finally, the Pt counter-electrode showed an efficiency of 2.95%, while the solar cell with polymer CE showed 2.18%.

2.2. Polymers in Mesoporous Layer of Photoanode

Currently, polymers are widely used to preparation of mesoporous oxide conductive layers to increase the porosity of material and thus the active surface area of the oxide. This chapter will briefly describe the polymers most commonly used to prepare the mesoporous metal oxide films. It is also worth noting that the increase in porosity improves yields a more effective penetration of the oxide substrate by the electrolyte, which will directly translate into an increase in short-circuit current. Additionally, and of particular importance, the increase in porosity will result in a larger number of dye molecules being able to anchor to the surface of the oxide semiconductor [70].

In order to create pores, polymers such as polystyrene, polyvinylpyrrolidone were used. In addition, copolymers are also used, which include P123 Pluronic (PEO$_{20}$–PPO$_{70}$–PEO$_{20}$), copolymer of PVC and POEM. In [31], a P123 copolymer was used without and with addition of polystyrene with different particle sizes (62, 130 and 250 nm). A series of cells differing in oxide substrates were prepared. The highest efficiencies for the liquid electrolyte were obtained for the cell containing only copolymer P123 (1.58%). A cell prepared from a substrate containing 130 nm PS nanoparticles (1.44%) showed a slightly lower efficiency. However, cells containing polymer electrolytes also based on the iodine redox couple were prepared. In the case of the polymer electrolyte, the highest efficiency was achieved by cells with substrates containing P123 and PS-130 (0.89%) and the lowest device efficiency was determined for a solar cell with a substrate containing only copolymer Pluronic P123 (0.42%). The subject of [33] concerned the influence of the obtained oxide substrate using polymers on the photovoltaic performance of the cell containing the dye N719. In this study, a TiO$_2$ paste was prepared using P123 to obtain a mesoporous layer, compared with a commercial P25 powder, and a TiO$_2$ substrate containing a mixture of the two pastes was prepared. It was the use of a mixture of the two pastes that led to the highest yields of 6.50%, compared to the commercial one (4.00%).

Park et al. [71] describe the comparison of mesoporous TiO$_2$ layers prepared from a commercial paste containing P25 and a paste prepared using a PVC-g-POEM copolymer to obtain a higher degree of porosity and a larger active surface area. The N719 dye was used to prepare the devices, which showed an efficiency of 5.36% when using a commercial paste, while the PCE values increased to 7.45% when using a mesoporous TiO$_2$ layer. A significant increase in the photocurrent density and fill factor was mainly observed. An important parameter from the point of view of developing the active surface is the number of adsorbed dye molecules. In the case of TiO$_2$ mesoporous layer, a significant increase in dye loading values was observed in the conventional layer of 143 and 95.1 mmol cm^{-2}, respectively.

2.3. Polymers as Dyes in DSSCs

Information on the use of polymers as dyes in DSSC-type cells is extremely rare in the literature. This can be seen in many review papers on the subject, in which no subsections on the subject appear. In preparing this work, a number of previous review papers were reviewed, where this information was omitted. This can be understood by the fact that finding this information in the literature is extremely difficult. However, it is worth noting that in the three cited papers, reasonably good yields were reported. Prakash and Subramanian [72] described the use of three poly(methacrylate)-based polymer sensitizers by employing phenothiazine (PPNPP), fluorine (FPNPP) and anthracene (APNPP). The anchor group for titanium oxide was NO_2. The photovoltaic parameters were recorded for devices prepared with and without the addition of the co-adsorbent CDCA (chenodeoxycholic acid). The highest efficiency was shown by the cell in which PPNPP was used together with CDCA (4.12%). There were significant increases in the PCE values relative to both the other two compounds and the cell without the addition of CDCA. Ramasamy et al. [73] obtained three new poly-(methacrylate) bearing push–pull-type pendants oxindole-phenothiazine with tetrazole anchoring acceptor used as sensitizers in devices. High efficiencies of the prepared cells were observed, reaching as high as 5.91% for the POTZP3 compound. Wang et al. [74] described new conjugated polymers based on poly(triphenylamine-phenothiazine) with carboxylic acid side groups. Four compounds, including three polymers, designated as PAT, PPAT4, PPAT5 and PPAT6, respectively, were presented. Additionally, the PAT compound was identical to polymer PPAT4 and when comparing PCE the polymer achieved a higher value (4.7%). The highest of the compounds studied in this work. After discussing these selected works, it can be seen that although reports on the use of polymers as dyes in DSSCs cells are few, their performance is promising.

2.4. Polymers in Gel Electrolyte to Quasi-Solid State DSSCs

Liquid electrolytes containing triiodide/iodide or Co^{2+}/Co^{3+} redox couples are now very commonly used. The use of liquid electrolytes, especially on a laboratory scale, is much simpler, quicker and cheaper than preparing a device with a solid electrolyte and applying a gold counter-electrode. However, the significant limitations of liquid electrolytes should be considered, such as volume change with temperature change, which significantly limits the operating conditions of the device. At high temperatures, the volume of the electrolyte increases considerably and evaporates, which makes it difficult to seal the solar cell; at low temperatures, the volume of the electrolyte decreases, which results in a decrease in the contact area between the electrodes and a decrease in the efficiency of the device. In addition, iodide electrolytes absorb part of the sunlight—the relatively low redox potential limits the available open-circuit voltage and causes corrosion of certain metals, making them unsuitable for use in a device such as silver [9,75,76]. Due to these limitations, solutions are being intensively sought to obtain similar PCE values to liquid electrolytes that also limit their sensitivity to changes in atmospheric conditions. For this reason, gel electrolytes, which obtain their consistency through the addition of polymers such as PEO, PEG, PAN or polyethylene glycol dimethyl ether (PGEDME), are now widely studied. All photovoltaic parameters of quasi-solid-state DSSCs based on the commercial Ru-dye are shown in Table 2.

In [34], the optimisation process of preparation and composition of the gel electrolyte was carried out to improve the performance of the device. The electrolyte contained triiodide/iodide redox couple, while the gel structure was obtained by using PEO and PGEDME. The study started with the choice of solvent, then focused on the amount of iodine in the electrolyte, ending with the addition of GuSCN. The study concluded that the most advantageous was the use of 0.2 M iodine and 0.1 M GuSCN which allowed to obtain V_{oc} = 793 mV, J_{sc} = 12.56 mA/cm^2, FF = 0.77 and PCE = 7.66%. Furthermore, it should be noted that the final cell efficiency obtained was slightly lower than when using liquid electrolyte (8.03%).

Table 2. Photovoltaic parameters of quasi-solid-state DSSC, based on the commercial Ru-dye.

Gel Electrolyte Structure	Solvent	V_{oc} (mV)	J_{sc} (mA/cm^2)	FF (-)	PCE (%)	Ref.
PEO + PGEDME/I$^-$/I$_3^-$ (0.1 M)	EtOH	730	12.91	0.66	6.34	[34]
	ACN	752	13.45	0.67	6.87	
	ACN/VN	785	11.56	0.76	6.88	
	ACN/3-MPN	785	12.05	0.76	7.16	
PEO + PGEDME/I$^-$/I$_3^-$ (0.2 M)	ACN/VN	785	12.49	0.75	7.39	
PEO + PGEDME/I$^-$/I$_3^-$ (0.4 M)		784	11.94	0.75	7.03	
PEO + PGEDME/I$^-$/I$_3^-$ (0.2 M) + GuSCN (0 M)		778	11.94	0.75	7.00	
PEO + PGEDME/I$^-$/I$_3^-$ (0.2 M) + GuSCN (0.05 M)		787	12.48	0.75	7.35	
PEO + PGEDME/I$^-$/I$_3^-$ (0.2 M) + GuSCN (0.1 M)		793	12.56	0.77	7.66	
PEO + PGEDME/I$^-$/I$_3^-$ (0.2 M) + GuSCN (0.2 M)		758	12.85	0.73	7.07	
PAN (I$^-$/I$_3^-$)	DMF	790	6.85	0.67	4.19	[35]
C11-AZO-C11/PAN (I$^-$/I$_3^-$)		780	11.96	0.75	6.28	
3 wt.% PAN-co-PBA (I$^-$/I$_3^-$)	ACN	593	9.86	0.64	3.77	[36]
5 wt.% PAN-co-PBA (I$^-$/I$_3^-$)		587	11.60	0.61	4.13	
7 wt.% PAN-co-PBA (I$^-$/I$_3^-$)		646	13.16	0.62	5.23	
9 wt.% PAN-co-PBA (I$^-$/I$_3^-$)		618	10.41	0.65	4.35	
10 wt.% PVdF-HFP (I$^-$/I$_3^-$)	ACN	6920	10.34	0.66	4.74	[77]
9 wt.% PVdF-HFP (I$^-$/I$_3^-$)		690	13.75	0.63	6.02	
8 wt.% PVdF-HFP (I$^-$/I$_3^-$)		670	12.04	0.62	5.03	
7 wt.% PVdF-HFP (I$^-$/I$_3^-$)		660	10.04	0.58	3.97	
pCMA-PGE (I$^-$/I$_3^-$)	PC:ACN	545	10.30	0.34	2.20	[78]
PVP-PGE (I$^-$/I$_3^-$)		640	6.67	0.60	3.00	
PVDF (I$^-$/I$_3^-$)	ACN	730	17.79	0.64	8.36	[79]
PDA@PVDF (I$^-$/I$_3^-$)		720	17.95	0.64	8.26	
esPME (I$^-$/I$_3^-$)	ACN	710	13.10	0.69	6.42	[80]
esCPME (2 wt.% PPy) (I$^-$/I$_3^-$)		720	13.90	0.70	7.02	

EtOH—ethanol, ACN—acetonitrile, VN—valeronitrile, 3-MPN—3-methoxypropionitrile, GuSCN—guanidine thiocyanate, DMF—N,N-Dimethylformamide, PC—propylene carbonate.

Huang et al. [35] prepared a DSSCs containing a PAN-based polymer electrolyte with iodine vapour redox, which was then doped with an azobenzene core compound. After the determination of photovoltaic parameters, a significant increase in the photocurrent density value and in the fill factor was found, thus changing the PCE from 4.19 to 6.28%. In addition, when compared to the liquid electrolyte, a negligible difference of 0.02% in favour of the liquid electrolyte was obtained. The influence of the effect of different amounts of copolymer on the photovoltaic performance obtained was studied by D. Kumar Shah et al. [36]. It was found that the addition of 7 wt.% PAN-co-PBA was the most beneficial, which caused a significant increase in both V_{oc} (646 mV), J_{sc} (13.16 mA/cm^2), and PCE (5.23%). The study focuses on determining the optimum concentration of poly(vinylidene fluoride-co-hexafluoropropylene) to prepare a polymer electrolyte. Concentrations of 7–10 wt.% were investigated. It was found that the most favourable performance was obtained for a device with a concentration of 9 wt.%, for which the efficiency was 6.02%.

The study described in [78] was devoted to the preparation of a polymer used to prepare a gel electrolyte. For this purpose, poly-3-(9H-carbazol-9-yl)propylmethacrylate (pCMA) was synthesised and then applied to a liquid electrolyte containing I$^-$/I$_3^-$ redox couple. A commercial polymer polyvinylpyrrolidone (PVP) was used for comparison and was also used to prepare the electrolyte. For the new electrolyte, the lower efficiencies were obtained than for the commercial polymer, 2.20 and 3.00%, respectively, while it is worth noting that the current density increased significantly from 6.67 (PVP) to 10.30 mA/cm^2 (pCMA). In another study [79], the effect of using a gel electrolyte obtained by adding prepared polyvinylidene fluoride (PVDF) or polydopamine- polyvinylidene fluoride (PDA@PVDF) to a commercial liquid electrolyte was studied. The polymer additives caused a decrease in the performance of the devices described by about one percentage point, which was mainly manifested by a decrease in the value of the short-circuit current density. Additionally, cells containing a gel electrolyte have been shown to have higher stability and less degradation over time. The work [80] concerned the preparation of gel electrolytes based on the electrospun polymer nanofibres (esPME) and composites (esCPME). PVdF-HFP fibres and PVdF-HFP composites with polypyrrole were obtained in the course of the research. The addition of polypyrrole resulted in an increase in cell efficiency of more than 7%. There was a slight improvement in the short-circuit current density (an increase of 0.8 mA/cm^2). In work [81], the effect of the addition of polylactic acid (PLA) was investigated when metal-free dye MK-2 was used in the solar cell. A significant increase in the density of the generated short-circuit current relative to the liquid electrolyte (10.2 and 2.7 mA/cm^2, respectively) was observed with PLA addition. This translated directly into an increase in device efficiency from 1.29 to 5.64%.

2.5. Conductive Polymers as Counter-Electrodes

The discovery of electrically conducting polymers has led scientists to delve deeper into a quest for replacing metals with their organic counterparts. Much progress has been made to replace the platinum metal counter-electrodes with organic polymers [82–84]. Counter-electrode being an indispensable component of DSSCs, catalyses the reduction of I_3^- to I^- by injecting electrons into the electrolyte and thus directly affects the device performance. Platinum metal is found to be the most suitable for this purpose due to its high conductivity and excellent electrocatalytic properties [85]. However, corrosion of the platinum, together with its high cost, makes it a less favoured choice for the counter-electrode, as it slowly deteriorates the stability of the device [86,87]. Use of expensive platinum metal as counter-electrode makes the device uneconomical. Hence, there is a need to replace the platinum metal electrodes with the noble-metal-free, low-cost, non-dissolving, electrically conducting and thermally stable materials that can alleviate the problems associated with platinum-based counter-electrodes.

Work continues in this field to develop such materials which can efficiently replace the metals-based electrodes. These include carbon materials [88,89], inorganic metal sulfides and oxides [90,91], conducting polymers [84,92–94], transition metal carbides and nitrides [95], alloys [95–97] and nanocomposites. The basic characteristics for a material to be used as counter-electrode in DSSCs include the optimum thickness of the active material, high electrocatalytic activity, porous structure, high surface area, good adhesion to substrate, and resistance against the corrosive electrolyte [98]. Among all other materials, conducting polymers have received special attention due to their low cost, facile synthesis and high electrical conductivity, along with other tunable properties. Here, we represent the most recent advances in the development of polymers for replacing metals as counter-electrodes.

2.5.1. Polypyrrole as Counter-Electrode

The high electrical conductivity and catalytic properties authenticate polypyrrole as a considerable material for fabricating counter-electrodes for DSSCs. Various attempts have been made to exploit the efficiency of polypyrrole as a counter-electrode. The ease of synthesis and versatility in its high-yielding synthetic routes, which include both chemical and electrochemical polymerization or vapor phase oxidation, allows researchers to get attracted to polypyrrole [99]. Sangiorgi et al. exploited novel molecularly imprinted polypyrrole as counter-electrodes for dye-sensitized solar cells. This molecular imprinting approach allowed them to enhance not only the catalytic property of polypyrrole, but the selectivity of the catalyst was also increased. They improved the power conversion efficiency up to 20% [100]. Khan et al. synthesized porous polypyrrole by a simple hydrothermal method and employed it as a counter-electrode in DSSC. By mixing a minute quantity of copper perchlorate and a zeolitic-imidazole framework in porous polypyrrole, they received the power conversion efficiency of 8.63% and 9.05%, respectively [37]. In the most recent work, Saberi Motlagh et al. reported the fabrication of novel counter-electrode from polypyrrole-coated on carbon fabric. Electro-polymerisation was carried out to synthesize polypyrrole-coated carbon fabric. They achieved a power conversion efficiency of 3.86% [101]. However, much work is going on using polypyrrole as counter-electrode and to achieve the results which can enable us to replace platinum electrodes with polypyrrole-based counter-electrodes.

2.5.2. Counter-Electrodes Based on Polypyrrole Nanocomposites

To improve the desired properties of polypyrrole, certain fillers are added to it or it is blended with appropriate materials. Various ways have been adopted to improve the required properties of polypyrrole so that it can be more efficiently used as a counter-electrode in DSSCs. Polypyrrole-covered graphene-based nanoplatelets are synthesized via electrochemical synthesis by Ohtani et al. The solar to electric power conversion efficiency of 4.30% was obtained, which is comparable to that of Pt-based counter-electrode (7.80%). η of PPy/GN-60 s containing DSSC was 3.3% which is even larger than DSSCs with Pt-based counter-electrode (3.00%) [102]. Another attempt to improve the performance of polypyrrole-based electrodes was done by Wu et al. Hybrid films from polyoxometalate doped polypyyrrole were prepared by electrochemical method and were exploited as counter-electrode in DSSCs. On average, the power conversion efficiency of these materials came out to be 6.19% [38]. Ahmed et al. incorporated polypyrrole with $SrTiO_3$ nanocubes via oxidative polymerization method. Scanning through the various concentrations of strontium titanate, power conversion efficiency of 2.52% was obtained for 50 percent loading of $SrTiO_3$. The incorporation of these particles into polypyrrole improved the efficiency of DSSCs from 1.29% by enhancing the surface area, electroactive response and catalytic property of the polypyrrole [103]. Rafique et al. reported the performance of DSSCs to 7.1% by synthesizing Cu-PPy-FWCNTs nanocomposites via dual step electrodeposition [104]. Relatively low conductivity and high charge-transfer resistance are two hurdles to paving the way for polypyrrole-based materials as counter-electrodes in DSSCs. The photovoltaic parameters of this type of solar cell are presented below in Table 3.

Table 3. Photovoltaic performance parameters of polypyrrole-based DSSCs counter-electrodes and their nanocomposites.

Counter-Electrodes	Fabrication Methods of PPy-Based Counter-Electrodes	V_{oc} (mV)	J_{sc} (mA cm^{-2})	FF (-)	PCE (%)	Ref.
Polypyrrole	Doctor Blade Technique	749	15.75	0.69	8.13	[37]
Poylpyrrole	Electropolymerization	727	10.20	0.42	3.12	[100]
Polypyrrole	Electropolymerization	630	12.00	0.51	3.86	[101]
PPy-POM	Electropolymerization	765	11.68	0.56	5.04	[38]
PPy-SrTiO$_3$	Doctor Blade Technique	671	10.45	0.36	2.52	[103]
PPy-MoS	In-situ Polymerization	708	18.90	0.62	8.28	[105]

2.5.3. Counter-Electrodes Based on Polyaniline

Intrinsically conducting polymers conceal potential members like polyaniline. In addition to polypyrrole, polyaniline has been widely used for the past three decades. Again, its facile synthesis, easy processability and tunable electric and redox properties make polyanilines unavoidable. Such properties are the result of three forms of polyaniline, viz. Leucomeraldine base (full reduced), Emeraldine base (half reduced form) and Pernigraniline base (full oxidized). Hence much of the research has been focused on fabricating counter-electrodes based on polyaniline.

Utami et al. synthesized nanostructured polyaniline using polyglyceryl-2-Dipolyhydroxystearate, a non-ionic surfactant. They used the nanostructured polyaniline to fabricate the counter-electrode for DSSCs. Through scanning of the surfactant concentration, they achieved the power conversion efficiency of 1.71% at 6% loading [106]. In another attempt, Krakus et al. exploited the polyaniline as the counter-electrode. Instead of the usual liquid electrolyte, they used cationic and anionic polymers as a quasi-solid electrolyte. The cationic quasi-gel electrolyte showed a higher electrolyte holding capability, elasticity and higher conductivity while the anionic one showed higher PCE. They were succeeded in achieving a high power conversion efficiency of 6.30% [39]. Furthermore, Jiao et al.

studied the cyclic voltametric behavior of polyaniline by growing a thin film of PANI on a plastic substrate. By employing these thin films as a counter-electrode in DSSC, they gained a power conversion efficiency of 7.27% [107]. However, there is a need to improve the performance of PANI-based counter-electrodes. Scientists are attempting this by adding fillers in polyaniline substrates.

2.5.4. Counter-Electrodes Based on PANI-Nanocomposites

Nano-sized fillers are usually added to polyaniline in order to improve both electrochemical and catalytic properties of PANI. The resulting materials usually show better properties than PANI itself. Here the most recent works on polyaniline nanocomposites for counter-electrodes in DSSCs are represented and all obtained photovoltaic parameters are gathered in Table 4.

Table 4. Photovoltaic performance parameters of polyaniline and its nanocomposites based on DSSCs counter-electrodes.

Counter-Electrode	Fabrication Methods of PANI-Based Electrodes	V_{oc} (mV)	J_{sc} (mA cm^{-2})	FF (-)	PCE (%)	Ref.
PANI	Doctor Blade Technique	645	20.8	0.41	4.20	[39]
PANI	Screen Printing Technique	630	5.10	0.48	1.71	[106]
PANI	Cyclic Voltametric-Electrochemical Method	740	15.34	0.64	7.27	[107]
Pristine PANI	Doctor Blade Technique	480	4.71	0.45	1.14	
H_2SO_4-doped PANI	Doctor Blade Technique	530	7.86	0.43	1.78	[40]
ALS-doped PANI	Doctor Blade Technique	603	10.84	0.43	2.79	
ALS-H_2SO_4-doped PANI	Doctor Blade Technique	603	15.13	0.53	4.54	
WO_3-PANI	Cyclic voltammetry Technique	685	18.00	0.55	6.78	[108]

Farooq et al. fabricated four different counter-electrodes using four novel polyaniline-based materials. They fabricated counter-electrodes using polyaniline, ammonium lauryl sulfate doped polyaniline, sulphuric acid doped polyaniline and binary doped polyaniline containing both components. They achieved a power conversion efficiency of 4.54% [40]. Zatirostami et al. prepared tungsten oxide containing polyaniline nanocomposites and used them to fabricate counter-electrodes for DSSC. The nanocomposites bore good electrocatalytic behavior and high electrical conductivity. They succeeded in achieving a 12.8% improved power conversion efficiency of 6.78% as compared to Pt-based counter-electrode DSSCs [108]. In the most recent work published, Ravichandran et al. fabricated the counter-electrodes from FeS_2 and achieved high FF value, higher Jsc and excellent electrocatalytic behavior towards electrolytes [109].

2.5.5. Counter-Electrodes Based on Poly(3,4-Ethylenedioxythiophene)

After polyaniline, another most widely used polymer in DSSCs as their counter-electrode is Poly(3,4-ethylenedioxythiophene), abbreviated as PEDOT. Bella et al. synthesized ammonium ion-bearing poly(3,4-ethylenedioxythiophene), which they used as counter-electrode in DSSC. Even without using any expensive rare earth metal, they succeeded in achieving the power conversion efficiency of 7.02% [110]. In another attempt, Pradhan et al. fabricated DSSC counter-electrodes from poly(3,4-ethylenedioxythiophene) films and evaluated the performance of PEDOT by the varying film thickness. They prepared the PEDOT films having 33 nm, 65 nm and 120 nm thickness. Studies showed that the film with a 33 nm thickness produced the highest power conversion efficiency of 10.39%, while conversion efficiency of 8.11% and 7.45% were used by films of 65 nm and 120 nm thicknesses, respectively. The results clearly indicated that the performance declined with an incline in film thickness [111]. In a most recent work presented by Venkatesan et al. the optimum poly(3,4-ethylenedioxythiophene) film thickness was found to be 90 nm. They studied the DSSCs performance using PEDOT counter-electrodes under different light intensities. As the light intensity decreased, the efficiency of the cells tested increased. Under standard 100 mW illumination, the PCE of the cell was about 2.5% [112].

2.5.6. Counter-Electrode Based on PEDOT-Nanocomposites

For obtaining more productive results using the PEDOT-counter-electrode, nanofillers are now added to the polymer substrate. Recently much work has been done to investigate the required catalytic and electrochemical properties of poly(3,4-ethylenedioxythiophene) nanocomposites. Mazloum-Ardakani et al. fabricated the DSSC counter-electrode from PEDOT-Ag/CuO nanocomposites. The cell showed the combined effect of poly(3,4-ethylenedioxythiophene), graphite and copper particles, thus obtaining an energy conversion efficiency of 9.06% [41]. Gaining confidence from the acquired results and in order to decrease the required amount of expensive platinum, Xu et al. fabricated counter-electrodes from transparent PEDOT film incorporated with Pt-nanoclusters. Enhancement in the electrochemical and catalytic properties in addition to improvement in film coverage, was observed. They prepared various counter-electrodes by varying the amounts of Pt-nanoclusters. The optimum concentration with the highest PCE was found to be in the Pt-10/PEDOT counter-electrode. By using the Pt-10/PEDOT electrode, the energy conversion efficiency was 6.77% when illuminating from the front side [113]. In a recent work by Gemeiner et al., various counter-electrodes were fabricated by screen-printing the poly(3,4-ethylenedioxythiophene): poly(styrene sulphonate). They doped the PEDOT:PSS with dimethyl sulfoxide, polyethylene glycol and ethylene glycol. Through optimizations, DSSC with a counter-electrode fabricated from PEDOT:PSS doped with 6 wt.% ethylene glycol showed a maximum power-converting efficiency of 3.12% [114].

Summing up the role of polymers in DSSCs, one can notice that polymers can be used in different forms and roles: as a flexible substrate, in the mesoporous layer of photoanode, dyes, in the gel electrolyte and as a counter-electrode.

In addition to the FTO-coated glass substrates, DSSC cells are increasingly being prepared on flexible polymer substrates made of PEN and also PET, which are coated with ITO. Considering another layer that forms the DSSC device, namely the conductive oxide layer, polymers are considered as materials used to receive the pores that develop the oxide surface. The most commonly used compounds are PS, PVP, as well as copolymers such as P123 Pluronic (PEO_{20}–PPO_{70}–PEO_{20}) and a copolymer of PVC and POEM.

This review presents only a few latest papers on the use of polymers as dyes in DSSCs, but it is worth noting that most papers of this type do not mention this at all, and looking at the cases cited, polymers may be materials worthy of attention in this aspect as well. The use of polymers containing either a phenothiazine derivative in the main or side chain gives very good results. The use of polymers as additives in electrolytes to obtain gel structures and improve their time stability is being explored far more extensively than for polymeric dyes. PEO, PAN, and copolymers such as PAN-co-PBA, pCMA-PGE, or PVP-PGE are the most commonly used. However, it is not possible to compare parameter values in this case and to indicate the best polymer due to the use of different dyes and preparation methods.

Excellent electronic and catalytic properties are crucial for a material to be used in DSSCs as counter-electrodes. Expensive and corrosive platinum-based counter-electrodes are now being replaced by low-cost, non-corrosive, highly efficient and easily synthesized organic counter-electrodes. Polypyrrole, Polyaniline, Poly(3,4-ethylenedioxythiophene) and Poly(3,4-propylenedioxythiophene) are the main candidates in this regard. A blend of Poly(3,4-ethylenedioxythiophene) and Polystyrene sulfonate (PEDOT:PSS) is proved to be efficient enough to be replaced by Platinum-based counter-electrode. PEDOT:PSS is the most catalytic counter-electrodes among all polymeric counter-electrodes. Still, most of the research is focused on the development of more efficient carbonaceous material, which not only reduces the production cost of the solar cell but also increases the photovoltaic parameters along with its long-term stability.

Different fabrication methods are being employed to synthesize materials for counter-electrodes with optimum required properties. Porous or network structured polymer material with high thickness can give better electrocatalytic properties. Electrochemical polymerization is an ideal technique to obtain polymeric materials with desired properties and dimensions. However, this technique has some limitations when applied on a large scale. Moreover, nanocomposites of these polymers are synthesized to further improve the available catalytic surface area and electronic properties of the material. For instance, depositing uniform films is a challenging task for PANI-based counter-electrodes. Hence, finding appropriate nanofillers improves the conductivity but also increases the surface area by acting as pore former without disturbing the film uniformity. The advantage of Polypyrrole-based counter-electrodes is that by carefully selecting the fabrication method and dopants, one can simultaneously obtain the benefits of high-catalytic and energy-storing properties of the polypyrrole.

Moreover, from the above-presented literature review of polymers-based counter-electrodes for DSSCs, we conclude that the conductive polymers with excellent electro-chemical and catalytic properties can selectively catalyze the redox reaction of the electrolyte and can further improve the photovoltaic parameters of the solar cells.

3. Polymers in Bulk-Heterojunction Solar Cells (BHJ)

BHJ solar cells are characterised by their layered structure, but it is important to note that the active layer is a donor and acceptor blend. The simplest architecture of both conventional and inverted systems is shown in Figure 2.

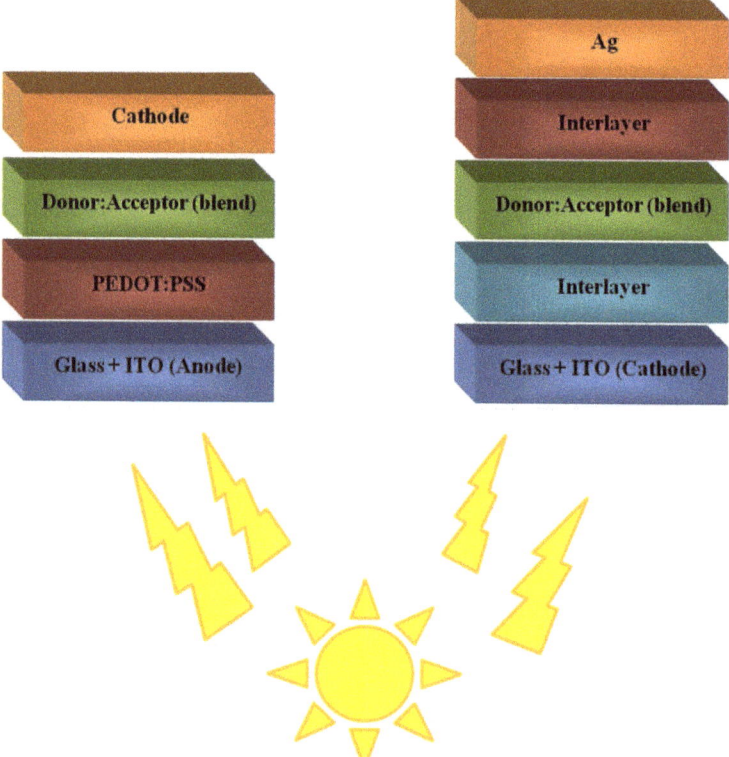

Figure 2. Schematic structure of a BHJ OSC with the conventional and inverted systems.

3.1. Polymers as Donors Materials

As for other types of solar cells, the power conversion efficiency is the most important parameter to assess the efficiency of a fabricated solar cell, which in turn depends directly upon short-circuit current density, open-circuit voltage and fill factor [42]. This fact prompts the researchers to look deeply into the factors which directly impact these parameters. Thus, plenty of work is being done to improve the PCE of bulk heterojunction solar cells by fine-tuning the (1) HOMO of donors to the deep-lying level and LUMO of the acceptor to low lying level for the improvement of V_{oc}, (2) incorporation of strong acceptors in D-A kind of donor systems, which alleviates the intramolecular charge transfers, (3) photon harvesting capability which in turn improves the external quantum efficiency and incident photon to current conversion efficiency, (4) use of superior charge carrying polymers as Donors and Acceptors to avoid energy loss due to charge recombination and (5) solution processability by increasing blending ability [43]. Out of these parameters, J_{sc} and V_{oc} are directly affected by the nature of the donor material. Hole-transporting ability and HOMO levels of donor directly impact the current density and open-circuit voltage, respectively. Hence it is inevitable to engineer the donor molecules that can efficiently improve the PCE of BHJ solar cells.

The main roles of the donor material are to absorb sunlight and transport the holes to the respective electrode. Certain requirements for organic semiconductors for employing them as donor materials in BHJ must be kept in mind. The first step is the exciton generation by absorbing solar light and most of the fraction of sunlight is absorbed by the donor material; therefore, it is necessary to optimize the absorption range of the donor materials to efficiently cover most of the solar spectrum. The use of low-bandgap donor materials, i.e., materials with a bandgap lower than 2 eV, is the key to this arduous task [44,115]. For instance, donor material having 1.1 eV bandgap can efficiently cover 77% part of AM 1.5 solar photon flux, while 1.9 eV bandgap donor material can cover only 30% of it [115]. Effective light absorption is achieved by employing π-conjugated polymers surely because of their superior absorption coefficient, i.e., 10^7 m^{-1} [116]. The thickness of the active layer in BHJ solar cells must be kept small (100 nm) so as to allow maximum exciton diffusion. It is due to the low charge carrier mobility of the polymers. The low thickness of the active layer makes the BHJ more cost-effective compared to the inorganic silicon-based solar cells, where the thickness of several micrometres is required. After exciton generation and their diffusion, the next step is the dissociation of excitons into separate charged particles, i.e., holes and electrons. The frontier orbital energy levels of the donor and acceptor molecules decide the efficiency of excitons dissociation [117,118]. The energy offset between the Acceptor LUMO and Donor LUMO should be between 0.1–1.4 eV for the dissociation of exciton into electrons and holes [119]. Herein, we review the most recent advances in the polymer donors for BHJ solar cells.

3.1.1. Wide-Bandgap Polymer Donors

The narrow absorption range and lower charge carrying ability had limited the wide applicability of poly(2-methoxy-5-(2'-ethyl-hexyloxy)-1,4-phenylene vinylene), which was the first ever polymer donor for organic solar cells [120]. Since then, a copious amount of research has been devoted to synthesizing an appropriate polymeric donor material having a wide range of spectral absorption by fine-tuning the bandgap. The PV parameters of the described devices are shown in Table 5.

Table 5. Photovoltaic parameters of devices containing wide-bandgap polymers.

Structure of Solar Cell	V_{oc} (mV)	J_{sc} (mA/cm^2)	FF (-)	PCE (%)	Ref.
ITO/PEDOT:PSS/W1:Y6 (1:1)/PDIN/Ag	890	25.36	0.68	15.39 (14.95) [a]	[121]
ITO/PEDOT:PSS/W1:Y6 (1:1.2)/PDIN/Ag	890	25.92	0.69	15.95 (15.69)	
ITO/PEDOT:PSS/W1:Y6 (1:1.4)/PDIN/Ag	890	25.06	0.71	15.87 (15.64)	
ITO/PEDOT:PSS/W1:Y6 (1:1.6)/PDIN/Ag	880	24.59	0.71	15.65 (15.35)	
ITO/PEDOT:PSS/PBDT-TTZ:N2200/PFN-Br/Ag	870	14.4	0.67	8.40	[122]
ITO/PEDOT:PSS/PBDT-TT:N2200/PFN-Br/Ag	750	2.0	0.46	0.70	
ITO/PEDOT:PSS/PBDT-TTz:PC61BM/PFN-Br/Ag	890	10.3	0.73	6.70	
ITO.PEDOT:PSS/PBTz:IT-4F/PFN-Br/Al	840	17.68	0.59	8.76	[123]
ITO/PEDOT:PSS/PTzTz:IT-4F/PFN-Br/Al	820	18.81	0.69	10.63	
ITO/PEDOT:PSS/D18:Y6 (1:0.8)/PDIN/Ag	861	27.16	0.72	16.98 (16.76) [a]	[15]
ITO/PEDOT:PSS/D18:Y6 (1:1.2)/PDIN/Ag	863	27.05	0.75	17.51 (17.34)	
ITO/PEDOT:PSS/D18:Y6 (1:1.6)/PDIN/Ag	865	27.31	0.75	17.84 (17.67)	
ITO/PEDOT:PSS/D18:Y6 (1:2)/PDIN/Ag	870	26.20	0.75	17.16 (16.87)	
ITO/PEDOT:PSS/D18:Y6/PDIN/Ag (170 nm) [b]	864	25.89	0.73	16.38 (16.34) [a]	
ITO/PEDOT:PSS/D18:Y6/PDIN/Ag (130 nm)	864	26.39	0.74	16.92 (16.77)	
ITO/PEDOT:PSS/D18:Y6/PDIN/Ag (112 nm)	866	27.16	0.75	17.65 (17.38)	
ITO/PEDOT:PSS/D18:Y6/PDIN/Ag (103 nm)	865	27.31	0.75	17.84 (17.67)	
ITO/PEDOT:PSS/D18:Y6/PDIN/Ag (91 nm)	869	26.75	0.76	17.73 (17.64)	
ITO/PEDOT:PSS/D18:Y6:PC61BM (1:1.6:0)/PDIN/Ag	862	26.09	0.76	17.23 (17.01) [b]	[124]
ITO/PEDOT:PSS/D18:Y6:PC61BM (1:1.6:0.1)/PDIN/Ag	865	26.33	0.76	17.42 (17.16)	
ITO/PEDOT:PSS/D18:Y6:PC61BM (1:1.6:0.2)/PDIN/Ag	870	26.48	0.77	17.89 (17.57)	
ITO/PEDOT:PSS/D18:Y6:PC61BM (1:1.6:0.4)/PDIN/Ag	874	25.70	0.75	16.94 (16.78)	
ITO/PEDOT:PSS/D18:Y6:PC61BM (1:1.6:0.6)/PDIN/Ag	882	25.64	0.71	16.05 (15.82)	
ITO/PEDOT:PSS/D18:Y6:PC61BM (1:1.6:0.2 [c])/PDIN/Ag	865	25.90	0.76	17.09 (16.92)	
ITO/PEDOT:PSS/D18:Y6:PC61BM (1:1.6:0 [d])/PDIN/Ag	865	27.31	0.75	17.84 (17.67)	
ITO/PEDOT:PSS/D18:Y6:PC61BM (1:1.6:0 [e])/PDIN/Ag	859	27.70	0.76	18.22 (18.01)	
ITO/PEDOT:PSS/D18:Y6:PC61BM/PDIN/Ag (90 nm) [b]	870	25.94	0.75	17.10 (16.97) [b]	
ITO/PEDOT:PSS/D18:Y6:PC61BM/PDIN/Ag (110 nm)	870	26.48	0.77	17.89 (17.57)	
ITO/PEDOT:PSS/D18:Y6:PC61BM/PDIN/Ag (130 nm)	864	26.44	0.75	17.12 (16.72)	
ITO/PEDOT:PSS/PTPD:Y6/PFN-Br/Ag	660	19.5	0.46	5.90	[125]
ITO/PEDOT:PSS/PBiTPD:Y6/PFN-Br/Ag	830	25.6	0.66	14.20	
ITO/PEDOT:PSS/P106:Y18-DMO/PFN/Al	870	22.78 (22.62) [f]	0.71	14.07 (13.91) [b]	[126]
ITO/PEDOT:PSS/P106:DBTBT-IC/PFN/Al	960	18.56 (18.41)	0.66	11.76 (11.59)	
ITO/PEDOT:PSS/P106:DBTBT-IC:Y18-DMO/PFN/Al	910	24.82 (24.66)	0.73	16.49 (16.32)	
ITO/ZnO/[PTB7-Th(1):Si-BDT(0):DCNBT-TPIC(0.6)/MoO$_3$/Ag	850	18.00 (18.07) [f]	0.64	10.11	[127]
ITO/ZnO/[PTB7-Th(0.8):Si-BDT(0.2):DCNBT-TPIC(0.6)/MoO$_3$/Ag	850	19.32 (19.74) [f]	0.65	11.20	
ITO/ZnO/[PTB7-Th(0.6):Si-BDT(0.4):DCNBT-TPIC(0.6)/MoO$_3$/Ag	860	22.32 (22.06) [f]	0.68	13.45	
ITO/ZnO/[PTB7-Th(0.4):Si-BDT(0.6):DCNBT-TPIC(0.6)/MoO$_3$/Ag	820	19.21 (19.23) [f]	0.66	10.88	
ITO/ZnO/[PTB7-Th(0.2):Si-BDT(0.8):DCNBT-TPIC(0.6)/MoO$_3$/Ag	820	16.00 (16.01) [f]	0.54	7.53	
ITO/ZnO/[PTB7-Th(0):Si-BDT(1):DCNBT-TPIC(0.6)/MoO$_3$/Ag	820	17.58 (17.59) [f]	0.65	9.92	

[a] Data in parentheses are averages for 10 cells; [b] Data in parentheses are the thickness of the active layer; [c] The active layer underwent CF solvent vapor annealing (SVA) for 5 min; [d] No CN additive; [e] No CN additive; the active layer underwent CF SVA for 5 min. Data from literature, [f] Estimated from the integration of EQE spectra.

The chemical structures of polymers used as wide bandgap materials are shown in Figure 3.

Wang et al. synthesized a 1,2-difluoro-4,5-bis(octyloxy)benzene wide-bandgap (2.16 eV) polymer W1 and used it with Y6 (BTP-4F non-fulerene electron acceptor) to fabricate a BHJ solar cell in the architecture of ITO/PEDOT:PSS/W1:Y6/PDIN/Ag. They succeeded in obtaining a PCE as high as 16.16% [121]. Cao et al. synthesized a wide-bandgap polymer PBDT-TTz based on thiazolothiazole motif and exploited the role of the imine substitution on the electronic charge transport and optical properties of the polymer. For comparison, a non-imine polymer PBDT-TT was also synthesized, and it was found that the imine substitution not only improved donor-acceptor miscibility, but also increased the face-on orientation and crystallinity of the donor phase. ITO/PEDOT:PSS/PBDT-TTZ:N2200/PFN-Br/Ag architecture was fabricated to measure the photovoltaic parameters. An imine-substituted polymer donor exhibited a power conversion efficiency of 8.4% as compared to non-imine polymers, which showed a 0.7% PCE value [122]. Changguo et al. synthesized two polymers based on 4,8-bis(5-(2-ethylhexyl)thiophen-2-yl)benzo[1,2-b:4,5-b']dithiophen, and a combination of thiazolothiazole and thiazoles. Donor polymers PTzTz and PBTz, when used along with a non-fluorine Acceptor (IT-4F) in the architecture of ITO.PEDOT:PSS/PTzTz:IT-4F/PFN-Br/Al and ITO.PEDOT:PSS/PBTz:IT-4F/PFN-Br/Al delivered energy conversion efficiencies of 10.63% and 8.76%, respectively [123]. Another wide-bandgap (1.98 eV) donor copolymer, as shown in Figure 3, was synthesized and employed in BHJ-solar cells by Qishi et al. By fabricating the cell in the ITO/PEDOT:PSS/D18:Y6/PDIN/Ag architecture provided a final PCE of 18.22% [15].

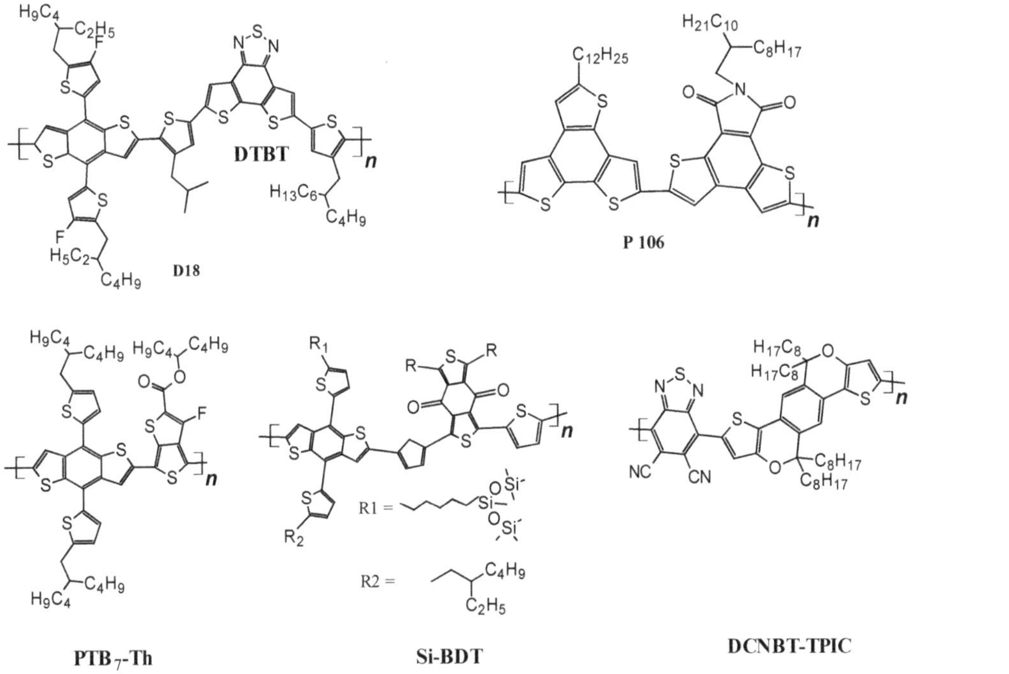

Figure 3. Chemical structures of wide bandgap polymers used in BHJ solar cells.

In a separate attempt, Jianqiang et al. fabricated a thick active layer of BHJ solar cells by adding PCBM into a D18-Y6 blend and used the architecture ITO/PEDOT:PSS/D18:Y6:PC61BM/PDIN/Ag. Increasing the thickness to 110 nm, they achieved a power conversion efficiency of 17.89% [124]. Advancing further, Zhao et al. reported bithieno[3,4-c]pyrrole-4,6-dione (PBiTPD), a donor based on the thieno[3,4-c]pyrrole-4,6-dione (TPD) motif. Using the solar cell structure ITO/PEDOT:PSS/PBiTPD:Y6/PFN-Br/Ag, the power conversion efficiency was 14.2% [125]. Keshtov et al. fabricated the binary and ternary BHJ solar cells by employing a D-A polymer P106 as a donor and two non-fluorine acceptors, Y18-DMO and DBTBT-IC. P106 contained 2-dodecylbenzo[1,2-b:3,4-b':6,5-b'']trithiophene (3TB) as a donor unit with dithieno [2,3-e;3'2'-g]isoindole-7,9 (8H) (DTID) as an acceptor unit. Two binary solar cells having the architectures of ITO/PEDOT:PSS/P106:DBTBT-IC/PFN/Al and ITO/PEDOT:PSS/P106:Y18-DMO/PFN/Al produced power conversion efficiencies of 11.76% and 14.07%, respectively, while the ternary solar cell with the structure ITO/PEDOT:PSS/P106: DBTBT-IC:Y18-DMO/PFN/Al obtained a power conversion efficiency of 16.49% [126]. In a more recent work on wide-bandgap donor polymers, Gokulnath et al. reported the fabrication of a ternary solar cell based on the siloxane-functionalized polymer Si-BDT. The ITO/ZnO/[PTB7-Th(0.6):Si-BDT(0.4):DCNBT-TPIC(0.6)/MoO_3/Ag architecture provided a power conversion efficiency of 13.45% [127].

3.1.2. Medium-Bandgap Polymer Donors

Thieno[3,4-b]thiophene or benzodithiohene-based polymers have produced satisfactory results when employed as donor materials in BHJ solar cells [128,129]. The photovoltaic parameters of the described devices are collected in Table 6 and the structures of the polymers are shown in Figure 4.

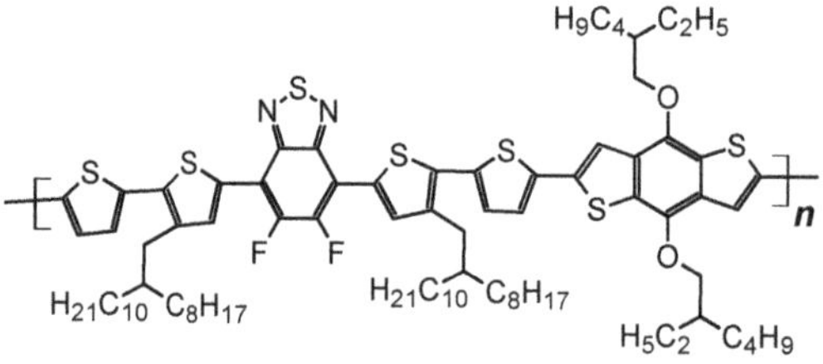

POBDFBT

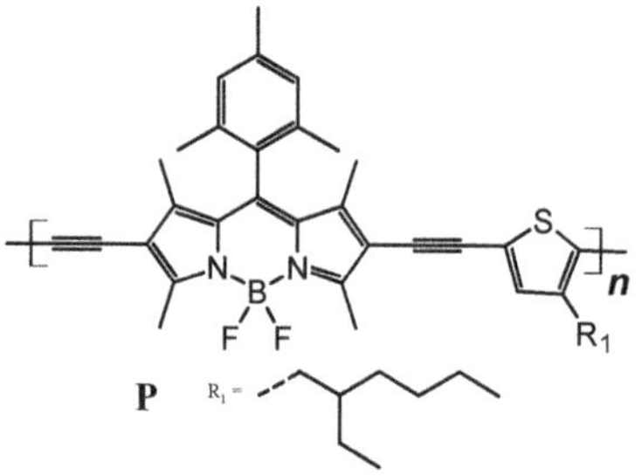

P

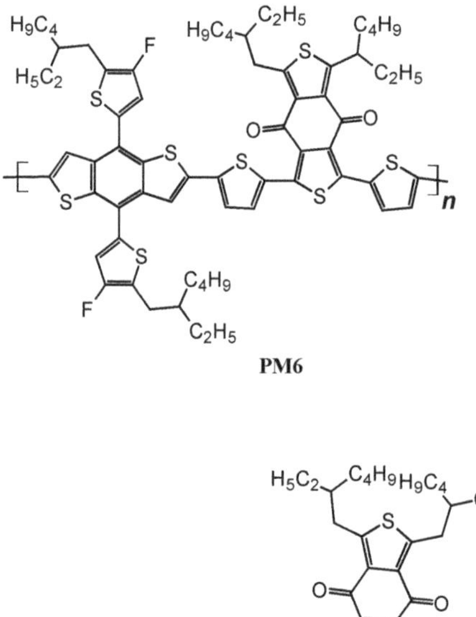

PM6

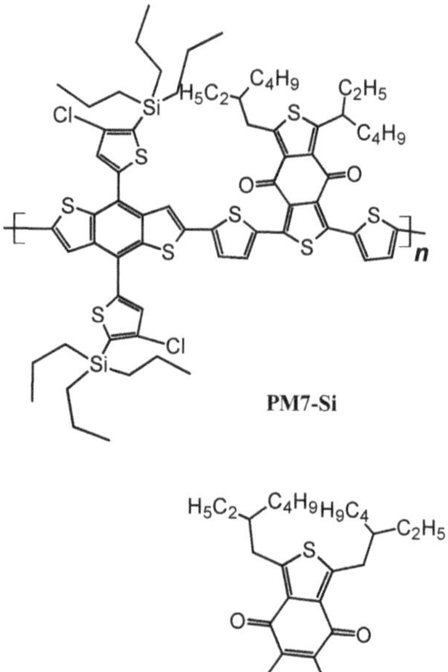

PM7-Si

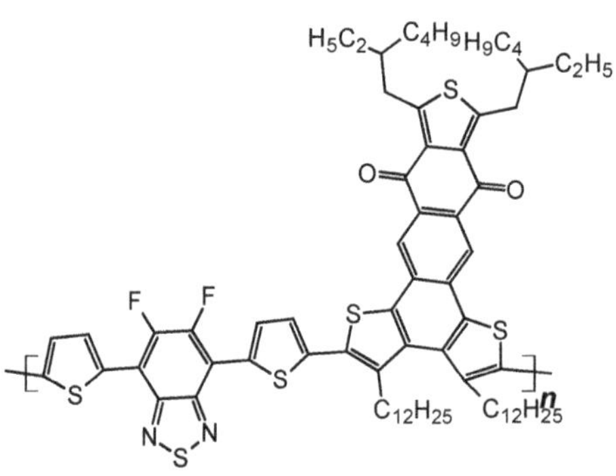

PBTZ2F-ATTD (P130)

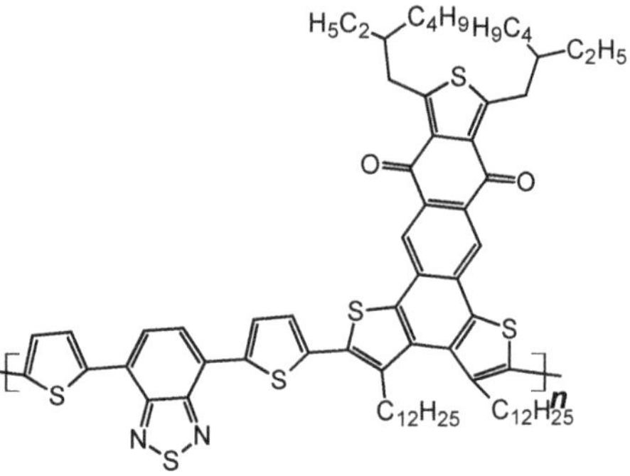

PBTZ-ATTD (P131)

Figure 4. Chemical structures of medium bandgap polymers used in BHJ solar cells.

Table 6. Photovoltaic parameters of devices containing medium-bandgap polymers.

Structure of Solar Cell	V_{oc} (mV)	J_{sc} (mA/cm^2)	FF (-)	PCE (%)	Ref.
ITO/PEDOT:PSS/POBDFBT(1):ITIC(1): PCBM(0)/PFN/Al	820	16.59	0.46	6.16	[130]
ITO/PEDOT:PSS/POBDFBT(1):ITIC(1): PCBM(0.5)/PFN/Al	780	12.7	0.64	6.26	
ITO/PEDOT:PSS/POBDFBT(1):ITIC(0.75): PCBM(0.75)/PFN/Al	760	13.8	0.61	6.39	
ITO/PEDOT:PSS/POBDFB(1):ITIC(0.5): PCBM(1)/PFN/Al	720	17.65	0.62	7.91	
ITO/PEDOT:PSS/POBDFB(1):ITIC(0.25): PCBM(1.25)/PFN/Al	790	13.78	0.61	6.66	
ITO/PEDOT:PSS/POBDFB(1):ITIC(0): PCBM(1.5)/PFN/Al	710	13.67	0.64	6.23	
ITO/PEDOT:PSS/P:ITIC-m/PFN/Al	1040	16.86	0.69	12.10	[131]
ITO/PEDOT:PSS/P:Y6/PFN/Al	940	19.72	0.71	13.16	
ITO/PEDOT:PSS/P:ITIC-m:Y6/PFN/Al	990	20.65	0.74	15.13	
ITO/PEDOT:PSS/PM6:MF1(0):Y6/PDIN/Al	843	25.11	0.75	15.93	[132]
ITO/PEDOT:PSS/PM6:MF1(10):Y6/PDIN/Al	853	25.68	0.77	17.22	
ITO/PEDOT:PSS/PM6:MF1(50):Y6/PDIN/Al	867	23.53	0.71	14.40	
ITO/PEDOT:PSS/PM6:MF1(100):Y6/PDIN/Al	914	16.67	0.79	12.09	
ITO/PEDOT:PSS/PM6:Y6/PDINO/Al (150)	860	24.3	0.73	15.3(15.2 ± 0.1)	[133]
ITO/PEDOT:PSS/PM6:Y6/PDINO/Al (150)	830	25.3	0.75	15.7(15.6 ± 0.1)	
ITO/PEDOT:PSS/PM6:Y6/PDINO/Al (200)	830	25.8	0.67	14.3(14.2 ± 0.1)	
ITO/PEDOT:PSS/PM6:Y6/PDINO/Al (250)	820	27.1	0.63	14.1(13.9 ± 0.2)	
ITO/PEDOT:PSS/PM6:Y6/PDINO/Al (300)	820	26.5	0.62	13.6(13.3 ± 0.3)	
ITO/ZnO/PM6:Y6/MoO$_3$/Ag (100)	820	25.2	0.76	15.7(15.5 ± 0.2)	
ITO/PEDOT:PSS/PM6:Y6/PDINO/Al	830	23.2	0.77	14.90	
ITO/PEDOT:PSS/PM6(1):Y6 (1.2): PC$_{71}$BM(0)/PDINO/Al	8450	24.89	0.74	15.75 (15.70)	[134]
ITO/PEDOT:PSS/PM6(1):Y6 (1.1): PC$_{71}$BM(0.1)/PDINO/Al	850	25.36	0.76	16.30 (16.26)	
ITO/PEDOT:PSS/PM6(1):Y6 (1.05): PC$_{71}$BM(0.15)/PDINO/Al	850	25.8	0.75	16.38 (16.32)	
ITO/PEDOT:PSS/PM6(1):Y6 (1.0): PC$_{71}$BM(0.2)/PDINO/Al	850	25.7	0.76	16.67 (16.61)	
ITO/PEDOT:PSS/PM6(1):Y6 (0.9): PC$_{71}$BM(0.3)/PDINO/Al	853	25.05	0.75	16.05 (16.0)	
ITO/PEDOT:PSS/PM6(1):Y6 (0.7): PC$_{71}$BM(0.5)/PDINO/Al	865	23.94	0.74	15.30 (15.23)	
ITO/PEDOT:PSS/PM6(1):Y6 (0.4): PC$_{71}$BM(0.8)/PDINO/Al	876	19.24	0.49	8.39 (8.27)	
ITO/PEDOT:PSS/PM6(1):Y6 (0.1): PC$_{71}$BM(1.2)/PDINO/Al	965	11.56	0.53	6.01 (5.94)	
ITO/PEDOT:PSS/PM6(1):PM7-Si(0):C9(1.2)/PFN-Br/Ag	841	26.36	0.76	17.0	[135]
ITO/PEDOT:PSS/PM6(0.9):PM7-Si(0.1):C9(1.2)/PFN-Br/Ag	864	26.35	0.77	17.7	
ITO/PEDOT:PSS/PM6(0):PM7-Si(1):C9(1.2)/PFN-Br/Ag	895	14.43	0.41	5.4	
ITO/PEDOT:PSS/P130:Y6/PFN/Al	890 (±5)	23.84 (±0.32)	0.72 (±0.05)	15.28 (±0.21)	[136]
ITO/PEDOT:PSS/P131:Y6/PFN/Al	780 (±3)	21.96 (0.22)	0.65 (±0.03)	11.13 (±0.18)	

Chen et al. described the enhanced photovoltaic performance by using a novel medium bandgap polymer as the donor for BHJ. A copolymer based on thiophene, diflurobenzothiadiazole (FBT) and benzodithiophene (BDT) motifs were synthesized and employed to fabricate ternary and binary BHJ solar cells. Non fluorine 3,9-bis(2-methylene-(3-(1,1-dicyanomethylene)-indanone))-5,5,11,11-tetrakis(4-hexylphenyl)-dithienol[2,3-d:2',3'-d']-s-indaceno[1,2-b:5,6-b']dithiophene (ITIC) and [6,6]-phenyl-C71-butyric acid methyl ester (PC71BM) were used as Acceptors.

The solar cells were fabricated in the following architectures: ITO/PEDOT:PSS/ POBDFBT:ITIC/PFN/Al, ITO/PEDOT:PSS/POBDFBT:PCBM/PFN/Al and ITO/PEDOT:PSS/ POBDFBT:PCBM:ITIC/PFN/Al. Power conversion efficiencies were calculated to be 6.16%, 6.23% and 7.91%, respectively [130]. Sharma et al. fabricated a ternary solar cell using BODIPY-thiophene-based conjugated polymer. The ternary cell was fabricated by mixing two polymers with two different acceptors ITIC-m and Y6, in a weight ratio of 1:0.3:1.2.

Fabricating architecture of the ternary solar cell was ITO/PEDOT:PSS/P:ITIC-m:Y6/ PFN/Al and it delivered a power conversion efficiency of 15.13%, which is higher as compared to the binary solar cells, i.e., 12.10% for P-ITIC-m (1:1.5) and 13.16% for P-Y6 (1:1.5) [131]. An and Qiaoshi et al. succeeded in achieving a PCE of 17.22 percent by employing ternary strategy of fabricating solar cells. For this PM6, a donor polymer was used along with Y6 and MF1 as acceptors in the architecture of ITO/PEDOT:PSS/active layer/PDIN/Al [132]. Separately, Yuan and Jun et al. used a ladder-type Y6 as acceptor and PM6 as a donor and succeeded in achieving a PCE of 15% [133]. Yan et al. fabricated ternary solar cells by incorporating PCBM as a third component in the PM6-Y6 binary mixture. The architecture used by them was ITO/PEDOT:PSS/PM6:Y6 (w and w/oPC$_{71}$BM)/PDINO/Al and they obtained the PCE values of 16.67% in rigid and 14.06% in flexible organic solar cells [134].

Another attempt to successfully employ medium bandgap polymers as donors in BHJ devices was made by Penget et al. They synthesized a D-A type polymer PM7-Si after modifying well-known PM6 by replacing the ethylhexyl group with alkylsilyl chains and fluorine atom with chorine. The fabrication of the ternary BHJ solar cell was made with a structure of ITO/PEDOT:PSS/PM6:PM7-Si:C9/PFN-Br/Ag. The obtained PCE of the ternary cell was 17.7% which was higher as compared to a binary cell based on PM6:C9 [135]. In a more recent work, a medium bandgap copolymer donor D-A_1-D-A_2 was synthesized where D is thiophene, A_2 is novel anthra[1,2-b:4,3,b':6,7-c″]trithiophene-8.12-dione (A3T) and A_1 is fluorinated benzothiadiazole in case of P130 or benzothiadiazole in case of P131. The architecture of the cell was ITO/PEDOT:PSS/P130 or P131:Y6/PFN/Al. The power conversion efficiencies are 15.28% and 11.13% for P130 and P131, respectively [136].

3.1.3. Narrow Bandgap Polymers

Researchers are doing continuous work on optimizing the bandgap width of the polymers to employ them efficiently as the donor materials in BHJ solar cells. The PV parameters of described devices are collected in Table 7.

Table 7. Photovoltaic parameters of solar cells contain narrow bandgap polymer donors.

Device Structure	Voc (mV)	Jsc (mA cm^{-2})	FF (-)	PCE (%)	Ref.
ITO/PEDOT:PSS/PffBT-DPP(1)/[70] PCBM(3)/MeIC(1)/ZrAcAc/Al	740	12.5	0.74	6.8	[137]
ITO/PEDOT:PSS/PffBT-DPP(1)/[70] PCBM(0)/MeIC(1)/ZrAcAc/Al	780	4.5	0.58	2.0	
ITO/PEDOT:PSS/PffBT-DPP(1)/[70] PCBM(2)/MeIC(1)/ZrAcAc/Al	760	16.1	0.73	9.0	
(ITO)/PEDOT:PSS/PTQ10:Y6/PFN-Br/Al	820 ± 1	23.9 ± 0.1	0.73	14.5 ± 0.1	[138]
ITO/PEDOT:PSS/P1(1):PC$_{71}$BM(2)/LiF/Al (500 rpm) ^{i}	770	5.76	0.43	1.92	[139]
ITO/PEDOT:PSS/P1(1):PC$_{71}$BM(3)/LiF/Al (500 rpm) ^{i}	770	7.32	0.39	2.21	
ITO/PEDOT:PSS/P1(1):PC$_{71}$BM(4)/LiF/Al (500 rpm) ^{i}	770	7.10	0.39	1.97	
ITO/PEDOT:PSS/P1(1):PC$_{71}$BM(3)/LiF/Al (500 rpm) ^{i}	580	3.07	0.30	0.55	
ITO/PEDOT:PSS/P1(1):PC$_{71}$BM(3)/LiF/Al (500 rpm) ^{i}	770	8.19	0.35	2.21	
ITO/PEDOT:PSS/P1(1):PC$_{71}$BM(3)/LiF/Al (350 rpm) ^{i}	790	7.29	0.41	2.36	
ITO/PEDOT:PSS/P1(1):PC$_{71}$BM(3)/LiF/Al (750 rpm) ^{i}	790	6.94	0.35	1.92	
ITO/PEDOT:PSS/P2(1):PC$_{71}$BM(2)/LiF/Al (500 rpm) ^{i}	710	5.27	0.55	2.07	
ITO/PEDOT:PSS/P2(1):PC$_{71}$BM(3)/LiF/Al (500 rpm) ^{i}	700	5.30	0.37	1.38	
ITO/PEDOT:PSS/P3(1):PC$_{71}$BM(1)/LiF/Al (500 rpm) ^{i}	750	2.50	0.49	0.92	
ITO/PEDOT:PSS/P3(1):PC$_{71}$BM(2)/LiF/Al (500 rpm) ^{i}	750	3.95	0.46	1.38	
ITO/PEDOT:PSS/P3(1):PC$_{71}$BM(3)/LiF/Al (500 rpm) ^{i}	750	3.85	0.49	1.43	
ITO/PEDOT:PSS/P3(1):PC$_{71}$BM(4)/LiF/Al (500 rpm) ^{i}	760	5.14	0.42	1.65	
ITO/PEDOT:PSS/P3(1):PC$_{71}$BM(4)/LiF/Al (500 rpm) ^{i}	740	7.13	0.34	1.83	
ITO/PEDOT:PSS/P3(1):PC$_{71}$BM(4)/LiF/Al (500 rpm) ^{i}	750	7.63	0.35	2.02	
ITO/PEDOT:PSS/P3(1):PC$_{71}$BM(4)/LiF/Al (350 rpm) ^{i}	770	7.59	0.41	2.45	
ITO/PEDOT:PSS/P3(1):PC$_{71}$BM(4)/LiF/Al (750 rpm) ^{i}	740	5.9	0.33	1.48	
ITO/PEDOT:PSS/PTT-EFQX:PCBM/PFN-Br/Ag	690	11.19	0.68	5.37	[140]
ITO/PEDOT:PSS/PT-DFBT-T-EFQX:PCBM/PFN-Br/Ag	870	5.62	0.54	2.69	
ITO/PEDOT:PSS/P(T2BDY−TBDT)/PNDIT-F3N−Br/Ag	780	12.07	0.47	4.40	[141]
ITO/PEDOT:PSS/P(TTzBDY−TBDT)/PNDIT-F3N−Br/Ag	800	7.71	0.40	2.49	
ITO/PEDOT:PSS/P(T2BDY−TBDT0.7−OBDT0.3)/PNDIT-F3N−**Br/Ag**	750	3.80	0.37	1.06	
ITO/PEDOT:PSS/P(TTzBDY−TBDT0.7−OBDT0.3)/PNDIT-F3N−**Br/Ag**	770	5.23	0.39	1.58	

^{i} Revolutions per minute.

Pan et al. reported diketopyrrolopyrrole (DPP)-based polymer PffBT-DPP has a narrow bandgap of 1.33 eV. One non-fullerene acceptor MeIC and one fullerene acceptor PCBM were employed to fabricate both binary and ternary solar cells with the architecture ITO/PEDOT:PSS/active layer/ZrAcAc/Al. The ternary device PffBT-DPP:PCBM:MeIC showed a power conversion efficiency of 9.0% while binary BHJ solar cells showed PCE of 6.8% and 2.0% for PffBT-DPP:PCBM and PffBT-DPP:MeIC respectively [137]. A PT10-based donor polymeric material was synthesized by Rech et al. and was used in the architecture of (ITO)/poly(3,4-ethylenedioxythiophene):polystyrenesulfonat(PEDOT:PSS)/PTQ10:Y6/PFN-Br/aluminum. They acquired the PCE of 15%. In [138], Caliskan et al. synthesized a donor material based on benzo dithiophene by attaching a 2-(2-octyldodecyl)selenophene ring at the fourth and eighth position of benzene ring in BDT. The structure of the solar cell was ITO/PEDOT:PSS/Polymer:PC$_{71}$BM/LiF/Al and the obtained PCEs were 2.36%, 2.07% and 2.45% for P1, P2 and P3, respectively [139]. Guo et al. synthesized narrow bandgap (1.6 eV) conjugated polymers based on bis(2-alkyl)-5,8-dibromo-6,7-difluoroquinoxaline-2,3-dicarboxylate (EF-Qx) unit. For better performance D-A (PTT-EFQX) and D-A$_1$-D-A$_2$(PT-DFBT-T-EFQX)-type materials were synthesized in the architecture of ITO/PEDOT:PSS/PTT-EFQX:PCBM/PFN-Br/Ag and ITO/PEDOT:PSS/PT-DFBT-T-EFQX:PCBM/PFN-Br/Ag respectively. PCE of the solar cells containing D-A type structure was found to be 5.37%, while D-A$_1$-D-A$_2$ type has 2.69% [140]. Can et al. recently synthesized low bandgap (1.30–1.35 eV) D-A copolymers 4,4-Difluoro-4-bora-3a,4a-diaza-s-indacene (BODIPY) as donor part and benzo[1,2-b:4,5-b']dithiophene (BDT) acting as acceptor. The highest conversion efficiency of 4.40% was shown by P(T2BDY−TBDT), having a very high current density of 12.07 mAcm^{-2} [141].

3.2. Polymers as Acceptor Materials

To a lesser extent, polymers are used as acceptors in BHJ cells. However, in the literature in recent years, few reports can be found on the use of polymers as acceptors (Table 8).

Table 8. Polymers used as an acceptor material in bulk-heterojunction solar cells.

Structure of Solar Cell	V_{oc} (mV)	J_{sc} (mA/cm^2)	FF (-)	PCE (%)	Ref.
ITO/PEDOT:PSS/PTzBISi:N2200/C60N/Ag CB- as print	930	2.76	0.43	1.01	[47]
ITO/PEDOT:PSS/PTzBISi:N2200/C60N/Ag CB-TA	890	3.98	0.48	1.57	
ITO/PEDOT:PSS/PTzBISi:N2200/C60N/Ag CB-TA+SVA	870	4.58	0.51	1.83	
ITO/PEDOT:PSS/PTzBISi:N2200/C60N/Ag MTHF-as print	890	15.41	0.70	9.01	
ITO/PEDOT:PSS/PTzBISi:N2200/C60N/Ag MTHF-TA	880	16.19	0.73	9.96	
ITO/PEDOT:PSS/PTzBISi:N2200/C60N/Ag MTHF-TA+SVA	880	17.62	0.76	11.25	
ITO/ZnO/PTB7-Th:NDP-V/V$_2$O$_5$/Al	740	17.07	0.67	8.59	[48]
ITO/ZnO/PTB7-Th:PDI-V/V$_2$O$_5$/Al	740	15.39	0.64	7.38	
ITO/ZnO/PEI/BSS0:PBDB-T/MoO$_3$/Ag	820	15.74	0.57	7.38	[49]
ITO/ZnO/PEI/BSS10:PBDB-T/MoO$_3$/Ag	860	18.55	0.64	10.10	
ITO/ZnO/PEI/BSS20:PBDB-T/MoO$_3$/Ag	860	17.07	0.65	9.58	
ITO/ZnO/PEI/BSS50:PBDB-T/MoO$_3$/Ag	850	17.50	0.65	9.69	
ITO/ZnO/PBDBT:PIID(CO) 2FT/MoO$_3$/Ag	640	8.30	0.50	2.65	[142]
ITO/ZnO/**PBDBT:PIID(CO) BTIA**/MoO$_3$/Ag	630	1.80	0.50	0.37	
ITO/PEDOT:PSS/PBDB-Tb-PYT/PDINN50/Ag (CF; area 5 mm^2)	919	16.90	0.46	7.18	[143]
ITO/PEDOT:PSS/PBDB-Tb-PYT/PDINN50/Ag (CF, 4% CN; area 5 mm^2)	916	19.60	0.63	11.32	
ITO/PEDOT:PSS/PBDB-Tb-PYT/PDINN50/Ag (CB, 4% CN; area 5 mm^2)	908	19.31	0.60	10.53	
ITO/PEDOT:PSS/PBDB-Tb-PYT/PDINN50/Ag (ODCB, 4% CN; area 5 mm^2)	917	18.67	0.59	10.08	
ITO/PEDOT:PSS/PBDB-Tb-PYT/PDINN50/Ag (THF, 4% CN; area 5 mm^2)	914	19.25	0.63	11.13	
ITO/PEDOT:PSS/PBDB-Tb-PYT/PDINN50/Ag (Toluene, 4% CN; area 5 mm^2)	912	19.38	0.62	11.07	
ITO/PEDOT:PSS/PBDB-Tb-PYT/PDINN50/Ag (CF, 4% CN; area 2.2 mm^2)	867	19.71	0.63	10.80	
ITO/PEDOT:PSS/PBDB-T:PYT/PDINN50/Ag (CF; area 5 mm^2)	883	22.70	0.72	14.57	

MTHF—2-methyltetrahydrofuran, TA—thermal annealing, SVA—solvent vapor annealing, CF—chloroform, CB—chlorobenzene, ODCB—o-dichlorobenzene, CN—1-chloronaphthalene.

Zhu et al. [47] prepared and described a series of solar cells with structure ITO/PEDOT:PSS/PTzBISi:N2200/C60N/Ag prepared under different conditions. N2200 polymer was used as an acceptor. Two solvents, such as chlorobenzene and 2-methyltetrahydrofuran were used, in addition to the thermal annealing and solvent vapour annealing. It has been shown that 2-methyltetrahydrofuran and thermal annealing together with solvent vapour annealing are the most favourable applications. For the solar cells prepared in this way, high photovoltaic parameters of V_{oc} = 880 mV, J_{sc} = 17.62 mA/cm^2, FF = 0.76 and PCE = 11.25% were obtained. The values given were averages and the maximum efficiency of the device was as high as 11.76%.

In [48], polymeric acceptors were used, which are naphthalene-diimide and perylenediimide derivatives. The prepared inverted structure devices showed high efficiencies of

8.59% (NDP-V) and 7.38 (PDI-V). When NDP-V was used, an increase in both Jsc and FF was observed (by 1.68 mA/cm^2 and 0.03, respectively). Nagesh et al. [49] prepared a series of inverted photovoltaic cells containing a copolymer as an acceptor. The fabricated devices differed in the ratio of NDI-biselenophene/NDI-selenophene copolymer repeating units. As a result of the research carried out, it was found that the most advantageous was the use of an NDI-biselenophene/NDI-selenophene copolymer with an equivalent proportion of 90:10 (BSS10); for this acceptor structure, yields of over 10% were obtained. In [143], a block copolymer containing donor and acceptor moieties (PBDB-Tb-PYT) was used. The effect of the solvent used (chloroform, chlorobenzene, o-dichlorobenzene, tetrahydrofuran, toluene) was studied. Moreover, the addition of 1-chloronaphthalene was used. By using the CN additive, a significant increase in both Jsc and FF and therefore, in efficiency, was observed (from 7.18 to 11.32%). The obtained photovoltaic performance results for the copolymer active layer were compared with the blend obtained by mixing the donor (PBDB) and acceptor (PYT), respectively. A PCE of 14.57% was recorded for the blends obtained by mixing the polymers in the active layer.

On the basis of the above-presented results from the latest scientific reports, we can summarize the role of polymeric materials in BHJ solar cells. The most important light-harvesting responsibility of the donor material in bulk heterojunction solar cells compels the researchers to choose the optimum bandgap materials in this regard. Inorganic silicon-based solar cells require the thick active donor layer and hence not only increase the production cost but also are based on a non-renewable silicon source. Therefore easily synthesized, low-cost, environmentally friendly and thermally stable polymeric donor materials are continuously increasing in demand. Low-bandgap donor polymers are optimum for bulk heterojunction solar cells because they absorb most parts of the solar spectrum and are thus efficient light absorbers. Non-fullerene acceptors are more compatible with the polymer donors because of the lowered LUMO levels and high extinction coefficient.

The wide visible light absorption range is the specialty of the polymers only. Poly(3-hexylthiophene) (P3HT) is the most widely used polymer donor with PCBM acceptor. Benzo[1,2-b:4,5-b′]dithiophene (BDT)-based donor polymers are among the best polymers used against fullerene and non-fullerene acceptors. Not only binary, but ternary strategies are employed to further increase the efficiency of all-polymer solar cells. PCEs greater than 11% have been achieved by fullerene-based acceptors, while PCEs greater than 17% have been accomplished by non-fullerene BHJ solar cells. Regioregular geometry of the polymers controls the polymeric chain supramolecular assembly and thus influences the charge transport properties. Developing novel synthetic methodologies has become crucial for controlling regioregular geometry of the polymers during copolymerization. From this literature review, it can be noted that good miscibility between acceptor and donor is another important parameter, which must be kept in mind to achieve high-performing bulk heterojunction solar cells. Side-chain engineering plays an important role in the electron-donating abilities of the donor material. Therefore, it is the need of the hour to develop facile and cost-effective synthetic methods for synthesizing polymer donors with optimum properties. Polymers used as acceptors in BHJs are far less common, although there is, of course, information on this in the literature. The most commonly used polymeric acceptors are N2200, PBDB-T or NDP-V, but it is impossible to compare the PV performance of these devices due to different preparation methods and architecture.

4. Polymers in Perovskite Solar Cells (PSCs)

Over the last two decades, the significant development of the sourcing electricity concept from solar energy is observed. The current research topic in photovoltaics is perovskite solar cells. The PSCs are cells of the latest technology, for which has been noted a very fast increase in efficiency (PCE) from 3.8% in 2009 to 25.2% in 2020, which may indicate that this type of cell will find commercial applications [12,144,145]. The perovskite solar cells are a hybrid system, a combination of organic and inorganic structures. A perovskite can be represented by a general formula ABX_3, where A is the organic ion (the most common is methylammonium ion $-[CH_3NH_3]^+$), B is Pb^{2+} ion, Sn^{2+} or Cd^{2+}, and X is a halogen ion I^-, Br^- or Cl^-. The perovskite is characterized by wide absorption of visible and near-infrared radiation, low binding energy exciton (~2 meV), and a direct bandgap. Additionally, perovskite materials show: (i) a long time carrier life (~270 ns), which generates the length of the diffusion path at the level of ~1 µm in thin layers and up to ~175 µm in single crystals, thus ensuring hassle-free transport of charge carriers through the absorber (perovskite) 300 nm thick (no recombination effect), (ii) high mobility load carrier (up to ~2320 cm^2 V^{-1} s^{-1}); and (iii) high dielectric constant (~18–70), which makes them ideal materials for photovoltaics [146]. A perovskite absorber, hole-transporting material (HTM), electron transporting material (ETM), and electrodes are all common components in PSC devices. Photo-generated electrons/holes in the perovskite absorber are transported to the ETM/HTM and selectively collected by the anode/cathode when a PSC is illuminated. Both n–i–p (traditional) and p–i–n (inverted) forms of PSCs can operate successfully due to perovskites' ambipolar charge transport feature [147].

This work is a presentation of the current achievements concerning the applications of polymers in perovskite solar cells, which will be shown in the following subsections of this publication. The conventional and inverted structures of PSCs are presented in Figure 5a, while the chemical structures of polymers used in PSCs are shown in Figure 5b.

(a)

Figure 5. *Cont.*

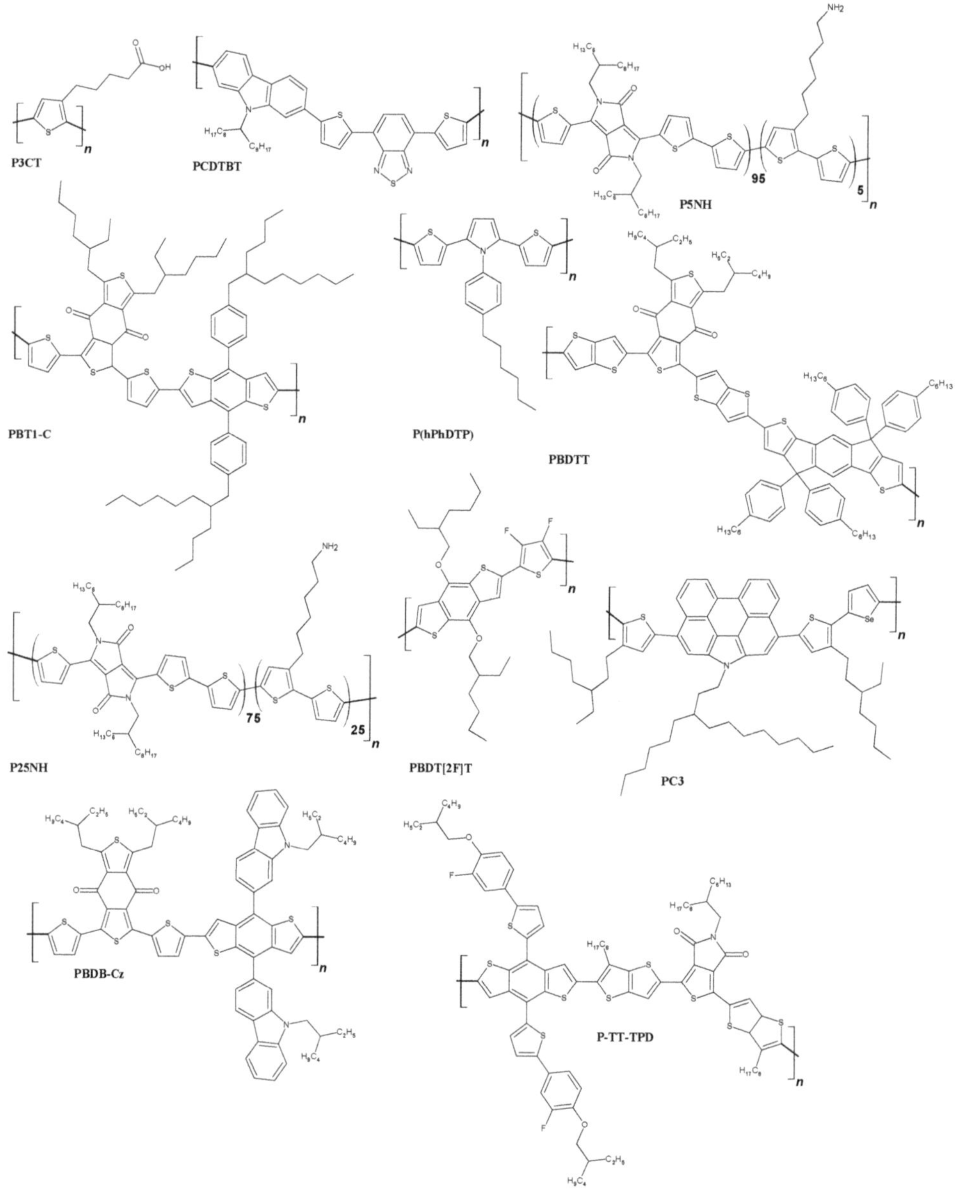

Figure 5. The conventional and inverted architecture of perovskite solar cells (**a**) and the chemical structures of polymers used in PSCs (**b**).

4.1. Polymers in Improving Perovskite Morphology

The large-scale development of perovskite solar cells requires high-quality failure-free perovskite foils with better surface coverage. Several solutions to this problem will be presented in this section.

Zhao et al. presented a polymerization-assisted grain growth (PAGG) technique for obtaining stable and efficient perovskite solar cells with $FA_{1-x}MA_xPbI_3$. DI (Dimethyl Itaconate) monomers were added to the PbI_2 precursor (1.0% molar ratio) to provide sufficient contact between the PbI_2 and their carbonyl groups (sequentially deposited approach). An in situ polymerization process was started during the subsequent PbI_2 annealing process, leaving the as-formed heavier polymers adhering to the grain boundaries with previously set contact. Due to the adequate polymer-PbI_2 interaction, there was a higher energy barrier for producing perovskite crystals when reacting with FAI (formamidinium iodide), resulting in more sizeable crystal grains. Furthermore, the carbonyl groups of polymers were led to the under coordinated Pb^{2+} and effectively diminished the trap density, whereby a PCE of 23.0% was obtained. Effective passivation, combined with the hydrophobic character of the polymer, significantly slowed the rate of deterioration, resulting in significant increases in stability [148].

Furthermore, Yousif and Agbolaghi investigated the potential application of the rGO and CNT precursors and their derivatives grafted with the rGO-g-PDDT and CNT-g-PDDT (irregioregular) and CNT-g-P3HT and CNT-g-P3HT (regioregular) polymers to improve the morphological, optical, and photovoltaic properties of FTO/b-TiO_2/mp-$TiO_2/CH_3NH_3PbI_3/spiro$-$OMeTAD/MoO_3/Ag$ perovskite devices (the ratio of carbonic materials to the perovskite was 1:15). The perovskite system behaviour (cell performance) was modified by the type of rGO or CNT (carbonic materials) and the regioregularity of grafts. The best results of PCE were obtained with CNT nanostructures grafted with P3HT backbones, which were 16.36% [149].

Yu et al. elaborated a new p-type p-conjugated ladder-like polymer P-Si (poly(3,30-(((2-(4,8-bis(5-(2-ethylhexyl)thiophen-2-yl)-6-methylbenzol[1,2-b:4,5-b0]dithiophen-2-yl)-5-methyl-1,4-phenylene)bis(oxy))bis(hexane-6,1-diyl))bis(1,1,1,3,5,5,5-heptamethyltrisiloxane)) for perovskite solar cells-based on SnO_2. This introduced of a small amount of P-Si into an antisolvent to improve the morphology and crystallinity of perovskite films. The P-Si (the HOMO energy level is -5.41 eV) could act as a hole-transport medium between the spiro-OMeTAD and the perovskite layer (enhanced hole transportation). As a result, the highest PCE of solar cell with P-Si (0.1 mg ml^{-1}) was achieved at 21.3% [150].

Fu et al. also applied a polymer in the anti-solvent process to passivate the defects of perovskite films and dominate the perovskite crystallization. The researchers used C60-PEG (fullerene end-capped polyethylene glycol). The application of C60-PEG also influenced the surface of the perovskite films. As a consequence, the highest PCE (17.71%) of the tested perovskite solar cells was registered [151].

Moreover, Chen et al. exploited a PBTI (poly(bithiophene imide)) in the anti-solvent step of the perovskite deposition process, resulting in effective passivation of the grain boundary defects and thus improvement of the tested devices performance. The PBTI (0.25 M) may be efficiently incorporated into grain boundaries (grain boundary defect passivation) cause of a vast lower in recombination losses and the ensuing increase in V_{oc} and PCE (20.57%) [152].

Yao et al. demonstrated that a polymer alloy of a PS (polystyrene) and a PMMA (poly (methyl methacrylate)) could profit the crystal growth and boost the flexibility of the perovskite solar cells. The polymer alloy (AMS, PS:PMMA, i.e., 1:2) was integrated with the perovskite layer ($CH_3NH_3PbI_3$) during the anti-solvent process. The additive of AMS may boost the grain size of perovskite crystals and suppress the crystallization of the absorber layer. As a result, the PSCs with AMS showed a PCE of 17.54% [153].

For the first time, Suwa et al. incorporated a small amount of a PTMA (poly(1-oxy-2,2,6,6-tetramethylpiperidin-4-yl methacrylate) into the perovskite layer, thereby increasing the durability of the perovskite. The superoxide anion radical generated following light irradiation on the layer was eliminated by PTMA, which could react with the perovskite molecule and degrade it into lead halide. The photovoltaic conversion efficiency of a cell made with a PTMA-incorporated perovskite layer (0.3 wt.% amount of the polymer vs. the perovskite) and a hole-transporting PTAA (polytriarylamine) layer was 18.8% [154].

Additionally, for the perovskite surface, Chen et al. used a PEA (poly(propylene glycol) bis(2-aminopropyl ether)) and applied grain boundary passivation. PEA's unshared ether–oxygen electron pair activates, forming a crosslinking complex with lead ions, thus lowering the trap state density and inhibiting non-radiative recombination in perovskite films. The PCE of the MAPbI$_3$-based cells with PEA was 18.87% (1 wt.% of PEA) [155].

In addition, Garai et al. designed and synthesized a PHIA (poly(p-phenylene)) as additives to the perovskite precursor solution. The side chains of the polymer were selectively functionalized, allowing it to be used in the effective trap passivation of perovskites. The PHIA polymer caused the production of perovskite films of a higher quality and with bigger grain sizes. The passivated device exhibited minimum charge collection at the interface, lower recombination and lesser traps, allowing for an improved charge transfer. As a result of the passivation, the device had a high PCE of 20.17% (0.50 mg mL^{-1} of PHIA) [156].

In contrast, Zarenezhad et al. utilized a PPy (polypyrrole) in the precursor solution to fabricate mixed halide devices. PPy was used as a conductive compound to ensure an enhanced electron-hole extraction and transfer. The PPy additive amended the layer quality by mitigating the growth of the perovskite crystals (the lower charge carrier recombination and efficient carrier extraction). The highest achieved PCE of perovskite solar cells (1 wt.% of PPy) was 13.2% [157].

Zhong et al. employed a mixture of a PVP (polyvinylpyrrolidone) and a PEG (polyethylene glycol) with an appropriate mass ratio in precursor solution to perfect the morphology of perovskite, optical and photovoltaic properties, and air stability of perovskite (CH$_3$NH$_3$PbI$_3$) films. After modifying the perovskite film with a polymer mixture (PVP and PEG), the crystallinity, uniformity, smoothness, compactness, and surface coverage of the perovskite film improved. The air stability of the tested PSCs could be imputed to the unique hygroscopicity of the polymer mixture. The bondings between polymer mixture and perovskite also contributed to the inhibition of ion migration and the synergistic stabilization of the perovskite structure [158].

4.2. Polymers as Hole-Transporting Materials (HTM)

Hole-transporting materials are essential elements of perovskite cells. Compounds acting as HTMs in PSCs should be of: (i) an appropriate level HOMO (i.e., Highest Occupied Molecular Orbital, which allows the band's energy valence perovskite material to be adjusted), (ii) high hole mobility, (iii) wide light absorption spectral range, (iv) photochemical stability, and (v) good layering ability [145,159]. The conjugated polymer HTMs have good stability and solution operability when compared to organic small molecule HTMs and inorganic HTMs. Photovoltaic parameters of PSCs with polymeric HTM are presented in Table 9.

Chawanpunyawat et al. developed an IDTB (poly(1,4-(2,5-bis((2-butyloctyloxy) phenylene)-2,7-(5,5,10,10-tetrakis(4-hexylphenyl)-5,10-dihydro-sindaceno[2,1-b:6,5-b'] dithiophene))) as dopant-free polymeric HTM. IDTB was shown a high mobility and an intensive interaction of the backbone to perovskites through IDTB's S/O atoms (a high holeextracting ability) and also an effective passivation of the defects in absorber layer. The prepared PSCs with IDTB as dopant-free HTM were attained PCE (19.38%) comparable to the devices with doped spiro-OMeTAD (2,2',7,7'-tetrakis[N,N-di(4-methoxyphenyl)amino]-9,9'-spirobifluorene) (18.22%) [160].

For the first time, Liao et al. applied a P3CT (poly[3-(4-carboxybutyl)thiophene-2,5-diyl]) as HTM in perovskite solar cells. The P3CT has demonstrated ideal dual functionality for device applications thanks to a plethora of carboxylic groups (−COOH) on the side chains. Molecules of the P3CT could firmly attach to the ITO surface, following it to achieve a work function that was similar to that of the perovskite active layer. To eliminate recombination defects, the Lewis base character of the −COOH group could efficiently passivate the under-coordinated Pb^{2+} ions at the HTL/perovskite interface. As a consequence of using the P3CT as HTL, a significant PCE of 21.09% was successfully produced [161].

Table 9. The collected photovoltaic parameters of PSCs based on the analysed HTMs.

Device Structure	Voc (mV)	Jsc (mA cm^{-2})	FF (-)	PCE (%)	Ref.
FTO/TiO$_2$/SnO$_2$/[Cs$_{0.05}$FA$_{0.8}$MA$_{0.15}$PbI$_{2.55}$Br$_{0.45}$]/IDTB/Au	1107	23.06	0.76	19.38	[160]
ITO/P3CT/[(FA$_{0.17}$MA$_{0.94}$PbI$_{3.11}$)$_{0.95}$(PbCl$_2$)$_{0.05}$]/C$_{60}$/ZrAcac/Ag	1120	22.88	0.82	21.09	[161]
FTO/TiO$_2$/[0.001 M FAI, 0.001 M PbI$_2$, 0.0002 M MABr, 0.0002 M PbBr$_2$ + CsI solution (1.5 M in DMSO)]/PBT1-C/-C/MoO$_3$/Ag	1030	22.10	0.79	19.06	[162]
ITO/SnO$_2$/[CH$_3$NH$_3$PbI$_3$]/PCDTBT/Ag	970	19.90	0.73	14.08	[163]
FTO/SnO$_2$/[0.001 M MAI,0.001 M PbI$_2$ + EACl (0.0002 M (15% molar ratio))]/PC3/Au	1110	23.50	0.80	20.80	[164]
ITO/SnO$_2$/[1.1 M PbI$_2$, 1.0 M FAI, 0.22 M PbBr$_2$, 0.2 M MABr + 1.5 M CsI]/PBDTT/MoO$_3$/Ag	1120	23.64	0.77	20.28	[165]
FTO/b-TiO$_2$/m-TiO$_2$/[CH$_3$NH$_3$PbI$_3$]/P(hPhDTP)/Ag	960	20.82	0.79	15.71	[166]
FTO/TiO$_2$/[0.0006 M PbI$_2$, 0.0001 M PbBr$_2$, 0.0001 M MABr, 0.0005 M FAI]/P-TT-TPD/Au	1040	21.68	0.73	16.82	[167]
FTO/SnO$_2$/[CH$_3$NH$_3$PbI$_3$]/PBDT[2F]T/Ag	1060	22.64	0.73	17.52	[168]
ITO/SnO$_2$/[1.1 M PbI$_2$, 1.0 M FAI, 0.22 M PbBr$_2$, 0.2 M MABr]/PBDB-Cz/MoO$_3$/Ag	1135	24.34	0.76	21.11	[169]
ITO/SnO$_2$/[0.26 M FAI,1.26 M PbI$_2$, 1.08 M MAI, 0.14 M PbCl$_2$]/P25NH/Ag	1049	19.81	0.83	17.30	[170]
ITO/SnO$_2$/[(MA$_{0.8}$FA$_{0.2}$)Pb(I$_{0.93}$Cl$_{0.07}$)$_3$]/P5NH/Ag	1041	20.95	0.83	18.10	[171]

Qi et al. proposed a PBT1-C obtained from the copolymerization between 1,3-bis(4-(2-ethylhexyl)thiophen-2-yl)-5,7-bis(2-alkyl)benzo[1,2-c:4,5-c']dithiophene-4,8-dione units and benzodithiophene. The PBT1-C was able to passivate the surface traps of the perovskite layer and was characterized by excellent hole mobility. Through its carbonyl (−CO) functional groups, PBT1-C might passivate under coordinated defective perovskites, reducing nonradiative recombination and enhancing charge extraction. The tested PSCs was shown a PCE of 19.06% [162].

Jeong et al. reported a PCDTBT (poly[N-9'-heptadecanyl-2,7-carbazole-alt-5,5-(4',7'-di-2-thienyl-2',1',3'-benzothiadiazole)]) as an efficient hole-transfer material (0.02 cm^2 V^{-1} s^{-1}). The greatest fracture energies in the perovskite devices were caused by PCDTBT fibrils produced at the grain boundaries of the perovskite layer. These energies have offered extrinsic reinforcement and shielding for improved mechanical and chemical stability. The PSCs with PCDTBT exhibited a PCE of 14.08% as well as significantly enhanced mechanical and air stability [163].

Yao et al. synthesized polymeric HTMs by inserting a phenanthrocarbazole unit into polymeric thiophene or selenophene chain (PC1, PC2, and PC3). The addition of a planar and broad phenanthrocarbazole unit was dramatically enhanced by the adjacent polymer strands' π−π stacking and interactions with the perovskite's surface. As a result, the PSC with PC3 as a dopant-free HTM had a stable PCE of 20.8% and a greatly increased lifetime [164].

Ma et al. explored a J71 (poly [[5,6-difluoro-2-(2-hexyldecyl)-2H-benzotriazole-4,7- diyl]-2,5-thiophenediyl [4,8-bis [5-(tripropylsilyl)-2-thienyl]benzo[1,2-b:4,5-b']dithiophene-2,6-diyl]-2,5-thiophenediyl]), PBDB-T (poly[(2,6-(4,8-bis(5-(2-ethylhexyl)thiophen-2-yl)benzo [1,2-b:4,5- b0]dithiophene)-co-(1,3-di (5-thiophen-2-yl)-5,7-bis(2-ethylhexyl)-benzo[1,2-c:4,5-c0]dithiophene-4,8-dione)]), and PM6 (poly [(2,6-(4,8-bis(5-(2-ethylhexyl-3-fluoro)thiophen-2-yl)-benzo [1,2-b:4,5-b0]dithiophene))-alt-(5,5-(10,30 -di-2-thienyl-50,70 -bis(2-ethylhexyl)benzo [10,20-c:40,50 -c0]dithiophene-4,8-dione)]) in PSCs. The alignment of the perovskite and HTM energy levels were crucial for hole extraction and recombination suppression at the interface. The fundamental techniques for obtaining a high-performance device were to increase the material carrier conveying capacities while retaining low-charge recombination [172].

You et al. developed PBDTT and PBTTT (D-A polymers) high-efficiency HTMs of PSCs. In the PBDTT and PBTTT, IDT or IDTT was used as the D unit, BDD served as the A unit, and thienothiophene acted as a π-bridge. The n-i-p PSCs incorporating these polymer HTMs displayed a highly promising device performance (PCE of around 20%). The devices with PBDTT functioned slightly better than PBTTT because of the superior solubility of IDT, which resulted in a smoother film and better perovskite/HTM/anode interfacial contact [165].

Shalan et al. investigated a new polymeric HTMs P(mPhDTP) (poly(1-(4-methoxyphenyl)-2,5-bis(5-methylthiophen-2-yl)-1H-pyrrole)), P(hPhDTP) (poly(1-(4-hexylphenyl)-2,5-bis(5-methylthiophen-2-yl)-1H-pyrrole)), P(hBT) (poly(3-hexyl-5,5′-dimethyl-2,3′-bithiophene)) and P(BT) (poly(5,5′-dimethyl-2,3′-bithiophene). These obtained polymers were discovered to be extremely soluble in a variety of halogenated and non-halogenated solvents, making them eco-friendly materials. The HOMO/LUMO band positions of the tested HTMs were aligned with those of perovskite, guaranteeing that holes were extracted from the $CH_3NH_3PbI_3$ to the HTM layer with appropriate driving force. The highest PCE (15.71%) was indicated by a device with p(hPhDTP) [166].

Kranthiraja et al. used a π-conjugated polymer P-TT-TPD (poly[4,8-bis(2-(4-(2-ethylhexyloxy)phenyl)-5-thienyl)benzo[1,2-b:4,5-b′]dithiophene-alt-1,3-bis(6-octylthieno[3,2-b]thiophen-2-yl)-5-(2-hexyldecyl)-4H-thieno[3,4-c]pyrrole-4,6(5H)-dione]) for PSCs. The device of P-TT-TPD had a PCE of 16.82% and 17.09% in dopant-free and tris(pentafluorophenyl) borane-doped PSCs, respectively, due to P-TT-TPD good solution processability, well-suited energy levels, its high mobility, better passivation and high-dipole moment difference between the ground and excited states [167].

Kong et al. studied F-substituted benzodithiophene copolymers PBDT[2H]T, PBDT[2F]T, PBDT(T)[2F]T as dopant-free efficient HTMs in PSCs. The PSC of PBDT[2F]T was shown a PCE of 17.52%. The experiments revealed that PBDT[2F]T as an HTM could extract holes while concurrently passivating surface traps, making it a strong rival to the doped spiro-OMeTAD. Furthermore, the hydrophobic character of PBDT[2F]T was provided greater ambient stability [168].

You et al. reported polymeric HTMs PBDB-O, PBDB-T (alkoxy and thiophene as the side chain of BDT appropriately), and PBDB-Cz (carbazole as the conjugated side chains of BDT). PBDB-Cz had the highest HOMO energy level, hole mobility, passivation effect, and effective interface modification, all of which helped improve the V_{oc}, J_{sc}, and FF in the devices. The PSC with PBDB-Cz was the best-performing PCE at 21.11% [169].

Liu et al. presented a novel polymer P25NH (DPP-based donor−acceptor) for application as a HTM in PSCs. The P25NH exhibited high mobility, better aggregation than P3HT and stability at low concentrations, and a perovskite surface passivation effect. All of these benefits resulted in devices with a dopant-free low concentration of the P25NH with a comparatively high PCE (17.3%) [170]. Furthermore, Liu et al. synthesized a new P5NH compound analogous to the previous polymer P25NH. The polymer P5NH was demonstrated to have higher mobility (5.13×10^{-2} cm^{-2} V^{-1} s^{-1}) than the reported P25NH (2.10×10^{-2} cm^{-2} V^{-1} s^{-1}). The fabricated PSC with the P5NH was achieved at an efficiency of 18.1% [170,171].

4.3. Polymers as Additives of Electron Transport Layers (ETL) and Electron-Transporting Materials (ETM)

The ETL collects electrons from the perovskite layer/s and transports them into the external circuit in perovskite solar cells. As a result, an ideal ETL material should have high electron mobility and a perovskite-like energy level.

Xiong et al. used a P3HT (poly(3-hexylthiophene)) as an additive to the electron transport layer of PCBM ([6,6]-phenyl-C61-butyric acid methyl ester). The addition of P3HT to PCBM could enhance the surface morphology of ETL as well as the moisture and water resistance of the ETL. The findings suggested that a small amount of P3HT did not result in a decreased power conversion efficiency (PCE) and could increase the PCBM aggregation, resulting in an improved ETL moisture and water resistance [173].

Jiang et al. applied a doping PCBM with F8BT (poly(9,9-dioctylfluorene-co-benzothiadiazole) as the electron transport layer. Doping with F8BT resulted in the creation of a smooth and uniform ETL, which was beneficial for electron-hole pair separation and hence increased the PSC performance. The power conversion efficiency of 15% of the PSC with 5 wt.% F8BT in PCBM was achieved (Figure 6) [174].

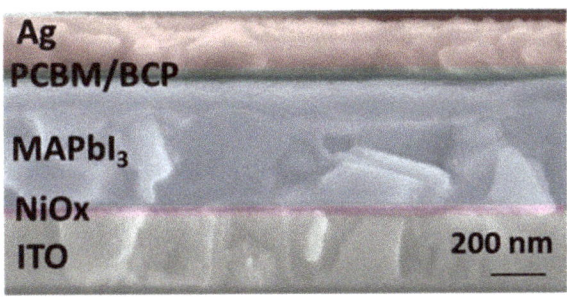

(a)

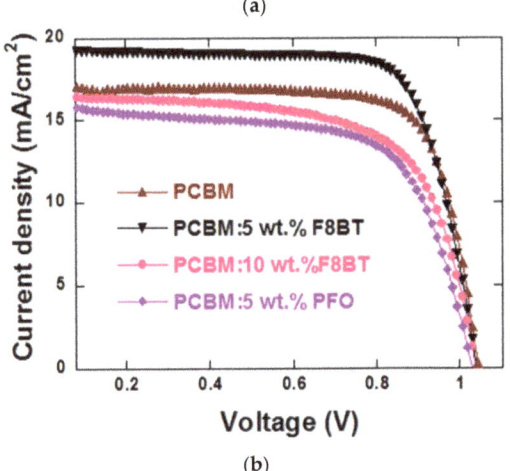

(b)

Figure 6. The sectional SEM image of PSC (**a**) and photocurrent density–voltage curves of the devices with F8BT in PCBM (**b**) [174].

Furthermore, You et al. introduced a biological polymer HP (heparin potassium) for stabilizing the ETL (SnO_2) dispersion and depositing arrangement of ETL. This method was discovered to enhance the interface contact between the ETL and the perovskite layer by generating vertically aligned crystal growth of mixed-cation perovskites. The planar PSCs based on SnO_2–HP had a PCE of over 23% (6 mg mL^{-1}) on rigid substrates and 19.47% on flexible substrates [175].

Liu et al. studied BCP (bathocuproine)/PMMA (poly(methyl methacrylate)) and BCP/PVP (polyvinylpyrrolidone) films as hole-blocking/electron-transporting interfacial layers. The storage stability of PSCs with BCP/PMMA was greatly improved over the PSCs with BCP, but the photovoltaic performance was marginally reduced when PMMA was added. The increased hydrophobicity and moisture resistance of the resultant BCP/PMMA layer ensured better storage stability. The PVP enhanced electron transport over the BCP-based interfacial layer to the cathode, resulting in greater current densities and power efficiency in the devices (Figure 7) [176].

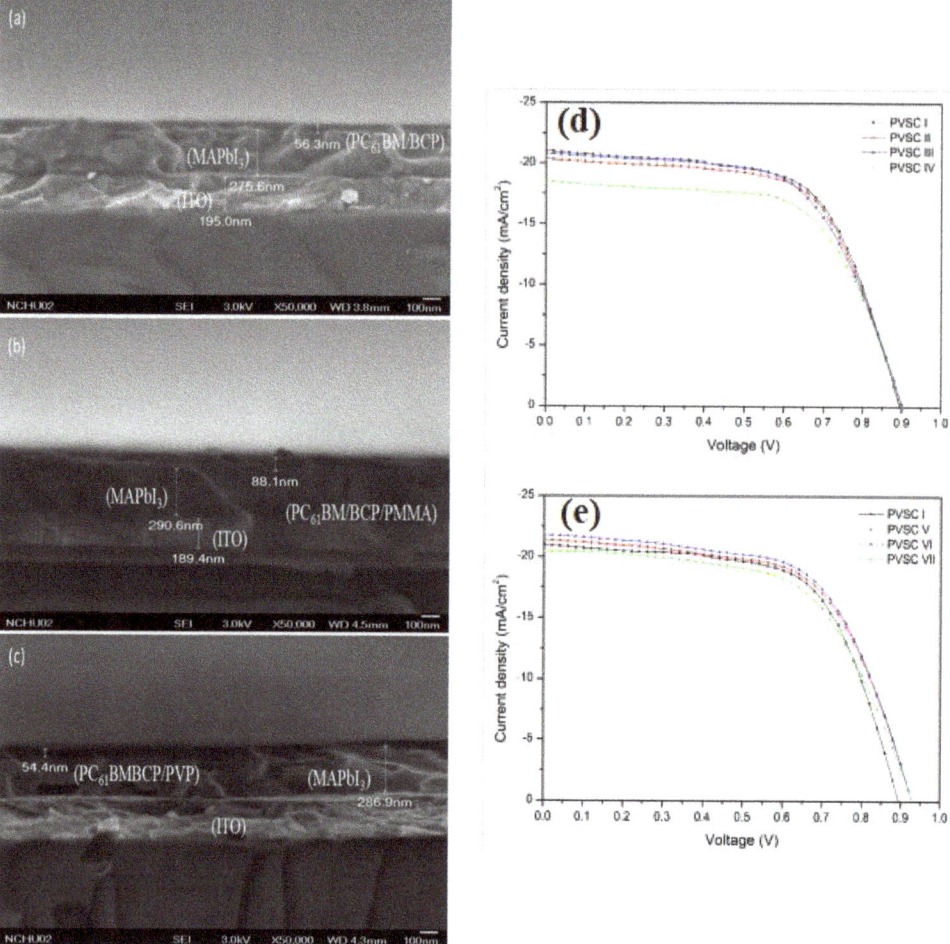

Figure 7. The cross-sectional SEM images of ITO/PEDOT:PSS/MAPbI$_3$/PC$_{61}$BM with BCP (**a**), BCP/PMMA (**b**), BCP/PVP (**c**) and the current density–potential plots of PSCs with BCP (PVSC I) (**d,e**), BCP/PMMA (PVSC II–IV) (**d**), BCP/PVP (PVSC V–VII)) (**e**) [176].

In addition, Said et al. investigated the impact of the sp^2-N substitution position in the main chains of the polymeric compounds on the photovoltaic properties of devices. They employed pBTT, pBTTz, and pSNT (naphthalenediimide-based n-type polymers) as ETLs in PSCs. Adding sp^2-N atoms to the donor thiophene units of pBTT resulted in pBTTz, which had somewhat lower electron mobility but greatly enhanced the PCE of PSCs. However, the PSC performance of pSNT with two extra sp^2-N atoms and very high electron mobility was significantly worse. Furthermore, the electron-rich sulfur atoms had a significant influence on the passivating of the under-coordinated Pb-atoms, as evidenced by the current density–voltage (J–V) hysteresis curves of the devices with pBTTz [177].

Tian et al. reported n-type conjugated polymers with fluoro- and amino-side chains (PN, PN-F25%, and PN-F50%) as ETM in a perovskite device of structure ITO/NiO$_x$/CH$_3$NH$_3$PbI$_{3-x}$Cl$_x$/PN or PN-F25% or PN-F50%/Ag. It was discovered that the amino side chains could provide good interface modification capabilities, while the fluoro side chains could supply hydrophobic qualities to these polymers. As a result, the bifunctional conjugated polymers successfully improved the performance of tested solar cells (17.5%), which was higher than the performance of devices with PC61BM (14.0%). Furthermore, the bifunctional ETMs were improved PSCs stability significantly [178].

Moreover, Elnaggar et al. tested a pyrrolo[3,4-c]pyrrole-1,4-dione-based n-type copolymer (P1, this polymer with the fullerene derivative [60]PCBM) as an electron transport material for PSCs. A conjugated polymer P1 and its composites with [60] PCBM provided reasonable efficiencies of 12–14%, respectively. Importantly, the use of the P1-PCBM [60] composite's ETL resulted in a significant increase in the operational stability of PSCs [179].

Yan et al. synthesized semiconducting copolymers NDI-Se, NDI-BiSe, and NDI-TriSe based naphthalene-diimide. The addition of a biselenophene or triselenophene unit to a polymer increased the polymer's planarity and delocalization, as well as conductivity. The perovskite solar cells of the ITO/NiO$_x$/perovskite/NDI-selenophene/Ag structure were prepared. The power conversion efficiency of 9.51% (NDI-Se), 7.66% (NDI-BiSe), and 14.00% (NDI-TriSe) were obtained [180].

These studies contributed to the development of new polymeric ETLs and additives to the ETL by providing useful design recommendations.

4.4. Polymeric Interlayer/s

Interface engineering has been shown to be an effective method for reducing defect density in organic–inorganic hybrid PSCs and is commonly utilized to improve their performance [181–188].

Ding et al. exhibited a PVAc (polyvinyl acetate) as an agent modifying the surface of perovskite (CsPbBr$_3$) film. The combination of O atoms in the carbonyl group (C=O) of PVAc with the positively charged under-coordinated Pb^{2+} and Cs$^+$ defect ions could contribute to the reduction of the CsPbBr$_3$ surface defects and improve the energy-level alignment between the carbon electrode (work function) and the valance band (VB) of perovskite. This results in reduced carrier recombination and energy loss at the perovskite/carbon contact, which effectively increased the V_{oc} and PCE [181].

Zhao et al. applied a DPP-DTT (poly(N-alkyldiketopyrrolo-pyrrole dithienylthieno[3,2-b]thiophene)) multifunctional passivation layer to receive the stable and highly efficient devices. By coordinate bonding between the atoms containing lone-pair electrons (sulphur, oxygen, and nitrogen) in DPP-DTT and the under the coordinate Pb atoms in perovskite, DPP-DTT acted as an efficient passivation layer to decrease defects on the perovskite surface. The DPP-DTT could function as a hole-selective layer because of its high hole mobility (~10 cm^2 V^{-1} s^{-1}) and acceptable valence band (VB) (−5.31 eV) between the perovskite (−5.67 eV) and spiro-OMeTAD (−5.22 eV) to efficiently increase hole extraction and transport. DPP-DTT as an ultra-hydrophobic agent, improved the perovskite stability [182].

Sharma et al. developed an n-type conjugated polymer with a naphthalene diimide core and a vinylene linker and oligo (ethylene glycol) (P2G) as a stable cathode interface layer (CIL). P2G was shown to be an effective CIL for lowering interfacial energy barriers in hybrid perovskite solar cells, with a PCE of 17.6% for MAPbI$_3$-based p-i-n planar devices vs. 15% for reference devices. Because of the effectiveness of P2G CIL, there appeared to be a potential technique for creating alcohol/water-soluble polymer interlayers with desirable electrical and electronic characteristics [183].

Zhou et al. reported a photoinitiation-crosslinked zwitterionic polymer (Dex-CB-MA) as an interfacial layer played an important role in the perovskite device performance. Dex-CB-MA (dextran with carboxybetaine modified by methacrylate) was used as the interfacial layer between the PEDOT:PSS (as hole extraction layer, HEL) and the perovskite layer and to improve the morphology of the perovskite film. The Dex-CB-MA was discovered to generate an effective charge-transfer process in perovskite solar cells. As a consequence, when compared to PSCs based on the PEDOT:PSS HEL, the PEDOT:PSS/Dex-CB-MA HEL perovskite solar cells using the PEDOT:PSS/Dex-CB-MA HEL showed a 30% increase in power conversion efficiency [184].

Zhao et al. demonstrated a polymer-based difluorobenzothiadiazole (PffBT4T-C9C13) as the interfacial material for planar PSCs. The PffBT4T-C9C13 was deposited between the HTL (spiro-OMeTAD) and the perovskite absorber by utilizing a well-refined deposition technique. When the prepared polymer was decorated, a uniform perovskite layer with a large grain was formed. The PffBT4T-C9C13 has passivated the surface defects of perovskite film and also protected the film from water corrosion. At the interface, the charge collection was adequately suppressed, which helped with charge extraction and transport. As a consequence, the ITO/TiO$_2$/MA$_{1-x}$FA$_x$PbI$_3$/PffBT4T-C9C13/spiro-MeOTAD/Ag structure of the device had the best power conversion efficiency of 19.37% (0.50 mg mL^{-1} of PffBT4T-C9C13) [185].

Liu et al. reported a conjugated polyelectrolyte PTFTS (poly[N-(4-sulfonatophenyl)-4,4′-diphenylamine-alt-N-(p-trifluoromethyl)phenyl-4,4′-diphenylamine] sodium salt) as the interlayer between GO (graphene oxide) and the perovskite. The polymers' interlayers were allowed for the identification of the recombination channels at the front-contact interface in the inverted-type planar PSCs. The sulfonate-charged PTFTS had the unanticipated benefit of causing strong contact between PTFTS as the binding force and GO, allowing for the development of a uniform interfacial layer onto GO using a simple wet-chemical method. As a result, the best PCE for the device with PTFTS was 18.39% [186].

Kim et al. utilized a PDMS (polydimethylsiloxane) interlayer between CuSCN (inorganic HTL) and the absorber to stabilize the perovskite deposition and working device. The PDMS successfully blocked a disintegration of perovskite at the surface throughout the upper layer's deposition. In addition, it was noticed that the polymer could form chemical bonds with CuSCN and perovskite as the cross-linking interlayer. This novel cross-linking layer alleviated the interfacial traps/defects in the solar cells and amended the hole-extraction property at the interface. The PDMS as a cross-linking interlayer was enabled to receive a highly efficient PSC with MAPbI$_3$/PDMS/CuSCN with a PCE of 19.04% [187].

Wang et al. synthesized a naphthalene imide dimer (2FBT2NDI) and applied it as interface material for inverted PSCs. With the introduction conjugated skeleton benzothiadiazole-dithiophene unit, two fluorine atoms could enhance intermolecular interactions and regulate the energy levels. The exploitation 2FBT2NDI as a polymeric interlayer suppressed the recombination of a charge trapped at the perovskite/ETL interface and improved the electron extraction and the efficiency of the device. The perovskite device with 2FBT2NDI exhibited the best PCE of 20.1% [188]. These works give precepts for implementing interface control and modification utilizing polymers.

The above-presented results prove that polymers can be utilized in PSCs to aid nucleation, control perovskite film crystallization, and improve device stability by developing different interactions with the perovskite films. The application of the DI monomer by adding it to the PbI$_2$ precursor (polymerization-assisted grain growth (PAGG) technique) allowed for the appropriate interaction of the polymer with PbI$_2$, thanks to which the PCE was obtained at the level of 23%, and the rate of perovskite degradation was slowed down [148]. Moreover, the utilization of the naphthalene imide dimer (2FBT2NDI) as interface material for reverse PSCs allowed for the inhibiting recombination of the charge trapped at the perovskite/ETL interface, improving the electron extraction as well as the device efficiency (20.1%) [188]. Additionally, polymers can also be shown high hole mobility, which makes their use possible as hole-transporting materials. The polymeric HTMs PBDB-Cz (carbazole as the conjugated side chains of BDT), was indicated a PCE of 21.11% [169]. Furthermore, biological polymer HP (heparin potassium) was introduced for stabilizing the electron transport layers' dispersion and depositing the arrangement of the ETL (the PCE of over 23%) [175]. Consequently, it is imperative to design the novel polymers used in perovskite solar cells to improve the stability and performance of the PSCs.

5. Summary and Conclusions

Photovoltaics is a strongly and rapidly growing branch of renewable energy sources. Many new materials are being used in solar cells. Polymeric materials are also widely used in devices that convert solar radiation into electricity. As presented in this review paper, polymeric compounds are widely used in many fields in photovoltaic cells due to their numerous advantages, which undoubtedly include the possibility of modifying their chemical structure and thus adjusting their physical and chemical properties to the given needs. Polymeric materials are also widely used in devices that convert solar radiation into electricity. As presented in this review paper, polymeric compounds are widely used in many fields in photovoltaic cells due to their numerous advantages, which undoubtedly include the possibility of modifying their chemical structure and thus adjusting their physical and chemical properties to the given needs. Given current trends and recent literature, more and more new polymeric materials are finding applications in photovoltaic cells, as seen, for example, in their use as dyes in DSSCs or HTMs in PSCs.

In summary, polymeric materials are increasingly used in a wide range of research and technological solutions and will certainly become more widely and extensively used in solar cells as well. As noted, polymers are used as the flexible transparent substrates for all types of photovoltaic devices discussed, as materials that impart gel character to electrolytes in DSSCs, counter-electrodes, materials responsible for the pore formation in inorganic oxides used in DSSCs and PSCs. They are widely used also in BHJ, mainly as donor materials, but numerous studies report that the substitution of acceptor fullerenes by polymers can also be found. It is also worth remembering to use polymers as intermediate or buffer layers, supporting the transport or separation of the generated charges to the appropriate electrodes. As previously mentioned, organic compounds and among them, polymers, are and will be widely used in new technologies for obtaining electric currents due to their relatively low costs of preparation, easy modification of the chemical structure, and thus easily obtain the required properties, as well as the possibility of manufacturing suitable layers from them.

As mentioned in the summaries of the individual sections, it is not easy or even entirely possible to compare and determine the best polymer on the basis of the collected results due to the differences in structures and methods of preparation of individual devices. In the case of the DSSC cells, apart from the use of the polymer as a dye or counter-electrode, it is very difficult to determine the direct effect on the recorded device parameters. Even in the case of a dye or counter-electrode, the preparation methods and conditions play a huge role, with very often different results in different publications. Therefore, it can only be stated that the following polymers are used PEN, PET as flexible substrates; PS, PVP, P123 used to obtain pores in the oxide material, PEO, PAN, PEG as materials to give a gel structure to the electrolyte and PANI, PPy, PTh, PEDOT, PEDOT:PSS as counter-electrodes. In the case of BHJ solar cells, high efficiencies are registered mainly for donors containing thiophene rings in the polymer, such as PTB-7, PTB-7-Th, D18, P106, among others. As for acceptors, polymeric compounds are used much less frequently, while high efficiencies are registered for polymers containing naphthalene moieties. In perovskite solar cells, polymers are used as additives to facilitate the nucleation and crystallization processes in the perovskite layer(s). By adding a DI monomer to the PbI_2 precursor, a PCE was obtained of 23.0%. Polymers can also be used as electrons (biological polymer HP, the PCE of over 23%), hole-transporting materials (PBDB-Cz, PCE = 21.11%), and interface layer(s) (2FBT2NDI, PCE = 20.1%).

Author Contributions: P.G. and B.J., conceptualization; P.G., A.K.P. and M.F.A., writing—original draft preparation; P.G., A.K.P. and M.F.A., visualization; P.G. and B.J., writing—review and editing; B.J., supervision. All authors have read and agreed to the published version of the manuscript.

Funding: This research received no external funding.

Institutional Review Board Statement: Not applicable.

Informed Consent Statement: Not applicable.

Data Availability Statement: Not applicable.

Conflicts of Interest: The authors declare no conflict of interest.

References

1. Grätzel, M. Recent advances in sensitized mesoscopic solar cells. *Acc. Chem. Res.* **2009**, *42*, 1788–1798. [CrossRef] [PubMed]
2. Khatibi, A.; Razi Astaraei, F.; Ahmadi, M.H. Generation and combination of the solar cells: A current model review. *Energy Sci. Eng.* **2019**, *7*, 305–322. [CrossRef]
3. Palm, J.; Probst, V.; Karg, F.H. Second generation CIS solar modules. *Sol. Energy* **2004**, *77*, 757–765. [CrossRef]
4. El Chaar, L.; Lamont, L.A.; El Zein, N. Review of photovoltaic technologies. *Renew. Sustain. Energy Rev.* **2011**, *15*, 2165–2175. [CrossRef]
5. Parida, B.; Iniyan, S.; Goic, R. A review of solar photovoltaic technologies. *Renew. Sustain. Energy Rev.* **2011**, *15*, 1625–1636. [CrossRef]
6. Polman, A.; Knight, M.; Garnett, E.C.; Ehrler, B.; Sinke, W.C. Photovoltaic materials: Present efficiencies and future challenges. *Science* **2016**, *352*, aad4424. [CrossRef]
7. NREL. Best Research-Cell Efficiencies: Rev. 04-06-2020. Best Res. Effic. Chart | Photovolt. Res. | NREL. 2020. Available online: https://www.nrel.gov/pv/cell-efficiency.html (accessed on 15 February 2022).
8. Roncali, J.; Leriche, P.; Blanchard, P. Molecular materials for organic photovoltaics: Small is beautiful. *Adv. Mater.* **2014**, *26*, 3821–3838. [CrossRef]
9. Ye, M.; Wen, X.; Wang, M.; Iocozzia, J.; Zhang, N.; Lin, C.; Lin, Z. Recent advances in dye-sensitized solar cells: From photoanodes, sensitizers and electrolytes to counter electrodes. *Mater. Today* **2015**, *18*, 155–162. [CrossRef]
10. Gnida, P.; Libera, M.; Pająk, A.; Schab-Balcerzak, E. Examination of the Effect of Selected Factors on the Photovoltaic Response of Dye-Sensitized Solar Cells. *Energy Fuels* **2020**, *34*, 14344–14355. [CrossRef]
11. Song, L.; Du, P.; Xiong, J.; Ko, F.; Cui, C. Efficiency enhancement of dye-sensitized solar cells by optimization of electrospun ZnO nanowire/nanoparticle hybrid photoanode and combined modification. *Electrochim. Acta* **2015**, *163*, 330–337. [CrossRef]
12. Nath, B.; Pradhan, B.; Panda, S.K. Optical tunability of lead free double perovskite Cs2AgInCl6: Via composition variation. *New J. Chem.* **2020**, *44*, 18656–18661. [CrossRef]
13. Ji, J.M.; Zhou, H.; Eom, Y.K.; Kim, C.H.; Kim, H.K. 14.2% Efficiency Dye-Sensitized Solar Cells by Co-sensitizing Novel Thieno[3,2-b]indole-Based Organic Dyes with a Promising Porphyrin Sensitizer. *Adv. Energy Mater.* **2020**, *10*, 2000124. [CrossRef]
14. Kakiage, K.; Aoyama, Y.; Yano, T.; Oya, K.; Fujisawa, J.I.; Hanaya, M. Highly-efficient dye-sensitized solar cells with collaborative sensitization by silyl-anchor and carboxy-anchor dyes. *Chem. Commun.* **2015**, *51*, 15894–15897. [CrossRef] [PubMed]
15. Liu, Q.; Jiang, Y.; Jin, K.; Qin, J.; Xu, J.; Li, W.; Xiong, J.; Liu, J.; Xiao, Z.; Sun, K.; et al. 18% Efficiency organic solar cells. *Sci. Bull.* **2020**, *65*, 272–275. [CrossRef]
16. Han, Y.W.; Jeon, S.J.; Lee, H.S.; Park, H.; Kim, K.S.; Lee, H.W.; Moon, D.K. Evaporation-Free Nonfullerene Flexible Organic Solar Cell Modules Manufactured by An All-Solution Process. *Adv. Energy Mater.* **2019**, *9*, 1902065. [CrossRef]
17. Kim, J.Y.; Park, S.; Lee, S.; Ahn, H.; Joe, S.Y.; Kim, B.J.; Son, H.J. Low-Temperature Processable High-Performance D–A-Type Random Copolymers for Nonfullerene Polymer Solar Cells and Application to Flexible Devices. *Adv. Energy Mater.* **2018**, *8*, 1801601. [CrossRef]
18. Gong, S.C.; Jang, S.K.; Ryu, S.O.; Jeon, H.; Park, H.H.; Chang, H.J. Post annealing effect of flexible polymer solar cells to improve their electrical properties. *Curr. Appl. Phys.* **2010**, *10*, e192–e196. [CrossRef]
19. Jung, H.S.; Han, G.S.; Park, N.G.; Ko, M.J. Flexible Perovskite Solar Cells. *Joule* **2019**, *3*, 1850–1880. [CrossRef]
20. Feng, J.; Zhu, X.; Yang, Z.; Zhang, X.; Niu, J.; Wang, Z.; Zuo, S.; Priya, S.; Liu, S.F.; Yang, D. Record Efficiency Stable Flexible Perovskite Solar Cell Using Effective Additive Assistant Strategy. *Adv. Mater.* **2018**, *30*, e1801418. [CrossRef]
21. Yang, D.; Yang, R.; Priya, S.; Liu, S.F. Recent Advances in Flexible Perovskite Solar Cells: Fabrication and Applications. *Angew. Chemie—Int. Ed.* **2019**, *58*, 4466–4483. [CrossRef]
22. Noorasid, N.S.; Arith, F.; Mustafa, A.N.; Azam, M.A.; Mahalingam, S.; Chelvanathan, P.; Amin, N. Current advancement of flexible dye sensitized solar cell: A review. *Optik* **2022**, *254*, 168089. [CrossRef]
23. Devadiga, D.; Selvakumar, M.; Shetty, P.; Santosh, M.S. The integration of flexible dye-sensitized solar cells and storage devices towards wearable self-charging power systems: A review. *Renew. Sustain. Energy Rev.* **2022**, *159*, 112252. [CrossRef]
24. Fan, X. Flexible dye-sensitized solar cells assisted with lead-free perovskite halide. *J. Mater. Res.* **2022**, *37*, 866–875. [CrossRef]
25. Khan, A.; Liang, C.; Huang, Y.T.; Zhang, C.; Cai, J.; Feng, S.P.; Li, W. Di Template-Electrodeposited and Imprint-Transferred Microscale Metal-Mesh Transparent Electrodes for Flexible and Stretchable Electronics. *Adv. Eng. Mater.* **2019**, *21*, 1801363. [CrossRef]
26. Mustafa, M.N.; Sulaiman, Y. Fully flexible dye-sensitized solar cells photoanode modified with titanium dioxide-graphene quantum dot light scattering layer. *Sol. Energy* **2020**, *212*, 332–338. [CrossRef]
27. Wante, H.P.; Yap, S.L.; Aidan, J.; Saikia, P. Efficiency Enhancement of Dye Sensitized Solar cells (DSSCs) by Atmospheric DBD Plasma Modification of Polyetherimide (PEI) Polymer Substrate. *J. Mater. Environ. Sci.* **2020**, *2020*, 713–722.

28. Hellert, C.; Wortmann, M.; Frese, N.; Grötsch, G.; Cornelißen, C.; Ehrmann, A. Adhesion of electrospun poly(Acrylonitrile) nanofibers on conductive and isolating foil substrates. *Coatings* **2021**, *11*, 249. [CrossRef]
29. Baiju, K.G.; Murali, B.; Subba Rao, R.; Jayanarayanan, K.; Kumaresan, D. Heat sink assisted elevated temperature sintering process of TiO_2 on polymer substrates for producing high performance flexible dye-sensitized solar cells. *Chem. Eng. Process.* **2020**, *149*, 107817. [CrossRef]
30. Hou, W.; Xiao, Y.; Han, G.; Zhou, H.; Chang, Y.; Zhang, Y. Preparation of mesoporous titanium dioxide anode by a film- and pore-forming agent for the dye-sensitized solar cell. *Mater. Res. Bull.* **2016**, *76*, 140–146. [CrossRef]
31. Bharwal, A.K.; Manceriu, L.; Alloin, F.; Iojoiu, C.; Dewalque, J.; Toupance, T.; Henrist, C. Tuning bimodal porosity in TiO_2 photoanodes towards efficient solid-state dye-sensitized solar cells comprising polysiloxane-based polymer electrolyte. *Microporous Mesoporous Mater.* **2019**, *273*, 226–234. [CrossRef]
32. Zukalovà, M.; Zukal, A.; Kavan, L.; Nazeeruddin, M.K.; Liska, P.; Grätzel, M. Organized mesoporous TiO_2 films exhibiting greatly enhanced performance in dye-sensitized solar cells. *Nano Lett.* **2005**, *5*, 1789–1792. [CrossRef]
33. Agarwala, S.; Kevin, M.; Wong, A.S.W.; Peh, C.K.N.; Thavasi, V.; Ho, G.W. Mesophase ordering of TiO_2 film with high surface area and strong light harvesting for dye-sensitized solar cell. *ACS Appl. Mater. Interfaces* **2010**, *2*, 1844–1850. [CrossRef]
34. Li, C.; Xin, C.; Xu, L.; Zhong, Y.; Wu, W. Components control for high-voltage quasi-solid state dye-sensitized solar cells based on two-phase polymer gel electrolyte. *Sol. Energy* **2019**, *181*, 130–136. [CrossRef]
35. Huang, Y.; Yang, F.; Tang, W.; Deng, Z.; Zhang, M.; Ruan, W. Azobenzene-dyed, nanofibrous microstructure for improving photothermal effect of polymer gel electrolyte. *Sol. Energy* **2021**, *230*, 1–9. [CrossRef]
36. Kumar Shah, D.; Son, Y.H.; Lee, H.R.; Shaheer Akhtar, M.; Kim, C.Y.; Yang, O.B. A stable gel electrolyte based on poly butyl acrylate (PBA)-co-poly acrylonitrile (PAN) for solid-state dye-sensitized solar cells. *Chem. Phys. Lett.* **2020**, *754*, 137756. [CrossRef]
37. Khan, S.A.; Li, L.; Zhao, D.; Chen, S. Increased power conversion efficiency of dye-sensitized solar cells with counter electrodes based on porous polypyrrole. *React. Funct. Polym.* **2020**, *148*, 104483. [CrossRef]
38. Wu, J.; Wu, S.; Sun, W. Electropolymerization and application of polyoxometalate-doped polypyrrole film electrodes in dye-sensitized solar cells. *Electrochem. Commun.* **2021**, *122*, 106879. [CrossRef]
39. Karakuş, M.Ö.; Yakışıklıer, M.E.; Delibaş, A.; Ayyıldız, E.; Çetin, H. Anionic and cationic polymer-based quasi-solid-state dye-sensitized solar cell with poly(aniline) counter electrode. *Sol. Energy* **2020**, *195*, 565–572. [CrossRef]
40. Farooq, S.; Tahir, A.A.; Krewer, U.; Shah, A.U.H.A.; Bilal, S. Efficient photocatalysis through conductive polymer coated FTO counter electrode in platinum free dye sensitized solar cells. *Electrochim. Acta* **2019**, *320*, 134544. [CrossRef]
41. Mazloum-Ardakani, M.; Arazi, R. Enhancement of photovoltaic performance using a novel photocathode based on poly(3,4-ethylenedioxythiophene)/Ag–CuO nanocomposite in dye-sensitized solar cells. *Comptes Rendus* **2020**, *23*, 105–115.
42. Sathiyan, G.; Sivakumar, E.K.T.; Ganesamoorthy, R.; Thangamuthu, R.; Sakthivel, P. Review of carbazole based conjugated molecules for highly efficient organic solar cell application. *Tetrahedron Lett.* **2016**, *57*, 243–252. [CrossRef]
43. Sakthivel, P.; Ban, T.W.; Kim, S.; Kim, S.; Gal, Y.S.; Chae, E.A.; Shin, W.S.; Moon, S.J.; Lee, J.C.; Jin, S.H. Synthesis and studies of methyl ester substituted thieno-o-quinodimethane fullerene multiadducts for polymer solar cells. *Sol. Energy Mater. Sol. Cells* **2013**, *113*, 13–19. [CrossRef]
44. Brabec, C.J.; Winder, C.; Sariciftci, N.S.; Hummelen, J.C.; Dhanabalan, A.; Van Hal, P.A.; Janssen, R.A.J. A low-bandgap semiconducting polymer for photovoltaic devices and infrared emitting diodes. *Adv. Funct. Mater.* **2002**, *12*, 709–712. [CrossRef]
45. Watson, B.W.; Meng, L.; Fetrow, C.; Qin, Y. Core/shell conjugated polymer/quantum dot composite nanofibers through orthogonal non-covalent interactions. *Polymers* **2016**, *8*, 408. [CrossRef]
46. Zhang, Z.; Qin, Y. Structurally Diverse Poly(thienylene vinylene)s (PTVs) with Systematically Tunable Properties through Acyclic Diene Metathesis (ADMET) and Postpolymerization Modification. *Macromolecules* **2016**, *49*, 3318–3327. [CrossRef]
47. Zhu, L.; Zhong, W.; Qiu, C.; Lyu, B.; Zhou, Z.; Zhang, M.; Song, J.; Xu, J.; Wang, J.; Ali, J.; et al. Aggregation-Induced Multilength Scaled Morphology Enabling 11.76% Efficiency in All-Polymer Solar Cells Using Printing Fabrication. *Adv. Mater.* **2019**, *31*, e1902899. [CrossRef]
48. Guo, Y.; Li, Y.; Awartani, O.; Han, H.; Zhao, J.; Ade, H.; Yan, H.; Zhao, D. Improved Performance of All-Polymer Solar Cells Enabled by Naphthodiperylenetetraimide-Based Polymer Acceptor. *Adv. Mater.* **2017**, *29*, 1700309. [CrossRef]
49. Kolhe, N.B.; Tran, D.K.; Lee, H.; Kuzuhara, D.; Yoshimoto, N.; Koganezawa, T.; Jenekhe, S.A. New Random Copolymer Acceptors Enable Additive-Free Processing of 10.1% Efficient All-Polymer Solar Cells with Near-Unity Internal Quantum Efficiency. *ACS Energy Lett.* **2019**, *4*, 1162–1170. [CrossRef]
50. Liao, J.F.; Wu, W.Q.; Zhong, J.X.; Jiang, Y.; Wang, L.; Kuang, D. Bin Enhanced efficacy of defect passivation and charge extraction for efficient perovskite photovoltaics with a small open circuit voltage loss. *J. Mater. Chem. A* **2019**, *7*, 9025–9033. [CrossRef]
51. Liu, G.; Liu, C.; Lin, Z.; Yang, J.; Huang, Z.; Tan, L.; Chen, Y. Regulated Crystallization of Efficient and Stable Tin-Based Perovskite Solar Cells via a Self-Sealing Polymer. *ACS Appl. Mater. Interfaces* **2020**, *12*, 14049–14056. [CrossRef]
52. Han, T.H.; Lee, J.W.; Choi, C.; Tan, S.; Lee, C.; Zhao, Y.; Dai, Z.; De Marco, N.; Lee, S.J.; Bae, S.H.; et al. Perovskite-polymer composite cross-linker approach for highly-stable and efficient perovskite solar cells. *Nat. Commun.* **2019**, *10*, 520. [CrossRef]
53. Fairfield, D.J.; Sai, H.; Narayanan, A.; Passarelli, J.V.; Chen, M.; Palasz, J.; Palmer, L.C.; Wasielewski, M.R.; Stupp, S.I. Structure and chemical stability in perovskite-polymer hybrid photovoltaic materials. *J. Mater. Chem. A* **2019**, *7*, 1687–1699. [CrossRef]

54. Chen, W.; Shi, Y.; Wang, Y.; Feng, X.; Djurišić, A.B.; Woo, H.Y.; Guo, X.; He, Z. N-type conjugated polymer as efficient electron transport layer for planar inverted perovskite solar cells with power conversion efficiency of 20.86%. *Nano Energy* **2020**, *68*, 104363. [CrossRef]
55. Yang, D.; Zhang, X.; Wang, K.; Wu, C.; Yang, R.; Hou, Y.; Jiang, Y.; Liu, S.; Priya, S. Stable Efficiency Exceeding 20.6% for Inverted Perovskite Solar Cells through Polymer-Optimized PCBM Electron-Transport Layers. *Nano Lett.* **2019**, *19*, 3313–3320. [CrossRef]
56. Syed, A.A.; Poon, C.Y.; Li, H.W.; Zhu, F. A sodium citrate-modified-PEDOT:PSS hole transporting layer for performance enhancement in inverted planar perovskite solar cells. *J. Mater. Chem. C* **2019**, *7*, 5260–5266. [CrossRef]
57. Zhang, L.; Zhou, X.; Zhong, X.; Cheng, C.; Tian, Y.; Xu, B. Hole-transporting layer based on a conjugated polyelectrolyte with organic cations enables efficient inverted perovskite solar cells. *Nano Energy* **2019**, *57*, 248–255. [CrossRef]
58. Zhang, F.; Yao, Z.; Guo, Y.; Li, Y.; Bergstrand, J.; Brett, C.J.; Cai, B.; Hajian, A.; Guo, Y.; Yang, X.; et al. Polymeric, cost-effective, dopant-free hole transport materials for efficient and stable perovskite solar cells. *J. Am. Chem. Soc.* **2019**, *141*, 19700–19707. [CrossRef]
59. Lee, J.; Kim, G.W.; Kim, M.; Park, S.A.; Park, T. Nonaromatic Green-Solvent-Processable, Dopant-Free, and Lead-Capturable Hole Transport Polymers in Perovskite Solar Cells with High Efficiency. *Adv. Energy Mater.* **2020**, *10*, 290. [CrossRef]
60. Zhang, L.; Liu, C.; Wang, X.; Tian, Y.; Jen, A.K.Y.; Xu, B. Side-Chain Engineering on Dopant-Free Hole-Transporting Polymers toward Highly Efficient Perovskite Solar Cells (20.19%). *Adv. Funct. Mater.* **2019**, *29*, 1904856. [CrossRef]
61. Wang, M.; Wang, H.; Li, W.; Hu, X.; Sun, K.; Zang, Z. Defect passivation using ultrathin PTAA layers for efficient and stable perovskite solar cells with a high fill factor and eliminated hysteresis. *J. Mater. Chem. A* **2019**, *7*, 26421–26428. [CrossRef]
62. Tian, J.; Xue, Q.; Tang, X.; Chen, Y.; Li, N.; Hu, Z.; Shi, T.; Wang, X.; Huang, F.; Brabec, C.J.; et al. Dual Interfacial Design for Efficient CsPbI2Br Perovskite Solar Cells with Improved Photostability. *Adv. Mater.* **2019**, *31*, 1901152. [CrossRef]
63. Liu, C.; Zhang, L.; Li, Y.; Zhou, X.; She, S.; Wang, X.; Tian, Y.; Jen, A.K.Y.; Xu, B. Highly Stable and Efficient Perovskite Solar Cells with 22.0% Efficiency Based on Inorganic–Organic Dopant-Free Double Hole Transporting Layers. *Adv. Funct. Mater.* **2020**, *30*, 1908462. [CrossRef]
64. Tan, F.; Tan, H.; Saidaminov, M.I.; Wei, M.; Liu, M.; Mei, A.; Li, P.; Zhang, B.; Tan, C.S.; Gong, X.; et al. In Situ Back-Contact Passivation Improves Photovoltage and Fill Factor in Perovskite Solar Cells. *Adv. Mater.* **2019**, *31*, e1807435. [CrossRef]
65. Weerasinghe, H.C.; Huang, F.; Cheng, Y.B. Fabrication of flexible dye sensitized solar cells on plastic substrates. *Nano Energy* **2013**, *2*, 174–189. [CrossRef]
66. Kang, M.G.; Park, N.G.; Ryu, K.S.; Chang, S.H.; Kim, K.J. A 4.2% efficient flexible dye-sensitized TiO_2 solar cells using stainless steel substrate. *Sol. Energy Mater. Sol. Cells* **2006**, *90*, 574–581. [CrossRef]
67. Hou, W.; Xiao, Y.; Han, G.; Lin, J.Y. The applications of polymers in solar cells: A review. *Polymers* **2019**, *11*, 143. [CrossRef]
68. Zardetto, V.; Di Giacomo, F.; Garcia-Alonso, D.; Keuning, W.; Creatore, M.; Mazzuca, C.; Reale, A.; Di Carlo, A.; Brown, T.M. Fully plastic dye solar cell devices by low-temperature UV-irradiation of both the mesoporous TiO_2 photo- and platinized counter-electrodes. *Adv. Energy Mater.* **2013**, *3*, 1292–1298. [CrossRef]
69. Fu, Y.; Lv, Z.; Hou, S.; Wu, H.; Wang, D.; Zhang, C.; Zou, D. TCO-free, flexible, and bifacial dye-sensitized solar cell based on low-cost metal wires. *Adv. Energy Mater.* **2012**, *2*, 37–41. [CrossRef]
70. Barbé, C.J.; Arendse, F.; Comte, P.; Jirousek, M.; Lenzmann, F.; Shklover, V.; Grätzel, M. Nanocrystalline titanium oxide electrodes for photovoltaic applications. *J. Am. Ceram. Soc.* **1997**, *80*, 3157–3171. [CrossRef]
71. Park, J.T.; Moon, J.; Choi, G.H.; Lim, S.M.; Kim, J.H. Facile graft copolymer template synthesis of mesoporous polymeric metal-organic frameworks to produce mesoporous TiO_2: Promising platforms for photovoltaic and photocatalytic applications. *J. Ind. Eng. Chem.* **2020**, *84*, 384–392. [CrossRef]
72. Prakash, G.; Subramanian, K. Interaction of pyridine π-bridge-based poly(methacrylate) dyes for the fabrication of dye-sensitized solar cells with the influence of different strength phenothiazine, fluorene and anthracene sensitizers as donor units with new anchoring mode. *New J. Chem.* **2018**, *42*, 17939–17949. [CrossRef]
73. Ramasamy, S.; Boopathy, M.; Johnsanthoshkumar, S.; Subramanian, K. Structural engineering of poly-(methacrylate) bearing push-pull type pendants oxindole-phenothiazine with tetrazole anchoring acceptor for efficient organic photovoltaic cells. *Polymer* **2017**, *115*, 128–136. [CrossRef]
74. Wang, G.; Wu, Y.; Ding, W.; Yu, G.; Hu, Z.; Wang, H.; Liu, S.; Zou, Y.; Pan, C. Photovoltaic performance of long-chain poly(triphenylamine-phenothiazine) dyes with a tunable π-bridge for dye-sensitized solar cells. *J. Mater. Chem. A* **2015**, *3*, 14217–14227. [CrossRef]
75. Nusbaumer, H.; Zakeeruddin, S.M.; Moser, J.E.; Grätzel, M. An alternative efficient redox couple for the dye-sensitized solar cell system. *Chemistry* **2003**, *9*, 3756–3763. [CrossRef]
76. Shakeel Ahmad, M.; Pandey, A.K.; Abd Rahim, N. Advancements in the development of TiO_2 photoanodes and its fabrication methods for dye sensitized solar cell (DSSC) applications. A review. *Renew. Sustain. Energy Rev.* **2017**, *77*, 89–108. [CrossRef]
77. Xiao, B.C.; Lin, L.Y. Tuning electrolyte configuration and composition for fiber-shaped dye-sensitized solar cell with poly(vinylidene fluoride-co-hexafluoropropylene) gel electrolyte. *J. Colloid Interface Sci.* **2020**, *571*, 126–133. [CrossRef]
78. Abdul Azeez, U.H.; Gunasekaran, A.; Sorrentino, A.; Syed, A.; Marraiki, N.; Anandan, S. Synthesis and characterization of poly-3-(9H-carbazol-9-yl)propylmethacrylate as a gel electrolyte for dye-sensitized solar cell applications. *Polym. Bull.* **2022**, *79*, 921–934. [CrossRef]

79. Cheng, F.; Wu, C.; Wang, S.; Wen, S. Polydopamine-Modified Electrospun Polyvinylidene Fluoride Nanofiber Based Flexible Polymer Gel Electrolyte for Highly Stable Dye-Sensitized Solar Cells. *ACS Omega* **2021**, *6*, 28663–28670. [CrossRef] [PubMed]
80. Elayappan, V.; Murugadoss, V.; Fei, Z.; Dyson, P.J.; Angaiah, S. Influence of polypyrrole incorporated electrospun poly (vinylidene fluoride-co-hexafluoropropylene) nanofibrous composite membrane electrolyte on the photovoltaic performance of dye sensitized solar cell. *Eng. Sci.* **2020**, *10*, 78–84. [CrossRef]
81. Gunasekaran, A.; Chen, H.Y.; Ponnusamy, V.K.; Sorrentino, A.; Anandan, S. Synthesis of high polydispersity index polylactic acid and its application as gel electrolyte towards fabrication of dye-sensitized solar cells. *J. Polym. Res.* **2021**, *28*, 252. [CrossRef]
82. Park, J.W.; Jang, J. Fabrication of graphene/free-standing nanofibrillar PEDOT/P(VDF-HFP) hybrid device for wearable and sensitive electronic skin application. *Carbon N. Y.* **2016**, *87*, 275–281. [CrossRef]
83. Zhou, L.; Yu, M.; Chen, X.; Nie, S.; Lai, W.Y.; Su, W.; Cui, Z.; Huang, W. Screen-Printed Poly(3,4-Ethylenedioxythiophene):Poly (Styrenesulfonate) Grids as ITO-Free Anodes for Flexible Organic Light-Emitting Diodes. *Adv. Funct. Mater.* **2018**, *28*, 1705955. [CrossRef]
84. Tang, Q.; Cai, H.; Yuan, S.; Wang, X. Counter electrodes from double-layered polyaniline nanostructures for dye-sensitized solar cell applications. *J. Mater. Chem. A* **2013**, *1*, 317–323. [CrossRef]
85. Calogero, G.; Calandra, P.; Irrera, A.; Sinopoli, A.; Citro, I.; Di Marco, G. A new type of transparent and low cost counter-electrode based on platinum nanoparticles for dye-sensitized solar cells. *Energy Environ. Sci.* **2011**, *4*, 1838–1844. [CrossRef]
86. Theerthagiri, J.; Senthil, R.A.; Buraidah, M.H.; Raghavender, M.; Madhavan, J.; Arof, A.K. Synthesis and characterization of (Ni1-xCox)Se2 based ternary selenides as electrocatalyst for triiodide reduction in dye-sensitized solar cells. *J. Solid State Chem.* **2016**, *238*, 113–120. [CrossRef]
87. Zhang, X.; Chen, X.; Zhang, K.; Pang, S.; Zhou, X.; Xu, H.; Dong, S.; Han, P.; Zhang, Z.; Zhang, C.; et al. Transition-metal nitride nanoparticles embedded in N-doped reduced graphene oxide: Superior synergistic electrocatalytic materials for the counter electrodes of dye-sensitized solar cells. *J. Mater. Chem. A* **2013**, *1*, 3340–3346. [CrossRef]
88. Jing, H.; Shi, Y.; Wu, D.; Liang, S.; Song, X.; An, Y.; Hao, C. Well-defined heteroatom-rich porous carbon electrocatalyst derived from biowaste for high-performance counter electrode in dye-sensitized solar cells. *Electrochim. Acta* **2018**, *281*, 646–653. [CrossRef]
89. Zheng, H.; Neo, C.Y.; Mei, X.; Qiu, J.; Ouyang, J. Reduced graphene oxide films fabricated by gel coating and their application as platinum-free counter electrodes of highly efficient iodide/triiodide dye-sensitized solar cells. *J. Mater. Chem.* **2012**, *22*, 14465–14474. [CrossRef]
90. Hou, W.; Xiao, Y.; Han, G. The dye-sensitized solar cells based on the interconnected ternary cobalt diindium sulfide nanosheet array counter electrode. *Mater. Res. Bull.* **2018**, *107*, 204–212. [CrossRef]
91. Yang, W.; Li, Z.; Xu, X.; Hou, L.; Tang, Y.; Deng, B.; Yang, F.; Wang, Y.; Li, Y. Atomic N-coordinated cobalt sites within nanomesh graphene as highly efficient electrocatalysts for triiodide reduction in dye-sensitized solar cells. *Chem. Eng. J.* **2018**, *349*, 782–790. [CrossRef]
92. Tai, Q.; Chen, B.; Guo, F.; Xu, S.; Hu, H.; Sebo, B.; Zhao, X.Z. In situ prepared transparent polyaniline electrode and its application in bifacial dye-sensitized solar cells. *ACS Nano* **2011**, *5*, 3795–3799. [CrossRef] [PubMed]
93. Jeon, S.S.; Kim, C.; Ko, J.; Im, S.S. Spherical polypyrrole nanoparticles as a highly efficient counter electrode for dye-sensitized solar cells. *J. Mater. Chem.* **2011**, *21*, 8146–8151. [CrossRef]
94. Tang, Z.; Wu, J.; Zheng, M.; Tang, Q.; Liu, Q.; Lin, J.; Wang, J. High efficient PANI/Pt nanofiber counter electrode used in dye-sensitized solar cell. *RSC Adv.* **2012**, *2*, 4062–4064. [CrossRef]
95. Li, G.R.; Song, J.; Pan, G.L.; Gao, X.P. Highly Pt-like electrocatalytic activity of transition metal nitrides for dye-sensitized solar cells. *Energy Environ. Sci.* **2011**, *4*, 1680–1683. [CrossRef]
96. Xiao, Y.; Han, G.; Li, Y.; Li, M.; Lin, J.Y. Three-dimensional hollow platinum-nickel bimetallic nanoframes for use in dye-sensitized solar cells. *J. Power Sources* **2015**, *278*, 149–155. [CrossRef]
97. Tang, Q.; Zhang, H.; Meng, Y.; He, B.; Yu, L. Dissolution Engineering of Platinum Alloy Counter Electrodes in Dye-Sensitized Solar Cells. *Angew. Chemie—Int. Ed.* **2015**, *54*, 11448–11452. [CrossRef]
98. Kim, S.S.; Nah, Y.C.; Noh, Y.Y.; Jo, J.; Kim, D.Y. Electrodeposited Pt for cost-efficient and flexible dye-sensitized solar cells. *Electrochim. Acta* **2006**, *51*, 3814–3819. [CrossRef]
99. Xia, J.; Chen, L.; Yanagida, S. Application of polypyrrole as a counter electrode for a dye-sensitized solar cell. *J. Mater. Chem.* **2011**, *21*, 4644–4649. [CrossRef]
100. Sangiorgi, N.; Sangiorgi, A.; Tarterini, F.; Sanson, A. Molecularly imprinted polypyrrole counter electrode for gel-state dye-sensitized solar cells. *Electrochim. Acta* **2019**, *305*, 322–328. [CrossRef]
101. Saberi Motlagh, M.; Mottaghitalab, V.; Rismanchi, A.; Rafieepoor Chirani, M.; Hasanzadeh, M. Performance modelling of textile solar cell developed by carbon fabric/polypyrrole flexible counter electrode. *Int. J. Sustain. Energy* **2022**, 1–21. [CrossRef]
102. Ohtani, Y.; Kumano, K.; Saneshige, M.; Takami, K.; Hoshi, H. Effect of electropolymerization duration on the structure and performance of polypyrrole/graphene nanoplatelet counter electrode for dye-sensitized solar cells. *J. Solid State Electrochem.* **2021**, *25*, 2107–2113. [CrossRef]
103. Ahmed, U.; Shahid, M.M.; Shahabuddin, S.; Rahim, N.A.; Alizadeh, M.; Pandey, A.K.; Sagadevan, S. An efficient platform based on strontium titanate nanocubes interleaved polypyrrole nanohybrid as counter electrode for dye-sensitized solar cell. *J. Alloys Compd.* **2021**, *860*, 158228. [CrossRef]

104. Rafique, S.; Rashid, I.; Sharif, R. Cost effective dye sensitized solar cell based on novel Cu polypyrrole multiwall carbon nanotubes nanocomposites counter electrode. *Sci. Rep.* **2021**, *11*, 14830. [CrossRef]
105. Yao, X.; He, B.; Cui, L.; Ti, J.; Chen, H.; Duan, Y.; Tang, Q. Polypyrrole-molybdenum sulfide complex as an efficient and transparent catalytic electrode for bifacial dye-sensitized solar cells. *Catal. Commun.* **2022**, *163*, 106403. [CrossRef]
106. Utami, A.N.; Reza, M.; Benu, D.P.; Fatya, A.I.; Yuliarto, B.; Suendo, V. Reverse micelle facilitated synthesis of nanostructured polyaniline as the counter electrode materials in dye-sensitized solar cells. *Polym. Technol. Mater.* **2020**, *59*, 1350–1358. [CrossRef]
107. Jiao, S.; Wen, J.; Zhou, Y.; Sun, Z.; Liu, Y.; Liu, R. Preparation and Property Studies of Polyaniline Film for Flexible Counter Electrode of Dye-Sensitized Solar Cells by Cyclic Voltammetry. *ChemistrySelect* **2021**, *6*, 230–233. [CrossRef]
108. Zatirostami, A. A new electrochemically prepared composite counter electrode for dye-sensitized solar cells. *Thin Solid Films* **2020**, *701*, 137926. [CrossRef]
109. Ravichandran, S.; Varthamanan, Y.; Akilandeswari; Elangoven, T.; Ragupathi, C.; Murugesan, S. Effect of polyaniline/FeS2 composite and usages of alternates counter electrode for dye-sensitized solar cells. *Mater. Today Proc.* **2021**, *49*, 2615–2619. [CrossRef]
110. Bella, F.; Porcarelli, L.; Mantione, D.; Gerbaldi, C.; Barolo, C.; Grätzel, M.; Mecerreyes, D. A water-based and metal-free dye solar cell exceeding 7% efficiency using a cationic poly(3,4-ethylenedioxythiophene) derivative. *Chem. Sci.* **2020**, *11*, 1485–1493. [CrossRef]
111. Pradhan, S.C.; Soman, S. Effect of thickness on charge transfer properties of conductive polymer based PEDOT counter electrodes in DSSC. *Results Surf. Interfaces* **2021**, *5*, 100030. [CrossRef]
112. Venkatesan, S.; Lin, W.-H.; Hsu, T.-H.; Teng, H.; Lee, Y.-L. Indoor Dye-Sensitized Solar Cells with Efficiencies Surpassing 26% Using Polymeric Counter Electrodes. *ACS Sustain. Chem. Eng.* **2022**, *10*, 2473–2483. [CrossRef]
113. Xu, T.; Cao, W.; Kong, D.; Qin, X.; Song, J.; Kou, K.; Chen, L.; Qiao, Q.; Huang, W. Enhanced catalytic property of transparent PEDOT counter electrodes for bifacial dye sensitized solar cells. *Mater. Today Commun.* **2020**, *25*, 101313. [CrossRef]
114. Gemeiner, P.; Pavlíčková, M.; Hatala, M.; Hvojnik, M.; Homola, T.; Mikula, M. The effect of secondary dopants on screen-printed PEDOT:PSS counter-electrodes for dye-sensitized solar cells. *J. Appl. Polym. Sci.* **2022**, *139*, 51929. [CrossRef]
115. Yeh, N.; Yeh, P. Organic solar cells: Their developments and potentials. *Renew. Sustain. Energy Rev.* **2013**, *21*, 421–431. [CrossRef]
116. Blom, P.W.M.; Mihailetchi, V.D.; Koster, L.J.A.; Markov, D.E. Device physics of polymer:Fullerene bulk heterojunction solar cells. *Adv. Mater.* **2007**, *19*, 1551–1566. [CrossRef]
117. Gao, F.; Inganäs, O. Charge generation in polymer-fullerene bulk-heterojunction solar cells. *Phys. Chem. Chem. Phys.* **2014**, *16*, 20291–20304. [CrossRef]
118. Dimitrov, S.D.; Durrant, J.R. Materials design considerations for charge generation in organic solar cells. *Chem. Mater.* **2014**, *26*, 616–630. [CrossRef]
119. Mayer, A.C.; Scully, S.R.; Hardin, B.E.; Rowell, M.W.; McGehee, M.D. Polymer-based solar cells. *Mater. Today* **2007**, *10*, 28–33. [CrossRef]
120. Yu, G.; Gao, J.; Hummelen, J.C.; Wudl, F.; Heeger, A.J. Internal Donor-Acceptor. *Science* **1995**, *270*, 1789–1791. [CrossRef]
121. Wang, T.; Qin, J.; Xiao, Z.; Meng, X.; Zuo, C.; Yang, B.; Tan, H.; Yang, J.; Yang, S.; Sun, K.; et al. A 2.16 eV bandgap polymer donor gives 16% power conversion efficiency. *Sci. Bull.* **2020**, *65*, 179–181. [CrossRef]
122. Cao, Z.; Chen, J.; Liu, S.; Qin, M.; Jia, T.; Zhao, J.; Li, Q.; Ying, L.; Cai, Y.P.; Lu, X.; et al. Understanding of Imine Substitution in Wide-Bandgap Polymer Donor-Induced Efficiency Enhancement in All-Polymer Solar Cells. *Chem. Mater.* **2019**, *31*, 8533–8542. [CrossRef]
123. Xue, C.; Tang, Y.; Liu, S.; Feng, H.; Li, S.; Xia, D. Achieving efficient polymer solar cells based on benzodithiophene-thiazole-containing wide band gap polymer donors by changing the linkage patterns of two thiazoles. *New J. Chem.* **2020**, *44*, 13100–13107. [CrossRef]
124. Qin, J.; Zhang, L.; Xiao, Z.; Chen, S.; Sun, K.; Zang, Z.; Yi, C.; Yuan, Y.; Jin, Z.; Hao, F.; et al. Over 16% efficiency from thick-film organic solar cells. *Sci. Bull.* **2020**, *65*, 1979–1982. [CrossRef]
125. Zhao, J.; Li, Q.; Liu, S.; Cao, Z.; Jiao, X.; Cai, Y.P.; Huang, F. Bithieno[3,4-c]pyrrole-4,6-dione-Mediated Crystallinity in Large-Bandgap Polymer Donors Directs Charge Transportation and Recombination in Efficient Nonfullerene Polymer Solar Cells. *ACS Energy Lett.* **2020**, *5*, 367–375. [CrossRef]
126. Keshtov, M.L.; Konstantinov, I.O.; Kuklin, S.A.; Zou, Y.; Agrawal, A.; Chen, F.C.; Sharma, G.D. Binary and Ternary Polymer Solar Cells Based on a Wide Bandgap D-A Copolymer Donor and Two Nonfullerene Acceptors with Complementary Absorption Spectral. *ChemSusChem* **2021**, *14*, 4731–4740. [CrossRef]
127. Gokulnath, T.; Feng, K.; Park, H.-Y.; Do, Y.; Park, H.; Gayathri, R.D.; Reddy, S.S.; Kim, J.; Guo, X.; Yoon, J.; et al. Facile Strategy for Third Component Optimization in Wide-Band-Gap π-Conjugated Polymer Donor-Based Efficient Ternary All-Polymer Solar Cells. *ACS Appl. Mater. Interfaces* **2022**, *14*, 11211–11221. [CrossRef]
128. Liang, Y.; Wu, Y.; Feng, D.; Tsai, S.T.; Son, H.J.; Li, G.; Yu, L. Development of new semiconducting polymers for high performance solar cells. *J. Am. Chem. Soc.* **2009**, *131*, 56–57. [CrossRef]
129. Liang, Y.; Xu, Z.; Xia, J.; Tsai, S.T.; Wu, Y.; Li, G.; Ray, C.; Yu, L. For the bright future-bulk heterojunction polymer solar cells with power conversion efficiency of 7.4%. *Adv. Mater.* **2010**, *22*, 135–138. [CrossRef]
130. Chen, Y.; You, G.; Zou, D.; Zhuang, Q.; Zhen, H.; Ling, Q. Enhanced photovoltaic performances via ternary blend strategy employing a medium-bandgap D-A type alternating copolymer as the single donor. *Sol. Energy* **2019**, *183*, 350–355. [CrossRef]

131. Sharma, G.D.; Bucher, L.; Desbois, N.; Gros, C.P.; Gupta, G.; Malhotra, P. Polymer solar cell based on ternary active layer consists of medium bandgap polymer and two non-fullerene acceptors. *Sol. Energy* **2020**, *207*, 1427–1433. [CrossRef]
132. An, Q.; Wang, J.; Gao, W.; Ma, X.; Hu, Z.; Gao, J.; Xu, C.; Hao, M.; Zhang, X.; Yang, C.; et al. Alloy-like ternary polymer solar cells with over 17.2% efficiency. *Sci. Bull.* **2020**, *65*, 538–545. [CrossRef]
133. Yuan, J.; Zhang, Y.; Zhou, L.; Zhang, G.; Yip, H.L.; Lau, T.K.; Lu, X.; Zhu, C.; Peng, H.; Johnson, P.A.; et al. Single-Junction Organic Solar Cell with over 15% Efficiency Using Fused-Ring Acceptor with Electron-Deficient Core. *Joule* **2019**, *3*, 1140–1151. [CrossRef]
134. Yan, T.; Song, W.; Huang, J.; Peng, R.; Huang, L.; Ge, Z. 16.67% Rigid and 14.06% Flexible Organic Solar Cells Enabled by Ternary Heterojunction Strategy. *Adv. Mater.* **2019**, *31*, 1–8. [CrossRef] [PubMed]
135. Peng, W.; Lin, Y.; Jeong, S.Y.; Firdaus, Y.; Genene, Z.; Nikitaras, A.; Tsetseris, L.; Woo, H.Y.; Zhu, W.; Anthopoulos, T.D.; et al. Using Two Compatible Donor Polymers Boosts the Efficiency of Ternary Organic Solar Cells to 17.7%. *Chem. Mater.* **2021**, *33*, 7254–7262. [CrossRef]
136. Keshtov, M.L.; Kuklin, S.A.; Khokhlov, A.R.; Xie, Z.; Alekseev, V.G.; Dahiya, H.; Singhal, R.; Sharma, G.D. New Medium Bandgap Donor D-A1-D-A2 Type Copolymers Based on Anthra[1,2-b: 4,3-b":6,7-c'''] Trithiophene-8,12-dione Groups for High-Efficient Non-Fullerene Polymer Solar Cells. *Macromol. Rapid Commun.* **2022**, *43*, 2100839. [CrossRef]
137. Pan, L.; Liu, T.; Wang, J.; Ye, L.; Luo, Z.; Ma, R.; Pang, S.; Chen, Y.; Ade, H.; Yan, H.; et al. Efficient Organic Ternary Solar Cells Employing Narrow Band Gap Diketopyrrolopyrrole Polymers and Nonfullerene Acceptors. *Chem. Mater.* **2020**, *32*, 7309–7317. [CrossRef]
138. Rech, J.J.; Neu, J.; Qin, Y.; Samson, S.; Shanahan, J.; Josey, R.F.; Ade, H.; You, W. Designing Simple Conjugated Polymers for Scalable and Efficient Organic Solar Cells. *ChemSusChem* **2021**, *14*, 3561–3568. [CrossRef]
139. Caliskan, M.; Erer, M.C.; Aslan, S.T.; Udum, Y.A.; Toppare, L.; Cirpan, A. Narrow band gap benzodithiophene and quinoxaline bearing conjugated polymers for organic photovoltaic applications. *Dye. Pigment.* **2020**, *180*, 108479. [CrossRef]
140. Guo, L.; Huang, X.; Luo, Y.; Liu, S.; Cao, Z.; Chen, J.; Luo, Y.; Li, Q.; Zhao, J.; Cai, Y.P. Novel narrow bandgap polymer donors based on ester-substituted quinoxaline unit for organic photovoltaic application. *Sol. Energy* **2021**, *220*, 425–431. [CrossRef]
141. Can, A.; Choi, G.-S.; Ozdemir, R.; Park, S.; Park, J.S.; Lee, Y.; Deneme, İ.; Mutlugun, E.; Kim, C.; Kim, B.J.; et al. Meso-π-Extended/Deficient BODIPYs and Low-Band-Gap Donor–Acceptor Copolymers for Organic Optoelectronics. *ACS Appl. Polym. Mater.* **2022**, *4*, 1991–2005. [CrossRef]
142. Cruciani, F.; Babics, M.; Liu, S.; Carja, D.; Mantione, D.; Beaujuge, P.M. N-Acylisoindigo Derivatives as Polymer Acceptors for "All-Polymer" Bulk-Heterojunction Solar Cells. *Macromol. Chem. Phys.* **2019**, *220*, 1900029. [CrossRef]
143. Wu, Y.; Guo, J.; Wang, W.; Chen, Z.; Chen, Z.; Sun, R.; Wu, Q.; Wang, T.; Hao, X.; Zhu, H.; et al. A conjugated donor-acceptor block copolymer enables over 11% efficiency for single-component polymer solar cells. *Joule* **2021**, *5*, 1800–1815. [CrossRef]
144. Aydin, E.; De Bastiani, M.; De Wolf, S. Defect and Contact Passivation for Perovskite Solar Cells. *Adv. Mater.* **2019**, *31*, e1900428. [CrossRef] [PubMed]
145. Zhao, X.; Wang, M. Organic hole-transporting materials for efficient perovskite solar cells. *Mater. Today Energy* **2018**, *7*, 208–220. [CrossRef]
146. Bakr, Z.H.; Wali, Q.; Fakharuddin, A.; Schmidt-Mende, L.; Brown, T.M.; Jose, R. Advances in hole transport materials engineering for stable and efficient perovskite solar cells. *Nano Energy* **2017**, *34*, 271–305. [CrossRef]
147. Yin, X.; Song, Z.; Li, Z.; Tang, W. Toward ideal hole transport materials: A review on recent progress in dopant-free hole transport materials for fabricating efficient and stable perovskite solar cells. *Energy Environ. Sci.* **2020**, *13*, 4057–4086. [CrossRef]
148. Zhao, Y.; Zhu, P.; Wang, M.; Huang, S.; Zhao, Z.; Tan, S.; Han, T.H.; Lee, J.W.; Huang, T.; Wang, R.; et al. A Polymerization-Assisted Grain Growth Strategy for Efficient and Stable Perovskite Solar Cells. *Adv. Mater.* **2020**, *32*, e1907769. [CrossRef]
149. Yousif, Q.A.; Agbolaghi, S. A Comparison Between Functions of Carbon Nanotube and Reduced Graphene Oxide and Respective Ameliorated Derivatives in Perovskite Solar Cells. *Macromol. Res.* **2020**, *28*, 425–432. [CrossRef]
150. Yu, B.; Zhang, L.; Wu, J.; Liu, K.; Wu, H.; Shi, J.; Luo, Y.; Li, D.; Bo, Z.; Meng, Q. Application of a new π-conjugated ladder-like polymer in enhancing the stability and efficiency of perovskite solar cells. *J. Mater. Chem. A* **2020**, *8*, 1417–1424. [CrossRef]
151. Fu, Q.; Xiao, S.; Tang, X.; Chen, Y.; Hu, T. Amphiphilic Fullerenes Employed to Improve the Quality of Perovskite Films and the Stability of Perovskite Solar Cells. *ACS Appl. Mater. Interfaces* **2019**, *11*, 24782–24788. [CrossRef]
152. Chen, W.; Wang, Y.; Pang, G.; Koh, C.W.; Djurišić, A.B.; Wu, Y.; Tu, B.; Liu, F.Z.; Chen, R.; Woo, H.Y.; et al. Conjugated Polymer-Assisted Grain Boundary Passivation for Efficient Inverted Planar Perovskite Solar Cells. *Adv. Funct. Mater.* **2019**, *29*, 1808855. [CrossRef]
153. Yao, Z.; Qu, D.; Guo, Y.; Huang, H. Grain boundary regulation of flexible perovskite solar cells via a polymer alloy additive. *Org. Electron.* **2019**, *70*, 205–210. [CrossRef]
154. Suwa, K.; Oyaizu, K.; Segawa, H.; Nishide, H. Anti-Oxidizing Radical Polymer-Incorporated Perovskite Layers and their Photovoltaic Characteristics in Solar Cells. *ChemSusChem* **2019**, *12*, 5207–5212. [CrossRef] [PubMed]
155. Chen, N.; Yi, X.; Zhuang, J.; Wei, Y.; Zhang, Y.; Wang, F.; Cao, S.; Li, C.; Wang, J. An Efficient Trap Passivator for Perovskite Solar Cells: Poly(propylene glycol) bis(2-aminopropyl ether). *Nano-Micro Lett.* **2020**, *12*, 177. [CrossRef]
156. Garai, R.; Gupta, R.K.; Tanwar, A.S.; Hossain, M.; Iyer, P.K. Conjugated Polyelectrolyte-Passivated Stable Perovskite Solar Cells for Efficiency beyond 20%. *Chem. Mater.* **2021**, *33*, 5709–5717. [CrossRef]
157. Zarenezhad, H.; Balkan, T.; Solati, N.; Halali, M.; Askari, M.; Kaya, S. Efficient carrier utilization induced by conductive polypyrrole additives in organic-inorganic halide perovskite solar cells. *Sol. Energy* **2020**, *207*, 1300–1307. [CrossRef]

158. Zhong, M.; Chai, L.; Wang, Y.; Di, J. Enhanced efficiency and stability of perovskite solar cell by adding polymer mixture in perovskite photoactive layer. *J. Alloys Compd.* **2021**, *864*, 158793. [CrossRef]
159. Wang, F.; Cao, Y.; Chen, C.; Chen, Q.; Wu, X.; Li, X.; Qin, T.; Huang, W. Materials toward the Upscaling of Perovskite Solar Cells: Progress, Challenges, and Strategies. *Adv. Funct. Mater.* **2018**, *28*, 1803753. [CrossRef]
160. Chawanpunyawat, T.; Funchien, P.; Wongkaew, P.; Henjongchom, N.; Ariyarit, A.; Ittisanronnachai, S.; Namuangruk, S.; Cheacharoen, R.; Sudyoadsuk, T.; Goubard, F.; et al. A Ladder-like Dopant-free Hole-Transporting Polymer for Hysteresis-less High-Efficiency Perovskite Solar Cells with High Ambient Stability. *ChemSusChem* **2020**, *13*, 5058–5066. [CrossRef]
161. Liao, Q.; Wang, Y.; Yao, X.; Su, M.; Li, B.; Sun, H.; Huang, J.; Guo, X. A Dual-Functional Conjugated Polymer as an Efficient Hole-Transporting Layer for High-Performance Inverted Perovskite Solar Cells. *ACS Appl. Mater. Interfaces* **2021**, *13*, 16744–16753. [CrossRef]
162. Qi, F.; Deng, X.; Wu, X.; Huo, L.; Xiao, Y.; Lu, X.; Zhu, Z.; Jen, A.K.Y. A Dopant-Free Polymeric Hole-Transporting Material Enabled High Fill Factor Over 81% for Highly Efficient Perovskite Solar Cells. *Adv. Energy Mater.* **2019**, *9*, 1902600. [CrossRef]
163. Jeong, S.; Lee, I.; Kim, T.S.; Lee, J.Y. An Interlocking Fibrillar Polymer Layer for Mechanical Stability of Perovskite Solar Cells. *Adv. Mater. Interfaces* **2020**, *7*, 2001425. [CrossRef]
164. Yao, Z.; Zhang, F.; Guo, Y.; Wu, H.; He, L.; Liu, Z.; Cai, B.; Guo, Y.; Brett, C.J.; Li, Y.; et al. Conformational and Compositional Tuning of Phenanthrocarbazole-Based Dopant-Free Hole-Transport Polymers Boosting the Performance of Perovskite Solar Cells. *J. Am. Chem. Soc.* **2020**, *142*, 17681–17692. [CrossRef] [PubMed]
165. You, G.; Zhuang, Q.; Wang, L.; Lin, X.; Zou, D.; Lin, Z.; Zhen, H.; Zhuang, W.; Ling, Q. Dopant-Free, Donor–Acceptor-Type Polymeric Hole-Transporting Materials for the Perovskite Solar Cells with Power Conversion Efficiencies over 20%. *Adv. Energy Mater.* **2020**, *10*, 1903146. [CrossRef]
166. Shalan, A.E.; Sharmoukh, W.; Elshazly, A.N.; Elnagar, M.M.; Al Kiey, S.A.; Rashad, M.M.; Allam, N.K. Dopant-free hole-transporting polymers for efficient, stable, and hysteresis-less perovskite solar cells. *Sustain. Mater. Technol.* **2020**, *26*, e00226. [CrossRef]
167. Kranthiraja, K.; Arivunithi, V.M.; Aryal, U.K.; Park, H.Y.; Cho, W.; Kim, J.; Reddy, S.S.; Kim, H.K.; Kang, I.N.; Song, M.; et al. Efficient and hysteresis-less perovskite and organic solar cells by employing donor-acceptor type π-conjugated polymer. *Org. Electron.* **2019**, *72*, 18–24. [CrossRef]
168. Kong, X.; Jiang, Y.; Wu, X.; Chen, C.; Guo, J.; Liu, S.; Gao, X.; Zhou, G.; Liu, J.M.; Kempa, K.; et al. Dopant-free F-substituted benzodithiophene copolymer hole-Transporting materials for efficient and stable perovskite solar cells. *J. Mater. Chem. A* **2020**, *8*, 1858–1864. [CrossRef]
169. You, G.; Li, L.; Wang, S.; Cao, J.; Yao, L.; Cai, W.; Zhou, Z.; Li, K.; Lin, Z.; Zhen, H.; et al. Donor–Acceptor Type Polymer Bearing Carbazole Side Chain for Efficient Dopant-Free Perovskite Solar Cells. *Adv. Energy Mater.* **2022**, *12*, 2102697. [CrossRef]
170. Liu, W.; Ma, Y.; Wang, Z.; Zhu, M.; Wang, J.; Khalil, M.; Wang, H.; Gao, W.; Fan, W.J.; Li, W.S.; et al. Improving the fill factor of perovskite solar cells by employing an amine-tethered diketopyrrolopyrrole-based polymer as the dopant-free hole transport layer. *ACS Appl. Energy Mater.* **2020**, *3*, 9600–9609. [CrossRef]
171. Liu, W.; Ma, Y.; Wang, Z.; Mu, Z.; Gao, W.; Fan, W.; Li, W.S.; Zhang, Q. Improving the hole transport performance of perovskite solar cells through adjusting the mobility of the as-synthesized conjugated polymer. *J. Mater. Chem. C* **2021**, *9*, 3421–3428. [CrossRef]
172. Ma, J.; Li, Y.; Li, J.; Qin, M.; Wu, X.; Lv, Z.; Hsu, Y.J.; Lu, X.; Wu, Y.; Fang, G. Constructing highly efficient all-inorganic perovskite solar cells with efficiency exceeding 17% by using dopant-free polymeric electron-donor materials. *Nano Energy* **2020**, *75*, 104933. [CrossRef]
173. Xiong, J.; Qi, Y.; Zhang, Q.; Box, D.; Williams, K.; Tatum, J.; Das, P.; Pradhan, N.R.; Dai, Q. Enhanced Moisture and Water Resistance in Inverted Perovskite Solar Cells by Poly(3-hexylthiophene). *ACS Appl. Energy Mater.* **2021**, *4*, 1815–1823. [CrossRef]
174. Jiang, M.; Niu, Q.; Tang, X.; Zhang, H.; Xu, H.; Huang, W.; Yao, J.; Yan, B.; Xia, R. Improving the performances of perovskite solar cells via modification of electron transport layer. *Polymers* **2019**, *11*, 147. [CrossRef] [PubMed]
175. You, S.; Zeng, H.; Ku, Z.; Wang, X.; Wang, Z.; Rong, Y.; Zhao, Y.; Zheng, X.; Luo, L.; Li, L.; et al. Multifunctional Polymer-Regulated SnO2 Nanocrystals Enhance Interface Contact for Efficient and Stable Planar Perovskite Solar Cells. *Adv. Mater.* **2020**, *32*, 2003990. [CrossRef]
176. Liu, G.Z.; Du, C.S.; Wu, J.Y.; Liu, B.T.; Wu, T.M.; Huang, C.F.; Lee, R.H. Enhanced photovoltaic properties of perovskite solar cells by employing bathocuproine/hydrophobic polymer films as hole-blocking/electron-transporting interfacial layers. *Polymers* **2021**, *13*, 42. [CrossRef]
177. Said, A.A.; Xie, J.; Wang, Y.; Wang, Z.; Zhou, Y.; Zhao, K.; Gao, W.B.; Michinobu, T.; Zhang, Q. Efficient Inverted Perovskite Solar Cells by Employing N-Type (D–A1–D–A2) Polymers as Electron Transporting Layer. *Small* **2019**, *15*, e1803339. [CrossRef]
178. Tian, L.; Hu, Z.; Liu, X.; Liu, Z.; Guo, P.; Xu, B.; Xue, Q.; Yip, H.L.; Huang, F.; Cao, Y. Fluoro- and Amino-Functionalized Conjugated Polymers as Electron Transport Materials for Perovskite Solar Cells with Improved Efficiency and Stability. *ACS Appl. Mater. Interfaces* **2019**, *11*, 5289–5297. [CrossRef]
179. Elnaggar, M.M.; Frolova, L.A.; Gordeeva, A.M.; Ustinova, M.I.; Laurenzen, H.; Akkuratov, A.V.; Nikitenko, S.L.; Solov'eva, E.A.; Luchkin, S.Y.; Fedotov, Y.S.; et al. Improving stability of perovskite solar cells using fullerene-polymer composite electron transport layer. *Synth. Met.* **2022**, *286*, 117028. [CrossRef]

180. Yan, W.; Wang, Z.; Gong, Y.; Guo, S.; Jiang, J.; Chen, J.; Tang, C.; Xia, R.; Huang, W.; Xin, H. Naphthalene-diimide selenophene copolymers as efficient solution-processable electron-transporting material for perovskite solar cells. *Org. Electron.* **2019**, *67*, 208–214. [CrossRef]
181. Ding, Y.; He, B.; Zhu, J.; Zhang, W.; Su, G.; Duan, J.; Zhao, Y.; Chen, H.; Tang, Q. Advanced Modification of Perovskite Surfaces for Defect Passivation and Efficient Charge Extraction in Air-Stable CsPbBr3 Perovskite Solar Cells. *ACS Sustain. Chem. Eng.* **2019**, *7*, 19286–19294. [CrossRef]
182. Zhao, H.; Yang, S.; Han, Y.; Yuan, S.; Jiang, H.; Duan, C.; Liu, Z.; Liu, S. A High Mobility Conjugated Polymer Enables Air and Thermally Stable CsPbI2Br Perovskite Solar Cells with an Efficiency Exceeding 15%. *Adv. Mater. Technol.* **2019**, *4*, 1900311. [CrossRef]
183. Sharma, A.; Singh, S.; Song, X.; Rosas Villalva, D.; Troughton, J.; Corzo, D.; Toppare, L.; Gunbas, G.; Schroeder, B.C.; Baran, D. A Nonionic Alcohol Soluble Polymer Cathode Interlayer Enables Efficient Organic and Perovskite Solar Cells. *Chem. Mater.* **2021**, *33*, 8602–8611. [CrossRef] [PubMed]
184. Zhou, S.; Zhu, T.; Zheng, L.; Zhang, D.; Xu, W.; Liu, L.; Cheng, G.; Zheng, J.; Gong, X. A zwitterionic polymer as an interfacial layer for efficient and stable perovskite solar cells. *RSC Adv.* **2019**, *9*, 30317–30324. [CrossRef] [PubMed]
185. Zhao, D.; Dai, S.; Li, M.; Wu, Y.; Zheng, L.; Wang, Y.; Yun, D.Q. Improved Efficiency and Stability of Perovskite Solar Cells Using a Difluorobenzothiadiazole-Based Interfacial Material. *ACS Appl. Energy Mater.* **2021**, *4*, 10646–10655. [CrossRef]
186. Liu, Z.; Li, S.; Wang, X.; Cui, Y.; Qin, Y.; Leng, S.; Xu, Y.X.; Yao, K.; Huang, H. Interfacial engineering of front-contact with finely tuned polymer interlayers for high-performance large-area flexible perovskite solar cells. *Nano Energy* **2019**, *62*, 734–744. [CrossRef]
187. Kim, J.; Lee, Y.; Yun, A.J.; Gil, B.; Park, B. Interfacial Modification and Defect Passivation by the Cross-Linking Interlayer for Efficient and Stable CuSCN-Based Perovskite Solar Cells. *ACS Appl. Mater. Interfaces* **2019**, *11*, 46818–46824. [CrossRef]
188. Wang, H.; Zhang, F.; Li, Z.; Zhang, J.; Lian, J.; Song, J.; Qu, J.; Wong, W.Y. Naphthalene imide dimer as interface engineering material: An efficient strategy for achieving high-performance perovskite solar cells. *Chem. Eng. J.* **2020**, *395*, 125062. [CrossRef]

Review

Hybrid Organic–Inorganic Perovskite Halide Materials for Photovoltaics towards Their Commercialization

Luke Jonathan [1], Lina Jaya Diguna [1], Omnia Samy [2], Muqoyyanah Muqoyyanah [3], Suriani Abu Bakar [3], Muhammad Danang Birowosuto [4,*] and Amine El Moutaouakil [2,*]

1. Department of Renewable Energy Engineering, Prasetiya Mulya University, Kavling Edutown I.1, Jl. BSD Raya Utama, BSD City, Tangerang 15339, Indonesia; 23301810001@student.prasetiyamulya.ac.id (L.J.); lina.diguna@prasetiyamulya.ac.id (L.J.D.)
2. Department of Electrical and Communication Engineering, College of Engineering, United Arab Emirates University, Al Ain P.O. Box 15551, United Arab Emirates; 202090009@uaeu.ac.ae
3. Department of Physics, Faculty of Science and Mathematics, Universiti Pendidikan Sultan Idris, Tanjung Malim 35900, Malaysia; anna.physics87@gmail.com (M.M.); suriani@fsmt.upsi.edu.my (S.A.B.)
4. Łukasiewicz Research Network—PORT Polish Center for Technology Development, Stabłowicka 147, 54-066 Wrocław, Poland
* Correspondence: muhammad.birowosuto@port.lukasiewicz.gov.pl (M.D.B.); a.elmoutaouakil@uaeu.ac.ae (A.E.M.)

Abstract: Hybrid organic–inorganic perovskite (HOIP) photovoltaics have emerged as a promising new technology for the next generation of photovoltaics since their first development 10 years ago, and show a high-power conversion efficiency (PCE) of about 29.3%. The power-conversion efficiency of these perovskite photovoltaics depends on the base materials used in their development, and methylammonium lead iodide is generally used as the main component. Perovskite materials have been further explored to increase their efficiency, as they are cheaper and easier to fabricate than silicon photovoltaics, which will lead to better commercialization. Even with these advantages, perovskite photovoltaics have a few drawbacks, such as their stability when in contact with heat and humidity, which pales in comparison to the 25-year stability of silicon, even with improvements are made when exploring new materials. To expand the benefits and address the drawbacks of perovskite photovoltaics, perovskite–silicon tandem photovoltaics have been suggested as a solution in the commercialization of perovskite photovoltaics. This tandem photovoltaic results in an increased PCE value by presenting a better total absorption wavelength for both perovskite and silicon photovoltaics. In this work, we summarized the advances in HOIP photovoltaics in the contact of new material developments, enhanced device fabrication, and innovative approaches to the commercialization of large-scale devices.

Keywords: power conversion efficiency; hybrid perovskite; tandem structure; photovoltaics; commercialization

1. Introduction

Nowadays, silicon photovoltaics (PVs) have successfully achieved a high power conversion efficiency (PCE) of 26.7% [1] and this nearly approaches the theoretical limit value (29.4%) [2], and has led the PV sale sector, with more than 90% of the market share [3]. However, its numerous drawbacks, such as the scarcity of its pure state, the requirement of high energy to separate the bonded oxygen in silicon dioxide, and their PCE stagnancy, which has kept it in the range of 25% for over 15 years, have limited silicon PVs for further development. Perovskite PVs with a higher PCE are a promising PV replacement in overcoming the limitations of silicon PVs. After a decade of development, a high PCE value of 29.3% [4] has been achieved by perovskite PVs, and this value is similar to the theoretical limit value for silicon PVs. Other advantages offered by perovskite PVs include its capability in absorbing the visible spectrum, simplicity and cost-effective production (about $2.5/cell) [5,6].

The meta-analysis study presented in Ref. [7] provides a comprehensive view of the total costs of different PV materials, where CdTe was found to have the least capital and lifecycle costs among the PVs. Another analysis, specifically on perovskite tandem costs, in [8] emphasizes the commercial potential of PV perovskite tandem cells. The structure and light absorption of the perovskite layer, which depends on the energy level and carrier transport properties, will affect the PV's PCE value. Therefore, most research has been focused on optimizing the morphology, energy level, and conductivity of perovskite PVs. Enhancing the structure, reducing defects, and increasing the grain size are several ways to improve the performance of perovskite PVs. These efforts have definitely been affected by the materials used in the synthesis process, and the selection of device fabrication methods [9,10].

With more research being focused on perovskites, HOIPs have been a focus since their emergence, and have shown numerous great characteristics, such as easy growth due to their cubic structures [9,10], a wide absorption range compatible with the solar spectrum [11], and a low exciton binding energy with a long carrier diffusion length [12]. All of these examples make HOIPs promising materials for emerging photovoltaic technology.

As an example of a HOIP, methylammonium lead halide (MAPbX$_3$, X = Cl$^-$, Br$^-$, or I$^-$) was the first material used for perovskite PV fabrication and resulted in a 3.8% PCE value, in 2009 [13]. In more than 10 years, this PCE value has increased to 29.8% by applying a smaller active area of 1 cm^2 and by continuously focusing on improving operational stability [4]. This rapid progress has triggered interest in transferring existing technology, from small-area to large-area perovskite, which is necessary for industrial expansion [14]. To meet the requirements of a high quality and large-area uniformity for perovskite films, several deposition methods, based on solution, vacuum [15–17], and solution–vacuum hybrid processes, have been developed and optimized. Recent works have been based on solution processes, such as spin-coating [18], spray-coating [19,20], blade coating [21,22], slot-die coating [23,24], softcover [25–27], and screen printing [28–30]. However, even with several achievements in terms of grain size and increases in efficiencies through different fabrication method, these are still not enough for successful commercialization of perovskite PVs due to the remaining challenges that need to be solved. The primary challenge here is to achieve long lifetimes with a good stability at the module level [31]. Although much progress has been made, it is still challenging for perovskite PVs to reach the most popular international standards (IEC61215:2016) for mature PV technologies.

This challenge has triggered further studies on joining the structure of perovskite and silicon PVs, so-called tandem PVs. This effort was undertaken in order to improve PV performance through a combined system that can efficiently absorb solar radiation in both the visible and the near-infrared region [32]. In addition, due to the possibility of bandgap tuning, the perovskites themselves can be combined into an all-perovskite tandem photovoltaic, with both the top and bottom cells using a perovskite absorber. All of these efforts are still very young technologies, but enable remarkable PCEs, close to—or even above—those of the best single-junction cells [33,34]. This potential has been recognized, not only by various research institutes, but also by several start-up companies, such as Oxford PV, Swift Solar, and Tandem PV, who have developed their own perovskite PVs for commercialization.

Previous reviews discussed the effectiveness of hole transport and molecular materials in PV cells, where we can obtain cells with PCEs of up to 25% [35,36]. Other detailed reviews dedicated to hybrid organic–inorganic halide perovskites have emphasized that HOIPs are high-quality materials for PVs [37,38]. Finally, a recent review discussed different perovskite fabrication methods to develop large-area PVs with moderate-to-high PCEs [39]. In this work, we present a review that summarized advances in perovskite to perovskite–silicon tandem PVs, as this leads to a better commercialization. Advancements in materials to improve characteristics, leading to large-scale devices that are prepared for commercialization, are discussed in detail. Finally, we also highlight future research directions to achieve advances in extensive commercialization.

2. Materials and Fabrication Methods
2.1. Materials

Perovskite is a type of PV that includes perovskite-structured compounds that use tin or lead halides to act as the base material for a light-harvesting active layer [40]. The material of each perovskite differs, as they can be hybrids of organic and inorganic materials. All hybrid organic–inorganic perovskite (HOIP) materials have the general chemical formula of ABX_3, as illustrated in Figure 1a. HOIP PVs are typically made using an organic/inorganic cation (A = methylammonium (MA) $CH_3NH_3^+$, formamidinium (FA) $CH_3(NH_2)_2^+$) [41], a divalent cation (B = Pb^{2+} or Sn^{2+}) [42], and an anion (X = Cl^-, Br^-, or I^-). The perovskite material, ABX_3, is sandwiched between electron transport layers (ETLs) and hole transport layers (HTLs), which are light absorbers. ETLs and HTLs have a pivotal role in charge transportation, separation, and recombination [43]. Examples of HTLs are shown in Figure 1b,c. Although a HOIP is a very attractive option for commercial applications, because this type of cell is very cheaply produced during a scale-up process [44], ETLs and HTLs face some challenges in terms of charge transfer [45]. One HTL used with a HOIP, such as spiro-OMeTAD (Figure 1b) has a low hole mobility and conductivity, but it does improve the optical and electrical properties of the HOIP [46]. Another HTL, like N4,N4′,N4″,N4‴-tetra([1,1-biphenyl]-4-yl)-[1,1:4,1-terphenyl]-4,4′-diamine (TaTm) in Figure 1c, allows efficient charge extraction, though there is a large misalignment of the highest occupied molecular orbit with the valence band of the HOIP [47]. The alignment of the valence and conduction bands of the perovskites with the transport layers of the HTLs and ETLs, respectively, is a critical factor for charge transfer and extraction. Some efforts have moved towards finding new materials for ETLs and HTLs, while others have investigated doping the layers to acquire enhanced properties [48]. Doping requires additives and solvents that may be hazardous, toxic, or harmful to the environment. Dopant-free structures have been developed to obtain better PCEs and stability without any additives [49,50]. An example for use as an ETL is C60 (Figure 1d) and it can achieve both the high stabilized power output and long-term operational stability of HOIP solar cells [51].

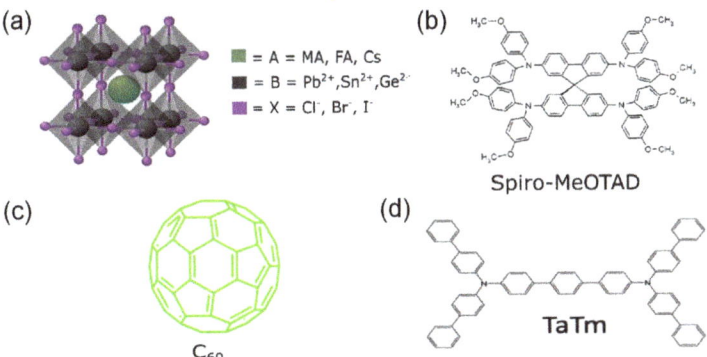

Figure 1. Common materials used in perovskite PVs: (**a**) crystal structure of $MAPbI_3$, $MAPbBr_3$, $FAPbI_3$, $FAPbBr_3$ [52]. Reproduced with permission from Ref. [52]. Copyright 2016 Advanced Science. (**b**) Hole transport layer (HTL) from Spiro–MeOTAD [53]. Reproduced with permission from Ref. [53]. Copyright 2017 Advanced Materials Interfaces. (**c**) HTL from TaTm [50]. Reproduced with permission from Ref. [50]. Copyright 2020 Frontiers in Chemistry. (**d**) Electron transport layer (ETL) from C60 [54]. Reproduced with permission from Ref. [54]. Copyright 2019 Scientific Reports.

Low-dimensional perovskites, such as quasi two-dimensional (2D) perovskites, are known for their stability, but they have a lower performance than that of three-dimensional (3D) ones [55]. They have unique optical and charge transport properties, but also a high open circuit voltage (V_{oc}) loss. These 2D halides can be used for photodetectors, lasers, resistive, solar cells, and LEDs. The 1D halide perovskites can be used in photodetectors and lasers; however, their use is challenging in other devices due to their rough structure and incomplete surface coverage, which result in a degraded performance. The 0D halide perovskite, or quantum dots, can be used in solar cells as they have a high photoluminescence quantum yield (up to 90%); however, their performance lags behind that of 3D halides because of their random orientation and the excess organic ligands on their surfaces, which reduce carrier mobility [56,57].

2.1.1. Methylammonium Lead Triiodide ($MAPbI_3$)

Metal halide perovskite ($MAPbX_3$, X = Cl^-, Br^-, or I^-) is the initial material used for perovskite PV fabrication and has a direct bandgap energy. The valence band is dominated by the p orbitals of the iodide, and the conduction band consists of Pb p orbitals; therefore, the transitions are p-to-p orbitals. Furthermore, the bandgap can be modified by varying A, B, and X. $MAPbI_3$ is an example of such a material and possesses several advantages, such as having an absorption of 800 nm, a direct bandgap of 1.57 eV [13], a large absorption coefficient of 1.5×10^4 cm^{-1} at 550 nm [58], a low-exciton binding energy of less than 10 meV [58], a very high charge carrier mobility of 66 cm^2/Vs, a large electron and hole diffusion length of over 1 µm, and potentially reaches over 100 µm [59], thus it is suitable to be applied as a perovskite material.

2.1.2. Formamidinium Lead Triiodide ($FAPbI_3$)

Formamidinium lead triiodide ($FAPbI_3$) has a bandgap energy of 1.48 eV near the optimal bandgap (1.34 eV), which is ideal for theoretical maximum device efficiency. Moreover, its excellent thermal- and photostability, in the form of black-phase $FAPbI_3$ (α-$FAPbI_3$), meets the requirements of perovskite PV devices. The earlier-reported $FAPbI_3$-based perovskite PVs have had their advantages of having a large charge-carrier diffusion and high short-circuit currents established [60]. Compared to $MAPbI_3$, $FAPbI_3$ is more thermally stable due to the variations in A-site cations, as an FA cation has a greater thermal stability compared to an MA cation [61]. As methylammonium cations are volatile, the MA^+ component will experience fragmentation with increased temperature, which leads to the quick decomposition of the $MAPbI_3$ crystal structure [62]. Moreover, the activation energies for thermal degradation of $FAPbI_3$ (115 ± 3 kj mol^{-1}) are higher than those of $MAPbI_3$ (93 ± 8 kj mol^{-1}), so $FAPbI_3$ has a better resistance to thermal decomposition [63]. In terms of humidity stability, both $MAPbI_3$ and $FAPbI_3$ can chemically decompose into lead iodide (PbI_2) under high moisture conditions; however, when exposed to moisture, the $FAPbI_3$ perovskite will undergo phase transition into an undesirable yellow non-perovskite (δ-$FAPbI_3$) [64,65]. Furthermore, the deposition and fabrication methods of $FAPbI_3$ films have difficulties in crystallizing the black perovskite phase.

2.1.3. Mixed-Cation Perovskite

Since there are concerns with $MAPbI_3$'s structural phase transitions and thermal stability, a combination of both $FAPbI_3$ and $MAPbI_3$ have been introduced with different ratios. With the perovskite composition of $FA_{1-x}MA_xPbI_3$, when x = 0–1, it shows different SEM characteristics. Based on the ratio of $FAPbI_3$ and $MAPbI_3$, the crystal structure shows some cracks that differ, based on the x composition. With $FA_{0.8}MA_{0.2}PbI_3$, there are several needle-like structures that can be obtained from the low MA contents, which lead to the formation of the δ-$FAPbI_3$ phase, while $FA_{0.4}MA_{0.6}PbI_3$ shows a better crystal structure with a lack of needle-like structures due to the increase in MA content, which results in the stabilization of α-$FAPbI_3$ [66]. The results showed an improvement in the optoelectronic properties and in the stability [67–69]. This was due to the larger ionic radius provided by FA and the dual-ammonia group that inhibits ion movement in the space of the PbI_6 octahedral [60]. However, a decrement of the open-circuit voltage (V_{oc}) value was observed when mixing FA and MA (with a V_{oc} of about 1.02 V) [70–72]. A slightly higher V_{oc} value of 1.03 V was then achieved by adding additional layers through surface passivation [69,73]. Moreover, a FAMA mixed cation perovskite will tend to form an undesirable yellow perovskite, so CsI is introduced into the FAMA mixed cation perovskite to guarantee the formation of black phase perovskite PV [67,73,74]. With the introduction of a third cation, in this case Cs with the mixture of MA and FA cation, it is recognized as a triple cation perovskite.

2.2. Fabrication Methods

The aim of PV devices is to generate high and efficient power to be used in large systems. However, controlling the morphology of perovskite is a difficulty in scaling large-area PVs, especially as this situation is not present in small-area PVs. Therefore, the method of fabrication must be carefully selected. The reported scalable deposition methods for perovskite PVs are solution-based and vapor-based deposition techniques.

The fabrication method is usually divided into one- or two-step depositions. In one-step deposition, both precursors' solutions are used, while for two-step deposition separating the layer deposition and the precursor solution to produce highly uniform and defect-free layers with a great morphology is performed. The layers obtained through one-step deposition has defects and, therefore, a higher recombination rate. Thus, two-step deposition offers better morphology control than one-step deposition through its deposition process [75].

2.2.1. Solution Processing-Based Method

Solution processing-based methods are the common methods for perovskite deposition. Various methods derived from solution processing include spin-coating, dip-coating, doctor blade, spray-coating, ink-jet printing, screen printing, drop-casting, and slot die coating.

2.2.2. Spin-Coating

Spin-coating is the simplest and most cost-effective method derived from solution processing methods. This method is done by spin-coating a precursor on a substrate followed by annealing the thin film to obtain a crystallized layer of perovskite, as shown in Figure 2a. The layer thickness can be optimized by changing the spin speed, acceleration, and spin-coating time. The advantage of using this technique is the good level of reproducibility and the morphology obtained for small-scale areas; however, the long processing duration and material waste limit this method for application for large-scale area perovskite fabrication [76].

2.2.3. Drop-Casting

Drop-casting is considered a cheap technique to produce PVs. The technique is based on dropping a volatile solvent on a substrate before it is evaporated and dried, as shown in Figure 2b. Layer thickness is dependent on the rate of evaporation, the drying process, and the volume and concentration of the solvent used for the dispersion. The wasted material in the case of drop-casting is less than that of spin-coating, which is counted as an advantage compared to the other method. However, the drawbacks of this method relate to the difficulty in controlling the layer thickness, which results is a non-uniform film formation which is thus ineffective for large-scale areas [77].

2.2.4. Spray-Coating

Spray-coating is a highly scalable deposition method for large-scale area perovskite PV fabrication. This technique has several advantages, such as rapid film deposition, as the spray head can move across a substrate at more than 5 m per minute, as shown in Figure 2c. Other advantages include the low cost of processing, and the capability for deposition on flexible and glass-based substrates for use in large-scale perovskite photovoltaics. Moreover, spray-coated films are characterized by their high thermal stability and they have better optoelectronic properties compared to spin-coated films due to a better charge transfer capability and a longer minority carrier lifetime [78]. However, the drawback of this method is the difficulty in achieving fully homogeneous layers due to there being unpredictable characteristics in the film. This problem occurs when different droplet sizes leave different patches sizes during the drying process, thus increasing the series resistance that affects the performance. These problems can be solved using a modified spraying technique, such as pen spray [79], electrostatic spray-coating [80], pulsed spray coating [81], and ultrasonic spray-coating [82].

2.2.5. Doctor Blade

Doctor blade is a simple coating system that uses a blade coater applicator—in this method, the height of the blade is adjusted depending on the substrate surface, as shown in Figure 2d—to produce a uniform film with a modifiable quality. Quality control is done by managing the evaporation rate, whether by adjusting the airflow over the substrate or heating the substrate to reach the boiling point of the solvent. A simple, environmentally friendly, vacuum free, low-cost deposition method and the capability to control film quality are some of the advantages offered by this method. Moreover, it is capable of overcoming performance degradation due to the presence of moisture and air [83]. This is caused by the slow crystallization rate that forms large-area grains that restricts the air and moisture permeability of the perovskite layer [84]. This method is suitable and stable and yields a great morphology for the deposited film.

2.2.6. Slot-Die Coating

Slot-die coating is a solution processed deposition method that applies solution to flat substrates, such as glass or metal, and is highly scalable for large-scale area perovskite PVs, as shown in Figure 2e. An advantage to this method is less, or no, material wastage compared to spin-coating techniques. Moreover, it also produces a uniform and controllable film thickness by controlling the amount of material that is fed into the process. Hence, it can be used on a commercial scale for perovskite PV production [85].

2.2.7. Ink-Jet Printing

Ink-jet printing is a common method used in the manufacturing of optoelectronic devices, field-effect transistors, and PVs. It is a non-contact technique that uses additive patterning. It is based on the selective ejection of ink from a chamber through a nozzle onto a substrate. A liquid droplet is ejected when an external bias is applied. This bias causes the chambers containing the liquid to contract, creating a shock wave in the liquid, and therefore ejection occurs, as shown in Figure 2f. The technique is considered fast, consumes less material, and can be used for large-scale production [86,87]. However, the main drawback is the possibility of a blocked nozzle because the used materials are poor solvents.

2.2.8. Vapor-Based Method

The vapor-based method is an alternative method for fabricating perovskite PVs and it has better film uniformity compared to the solution-processing-based methods. This method is mostly used on an industrial scale for glazing, liquid crystal displays, and in the thin-film solar industry. The absorption efficiency of solar cells depends on the thickness of the perovskite layer. Thin layers absorb less sunlight, whereas thick films take a great deal of time to operate, as electrons and holes take significant time to reach their contacts. If the film is not uniform in the overall area, there will be direct contact with the electron transport material (ETM) and the hole transport material (HTM), resulting in a lower FF and V_{oc}. Vapor-based deposition surpasses other techniques due to its ability to produce large-scale multi-stacked thin films with a uniform thickness. However, vapor-based deposition requires a vacuum to increase the mean free path between collisions to produce highly uniform and pure thin films. The vapor-based method is divided into two categories, physical and chemical.

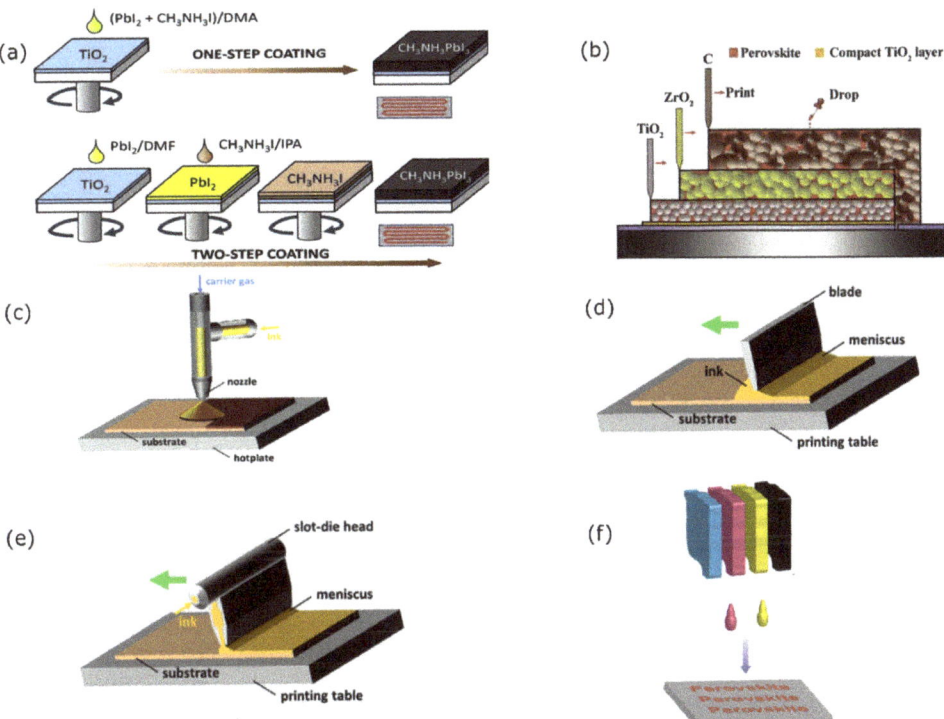

Figure 2. Different solution processing-based method for perovskite PV fabrication: (**a**) spin-coating [88], (**b**) drop-casting [88]. Reproduced with permission from Ref. [88]. Copyright 2017 Journal of Materiomics. (**c**) Spray-coating, (**d**) doctor blade, and (**e**) slot-die coating [89]. Reproduced with permission from Ref. [89]. Copyright 2019 Advanced Materials. (**f**) Ink-jet printing [90]. Reproduced with permission from Ref. [90]. Copyright 2016 Hal.

2.2.9. Chemical Vapor Deposition

Chemical vapor deposition (CVD) is a deposition method that produces highly scalable and pinhole-free large-scale perovskite PVs. A co-evaporation technique is used to deposit the perovskite layers with the help of two different precursors. The precursors are heated, mixed, and then moved to another substrate at a lower temperature, using a carrier gas (argon) to form highly uniform films, which are pinhole-free and have larger grain sizes and long carrier lifetimes, as shown in Figure 3a. CVD has been mostly deployed for fabricating perovskite layers to prevent the drawbacks of using low amounts of materials and the difficulty in controlling the flux deposition [91–93]. Moreover, while the use of CVD will produce a high material yield ratio and highly scalable for perovskite layer depositions [94], it requires a vacuum and uniform co-evaporation of the sample material, which is a challenge at an industrial scale.

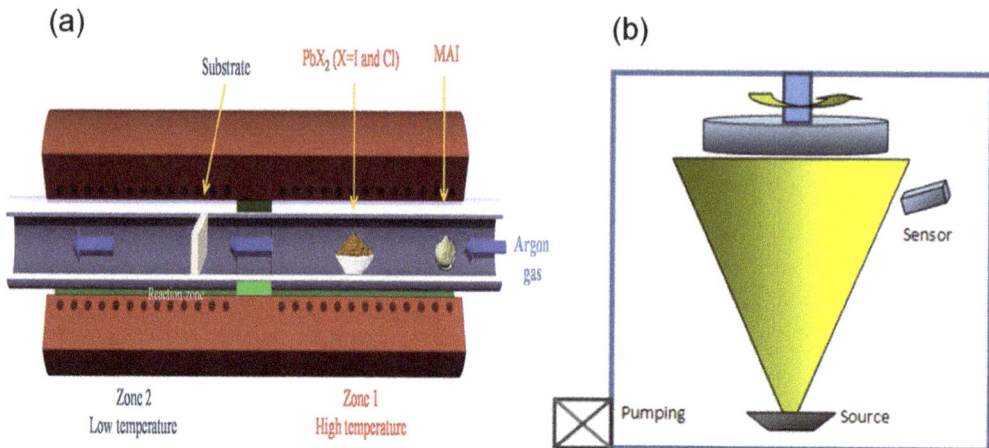

Figure 3. Type of vapor-based fabrication method: (**a**) chemical vapor deposition (CVD) [91]. Reproduced with permission from Ref. [91]. Copyright 2015 Scientific Reports. (**b**) Physical vapor deposition (PVD) [95]. Reproduced with permission from Ref. [95]. Copyright 2016 Scientific Reports.

2.2.10. Physical Vapor Deposition

Physical vapor deposition (PVD) is a simple and non-reactive deposition process where a whole substrate surface area is covered, and the resulting layer has a high stability against moisture. Utilizing this method, a deposited film shows a uniform and smooth surface, with less defects and good crystallization layers. Figure 3b shows a single-source PVD for $MAPbI_3$, where the temperature of the source is raised rapidly until the $MAPbI_3$ evaporates without a chemical reaction and is then deposited on the substrate. The advantages of using PVD are the full surface coverage, great grain structure, high crystallization, scalability for mass production [95], and controllable film quality, thickness, and morphology, which makes PVD preferable to solution processing-based methods.

Table 1 shows various materials and their methods for HOIP PV fabrication, complete with the PCE values. With each material, for HOIP photovoltaics, improved, new and excellent perovskite materials with superior stability, light-absorption, charge mobility, and lifetimes were produced. The optimization of materials and structures is one of the keys to improving the photoelectric conversion efficiency [96]. The optimization of materials is shown in the work of Saliba et al. [67], via spin coating with different material compositions of $Cs_{0.05}(MA_{0.17}FA_{0.83})_{(0.95)}Pb(I_{0.83}Br_{0.17})_3$ and $Cs_{0.1}(MA_{0.17}FA_{0.83})_{(0.90)}Pb(I_{0.83}Br_{0.17})_3$ with the increase in the fill factor, from 74 to 77%. This is due to $Cs_{0.1}(MA_{0.17}FA_{0.83})_{(0.90)}Pb(I_{0.83}Br_{0.17})_3$ having more uniform grains with each increase in Cs value, obtaining a better charge transport, which explains the higher fill factor value, which is consistent with works that used $Cs_{0.15}(MA_{0.17}FA_{0.83})_{(0.85)}Pb(I_{0.83}Br_{0.17})_3$ as the perovskite material.

The optimization of device structure is shown in several works on the topic of fabricating $MAPbI_3$ perovskite photovoltaics using different fabrication methods. Among $MAPbI_3$ perovskite photovoltaic fabrication methods, spin coating shows the lowest PCE as the uniform structure is capable of obtaining a grain size of 100 nm [97]. Doctor blade and slot-die coating show a better PCE among all $MAPbI_3$ perovskite photovoltaics with a more uniform grain size and thicknesses of 200 μm [98] and 5 μm [99], respectively, which are slightly better compared to spin coating due to the controllable fabrication methods.

Table 1. Various materials and methods for HOIP PV fabrications. Voc, Jsc, and FF are open-circuit voltage, short-circuit current density, and fill factor, respectively.

Coating Method	Material	Voc (V)	Jsc (mA cm^{-2})	FF (%)	Size (cm^2)	PCE (%)	Ref.
Spin-coating	FAPbI$_3$	1.06	24.7	77.5	~1	20.2	[100]
Spin-coating	Cs$_{0.05}$(MA$_{0.17}$FA$_{0.83}$)$_{(0.95)}$Pb(I$_{0.83}$Br$_{0.17}$)$_3$	1.109	22.7	74.0	~1	18.6	[67]
Spin-coating	Cs$_{0.1}$(MA$_{0.17}$FA$_{0.83}$)$_{(0.90)}$Pb(I$_{0.83}$Br$_{0.17}$)$_3$	1.13	22.0	77.0	~1	19.1	[67]
Ink-jet printing	Cs$_{0.1}$(FA$_{0.83}$MA$_{0.17}$)$_{0.9}$Pb(Br$_{0.17}$I$_{0.83}$)$_3$	1.11	23.1	82	2.3	20.7	[101]
Spin-coating	Cs$_{0.15}$(MA$_{0.17}$FA$_{0.83}$)$_{(0.85)}$Pb(I$_{0.83}$Br$_{0.17}$)$_3$	1.088	19.4	69.3	~1	14.6	[67]
Spin-coating	(FAPbI$_3$)$_{0.95}$(MAPbBr$_3$)$_{0.05}$	1.14	24.9	81	~0.094	23.2	[102]
Doctor blade	MAPbI$_3$	1.12	22.6	81	0.075	20.3	[98]
Doctor blade	MAPbI$_3$	1.10	22.7	81	0.08	20.2	[103]
Slot-die coating	MAPbI$_3$	1.03	22.1	74	~232.3	16.8	[99]
Spin-coating	MAPbI$_3$	1.08	20.7	68	0.16	15.2	[97]
Ink-jet printing	MAPbI$_3$	1.08	22.66	76.2	0.04	18.6	[104]
Slot-die coating	MAPbI$_{3-x}$Cl$_x$	1.06	21.7	~78	0.06	18.0	[105]
Slot-die coating	MAPbI$_{3-x}$Cl$_x$	1.09	22.38	74.7	0.096	18.3	[106]
Spray coating	MAPbI$_{3-x}$Cl$_x$	1.10	21.4	77.6	0.08	18.3	[20]
Spin-coating	MAPb(I$_{0.85}$Br$_{0.15}$)$_3$	1.07	21.5	68	0.076	15.4	[107]
Spray-coating	CsI$_{0.05}$((FAPbI$_3$)$_{0.85}$(MAPbBr$_3$)$_{0.15}$)$_{0.95}$	1.10	22.3	73	0.16	17.8	[19]
Screen printing	(AB)$_2$(MA)$_{49}$Pb$_{50}$I$_{151}$	0.94	23.4	71	0.8	15.6	[30]

Based on Table 1, it can be concluded that the best material for perovskite PV fabrication is (FAPbI$_3$)$_{0.95}$(MAPbBr$_3$)$_{0.05}$ [102]. Among the fabrication methods, the doctor blade method can be considered as the best method since it resulted in the highest PCE value. Such a method may improve the PCE of (FAPbI$_3$)$_{0.95}$(MAPbBr$_3$)$_{0.05}$, which already reaches 23.2% with spin coating [102].

3. Materials and Device Characterizations

This section describes the material characteristics that should be met for perovskite PVs, including the structural and morphological properties. XRD analysis can be used to detect a crystallite structure from the layer formation [108]. Based on synthesis conditions, there are three types of MAPbI$_3$ structure. At T < 163 K, it is in an orthorhombic phase, at 163 K < T < 327.3 K, it is in a tetragonal phase and beyond 327.3 K, it will begin to form a cubic phase [109]. Substantial changes occur at 60 °C, as it shifts from the tetragonal to the cubic phase, with better grain sizes, based on the XRD results from several research studies, as multiple new peaks are present compared to the tetragonal phase [110–112].

The properties of HOIPs are also affected by organic monocation. Adding a bulky hydrophobic organic cation to the perovskite lattice can prevent moisture intrusion [113]. Because of this structure, the bandgap is large and tunable. Moreover, increasing the chain length of the organic monocation allows the prevention of oxidation of Sn^{2+} to Sn^{4+} [114].

Empirically, the structural stability of 3D perovskites can be predicted using the Goldschmidt tolerance factor, as the rule of thumb, for the ABX$_3$ crystal structure, $t = \frac{(r_a+r_x)}{\sqrt{2}(r_b+r_x)}$, which is calculated based on the radii of the A, B and X ions. The ideal t value for a stable perovskite would be $0.8 < t < 1.0$, so the organic monocation value has to be adjusted to obtain the necessary t value [115]. However, the perovskite ligand size can have an influence on this value [116].

A key example to improve perovskite PV performance would be by using XRD techniques that will detect the best crystal structures in the formation of layers of a sample perovskite Cs$_x$M, with x = 0, 5, 10, 15%, as shown in the XRD spectra of Figure 4. At about 14°, all Cs$_x$M compositions with varying x values show the same perovskite peak. For Cs$_0$M, there are small peaks observed at about 11.6° and 12.7° that represents the undesirable yellow non-perovskite phase (δ-FAPbI$_3$) and the PbI$_2$ that was used in its fabrication, respectively. As Cs amounts of 5, 10, 15% in the XRD spectra did not show these specific peaks, this would indicate that the mixed perovskite Cs$_0$M shows an incomplete conversion of FA perovskite to the black-phase perovskite (α-FAPbI$_3$).

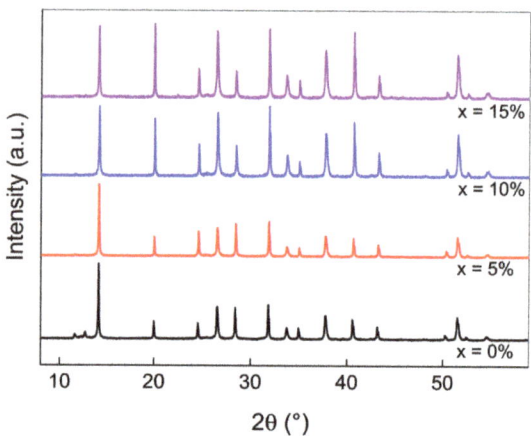

Figure 4. XRD characterization of Cs$_x$M compounds. XRD spectra of perovskite with addition of Cs Cs$_x$(MA$_{0.17}$FA$_{0.83}$)$_{(1-x)}$Pb(I$_{0.83}$Br$_{0.17}$)$_3$, abbreviated as Cs$_x$M, where M stands for "mixed perovskite". Cs$_x$M with x = 0, 5, 10, 15%. Reproduced with permission from Ref. [67]. Copyright 2016 Energy and Environmental Science.

Other than XRD characterization, with the different methods implemented for the fabrication of perovskite PVs, the resulting devices will also have different crystal structures relative to their fabrication methods [117]. The morphology of the perovskite is another factor to improve perovskite PV performance. Numerous work have been done on optimizing morphology by improving the fabrication methods [84,93] and by applying additional additives [118,119]. As shown in Figure 5a,b, different crystal grain structures were observed from using different fabrication methods (blow-dry and spin-coating).

The thicknesses of the layers differ between those fabricated using blow-drying and those using spin coating. For example, the thickness of the MAPbI$_3$ and spiro-MeOTAD deposited by blow-drying is thinner than that fabricated by spin-coating. The blow-dried spiro-MeOTAD layer has been shown to be more compact and smaller than the spin-coated spiro-MeOTAD layer. Moreover, blow-dried MAPbI$_3$ merged better with mp-TiO$_2$ than the spin-coated MAPbI$_3$, as shown in Figure 5c,d. As a result, devices with blow-dried MAPbI$_3$ consistently achieve higher V_{OC}, J_{SC} and PCE values than spin-coated ones.

Based on the work of Zhou et al. [120], the characteristics of the grain boundaries, grains, and material composition will have an effect on carrier transport, emissions, ionic diffusion, and performance. The simple crystal structures of the 3D halide PVs, such as MAPbI$_3$ and FAPbI$_3$, were determined using XRD or SEM analysis.

The presence of defects and grain boundaries in the PV film will have a detrimental effect on PV performance. Defect passivation is a strategy developed to enhance device efficiency and stability at the same time [121–123]. As shown in the work of Medjahed et al. [124], adding an additive in the presence of chlorine during the MAPbI$_3$ synthesis has a positive advantage and impact on the growth and morphology (improvement on crystallite size and structure) of the obtained fabrication. Moreover, it also improves the electrical properties of the material, such as the electron diffusion length, carrier lifetime, and a high PCE value of 11.4%. It was also revealed, using in situ XRD, that the presence of two areas of MAPbI$_3$ showed the presence of MAPbCl$_3$. There were crystalline phases that could not be identified, but did not correspond to a chlorine-based intermediate phase, but rather the formation of MAPbCl$_3$ in the perovskite formation from PbCl$_2$. It can be concluded that MAPbI$_3$ forms as annealing begins, MAPbCl$_3$ disappears gradually, and that MAPbI$_3$ decreases as the formation of PbI$_2$ into MAPbI$_3$ is completed, which corresponds with the work done by Saliba et al., who showed that XRD is able to analyze this information based on the patterns of the formation of the perovskite.

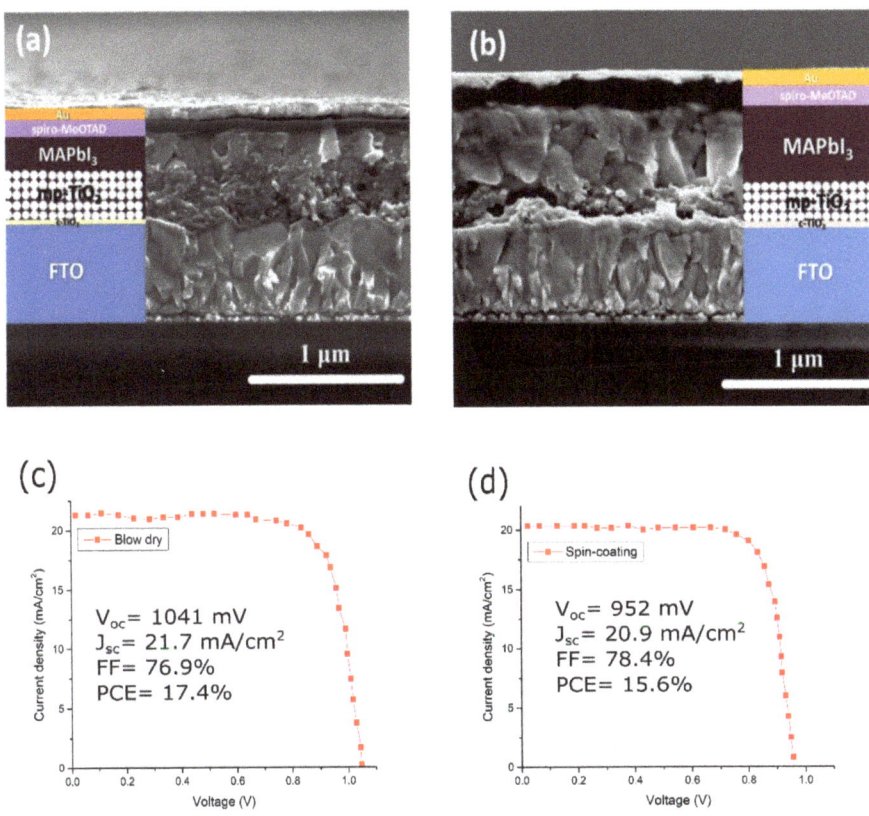

Figure 5. SEM images and J-V curve of the fabricated devices by using: (a,c) blow-dry and (b,d) spin-coating methods [117]. Reproduced with permission from Ref. [117]. Copyright 2017 Solar Energy Materials and Solar Cells.

Feng et al. also demonstrated another additive enhancement to the morphology in a one-step solution-processed perovskite that utilizes methanol as shown in Figure 6 [125]. It was observed that, by adding a proper amount (5 vol%) of methanol solution to the perovskite (MAPbI$_3$) precursor solution, the morphology, crystallization, optical and electrical properties of the perovskite layer were enhanced. SEM analysis was performed to investigate the morphology of the perovskite without methanol, and is shown in Figure 5a, and with (5 vol%) methanol solution, shown in Figure 5b. Under SEM, the perovskite film without methanol shows pinholes and grain boundaries. These pinholes and grain boundaries will impede charge transport, which easily induces the recombination between electron and hole, thus reducing the overall PV performance. However, the addition of methanol significantly enhances the morphology since it has rougher surface, larger grain sizes, and fewer grain boundaries compared to perovskite without methanol. Since the grain size in the vertical direction, as shown under SEM, is equivalent to the thickness of the perovskite, this indicates that the charge carriers can transport efficiently across the perovskite film and reach the electrodes before recombination occurs. The fabricated perovskite using methanol shows a higher PCE of 19.51% compared to the fabricated perovskite without methanol (16.53%). Moreover, it also has a better stability since it still shows a high PCE in the dark under ambient temperature for 30 days.

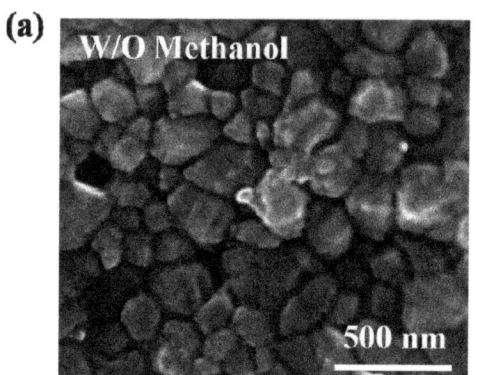 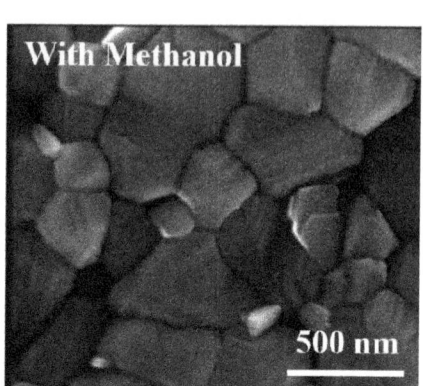

Figure 6. SEM images of the perovskite PV fabricated: (**a**) without and (**b**) with (5 vol%) methanol [125]. Reproduced with permission from Ref. [125]. Copyright 2019 Electrochimica Acta.

While the addition of additives may have improved the morphology of the HOIP, its lifetime and performance were not satisfactory for commercial and industrial applications. This will require further enhancements to the HOIP-like tandem structure to improve the overall lifetime and performance.

3.1. Perovskite Tandem Photovoltaics

Perovskite tandem PVs have been suggested as a method to improve the overall performance, stability, and lifetime of perovskite PVs. A perovskite tandem PV usually consist of a cell—either silicon, perovskite, or copper indium gallium selenide (CIGS)—overlaid by a perovskite PV [126], to increase the efficiencies beyond a single junction limit [127], without adding a substantial cost during production [128,129]. Typical single-junction PV cells do not make use of 67% of the solar energy they receive, because of the weak absorbance capabilities of the semiconductors. Semiconductors can only absorb photons with energy is above their bandgap energy (E_g) and they generate energy equal to E_g, where the rest of the energy are lost through thermalization as heat. This will severely affect the PCE of a PV because it corresponds to the V_{oc} and J_{sc} of the PVs. As one of solutions, tandem photovoltaics can address this problem [130].

Tandem PVs use stacks of materials with different bandgaps, where materials with larger bandgaps are put at the top of a cell and those with small bandgaps are at the bottom. High-energy photons are absorbed by the upper materials, while low-energy photons are not lost, in this case, but rather are absorbed by the lower stack materials, making use of most of the incident energy. One of the most known structures of tandem cells is the double-junction tandem device. They have two different configurations: two-terminal (2-T) and four-terminal (4-T) tandem, according to the stacking method used [130].

The 2-T tandem cells are synthesized by stacking a transparent front electrode, with a front cell and an opaque rear electrode, with the rear cell being one substrate where an interconnection layer (ICL) separates them, as shown in Figure 7a. The recombination of the photogenerated carriers from either sub-cell takes place in the ICL. On the other hand, 4-T tandem cells are made of two separate devices with two separate electrodes, linked together through a dichromatic mirror, as shown in Figure 7b. However, due to the additional electrodes, the optical loss will result in a more expensive cost compared to 2-T tandem PVs. Moreover, 2-T tandem PVs are cheaper to fabricate, though it is harder to fabricate 2-T or monolithic tandem PVs compared to 4-T tandem PVs for general applications [130].

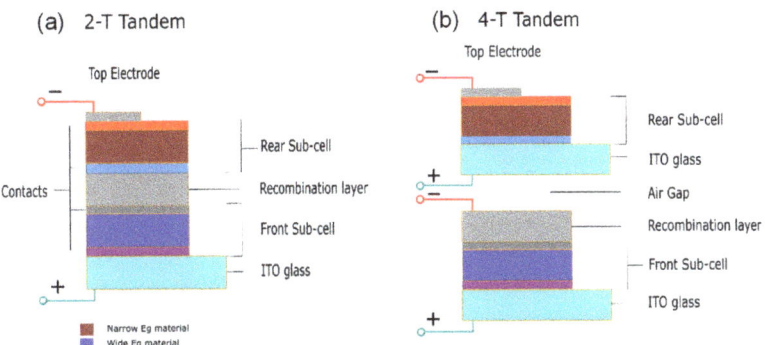

Figure 7. Schematic diagram of (**a**) 2-T and (**b**) 4-T tandem perovskite.

Based on the values shown in Table 2, Voc, Jsc, and FF have similar in values with the 4-T Si/MAPbI$_3$ configuration [131–133] and the values of Voc, Jsc, and FF in these works are about 1.1 V, 21 mA cm^{-2}, and 75–80%, respectively.

Table 2. Some reports of perovskite tandem PVs. Voc, Jsc, and FF are open-circuit voltage, short-circuit current density, and fill factor, respectively.

Type	Tandem	Perovskite Material	Voc (V)	Jsc (mA cm^{-2})	FF (%)	Size (cm^2)	PCE (%)	Ref.
4-T	Si	MAPbI$_3$	1.08	20.6	74.1	0.075	23.0	[131]
4-T	Si	MAPbI$_3$	1.098	21.0	74.1	1.10	26.7	[132]
4-T	Si	MAPbI$_3$	1.156	19.8	79.9	1.00	27.0	[133]
4-T	Si	MAPbI$_3$	1.69	15.9	77.6	0.17	21.2	[134]
2-T	Si	Cs$_{0.05}$(FA$_{0.77}$MA$_{0.23}$)$_{0.95}$Pb(I$_{0.77}$Br$_{0.23}$)$_3$	1.87	19.37	79.9	1.06	29.3	[4]
2-T	Si	Cs$_{0.05}$(MA$_{0.17}$FA$_{0.83}$)$_{0.95}$Pb(I$_{0.83}$Br$_{0.17}$)$_3$	1.792	19.02	74.6	1.088	25.2	[135]
2-T	Si	Cs$_{0.15}$(MA$_{0.17}$FA$_{0.83}$)$_{0.85}$Pb(I$_{0.7}$Br$_{0.3}$)$_3$	1.80	17.8	79.4	~1.00	25.4	[136]
4-T	CIGS	Cs$_{0.09}$FA$_{0.77}$MA$_{0.14}$Pb(I$_{0.86}$Br$_{0.14}$)$_3$	1.77	17.3	73.1	0.04	22.4	[137]
4-T	CIGS	Cs$_{0.05}$(MA$_{0.17}$FA$_{0.83}$)Pb$_{1.1}$(I$_{0.83}$Br$_{0.17}$)$_3$	1.59	18.0	75.7	0.78	21.6	[138]
2-T	CIGS	Cs$_{0.05}$(MA$_{0.17}$FA$_{0.83}$)$_{0.95}$Pb(I$_{0.83}$Br$_{0.17}$)$_3$	1.68	19.2	71.9	1.03	23.3	[33]

This occurs similarly in the 4-T CIGS/Cs$_{0.05}$(MA$_{0.17}$FA$_{0.83}$)Pb$_{1.1}$(I$_{0.83}$Br$_{0.17}$)$_3$ configuration [138] and the 2-T CIGS/Cs$_{0.05}$(MA$_{0.17}$FA$_{0.83}$)$_{0.95}$Pb(I$_{0.83}$Br$_{0.17}$)$_3$ configuration [33], since these works obtained similar values for Voc, Jsc, and FF of about 1.6–1.8 V, 17.0–19.2 mA cm^{-2}, and 72.0–75.7%, respectively.

In comparison to the PCE values in Table 2, the 2-T configurations of CIGS/Cs$_{0.05}$(MA$_{0.17}$FA$_{0.83}$)$_{0.95}$Pb(I$_{0.83}$Br$_{0.17}$)$_3$ [33,138] are better by 1.08-fold compared to those of the 4-T configurations. The 4-T configurations experience optical loss, which is found in the encapsulation and absorption in the transparent conductive electrode [139], which affects the PCE. The 2-T configurations experience series resistance loss in the large-area modules [140].

Further analysis is required to determine the increase in the PCE of the devices through development and improvements using tandem technologies. The configurations in Table 2 for tandem PVs refer to the materials in Table 1. The best performing PCE is from the Cs$_{0.05}$(MA$_{0.17}$FA$_{0.83}$)$_{(0.95)}$Pb(I$_{0.83}$Br$_{0.17}$)$_3$ configuration in Table 1, and shows a PCE of 18.6% [81]; the best performing PCE from the 2-T Si/Cs$_{0.05}$(FA$_{0.77}$MA$_{0.23}$)$_{0.95}$Pb(I$_{0.77}$Br$_{0.23}$)$_3$ configuration has a PCE of 29.3% [4] with an improvement achieved through tandem technologies using silicon and also through the use of an appropriate carbazole-based layer to efficiently extract the holes. Silicon was used as the bottom cell of the tandem since it has the properties of absorbing solar radiation from the near infrared region of the absorption spectrum. There is also another development in tandem PVs between Cs$_{0.05}$(MA$_{0.17}$FA$_{0.83}$)$_{0.95}$Pb(I$_{0.83}$Br$_{0.17}$)$_3$ perovskite and CIGS, and the PCE value of the

$Cs_{0.05}(MA_{0.17}FA_{0.83})_{0.95}Pb(I_{0.83}Br_{0.17})_3$ tandem [33] increases by 1.25-fold compared to that of the intrinsic one [67].

The CIGS allows for a tunable bandgap, and it varies based on the temperature. A tunable bandgap was obtained from 1.1 to 2.3 eV by interchanging the cations, metals, and halides. These allow for the tandem PV to absorb photons that have an energy above the bandgap, so a higher bandgap will require a higher energy for the absorption spectrum. The best configurations for perovskite tandem PVs came from silicon and CIGS. In fact, another perovskite–perovskite tandem PV has been recorded, with a high PCE of 24.4% [141] and there are few reports presently beating this record. Therefore, the best reported tandem perovskite photovoltaic belongs to the 2-T Si/$Cs_{0.05}(FA_{0.77}MA_{0.23})_{0.95}Pb(I_{0.77}Br_{0.23})_3$ configuration with a PCE of 29.3% [4].

For further commercialization of perovskite PVs, large area modules should be developed since the previous studies were conducted for a small area (1 cm^2). This development needs to be carried out to determine if there is any improvement in the crystal grain structure and PCE value when the module size is increased.

3.2. Large Scale Modules

Several issues need to be addressed for further perovskite PV commercialization: (i) a thin-film deposition method that is scalable and reproducible; (ii) high stability and long lifetime; and (iii) low or less toxic materials for large-area devices.

The rapid increment in PCE performance for perovskite PV has developed at a quick pace compared to other solar technologies. However, several high PCE values have been obtained with very small areas (~1.0 cm^2). To fabricate a large area with a high-quality perovskite (good uniformity and few structural defects), several deposition methods have been developed, such as spin-coating, slot-die coating, doctor blade, and vacuum deposition. In 2015, Chen et al. [142] fabricated perovskite using one-step spin-coating with an active area of 1.02 cm^2 and obtained a PCE of 15%. In 2016, Qiu et al. [143] fabricated a large-area perovskite with an active area of 1 cm^2 and obtained a PCE of 13.6%.

While spin-coating is widely used in the deposition of small-area thin films in laboratories, it might not be suitable for industrial large-area fabrications. Spin-coated films are not uniform throughout the area [144], require a large amount of solution mixture of materials, and there is significant wasted solvent, which increases the cost of fabrication. Its performance also reduced due to the increase in series resistance and the decrease in film quality [145,146]. As in the case of Hossain et al. [147], since spin-coating is not suitable for the fabrication of perovskite solar cell on a silicon solar cell, so some methods have to be replaced by PVD or CVD.

In 2018, Zheng et al. [148] were able to achieve 21.8% PCE using CVD on a 16 cm^2 monolithic $(FAPbI_3)_{0.83}(MAPbBr_3)_{0.17}$ perovskite–silicon tandem PVs. The final product of the PVs has an enhanced V_{oc} with a J_{sc} that ranged from 15.6 to 16.2 mA/cm^2 and an FF of 78%. The supporting information regarding further improvements that could be achieved for this device would be the spiro-OMeTAD stack being replaced with a high refractive index inorganic hole transport layer to eliminate unnecessary wavelength absorptions. To improve it for commercial use in larger area of 6″ × 6″, the PDMS layer can be replaced with thin glass and the spin-coating process can be replaced with something less expensive, such as the doctor blade or spray-coating methods, which will be discussed in further works.

After conducting their previous work, in 2019, Zheng et al. [149] showed another improvement that increased the PCE value. They decided to use micron green-emitting $(Ba,Sr)_2SiO_4:Eu^{2+}$ phosphor—a cheap material that is commercially available and is mostly used in light-emitting diodes—as an antireflective down shifting material on PDMS that acts as the hydrophobic layer of the silicon–perovskite tandem PVs with an area of 4 cm^2, which achieved a PCE of 23%. SEM images of the $(Ba,Sr)_2SiO_4:Eu^{2+}$ sample on PDMS showed a high-quality crystal size of about 10–20 μm. Moreover, the results of energy-dispersive X-ray spectroscopy (EDX), not only confirmed the homogenous distributions of Sr, Ba, Si, and O within the specimen, but were also able to detect the Eu content.

Figure 8 shows schematic diagrams of the (a) previous and (b) current large-scale modules of perovskite photovoltaics.

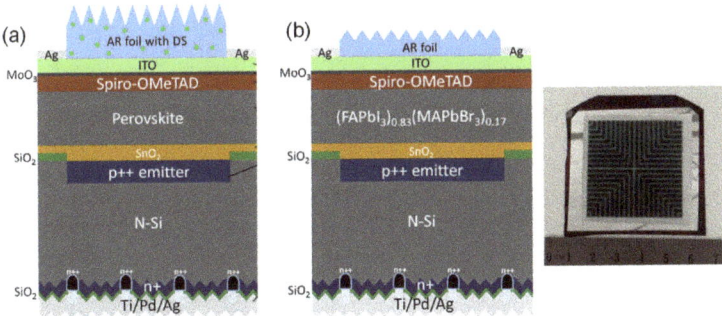

Figure 8. (a) Schematic of perovskite–silicon tandem homojunction photovoltaic with downshifting AR PDMS layer. From Ref. [149]. Copyright 2019 ACS Energy Letters. (b) Schematic of monolithic (FAPbI$_3$)$_{0.83}$(MAPbBr$_3$)$_{0.17}$ perovskite/rear-textured-homo-junction-silicon tandem photovoltaics. Reproduced with permission from Ref. [148]. Copyright 2018 ACS Energy Letters.

After further investigation, with a higher concentration of (Ba,Sr)$_2$SiO$_4$:Eu^{2+} phosphor on top of the perovskite–silicon PV, with initial conditions of J_{sc} of 14.1 mA/cm^2, V_{oc} of 1.73 V, FF of 82%, and a PCE of 20.1%, the result shows that, while the open-circuit voltage remains unchanged or at a constant level, there was an improvement in J_{sc} that was caused by the increased broad absorption of the cell with a reduced front reflection and increased light trapping. Eventually the best choice was obtained and had an area of 4 cm^2 with conditions of J_{sc} of 16.4 mA/cm^2, V_{oc} of 1.73 V, FF of 81%, and a PCE of 23%.

Although works on monolithic perovskite–silicon tandem PVs showed progress in terms of PCE values (about 22–23%), there are several issues that need to be solved for further large-scale fabrication. These issues include stability improvements, the degradation rate, commercialization costs, and the use of environmentally friendly materials.

3.3. Improvements to Perovskite Material and Its Tandem Structures

A recent investigation, conducted by Green et al. [1] on perovskite–silicon tandem PVs, states that a high PCE value of 29.15% had been achieved with an area of 1.060 cm^2. Though it is a small area, in the future, people will continue to develop perovskite–silicon tandem photovoltaics, which will progress further so that it is possible to achieve 4 cm^2 perovskite–silicon tandem photovoltaic with an efficiency of 29.15%, if there is a breakthrough on how to further increase the broad absorption of a cell with reduced reflection but while increase the light trapping.

Furthermore, since it is difficult to deposit a large area of continuous perovskite film using the previously described traditional methods, other methods should be improved to prepare high-quality and large-area perovskite PVs for commercial production in the future. Moreover, from the perspective of green energy, the Pb employed in perovskite PVs is highly toxic, which will hinder the industrial promotion and development of perovskite PVs. Therefore, it is necessary to find a low-toxicity or non-toxic material to replace Pb in the future [96].

Realistically, the halide of Pb is 10 times more dangerous than the Pb that already exists on Earth [150–153]. Several lines of research indicate that contamination due to leaks of lead ions into soil and water resources is a permanent affliction and generates harmful effects on humans, animals, and plant survivability [154–163]. To decrease and reduce its toxicity level, Pb-free [164–166] (or at the very least less Pb content) perovskite–silicon PVs have been researched using a safer option by mixing chalcogen and halogen anions [167] or using tin or (Sn)-based perovskite PVs [168–170]. Recently, it was determined that there is an all-perovskite Pb–Sn tandem photovoltaic with a PCE of 26.4% [171], showing a value that potentially exceeds that of the best-performing single-junction perovskite solar cells, which are capable of retaining more than 90% of their initial performance after 600 h of operation at maximum power under one-sun illumination under ambient conditions. Recent research [172], developed a low-cost device made of sulfonic acid-based lead-adsorbing resin, which freezes lead ions into a scaffold and prevents their leakage when the perovskites are exposed to rainfall. This new device does not affect efficiency or scaling and the structure can be scaled up to 60.8 cm^2.

However, Sn-based perovskite PVs have lower efficiencies and faster degradation than Pb-based perovskite PVs due to their phase fluctuations and they easily oxidize from Sn^{2+} to Sn^{4+}. In the case of Sn-based perovskite PVs being unable to breach the efficiency limit of Pb-based perovskite PVs, new approaches had been created to prevent Pb leakage in perovskite PVs by trapping Pb with cation-exchange resins that are abundant in Ca^{2+} and Mg^{2+} [173].

In addition, the cost barrier for perovskite–silicon cells was identified as being due to expensive organic charge transport materials, such as spiro-OMeTAD, PTAA, and PC$_{60}$BM [174–177]. The cost of organic charge transport, especially for one with a higher quality that can yield a better performance, is very expensive due to the intricate nature of the fabrication steps, in addition to the additional costs that might be derived from better or higher purity in terms of the formation of crystal grains. As an alternative to using organic charge transport materials, inorganic charge transport materials, such as NiO, CuSCN, SnO$_2$, and Nb$_2$O$_5$, which are much cheaper in comparison to organic charge transport materials, have been successfully developed for some perovskite PVs [101,178–180].

Further improvements concern the degradation of the perovskite PVs. PVs are always exposed to various degradation sources, such as humidity, oxygen, heat, and ultraviolet light. To ensure the lifetime of perovskite–silicon tandem PVs, their stability needs to be tested and improved against degradation. Among perovskite PVs, methyl ammonium lead triiodide is easily degraded by humidity and heat, in comparison to perovskite PVs that are based on FAPbI$_3$ [60] or CsPbI$_3$ [181], which proved to have a higher resistance to thermal decomposition—which can be solved by mixing the cations and the halogen anions, which improves the thermal and crystal structure stability. As an example of mixing halide anions, bromide-based perovskite PVs show better resistance to degradation under humidity and heat when compared to iodide-based perovskite PVs as they developed a 2D or 3D hetero-structure in the perovskite PVs [182–184].

It is also suggested that organic charge transport materials, such as spiro-OMeTAD, PTAA, and PC$_{60}$BM, are easily disintegrated by humidity and oxygen, so they require a higher level of encapsulation to prevent the elements from disintegrating the organic charge transport materials. An alternative solution would be to use inorganic charge transport materials, which are beneficial in terms of their stability due to their basic properties. In addition, a densely formed inorganic layer can act as a diffusion barrier to prevent organics or iodine species escaping from the lattice and reacting with the top metal electrode [185,186]. It has been reported in the literature that a semi-transparent perovskite PV with a dense charge transport layer and a transparent electrode endured through thermal cycling, damp heat, and UV stress tests [185–187]. Moreover, low-temperature, glass–glass encapsulation techniques, using high-performance polyisobutylene (PIB) on planar perovskite solar cells, have been reported using three different electrical configurations and methods are as shown in Figure 9 [188]. In method 1, PIB is put on the top of a thin gold film that acts as a positive electrical feedthrough for the cell. In methods 2 and 3, the FTO layer provides the electrical feedthroughs. It is worth noting that the PIB in methods 1 and 2 plays the role of an edge seal, but it blankets the entire area under the glass in method 3.

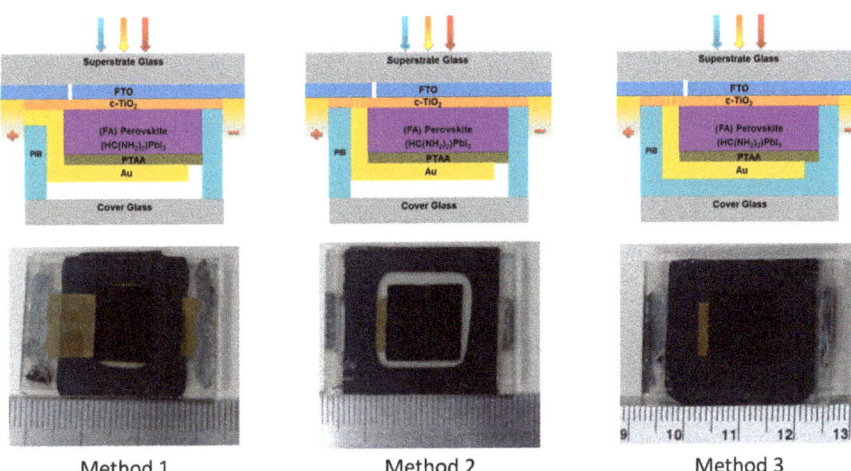

Figure 9. Cross-sectional schematics with the photographs of perovskite solar cells encapsulated by three different methods [188]. From Ref. [188]. Copyright 2017 ACS Applied Materials and Interfaces.

Lead-free PVs are also being investigated as potential improvements to PVs [157,164], as these same materials were investigated as scintillators [114,189]. Several improvements regarding their material quality, degradation have been considered, as have techniques to fabricate them. To further improve the commercialization rate in the future, certain costs that are more expensive, such as organic charge transport materials, must be replaced. A better alternative would be the use of inorganic charge transport materials to reduce costs. While these methods will reduce the cost for commercialization, the most harmful content, which is the lead, will need to be replaced, reduced, or they will need to be made using Pb-free methods, discussed above, to reduce harm to the environment.

Several improvements have been achieved for perovskite–silicon tandem PVs, such as a longer lifetime, lower cost, and less-harmful chemicals being used. These achievements have inspired some companies to produce and commercialize perovskite–silicon tandem PVs.

3.4. Commercialization

For future commercialization, although there are several challenges to be faced in fabricating perovskite PVs, some companies are in the early stages of developing perovskite PVs, e.g., Saule technologies [190] and Quantum Solution [191]. These developments are driven by perovskite's wide bandgap, low-cost, and simple fabrication methods. For Saule technologies, they launched the first industrial production line of solar panels in May 2021 in Poland, based on groundbreaking perovskite technology. They are making sheets of solar panels using a novel inkjet printing procedure invented by the founder, Olga Malinkiewicz [190].

However, the long-term performance of such a PV is an issue to pursue. As an example, the longest lifetime for the prototype was 6000 h under continuous one-sun illumination before degrading beyond 80% of its initial performance [192]. Interfacial recombination can be a severely degrading performance issue, but some methods are used to inhibit this. In [193], a 2D octyl-diammonium lead iodide interlayer was used to decrease recombination losses and obtain a PCE of 22.27% in tandem solar cells. A 2D/3D perovskite interface in [194] suppressed interface losses with a PCE $\approx$ 21%. As an alternative, perovskite–silicon tandem PV has a better lifetime of 300 h of operation, and retained 95% of its initial efficiency without encapsulant, and performs at a PCE of 29.3% in comparison to the theoretical limit of 29% for the silicon PV [195]. Therefore, the perovskite–silicon tandem PV is a prospective option for future commercialization; particularly, OxfordPV has pioneered producing their heterojunction silicon PVs to enhance their PV cells [196].

3.5. Summary

HOIPs have come a long way since their predecessors, with a current standing performance (or PCE) of about 29.3%. In this review, we highlighted the progress of HOIP materials and large-area fabrication techniques, in detail, and provided several comparisons between the techniques, and materials used in the fabrication of solar cells through their power conversion efficiencies. The fabrication techniques were also covered along with the advantages and disadvantages of using certain materials, which further enhance the properties of perovskite–silicon tandem solar cells; as shown using SEM. A scheme of this is shown in Figure 10, and it shows how far the progress of research towards commercialization and market must go. Although there has been feedback and improvements for materials, fabrication, characterization, and large-scale modules, commercialization is just starting. To enter the market of PVs, the other aspects, such as the total cost (although the material is cheap enough) and the aging of perovskite materials relative to those made using silicon, need to be addressed. Advances in other materials for optoelectronic devices such as graphene, MoS_2 and compound semiconductors [197–212] are having a huge impact and benefitting the progress in the field of PVs either through the process techniques, or through the marketization strategies. In the end, the perovskite–silicon tandem solar cells have been introduced as the next generation of PVs and will replace the current technology of silicon PVs (in comparison of electric costs) [6].

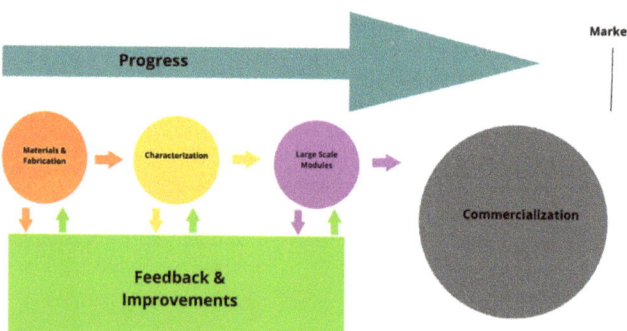

Figure 10. Scheme of the progress of perovskite tandem PVs.

Author Contributions: Conceptualization, L.J.D., M.D.B. and A.E.M.; methodology, L.J., L.J.D. and M.D.B.; validation, M.D.B. and A.E.M.; formal analysis, L.J. and L.J.D.; investigation, L.J. and L.J.D.; resources, L.J. and L.J.D.; data curation, L.J. and L.J.D. and O.S.; writing—original draft preparation, L.J., L.J.D., M.M. and S.A.B.; writing—review and editing, O.S., L.J.D., M.D.B. and A.E.M.; visualization, L.J. and O.S.; supervision, M.D.B. and A.E.M.; project administration, M.D.B. and A.E.M.; funding acquisition, L.J.D. and A.E.M. All authors have read and agreed to the published version of the manuscript.

Funding: This work was partly funded by Universitas Prasetiya Mulya, Indonesia Toray Science Foundation, the Indonesian Ministry for Research, Technology and Higher Education through INSINAS RISET PRATAMA scheme (No. 47/INS-1/PPK/E4/2018 and No. 35/INS-1/PPK/E4/2019) and the United Arab Emirates University UPAR project, grant number 31N393.

Informed Consent Statement: Not applicable.

Conflicts of Interest: The authors declare no conflict of interest.

References

1. Green, M.; Dunlop, E.; Hohl-Ebinger, J.; Yoshita, M.; Kopidakis, N.; Hao, X. Solar cell efficiency tables (version 57). *Prog. Photovolt. Res. Appl.* **2020**, *29*, 3–15. [CrossRef]
2. Richter, A.; Hermle, M.; Glunz, S.W. Reassessment of the Limiting Efficiency for Crystalline Silicon Solar Cells. *IEEE J. Photovolt.* **2013**, *3*, 1184–1191. [CrossRef]
3. Jäger, K.; Sutter, J.; Hammerschmidt, M.; Schneider, P.-I.; Becker, C. Prospects of light management in perovskite/silicon tandem solar cells. *Nanophotonics* **2021**, *10*, 1991–2000. [CrossRef]
4. Al-Ashouri, A.; Köhnen, E.; Li, B.; Magomedov, A.; Hempel, H.; Caprioglio, P.; Márquez, J.A.; Vilches, A.B.M.; Kasparavicius, E.; Smith, J.A.; et al. Monolithic perovskite/silicon tandem solar cell with >29% efficiency by enhanced hole extraction. *Science* **2020**, *370*, 1300–1309. [CrossRef]
5. Vidal, R.; Alberola-Borràs, J.-A.; Sánchez-Pantoja, N.; Mora-Seró, I. Comparison of Perovskite Solar Cells with other Photovoltaics Technologies from the Point of View of Life Cycle Assessment. *Adv. Energy Sustain. Res.* **2021**, *2*, 2000088. [CrossRef]
6. Chang, N.L.; Zheng, J.; Wu, Y.; Shen, H.; Qi, F.; Catchpole, K.; Ho-Baillie, A.W.Y.; Egan, R.J. A bottom-up cost analysis of silicon–perovskite tandem photovoltaics. *Prog. Photovolt. Res. Appl.* **2020**, *29*, 401–413. [CrossRef]
7. Dale, M. A Comparative Analysis of Energy Costs of Photovoltaic, Solar Thermal, and Wind Electricity Generation Technologies. *Appl. Sci.* **2013**, *3*, 325–337. [CrossRef]
8. Li, Z.; Zhao, Y.; Wang, X.; Sun, Y.; Zhao, Z.; Li, Y.; Zhou, H.; Chen, Q.; Li, Z.; Zhao, Y.; et al. Cost Analysis of Perovskite Tandem Photovoltaics. *Joule* **2018**, *2*, 1559–1572. [CrossRef]
9. Seok, S.I.; Grätzel, M.; Park, N.-G. Methodologies toward Highly Efficient Perovskite Solar Cells. *Small* **2018**, *14*, e1704177. [CrossRef]
10. Chen, B.; Song, J.; Dai, X.; Liu, Y.; Rudd, P.N.; Hong, X.; Huang, J. Synergistic Effect of Elevated Device Temperature and Excess Charge Carriers on the Rapid Light-Induced Degradation of Perovskite Solar Cells. *Adv. Mater.* **2019**, *31*, e1902413. [CrossRef]
11. De Wolf, S.; Holovsky, J.; Moon, S.-J.; Löper, P.; Niesen, B.; Ledinsky, M.; Haug, F.-J.; Yum, J.-H.; Ballif, C. Organometallic Halide Perovskites: Sharp Optical Absorption Edge and Its Relation to Photovoltaic Performance. *J. Phys. Chem. Lett.* **2014**, *5*, 1035–1039. [CrossRef] [PubMed]
12. Xing, G.; Mathews, N.; Sun, S.; Lim, S.S.; Lam, Y.M.; Grätzel, M.; Mhaisalkar, S.G.; Sum, T.C. Long-Range Balanced Electron- and Hole-Transport Lengths in Organic-Inorganic $CH_3NH_3PbI_3$. *Science* **2013**, *342*, 344–347. [CrossRef] [PubMed]
13. Kojima, A.; Teshima, K.; Shirai, Y.; Miyasaka, T. Organometal Halide Perovskites as Visible-Light Sensitizers for Photovoltaic Cells. *J. Am. Chem. Soc.* **2009**, *131*, 6050–6051. [CrossRef] [PubMed]
14. Kim, D.I.; Lee, J.W.; Jeong, R.H.; Boo, J.-H. A High-efficiency and Stable Perovskite Solar Cell Fabricated in Ambient Air Using a Polyaniline Passivation Layer. *Sci. Rep.* **2022**, *12*, 697. [CrossRef] [PubMed]
15. Roß, M.; Gil-Escrig, L.; Al-Ashouri, A.; Tockhorn, P.; Jošt, M.; Rech, B.; Albrecht, S. Co-Evaporated p-i-n Perovskite Solar Cells beyond 20% Efficiency: Impact of Substrate Temperature and Hole-Transport Layer. *ACS Appl. Mater. Interfaces* **2020**, *12*, 39261–39272. [CrossRef] [PubMed]
16. Feng, J.; Jiao, Y.; Wang, H.; Zhu, X.; Sun, Y.; Du, M.; Cao, Y.; Yang, D.; Liu, S. High-throughput large-area vacuum deposition for high-performance formamidine-based perovskite solar cells. *Energy Environ. Sci.* **2021**, *14*, 3035–3043. [CrossRef]
17. Li, J.; Dewi, H.A.; Wang, H.; Lew, J.H.; Mathews, N.; Mhaisalkar, S.; Bruno, A. Design of Perovskite Thermally Co-Evaporated Highly Efficient Mini-Modules with High Geometrical Fill Factors. *Sol. RRL* **2020**, *4*, 2070121. [CrossRef]
18. Park, N.-G.; Zhu, K. Scalable fabrication and coating methods for perovskite solar cells and solar modules. *Nat. Rev. Mater.* **2020**, *5*, 333–350. [CrossRef]
19. Bishop, J.E.; Smith, J.A.; Greenland, C.; Kumar, V.; Vaenas, N.; Game, O.S.; Routledge, T.J.; Wong-Stringer, M.; Rodenburg, C.; Lidzey, D.G. High-Efficiency Spray-Coated Perovskite Solar Cells Utilizing Vacuum-Assisted Solution Processing. *ACS Appl. Mater. Interfaces* **2018**, *10*, 39428–39434. [CrossRef]
20. Heo, J.H.; Lee, M.H.; Jang, M.H.; Im, S.H. Highly efficient $CH_3NH_3PbI_3-xCl_x$ mixed halide perovskite solar cells prepared by re-dissolution and crystal grain growth via spray coating. *J. Mater. Chem. A* **2016**, *4*, 17636–17642. [CrossRef]
21. Chen, B.; Yu, Z.J.; Manzoor, S.; Wang, S.; Weigand, W.; Yu, Z.; Yang, G.; Ni, Z.; Dai, X.; Holman, Z.C.; et al. Blade-Coated Perovskites on Textured Silicon for 26%-Efficient Monolithic Perovskite/Silicon Tandem Solar Cells. *Joule* **2020**, *4*, 850–864. [CrossRef]
22. Wu, R.; Wang, C.; Jiang, M.; Liu, C.; Liu, D.; Li, S.; Kong, Q.; He, W.; Zhan, C.; Zhang, F.; et al. Progress in blade-coating method for perovskite solar cells toward commercialization. *J. Renew. Sustain. Energy* **2021**, *13*, 012701. [CrossRef]
23. Subbiah, A.S.; Isikgor, F.H.; Howells, C.T.; De Bastiani, M.; Liu, J.; Aydin, E.; Furlan, F.; Allen, T.G.; Xu, F.; Zhumagali, S.; et al. High-Performance Perovskite Single-Junction and Textured Perovskite/Silicon Tandem Solar Cells via Slot-Die Coating. *ACS Energy Lett.* **2020**, *5*, 3034–3040. [CrossRef]
24. Patidar, R.; Burkitt, D.; Hooper, K.; Richards, D.; Watson, T. Slot-die coating of perovskite solar cells: An overview. *Mater. Today Commun.* **2019**, *22*, 100808. [CrossRef]
25. He, J.; Bi, E.; Tang, W.; Wang, Y.; Yang, X.; Chen, H.; Han, L. Low-Temperature Soft-Cover-Assisted Hydrolysis Deposition of Large-Scale TiO_2 Layer for Efficient Perovskite Solar Modules. *Nano-Micro Lett.* **2018**, *10*, 49. [CrossRef]
26. Ye, F.; Tang, W.; Xie, F.; Yin, M.; He, J.; Wang, Y.; Chen, H.; Qiang, Y.; Yang, X.; Han, L. Low-Temperature Soft-Cover Deposition of Uniform Large-Scale Perovskite Films for High-Performance Solar Cells. *Adv. Mater.* **2017**, *29*, 1701440. [CrossRef] [PubMed]

27. Ye, F.; Chen, H.; Xie, F.; Tang, W.; Yin, M.; He, J.; Bi, E.; Wang, Y.; Yang, X.; Han, L. Soft-cover deposition of scaling-up uniform perovskite thin films for high cost-performance solar cells. *Energy Environ. Sci.* **2016**, *9*, 2295–2301. [CrossRef]
28. Kamino, B.A.; Paviet-Salomon, B.; Moon, S.-J.; Badel, N.; Levrat, J.; Christmann, G.; Walter, A.; Faes, A.; Ding, L.; Leon, J.J.D.; et al. Low-Temperature Screen-Printed Metallization for the Scale-Up of Two-Terminal Perovskite–Silicon Tandems. *ACS Appl. Energy Mater.* **2019**, *2*, 3815–3821. [CrossRef]
29. Poshan Kumar Reddy, K.; Rameez, M.; Wang, T.-T.; Wang, K.; Yan-Ru Lin, E.; Lin, M.-C.; Wei-Guang Diau, E.; Hung, C.-H.; Chueh, Y.-L.; Pande, K.P.; et al. Screen-Printed Hole Transport Material-Free Perovskite Solar Cell for Water Splitting Incorporating Cu-NiCo2O4 Catalyst. *Mater. Lett.* **2022**, *313*, 131838. [CrossRef]
30. Hu, Y.; Zhang, Z.; Mei, A.; Jiang, Y.; Hou, X.; Wang, Q.; Du, K.; Rong, Y.; Zhou, Y.; Xu, G.; et al. Improved Performance of Printable Perovskite Solar Cells with Bifunctional Conjugated Organic Molecule. *Adv. Mater.* **2018**, *30*, 1706759. [CrossRef]
31. Wang, D.; Wright, M.; Elumalai, N.K.; Uddin, A. Stability of perovskite solar cells. *Sol. Energy Mater. Sol. Cells* **2016**, *147*, 255–275. [CrossRef]
32. Yeom, K.M.; Kim, S.U.; Woo, M.Y.; Noh, J.H.; Im, S.H. Recent Progress in Metal Halide Perovskite-Based Tandem Solar Cells. *Adv. Mater.* **2020**, *32*, 2002228. [CrossRef] [PubMed]
33. Al-Ashouri, A.; Magomedov, A.; Roß, M.; Jošt, M.; Talaikis, M.; Chistiakova, G.; Bertram, T.; Márquez, J.A.; Köhnen, E.; Kasparavičius, E.; et al. Conformal monolayer contacts with lossless interfaces for perovskite single junction and monolithic tandem solar cells. *Energy Environ. Sci.* **2019**, *12*, 3356–3369. [CrossRef]
34. Oxford, P.V. Oxford PV Perovskite Solar Cell Achieves 28% Efficiency. Available online: https://www.oxfordpv.com/news/oxford-pv-perovskite-solar-cell-achieves-28-efficiency (accessed on 28 July 2021).
35. Calió, L.; Kazim, S.; Grätzel, M.; Ahmad, S. Hole-Transport Materials for Perovskite Solar Cells. *Angew. Chem. Int. Ed.* **2016**, *55*, 14522–14545. [CrossRef] [PubMed]
36. Vasilopoulou, M.; Fakharuddin, A.; Coutsolelos, A.G.; Falaras, P.; Argitis, P.; Yusoff, A.R.B.M.; Nazeeruddin, M.K. Molecular materials as interfacial layers and additives in perovskite solar cells. *Chem. Soc. Rev.* **2020**, *49*, 4496–4526. [CrossRef] [PubMed]
37. Mukherjee, R. Review on Hybrid Organic-Inorganic Halide Perovskite. *J. Mol. Clin. Med.* **2020**, *7*, 5.
38. Brenner, T.M.; Egger, D.A.; Kronik, L.; Hodes, G.; Cahen, D. Hybrid organic—inorganic perovskites: Low-cost semiconductors with intriguing charge-transport properties. *Nat. Rev. Mater.* **2016**, *1*, 15007. [CrossRef]
39. Hamukwaya, S.L.; Hao, H.; Zhao, Z.; Dong, J.; Zhong, T.; Xing, J.; Hao, L.; Mashingaidze, M.M. A Review of Recent Developments in Preparation Methods for Large-Area Perovskite Solar Cells. *Coatings* **2022**, *12*, 252. [CrossRef]
40. Yan, C.; Huang, J.; Li, D.D.; Li, G. Recent progress of metal-halide perovskite-based tandem solar cells. *Mater. Chem. Front.* **2021**, *5*, 4538–4564. [CrossRef]
41. Zhang, Y.; Kirs, A.; Ambroz, F.; Lin, C.-T.; Bati, A.S.R.; Parkin, I.P.; Shapter, J.G.; Batmunkh, M.; Macdonald, T.J. Ambient Fabrication of Organic–Inorganic Hybrid Perovskite Solar Cells. *Small Methods* **2021**, *5*, 2000744. [CrossRef]
42. Sun, M.; Zhang, F.; Liu, H.; Li, X.; Xiao, Y.; Wang, S. Tuning the crystal growth of perovskite thin-films by adding the 2-pyridylthiourea additive for highly efficient and stable solar cells prepared in ambient air. *J. Mater. Chem. A* **2017**, *5*, 13448–13456. [CrossRef]
43. Kumar, A.; Ojha, S.K.; Vyas, N.; Ojha, A.K. Designing Organic Electron Transport Materials for Stable and Efficient Performance of Perovskite Solar Cells: A Theoretical Study. *ACS Omega* **2021**, *6*, 7086–7093. [CrossRef] [PubMed]
44. Petrović, M.; Chellappan, V.; Ramakrishna, S. Perovskites: Solar cells & engineering applications—Materials and device developments. *Sol. Energy* **2015**, *122*, 678–699. [CrossRef]
45. Kim, T.; Lim, J.; Song, S. Recent Progress and Challenges of Electron Transport Layers in Organic–Inorganic Perovskite Solar Cells. *Energies* **2020**, *13*, 5572. [CrossRef]
46. Zuo, X.; Chang, K.; Zhao, J.; Xie, Z.; Tang, H.; Li, B.; Chang, Z. Integrated Organic–Inorganic Hole Transport Layer for Efficient and Stable Perovskite Solar Cells. *J. Mater. Chem. A* **2016**, *4*, 51–58. [CrossRef]
47. Shao, S.; Loi, M.A. The Role of the Interfaces in Perovskite Solar Cells. *Adv. Mater. Interfaces* **2020**, *7*, 1901469. [CrossRef]
48. Haque, M.A.; Villalva, D.R.; Hernandez, L.H.; Tounesi, R.; Jang, S.; Baran, D. Role of Dopants in Organic and Halide Perovskite Energy Conversion Devices. *Chem. Mater.* **2021**, *33*, 8147–8172. [CrossRef]
49. Kranthiraja, K.; Gunasekar, K.; Kim, H.; Cho, A.-N.; Park, N.-G.; Kim, S.; Kim, B.J.; Nishikubo, R.; Saeki, A.; Song, M.; et al. High-Performance Long-Term-Stable Dopant-Free Perovskite Solar Cells and Additive-Free Organic Solar Cells by Employing Newly Designed Multirole π-Conjugated Polymers. *Adv. Mater.* **2017**, *29*, 1700183. [CrossRef]
50. Pham, H.D.; Yang, T.C.; Jain, S.M.; Wilson, G.J.; Sonar, P. Development of Dopant-Free Organic Hole Transporting Materials for Perovskite Solar Cells. *Adv. Energy Mater.* **2020**, *10*, 1903326. [CrossRef]
51. Wojciechowski, K.; Leijtens, T.; Siprova, S.; Schlueter, C.; Hörantner, M.T.; Wang, J.T.-W.; Li, C.-Z.; Jen, A.K.-Y.; Lee, T.-L.; Snaith, H.J. C60 as an Efficient n-Type Compact Layer in Perovskite Solar Cells. *J. Phys. Chem. Lett.* **2015**, *6*, 2399–2405. [CrossRef]
52. Chen, Y.; He, M.; Peng, J.; Sun, Y.; Liang, Z. Structure and Growth Control of Organic–Inorganic Halide Perovskites for Optoelectronics: From Polycrystalline Films to Single Crystals. *Adv. Sci.* **2016**, *3*, 1500392. [CrossRef] [PubMed]
53. Hawash, Z.; Ono, L.K.; Qi, Y. Recent Advances in Spiro-MeOTAD Hole Transport Material and Its Applications in Organic–Inorganic Halide Perovskite Solar Cells. *Adv. Mater. Interfaces* **2018**, *5*, 1700623. [CrossRef]
54. Hasanzadeh, A.; Khataee, A.; Zarei, M.; Zhang, Y. Two-electron oxygen reduction on fullerene C60-carbon nanotubes covalent hybrid as a metal-free electrocatalyst. *Sci. Rep.* **2019**, *9*, 13780. [CrossRef] [PubMed]

55. He, T.; Li, S.; Jiang, Y.; Qin, C.; Cui, M.; Qiao, L.; Xu, H.; Yang, J.; Long, R.; Wang, H.; et al. Reduced-dimensional perovskite photovoltaics with homogeneous energy landscape. *Nat. Commun.* **2020**, *11*, 1672. [CrossRef] [PubMed]
56. Zhu, T.; Gong, X. Low-dimensional perovskite materials and their optoelectronics. *InfoMat* **2021**, *3*, 1039–1069. [CrossRef]
57. Hong, K.; Van Le, Q.; Kim, S.Y.; Jang, H.W. Low-dimensional halide perovskites: Review and issues. *J. Mater. Chem. C* **2018**, *6*, 2189–2209. [CrossRef]
58. Savenije, T.J.; Ponseca, C.S., Jr.; Kunneman, L.; Abdellah, M.; Zheng, K.; Tian, Y.; Zhu, Q.; Canton, S.E.; Scheblykin, I.G.; Pullerits, T.; et al. Thermally Activated Exciton Dissociation and Recombination Control the Carrier Dynamics in Organometal Halide Perovskite. *J. Phys. Chem. Lett.* **2014**, *5*, 2189–2194. [CrossRef]
59. Dong, Q.; Fang, Y.; Shao, Y.; Mulligan, P.; Qiu, J.; Cao, L.; Huang, J. Solar Cells. Electron-hole diffusion lengths > 175 μm in solution-grown CH$_3$NH$_3$PbI$_3$ single crystals. *Science* **2015**, *347*, 967–970. [CrossRef]
60. Eperon, G.E.; Stranks, S.D.; Menelaou, C.; Johnston, M.B.; Herz, L.M.; Snaith, H.J. Formamidinium lead trihalide: A broadly tunable perovskite for efficient planar heterojunction solar cells. *Energy Environ. Sci.* **2014**, *7*, 982–988. [CrossRef]
61. Juarez-Perez, E.J.; Ono, L.K.; Qi, Y. Thermal degradation of formamidinium based lead halide perovskites into sym-triazine and hydrogen cyanide observed by coupled thermogravimetry-mass spectrometry analysis. *J. Mater. Chem. A* **2019**, *7*, 16912–16919. [CrossRef]
62. Conings, B.; Drijkoningen, J.; Gauquelin, N.; Babayigit, A.; D'Haen, J.; D'Olieslaeger, L.; Ethirajan, A.; Verbeeck, J.; Manca, J.; Mosconi, E.; et al. Intrinsic Thermal Instability of Methylammonium Lead Trihalide Perovskite. *Adv. Energy Mater.* **2015**, *5*, 1500477. [CrossRef]
63. Haeger, T.; Heiderhoff, R.; Riedl, T. Thermal properties of metal-halide perovskites. *J. Mater. Chem. C* **2020**, *8*, 14289–14311. [CrossRef]
64. Park, Y.H.; Jeong, I.; Bae, S.; Son, H.J.; Lee, P.; Lee, J.; Lee, C.-H.; Ko, M.J. Inorganic Rubidium Cation as an Enhancer for Photovoltaic Performance and Moisture Stability of HC(NH$_2$)$_2$ PbI$_3$ Perovskite Solar Cells. *Adv. Funct. Mater.* **2017**, *27*, 1605988. [CrossRef]
65. Song, Z.; Abate, A.; Watthage, S.C.; Liyanage, G.K.; Phillips, A.B.; Steiner, U.; Graetzel, M.; Heben, M.J. Perovskite Solar Cell Stability in Humid Air: Partially Reversible Phase Transitions in the PbI$_2$-CH$_3$NH$_3$I-H$_2$O System. *Adv. Energy Mater.* **2016**, *6*, 1600846. [CrossRef]
66. Slimi, B.; Mollar, M.; ben Assaker, I.; Kriaa, I.; Chtourou, R.; Marí, B. Perovskite FA1-xMAxPbI3 for Solar Cells: Films Formation and Properties. *Energy Proc.* **2016**, *102*, 87–95. [CrossRef]
67. Saliba, M.; Matsui, T.; Seo, J.-Y.; Domanski, K.; Correa-Baena, J.-P.; Nazeeruddin, M.K.; Zakeeruddin, S.M.; Tress, W.; Abate, A.; Hagfeldt, A.; et al. Cesium-containing triple cation perovskite solar cells: Improved stability, reproducibility and high efficiency. *Energy Environ. Sci.* **2016**, *9*, 1989–1997. [CrossRef]
68. Luo, D.; Yang, W.; Wang, Z.; Sadhanala, A.; Hu, Q.; Su, R.; Shivanna, R.; Trindade, G.F.; Watts, J.F.; Xu, Z.; et al. Enhanced photovoltage for inverted planar heterojunction perovskite solar cells. *Science* **2018**, *360*, 1442–1446. [CrossRef]
69. Zheng, X.; Chen, B.; Dai, J.; Fang, Y.; Bai, Y.; Lin, Y.; Wei, H.; Zeng, X.C.; Huang, J. Defect passivation in hybrid perovskite solar cells using quaternary ammonium halide anions and cations. *Nat. Energy* **2017**, *2*, 17102. [CrossRef]
70. Jiang, Q.; Zhang, L.; Wang, H.; Yang, X.; Meng, J.; Liu, H.; Yin, Z.; Wu, J.; Zhang, X.; You, J. Enhanced electron extraction using SnO2 for high-efficiency planar-structure HC(NH2)2PbI3-based perovskite solar cells. *Nat. Energy* **2017**, *2*, 16177. [CrossRef]
71. Liu, C.; Zhu, R.; Ng, A.; Ren, Z.; Cheung, S.H.; Du, L.; So, S.K.; Zapien, J.A.; Djurišić, A.B.; Phillips, D.L.; et al. Investigation of high performance TiO2nanorod array perovskite solar cells. *J. Mater. Chem. A* **2017**, *5*, 15970–15980. [CrossRef]
72. Kim, H.-S.; Lee, C.-R.; Im, J.-H.; Lee, K.-B.; Moehl, T.; Marchioro, A.; Moon, S.-J.; Humphry-Baker, R.; Yum, J.-H.; Moser, J.E.; et al. Lead Iodide Perovskite Sensitized All-Solid-State Submicron Thin Film Mesoscopic Solar Cell with Efficiency Exceeding 9%. *Sci. Rep.* **2012**, *2*, 591. [CrossRef] [PubMed]
73. Chen, R.; Bu, T.; Li, J.; Li, W.; Zhou, P.; Liu, X.; Ku, Z.; Zhong, J.; Peng, Y.; Huang, F.; et al. Efficient and Stable Inverted Planar Perovskite Solar Cells Using a Triphenylamine Hole-Transporting Material. *ChemSusChem* **2018**, *11*, 1467–1473. [CrossRef] [PubMed]
74. Lee, J.-W.; Kim, D.-H.; Kim, H.-S.; Seo, S.-W.; Cho, S.M.; Park, N.-G. Formamidinium and Cesium Hybridization for Photo- and Moisture-Stable Perovskite Solar Cell. *Adv. Energy Mater.* **2015**, *5*, 1501310. [CrossRef]
75. Im, J.-H.; Kim, H.-S.; Park, N.-G. Morphology-photovoltaic property correlation in perovskite solar cells: One-step versus two-step deposition of CH3NH3PbI3. *APL Mater.* **2014**, *2*, 081510. [CrossRef]
76. Razza, S.; Castro-Hermosa, S.A.; Di Carlo, A.; Brown, T.M. Research Update: Large-area deposition, coating, printing, and processing techniques for the upscaling of perovskite solar cell technology. *Appl. Phys. Lett.* **2016**, *4*, 091508. [CrossRef]
77. Chang, C.-Y.; Huang, Y.-C.; Tsao, C.-S.; Su, W.-F. Formation Mechanism and Control of Perovskite Films from Solution to Crystalline Phase Studied by in Situ Synchrotron Scattering. *ACS Appl. Mater. Interfaces* **2016**, *8*, 26712–26721. [CrossRef]
78. Habibi, M.; Rahimzadeh, A.; Bennouna, I.; Eslamian, M. Defect-Free Large-Area (25 cm2) Light Absorbing Perovskite Thin Films Made by Spray Coating. *Coatings* **2017**, *7*, 42. [CrossRef]
79. Ramesh, M.; Boopathi, K.M.; Huang, T.-Y.; Huang, Y.-C.; Tsao, C.-S.; Chu, C.-W. Using an Airbrush Pen for Layer-by-Layer Growth of Continuous Perovskite Thin Films for Hybrid Solar Cells. *ACS Appl. Mater. Interfaces* **2015**, *7*, 2359–2366. [CrossRef]

80. Chandrasekhar, P.S.; Kumar, N.; Swami, S.K.; Dutta, V.; Komarala, V.K. Fabrication of perovskite films using an electrostatic assisted spray technique: The effect of the electric field on morphology, crystallinity and solar cell performance. *Nanoscale* **2016**, *8*, 6792–6800. [CrossRef]
81. Habibi, M.; Ahmadian-Yazdi, M.-R.; Eslamian, M. Optimization of spray coating for the fabrication of sequentially deposited planar perovskite solar cells. *J. Photon. Energy* **2017**, *7*, 22003. [CrossRef]
82. Shen, P.-S.; Chiang, Y.-H.; Li, M.-H.; Guo, T.-F.; Chen, P. Research Update: Hybrid organic-inorganic perovskite (HOIP) thin films and solar cells by vapor phase reaction. *APL Mater.* **2016**, *4*, 91509. [CrossRef]
83. Razza, S.; Di Giacomo, F.; Matteocci, F.; Cinà, L.; Palma, A.L.; Casaluci, S.; Cameron, P.; D'Epifanio, A.; Licoccia, S.; Reale, A.; et al. Perovskite solar cells and large area modules (100 cm^2) based on an air flow-assisted PbI$_2$ blade coating deposition process. *J. Power Sour.* **2015**, *277*, 286–291. [CrossRef]
84. Abbas, M.; Zeng, L.; Guo, F.; Rauf, M.; Yuan, X.-C.; Cai, B. A Critical Review on Crystal Growth Techniques for Scalable Deposition of Photovoltaic Perovskite Thin Films. *Materials* **2020**, *13*, 4851. [CrossRef] [PubMed]
85. Qin, T.; Huang, W.; Kim, J.-E.; Vak, D.; Forsyth, C.; McNeill, C.R.; Cheng, Y.-B. Amorphous hole-transporting layer in slot-die coated perovskite solar cells. *Nano Energy* **2017**, *31*, 210–217. [CrossRef]
86. Li, S.-G.; Jiang, K.-J.; Su, M.-J.; Cui, X.-P.; Huang, J.-H.; Zhang, Q.-Q.; Zhou, X.-Q.; Yang, L.-M.; Song, Y.-L. Inkjet printing of CH3NH3PbI3 on a mesoscopic TiO2 film for highly efficient perovskite solar cells. *J. Mater. Chem. A* **2015**, *3*, 9092–9097. [CrossRef]
87. Singh, M.; Haverinen, H.M.; Dhagat, P.; Jabbour, G.E. Inkjet Printing—Process and Its Applications. *Adv. Mater.* **2010**, *22*, 673–685. [CrossRef]
88. Yang, Z.; Zhang, S.; Li, L.; Chen, W. Research progress on large-area perovskite thin films and solar modules. *J. Materiomics* **2017**, *3*, 231–244. [CrossRef]
89. Howard, I.A.; Abzieher, T.; Hossain, I.M.; Eggers, H.; Schackmar, F.; Ternes, S.; Richards, B.S.; Lemmer, U.; Paetzold, U.W. Coated and Printed Perovskites for Photovoltaic Applications. *Adv. Mater.* **2018**, *31*, e1806702. [CrossRef]
90. Jiang, Z.; Bag, M.; Renna, L.; Jeong, S.P.; Rotello, V.; Venkataraman, D. Aqueous-Processed Perovskite Solar Cells Based on Reactive Inkjet Printing. *hal* **2016**, hal-01386295. Available online: https://hal.archives-ouvertes.fr/hal-01386295 (accessed on 28 February 2022).
91. Tavakoli, M.M.; Gu, L.; Gao, Y.; Reckmeier, C.; He, J.; Rogach, A.L.; Yao, Y.; Fan, Z. Fabrication of efficient planar perovskite solar cells using a one-step chemical vapor deposition method. *Sci. Rep.* **2015**, *5*, 14083. [CrossRef]
92. Chen, C.-W.; Kang, H.-W.; Hsiao, S.-Y.; Yang, P.-F.; Chiang, K.-M.; Lin, H.-W. Efficient and Uniform Planar-Type Perovskite Solar Cells by Simple Sequential Vacuum Deposition. *Adv. Mater.* **2014**, *26*, 6647–6652. [CrossRef]
93. Chen, Q.; Zhou, H.; Hong, Z.; Luo, S.; Duan, H.-S.; Wang, H.-H.; Liu, Y.; Li, G.; Yang, Y. Planar Heterojunction Perovskite Solar Cells via Vapor-Assisted Solution Process. *J. Am. Chem. Soc.* **2014**, *136*, 622–625. [CrossRef] [PubMed]
94. Shen, P.-S.; Chen, J.-S.; Chiang, Y.-H.; Li, M.-H.; Guo, T.-F.; Chen, P. Low-Pressure Hybrid Chemical Vapor Growth for Efficient Perovskite Solar Cells and Large-Area Module. *Adv. Mater. Interfaces* **2016**, *3*, 1500849. [CrossRef]
95. Fan, P.; Gu, D.; Liang, G.-X.; Luo, J.-T.; Chen, J.-L.; Zheng, Z.-H.; Zhang, D.-P. High-performance perovskite CH3NH3PbI3 thin films for solar cells prepared by single-source physical vapour deposition. *Sci. Rep.* **2016**, *6*, 29910. [CrossRef] [PubMed]
96. Zhou, D.; Zhou, T.; Tian, Y.; Zhu, X.; Tu, Y. Perovskite-Based Solar Cells: Materials, Methods, and Future Perspectives. *J. Nanomater.* **2018**, *2018*, 8148072. [CrossRef]
97. Zhou, Y.; Yang, M.; Wu, W.; Vasiliev, A.L.; Zhu, K.; Padture, N.P. Room-Temperature Crystallization of Hybrid-Perovskite Thin Films via Solvent–Solvent Extractionfor High-Performance Solar Cells. *J. Mater. Chem. A* **2015**, *3*, 8178. [CrossRef]
98. Deng, Y.; Zheng, X.; Bai, Y.; Wang, Q.; Zhao, J.; Huang, J. Surfactant-controlled ink drying enables high-speed deposition of perovskite films for efficient photovoltaic modules. *Nat. Energy* **2018**, *3*, 560–566. [CrossRef]
99. Di Giacomo, F.; Shanmugam, S.; Fledderus, H.; Bruijnaers, B.J.; Verhees, W.J.H.; Dorenkamper, M.S.; Veenstra, S.C.; Qiu, W.; Gehlhaar, R.; Merckx, T.; et al. Up-scalable sheet-to-sheet production of high efficiency perovskite module and solar cells on 6-in. substrate using slot die coating. *Sol. Energy Mater. Sol. Cells* **2018**, *181*, 53–59. [CrossRef]
100. Yang, W.S.; Noh, J.H.; Jeon, N.J.; Kim, Y.C.; Ryu, S.; Seo, J.; Seok, S.I. High-performance Photovoltaic Perovskite Layers Fabricated Through Intramolecular Exchange. *Science* **2015**, *348*, 1234–1237. [CrossRef]
101. Abzieher, T.; Moghadamzadeh, S.; Schackmar, F.; Eggers, H.; Sutterlüti, F.; Farooq, A.; Kojda, D.; Habicht, K.; Schmager, R.; Mertens, A.; et al. Electron-Beam-Evaporated Nickel Oxide Hole Transport Layers for Perovskite-Based Photovoltaics. *Adv. Energy Mater.* **2019**, *9*, 1802995. [CrossRef]
102. Jeon, N.J.; Na, H.; Jung, E.H.; Yang, T.-Y.; Lee, Y.G.; Kim, G.; Shin, H.-W.; Seok, S.I.; Lee, J.; Seo, J. A fluorene-terminated hole-transporting material for highly efficient and stable perovskite solar cells. *Nat. Energy* **2018**, *3*, 682–689. [CrossRef]
103. Wu, W.-Q.; Wang, Q.; Fang, Y.; Shao, Y.; Tang, S.; Deng, Y.; Lu, H.; Liu, Y.; Li, T.; Yang, Z.; et al. Molecular doping enabled scalable blading of efficient hole-transport-layer-free perovskite solar cells. *Nat. Commun.* **2018**, *9*, 1625. [CrossRef]
104. Li, P.; Liang, C.; Bao, B.; Li, Y.; Hu, X.; Wang, Y.; Zhang, Y.; Li, F.; Shao, G.; Song, Y. Inkjet manipulated homogeneous large size perovskite grains for efficient and large-area perovskite solar cells. *Nano Energy* **2018**, *46*, 203–211. [CrossRef]
105. Whitaker, J.B.; Kim, D.H.; Larson, B.W.; Zhang, F.; Berry, J.J.; van Hest, M.F.A.M.; Zhu, K. Scalable slot-die coating of high-performance perovskite solar cells. *Sustain. Energy Fuels* **2018**, *2*, 2442–2449. [CrossRef]
106. Kim, Y.Y.; Park, E.Y.; Yang, T.-Y.; Noh, J.H.; Shin, T.J.; Jeon, N.J.; Seo, J. Fast two-step deposition of perovskite via mediator extraction treatment for large-area, high-performance perovskite solar cells. *J. Mater. Chem. A* **2018**, *6*, 12447–12454. [CrossRef]

107. Liu, M.; Johnston, M.B.; Snaith, H.J. Efficient planar heterojunction perovskite solar cells by vapour deposition. *Nature* **2013**, *501*, 395–398. [CrossRef] [PubMed]
108. Si, H.; Zhang, Z.; Liao, Q.; Zhang, G.; Ou, Y.; Zhang, S.; Wu, H.; Wu, J.; Kang, Z.; Zhang, Y. A-Site Management for Highly Crystalline Perovskites. *Adv. Mater.* **2020**, *32*, 1904702. [CrossRef] [PubMed]
109. Singh, S.; Li, C.; Panzer, F.; Narasimhan, K.L.; Graeser, A.; Gujar, T.P.; Köhler, A.; Thelakkat, M.; Huettner, S.; Kabra, D. Effect of Thermal and Structural Disorder on the Electronic Structure of Hybrid Perovskite Semiconductor $CH_3NH_3PbI_3$. *J. Phys. Chem. Lett.* **2016**, *7*, 3014–3021. [CrossRef]
110. Pratiwi, Z.R.; Nuraeni, L.; Aimon, A.H.; Iskandar, F. Morphology Control of MAPbI3 Perovskite Thin Film as An Active Layer of Solar Cells. *IOP Conf. Ser. Mater. Sci. Eng.* **2018**, *395*, 012010. [CrossRef]
111. Kumar, G.R.; Savariraj, A.D.; Karthick, S.N.; Selvam, S.; Balamuralitharan, B.; Kim, H.-J.; Viswanathan, K.K.; Vijaykumar, M.; Prabakar, K. Phase transition kinetics and surface binding states of methylammonium lead iodide perovskite. *Phys. Chem. Chem. Phys.* **2016**, *18*, 7284–7292. [CrossRef]
112. Baikie, T.; Fang, Y.; Kadro, J.M.; Schreyer, M.; Wei, F.; Mhaisalkar, S.G.; Graetzel, M.; White, T.J. Synthesis and crystal chemistry of the hybrid perovskite (CH3NH3)PbI3 for solid-state sensitised solar cell applications. *J. Mater. Chem. A* **2013**, *1*, 5628–5641. [CrossRef]
113. Yang, S.; Wang, Y.; Liu, P.; Cheng, Y.-B.; Zhao, H.; Yang, H. Functionalization of perovskite thin films with moisture-tolerant molecules. *Nat. Energy* **2016**, *1*, 15016. [CrossRef]
114. Diguna, L.J.; Kaffah, S.; Mahyuddin, M.H.; Arramel; Maddalena, F.; Bakar, S.A.; Aminah, M.; Onggo, D.; Witkowski, M.E.; Makowski, M.; et al. Scintillation in $(C_6H_5CH_2NH_3)_2SnBr_4$: Green-emitting lead-free perovskite halide materials. *RSC Adv.* **2021**, *11*, 20635–20640. [CrossRef]
115. Jin, S. Can We Find the Perfect A-Cations for Halide Perovskites? *ACS Energy Lett.* **2021**, *6*, 3386–3389. [CrossRef]
116. Arramel, A.; Fauzi, A.D.; Yin, X.; Tang, C.S.; Mahyuddin, M.H.; Sahdan, M.F.; Aminah, M.; Onggo, D.; Shukri, G.; Diao, C.; et al. Ligand size effects in two-dimensional hybrid copper halide perovskites crystals. *Commun. Mater.* **2021**, *2*, 1–12. [CrossRef]
117. Zheng, J.; Zhang, M.; Lau, C.F.J.; Deng, X.; Kim, J.; Ma, Q.; Chen, C.; Green, M.A.; Huang, S.; Ho-Baillie, A.W.Y. Spin-coating free fabrication for highly efficient perovskite solar cells. *Sol. Energy Mater. Sol. Cells* **2017**, *168*, 165–171. [CrossRef]
118. Liang, P.-W.; Liao, C.-Y.; Chueh, C.-C.; Zuo, F.; Williams, S.T.; Xin, X.-K.; Lin, J.-J.; Jen, A.K.-Y. Additive Enhanced Crystallization of Solution-Processed Perovskite for Highly Efficient Planar-Heterojunction Solar Cells. *Adv. Mater.* **2014**, *26*, 3748–3754. [CrossRef]
119. Zuo, C.; Ding, L. An 80.11% FF record achieved for perovskite solar cells by using the NH_4Cl additive. *Nanoscale* **2014**, *6*, 9935–9938. [CrossRef]
120. Zhou, Y.; Zhou, H.; Deng, J.; Cha, W.; Cai, Z. Decisive Structural and Functional Characterization of Halide Perovskites with Synchrotron. *Matter* **2020**, *2*, 360–377. [CrossRef]
121. Tailor, N.K.; Abdi-Jalebi, M.; Gupta, V.; Hu, H.; Dar, M.I.; Li, G.; Satapathi, S. Recent progress in morphology optimization in perovskite solar cell. *J. Mater. Chem. A* **2020**, *8*, 21356–21386. [CrossRef]
122. Zhu, J.; He, B.; Gong, Z.; Ding, Y.; Zhang, W.; Li, X.; Zong, Z.; Chen, H.; Tang, Q. Grain Enlargement and Defect Passivation with Melamine Additives for High Efficiency and Stable CsPbBr 3 Perovskite Solar Cells. *ChemSusChem* **2020**, *13*, 1834–1843. [CrossRef]
123. Lei, H.; Dai, P.; Wang, X.; Pan, Z.; Guo, Y.; Shen, H.; Chen, J.; Xie, J.; Zhang, B.; Zhang, S.; et al. Perovskite Solar Cells: In Situ Defect Passivation with Silica Oligomer for Enhanced Performance and Stability of Perovskite Solar Cells (Adv. Mater. Interfaces 2/2020). *Adv. Mater. Interfaces* **2020**, *7*, 2070013. [CrossRef]
124. Medjahed, A.A.; Dally, P.; Zhou, T.; Lemaitre, N.; Djurado, D.; Reiss, P.; Pouget, S. Unraveling the Formation Mechanism and Ferroelastic Behavior of MAPbI3 Perovskite Thin Films Prepared in the Presence of Chloride. *Chem. Mater.* **2020**, *32*, 3346–3357. [CrossRef]
125. Feng, M.; You, S.; Cheng, N.; Du, J. High quality perovskite film solar cell using methanol as additive with 19.5% power conversion efficiency. *Electrochim. Acta* **2019**, *293*, 356–363. [CrossRef]
126. De Vos, A. Detailed balance limit of the efficiency of tandem solar cells. *J. Phys. D Appl. Phys.* **1980**, *13*, 839–846. [CrossRef]
127. Shockley, W.; Queisser, H.J. Detailed Balance Limit of Efficiency of p-n Junction Solar Cells. *J. Appl. Phys.* **1961**, *32*, 510–519. [CrossRef]
128. Sofia, S.E.; Wang, H.; Bruno, A.; Cruz-Campa, J.L.; Buonassisi, T.; Peters, I.M. Roadmap for cost-effective, commercially-viable perovskite silicon tandems for the current and future PV market. *Sustain. Energy Fuels* **2019**, *4*, 852–862. [CrossRef]
129. Chen, B.; Ren, N.; Li, Y.; Yan, L.; Mazumdar, S.; Zhao, Y.; Zhang, X. Insights into the Development of Monolithic Perovskite/Silicon Tandem Solar Cells. *Adv. Energy Mater.* **2022**, *12*, 2003628. [CrossRef]
130. Wang, R.; Huang, T.; Xue, J.; Tong, J.; Zhu, K.; Yang, Y. Prospects for metal halide perovskite-based tandem solar cells. *Nat. Photon.* **2021**, *15*, 411–425. [CrossRef]
131. Chen, B.; Bai, Y.; Yu, Z.J.; Li, T.; Zheng, X.; Dong, Q.; Shen, L.; Boccard, M.; Gruverman, A.; Holman, Z.C.; et al. Efficient Semitransparent Perovskite Solar Cells for 23.0%-Efficiency Perovskite/Silicon Four-Terminal Tandem Cells. *Adv. Energy Mater.* **2016**, *6*, 1601128. [CrossRef]
132. Quiroz, C.O.R.; Shen, Y.; Salvador, M.; Forberich, K.; Schrenker, N.; Spyropoulos, G.D.; Heumüller, T.; Wilkinson, B.; Kirchartz, T.; Spiecker, E.; et al. Balancing electrical and optical losses for efficient 4-terminal Si–perovskite solar cells with solution processed percolation electrodes. *J. Mater. Chem. A* **2018**, *6*, 3583–3592. [CrossRef]

133. Wang, Z.; Zhu, X.; Zuo, S.; Chen, M.; Zhang, C.; Wang, C.; Ren, X.; Yang, Z.; Liu, Z.; Xu, X.; et al. 27%-Efficiency Four-Terminal Perovskite/Silicon Tandem Solar Cells by Sandwiched Gold Nanomesh. *Adv. Funct. Mater.* **2020**, *30*, 1908298. [CrossRef]
134. Werner, J.; Weng, C.-H.; Walter, A.; Fesquet, L.; Seif, J.P.; De Wolf, S.; Niesen, B.; Ballif, C. Efficient Monolithic Perovskite/Silicon Tandem Solar Cell with Cell Area >1 cm^2. *J. Phys. Chem. Lett.* **2016**, *7*, 161–166. [CrossRef] [PubMed]
135. Mazzarella, L.; Lin, Y.-H.; Kirner, S.; Morales-Vilches, A.B.; Korte, L.; Albrecht, S.; Crossland, E.; Stannowski, B.; Case, C.; Snaith, H.J.; et al. Infrared Light Management Using a Nanocrystalline Silicon Oxide Interlayer in Monolithic Perovskite/Silicon Heterojunction Tandem Solar Cells with Efficiency above 25%. *Adv. Energy Mater.* **2019**, *9*, 1803241. [CrossRef]
136. Chen, B.; Yu, Z.J.; Liu, K.; Zheng, X.; Liu, Y.; Shi, J.; Spronk, D.; Rudd, P.N.; Holman, Z.C.; Huang, J. Grain Engineering for Perovskite/Silicon Monolithic Tandem Solar Cells with Efficiency of 25.4%. *Joule* **2019**, *3*, 177–190. [CrossRef]
137. Han, Q.; Hsieh, Y.-T.; Meng, L.; Wu, J.-L.; Sun, P.; Yao, E.-P.; Chang, S.-Y.; Bae, S.-H.; Kato, T.; Bermudez, V.; et al. High-performance perovskite/Cu(In,Ga)Se$_2$ monolithic tandem solar cells. *Science* **2018**, *361*, 904–908. [CrossRef]
138. Jošt, M.; Bertram, T.; Koushik, D.; Marquez, J.A.; Verheijen, M.A.; Heinemann, M.D.; Köhnen, E.; Al-Ashouri, A.; Braunger, S.; Lang, F.; et al. 21.6%-Efficient Monolithic Perovskite/Cu(In,Ga)Se2 Tandem Solar Cells with Thin Conformal Hole Transport Layers for Integration on Rough Bottom Cell Surfaces. *ACS Energy Lett.* **2019**, *4*, 583–590. [CrossRef]
139. Singh, M.; Santbergen, R.; Syifai, I.; Weeber, A.; Zeman, M.; Isabella, O. Comparing optical performance of a wide range of perovskite/silicon tandem architectures under real-world conditions. *Nanophotonics* **2021**, *10*, 2043–2057. [CrossRef]
140. Todorov, T.; Gunawan, O.; Guha, S. A road towards 25% efficiency and beyond: Perovskite tandem solar cells. *Mol. Syst. Des. Eng.* **2016**, *1*, 370–376. [CrossRef]
141. Lin, R.; Xiao, K.; Qin, Z.; Han, Q.; Zhang, C.; Wei, M.; Saidaminov, M.I.; Gao, Y.; Xu, J.; Xiao, M.; et al. Monolithic all-perovskite tandem solar cells with 24.8% efficiency exploiting comproportionation to suppress Sn(ii) oxidation in precursor ink. *Nat. Energy* **2019**, *4*, 864–873. [CrossRef]
142. Chen, W.; Wu, Y.; Yue, Y.; Liu, J.; Zhang, W.; Yang, X.; Chen, H.; Bi, E.; Ashraful, I.; Grätzel, M.; et al. Efficient and stable large-area perovskite solar cells with inorganic charge extraction layers. *Science* **2015**, *350*, 944–948. [CrossRef]
143. Qiu, W.; Merckx, T.; Jaysankar, M.; de la Huerta, C.M.; Rakocevic, L.; Zhang, W.; Paetzold, U.W.; Gehlhaar, R.; Froyen, L.; Poortmans, J.; et al. Pinhole-free perovskite films for efficient solar modules. *Energy Environ. Sci.* **2016**, *9*, 484–489. [CrossRef]
144. Yuan, Y.; Giri, G.; Ayzner, A.L.; Zoombelt, A.P.; Mannsfeld, S.C.B.; Chen, J.; Nordlund, D.; Toney, M.F.; Huang, J.; Bao, Z. Ultra-high mobility transparent organic thin film transistors grown by an off-centre spin-coating method. *Nat. Commun.* **2014**, *5*, 3005. [CrossRef] [PubMed]
145. Ding, B.; Li, Y.; Huang, S.-Y.; Chu, Q.-Q.; Li, C.-X.; Li, C.-J.; Yang, G.-J. Material nucleation/growth competition tuning towards highly reproducible planar perovskite solar cells with efficiency exceeding 20%. *J. Mater. Chem. A* **2017**, *5*, 6840–6848. [CrossRef]
146. Yang, M.; Zhou, Y.; Zeng, Y.; Jiang, C.-S.; Padture, N.P.; Zhu, K. Square-Centimeter Solution-Processed Planar CH3NH3PbI3Perovskite Solar Cells with Efficiency Exceeding 15%. *Adv. Mater.* **2015**, *27*, 6363–6370. [CrossRef] [PubMed]
147. Hossain, M.I.; Qarony, W.; Jovanov, V.; Tsang, Y.H.; Knipp, D. Nanophotonic design of perovskite/silicon tandem solar cells. *J. Mater. Chem. A* **2018**, *6*, 3625–3633. [CrossRef]
148. Zheng, J.; Mehrvarz, H.; Ma, F.-J.; Lau, C.F.J.; Green, M.A.; Huang, S.; Ho-Baillie, A.W.Y. 21.8% Efficient Perovskite/Homo-Junction-Silicon Tandem Solar Cell on 16 cm2. *ACS Energy Lett.* **2018**, *3*, 2299–2300. [CrossRef]
149. Zheng, J.; Mehrvarz, H.; Liao, C.; Bing, J.; Cui, X.; Li, Y.; Gonçales, V.R.; Lau, C.F.J.; Lee, D.S.; Li, Y.; et al. Large-Area 23%-Efficient Monolithic Perovskite/Homojunction-Silicon Tandem Solar Cell with Enhanced UV Stability Using Down-Shifting Material. *ACS Energy Lett.* **2019**, *4*, 2623–2631. [CrossRef]
150. Zhai, Y.; Wang, Z.; Wang, G.; Peijnenburg, W.J.G.M.; Vijver, M.G. The fate and toxicity of Pb-based perovskite nanoparticles on soil bacterial community: Impacts of pH, humic acid, and divalent cations. *Chemosphere* **2020**, *249*, 126564. [CrossRef]
151. Schileo, G.; Grancini, G. Lead or no lead? Availability, toxicity, sustainability and environmental impact of lead-free perovskite solar cells. *J. Mater. Chem. C* **2021**, *9*, 67–76. [CrossRef]
152. Su, P.; Liu, Y.; Zhang, J.; Chen, C.; Yang, B.; Zhang, C.; Zhao, X. Pb-Based Perovskite Solar Cells and the Underlying Pollution behind Clean Energy: Dynamic Leaching of Toxic Substances from Discarded Perovskite Solar Cells. *J. Phys. Chem. Lett.* **2020**, *11*, 2812–2817. [CrossRef]
153. Zhang, Q.; Hao, F.; Li, J.; Zhou, Y.; Wei, Y.; Lin, H. Perovskite solar cells: Must lead be replaced–And can it be done? *Sci. Technol. Adv. Mater.* **2018**, *19*, 425–442. [CrossRef] [PubMed]
154. Hailegnaw, B.; Kirmayer, S.; Edri, E.; Hodes, G.; Cahen, D. Rain on Methylammonium Lead Iodide Based Perovskites: Possible Environmental Effects of Perovskite Solar Cells. *J. Phys. Chem. Lett.* **2015**, *6*, 1543–1547. [CrossRef]
155. Li, J.; Duan, J.; Yang, X.; Duan, Y.; Yang, P.; Tang, Q. Review on recent progress of lead-free halide perovskites in optoelectronic applications. *Nano Energy* **2021**, *80*, 105526. [CrossRef]
156. Wang, R.; Wang, J.; Tan, S.; Duan, Y.; Wang, Z.-K.; Yang, Y. Opportunities and Challenges of Lead-Free Perovskite Optoelectronic Devices. *Trends Chem.* **2019**, *1*, 368–379. [CrossRef]
157. Giustino, F.; Snaith, H.J. Toward Lead-Free Perovskite Solar Cells. *ACS Energy Lett.* **2016**, *1*, 1233–1240. [CrossRef]
158. Shi, Z.; Guo, J.; Chen, Y.; Li, Q.; Pan, Y.; Zhang, H.; Xia, Y.; Huang, W. Lead-Free Organic-Inorganic Hybrid Perovskites for Photovoltaic Applications: Recent Advances and Perspectives. *Adv. Mater.* **2017**, *29*, 1605005. [CrossRef] [PubMed]
159. Yang, S.; Fu, W.; Zhang, Z.; Chen, H.; Li, C.-Z. Recent advances in perovskite solar cells: Efficiency, stability and lead-free perovskite. *J. Mater. Chem. A* **2017**, *5*, 11462–11482. [CrossRef]

160. Xu, P.; Chen, S.; Xiang, H.-J.; Gong, X.-G.; Wei, S.-H. Influence of Defects and Synthesis Conditions on the Photovoltaic Performance of Perovskite Semiconductor CsSnI$_3$. *Chem. Mater.* **2014**, *26*, 6068–6072. [CrossRef]
161. Hoefler, S.F.; Trimmel, G.; Rath, T. Progress on lead-free metal halide perovskites for photovoltaic applications: A review. *Monatsh. Chem.* **2017**, *148*, 795–826. [CrossRef]
162. Lyu, M.; Yun, J.-H.; Chen, P.; Hao, M.; Wang, L. Addressing Toxicity of Lead: Progress and Applications of Low-Toxic Metal Halide Perovskites and Their Derivatives. *Adv. Energy Mater.* **2017**, *7*, 1602512–1602537. [CrossRef]
163. Ming, W.; Shi, H.; Du, M.-H. Large dielectric constant, high acceptor density, and deep electron traps in perovskite solar cell material CsGeI$_3$. *J. Mater. Chem. A* **2016**, *4*, 13852–13858. [CrossRef]
164. Wang, M.; Wang, W.; Ma, B.; Shen, W.; Liu, L.; Cao, K.; Chen, S.; Huang, W. Lead-Free Perovskite Materials for Solar Cells. *Nano-Micro Lett.* **2021**, *13*, 62. [CrossRef] [PubMed]
165. Shalan, A.E.; Kazim, S.; Ahmad, S. Lead Free Perovskite Materials: Interplay of Metals Substitution for Environmentally Compatible Solar Cells Fabrication. *ChemSusChem* **2019**, *12*, 4116–4139. [CrossRef]
166. Zhou, J.; An, K.; He, P.; Yang, J.; Zhou, C.; Luo, Y.; Kang, W.; Hu, W.; Feng, P.; Zhou, M.; et al. Solution-Processed Lead-Free Perovskite Nanocrystal Scintillators for High-Resolution X-Ray CT Imaging. *Adv. Opt. Mater.* **2021**, *9*, 2002144. [CrossRef]
167. Hong, F.; Saparov, B.; Meng, W.; Xiao, Z.; Mitzi, D.B.; Yan, Y. Viability of Lead-Free Perovskites with Mixed Chalcogen and Halogen Anions for Photovoltaic Applications. *J. Phys. Chem. C* **2016**, *120*, 6435–6441. [CrossRef]
168. Zhang, X.; Wang, W.; Xu, B.; Liu, H.; Shi, H.; Dai, H.; Zhang, X.; Chen, S.; Wang, K.; Sun, X.W. Less-Lead Control toward Highly Efficient Formamidinium-Based Perovskite Light-Emitting Diodes. *ACS Appl. Mater. Interfaces* **2018**, *10*, 24242–24248. [CrossRef] [PubMed]
169. Soleimanioun, N.; Rani, M.; Sharma, S.; Kumar, A.; Tripathi, S.K. Binary metal zinc-lead perovskite built-in air ambient: Towards lead-less and stable perovskite materials. *Sol. Energy Mater. Sol. Cells* **2019**, *191*, 339–344. [CrossRef]
170. Shao, S.; Liu, J.; Portale, G.; Fang, H.-H.; Blake, G.R.; Brink, G.H.T.; Koster, L.J.A.; Loi, M.A. Highly Reproducible Sn-Based Hybrid Perovskite Solar Cells with 9% Efficiency. *Adv. Energy Mater.* **2017**, *8*, 1702019. [CrossRef]
171. Lin, R.; Xu, J.; Wei, M.; Wang, Y.; Qin, Z.; Liu, Z.; Wu, J.; Xiao, K.; Chen, B.; Park, S.M.; et al. All-perovskite tandem solar cells with improved grain surface passivation. *Nature* **2022**, *603*, 73–78. [CrossRef]
172. Chen, S.; Deng, Y.; Xiao, X.; Xu, S.; Rudd, P.N.; Huang, J. Preventing lead leakage with built-in resin layers for sustainable perovskite solar cells. *Nat. Sustain.* **2021**, *4*, 636–643. [CrossRef]
173. Chen, S.; Deng, Y.; Gu, H.; Xu, S.; Wang, S.; Yu, Z.; Blum, V.; Huang, J. Trapping lead in perovskite solar modules with abundant and low-cost cation-exchange resins. *Nat. Energy* **2020**, *5*, 1003–1011. [CrossRef]
174. Kim, C.U.; Jung, E.D.; Noh, Y.W.; Seo, S.K.; Choi, Y.; Park, H.; Song, M.H.; Choi, K.J. Strategy for large-scale monolithic Perovskite /Silicon tandem solar cell: A review of recent progress. *EcoMat* **2021**, *3*, e12084. [CrossRef]
175. Aitola, K.; Domanski, K.; Correa-Baena, J.-P.; Sveinbjörnsson, K.; Saliba, M.; Abate, A.; Grätzel, M.; Kauppinen, E.; Johansson, E.M.J.; Tress, W.; et al. High Temperature-Stable Perovskite Solar Cell Based on Low-Cost Carbon Nanotube Hole Contact. *Adv. Mater.* **2017**, *29*, 1606398. [CrossRef]
176. Kim, Y.; Jung, E.H.; Kim, G.; Kim, D.; Kim, B.J.; Seo, J. Sequentially Fluorinated PTAA Polymers for Enhancing V OC of High-Performance Perovskite Solar Cells. *Adv. Energy Mater.* **2018**, *8*, 1801668. [CrossRef]
177. Zhou, L.; Chang, J.; Liu, Z.; Sun, X.; Lin, Z.; Chen, D.; Zhang, C.; Zhang, J.; Hao, Y. Enhanced planar perovskite solar cell efficiency and stability using a perovskite/PCBM heterojunction formed in one step. *Nanoscale* **2018**, *10*, 3053–3059. [CrossRef] [PubMed]
178. Yang, I.S.; Sohn, M.R.; Sung, S.D.; Kim, Y.J.; Yoo, Y.J.; Kim, J.; Lee, W.I. Formation of pristine CuSCN layer by spray deposition method for efficient perovskite solar cell with extended stability. *Nano Energy* **2017**, *32*, 414–421. [CrossRef]
179. Yun, A.J.; Kim, J.; Hwang, T.; Park, B. Origins of Efficient Perovskite Solar Cells with Low-Temperature Processed SnO2 Electron Transport Layer. *ACS Appl. Energy Mater.* **2019**, *2*, 3554–3560. [CrossRef]
180. Feng, J.; Yang, Z.; Yang, D.; Ren, X.; Zhu, X.; Jin, Z.; Zi, W.; Wei, Q.; Liu, S. E-beam evaporated Nb2O5 as an effective electron transport layer for large flexible perovskite solar cells. *Nano Energy* **2017**, *36*, 1–8. [CrossRef]
181. Wang, K.; Jin, Z.; Liang, L.; Bian, H.; Bai, D.; Wang, H.; Zhang, J.; Wang, Q.; Liu, S. All-inorganic cesium lead iodide perovskite solar cells with stabilized efficiency beyond 15%. *Nat. Commun.* **2018**, *9*, 4544. [CrossRef]
182. Grancini, G.; Roldán-Carmona, C.; Zimmermann, I.; Mosconi, E.; Lee, X.; Martineau, D.; Narbey, S.; Oswald, F.; De Angelis, F.; Graetzel, M.; et al. One-Year stable perovskite solar cells by 2D/3D interface engineering. *Nat. Commun.* **2017**, *8*, 15684. [CrossRef]
183. Gharibzadeh, S.; Nejand, B.A.; Jakoby, M.; Abzieher, T.; Hauschild, D.; Moghadamzadeh, S.; Schwenzer, J.A.; Brenner, P.; Schmager, R.; Haghighirad, A.A.; et al. Record Open-Circuit Voltage Wide-Bandgap Perovskite Solar Cells Utilizing 2D/3D Perovskite Heterostructure. *Adv. Energy Mater.* **2019**, *9*, 1803699. [CrossRef]
184. Chen, P.; Bai, Y.; Wang, S.; Lyu, M.; Yun, J.-H.; Wang, L. In Situ Growth of 2D Perovskite Capping Layer for Stable and Efficient Perovskite Solar Cells. *Adv. Funct. Mater.* **2018**, *28*, 1706923. [CrossRef]
185. Cheacharoen, R.; Boyd, C.C.; Burkhard, G.F.; Leijtens, T.; Raiford, J.A.; Bush, K.A.; Bent, S.F.; McGehee, M.D. Encapsulating perovskite solar cells to withstand damp heat and thermal cycling. *Sustain. Energy Fuels* **2018**, *2*, 2398–2406. [CrossRef]
186. Cheacharoen, R.; Bush, K.A.; Rolston, N.; Harwood, D.; Dauskardt, R.H.; McGehee, M.D. Damp Heat, Temperature Cycling and UV Stress Testing of Encapsulated Perovskite Photovoltaic Cells. In Proceedings of the 2018 IEEE 7th World Conference on Photovoltaic Energy Conversion (WCPEC) (a Joint Conference of 45th IEEE PVSC, 28th PVSEC & 34th EU PVSEC), Waikoloa, HI, USA, 10–15 June 2018; pp. 3498–3502. [CrossRef]

187. Boyd, C.C.; Cheacharoen, R.; Bush, K.A.; Prasanna, R.; Leijtens, T.; McGehee, M.D. Barrier Design to Prevent Metal-Induced Degradation and Improve Thermal Stability in Perovskite Solar Cells. *ACS Energy Lett.* **2018**, *3*, 1772–1778. [CrossRef]
188. Shi, L.; Young, T.L.; Kim, J.; Sheng, Y.; Wang, L.; Chen, Y.; Feng, Z.; Keevers, M.J.; Hao, X.; Verlinden, P.J.; et al. Accelerated Lifetime Testing of Organic–Inorganic Perovskite Solar Cells Encapsulated by Polyisobutylene. *ACS Appl. Mater. Interfaces* **2017**, *9*, 25073–25081. [CrossRef]
189. Hardhienata, H.; Ahmad, F.; Arramel; Aminah, M.; Onggo, D.; Diguna, L.J.; Birowosuto, M.D.; Witkowski, M.E.; Makowski, M.; Drozdowski, W. Optical and x–ray scintillation properties of X2MnCl4 (X = PEA, PPA) perovskite crystals. *J. Phys. D Appl. Phys.* **2020**, *53*, 455303. [CrossRef]
190. Malinkiewicz, O.; Yella, A.; Lee, Y.H.; Espallargas, G.M.; Graetzel, M.; Nazeeruddin, M.K.; Bolink, H.J. Perovskite solar cells employing organic charge-transport layers. *Nat. Photon.* **2013**, *8*, 128–132. [CrossRef]
191. Rong, Y.; Hu, Y.; Mei, A.; Tan, H.; Saidaminov, M.I.; Seok, S.I.; McGehee, M.D.; Sargent, E.H.; Han, H. Challenges for commercializing perovskite solar cells. *Science* **2018**, *361*, eaat8235. [CrossRef]
192. Krishna, A.; Zhang, H.; Zhou, Z.; Gallet, T.; Dankl, M.; Ouellette, O.; Eickemeyer, F.T.; Fu, F.; Sanchez, S.; Mensi, M.; et al. Nanoscale interfacial engineering enables highly stable and efficient perovskite photovoltaics. *Energy Environ. Sci.* **2021**, *14*, 5552–5562. [CrossRef]
193. Wang, D.; Guo, H.; Wu, X.; Deng, X.; Li, F.; Li, Z.; Lin, F.; Zhu, Z.; Zhang, Y.; Xu, B.; et al. Interfacial Engineering of Wide-Bandgap Perovskites for Efficient Perovskite/CZTSSe Tandem Solar Cells. *Adv. Funct. Mater.* **2021**, *32*, 2107359. [CrossRef]
194. Sutanto, A.A.; Caprioglio, P.; Drigo, N.; Hofstetter, Y.J.; Garcia-Benito, I.; Queloz, V.I.E.; Neher, D.; Nazeeruddin, M.K.; Stolterfoht, M.; Vaynzof, Y.; et al. 2D/3D perovskite engineering eliminates interfacial recombination losses in hybrid perovskite solar cells. *Chem* **2021**, *7*, 1903–1916. [CrossRef]
195. Hossain, M.I.; Qarony, W.; Ma, S.; Zeng, L.; Knipp, D.; Tsang, Y.H. Perovskite/Silicon Tandem Solar Cells: From Detailed Balance Limit Calculations to Photon Management. *Nano-Micro Lett.* **2019**, *11*, 1–24. [CrossRef] [PubMed]
196. Oxford, P.V. Tandem Cell Production. Available online: https://www.oxfordpv.com/tandem-cell-production (accessed on 10 August 2021).
197. Samy, O.; Zeng, S.; Birowosuto, M.D.; El Moutaouakil, A. A Review on MoS2 Properties, Synthesis, Sensing Applications and Challenges. *Crystals* **2021**, *11*, 355. [CrossRef]
198. Samy, O.; Birowosuto, D.; El Moutaouakil, A. A Short Review on Molybdenum Disulfide (MoS2) Applications and Challenges. In Proceedings of the 2021 6th International Conference on Renewable Energy: Generation and Applications (ICREGA), Al Ain, United Arab Emirates, 2–4 February 2021; pp. 220–222.
199. Samy, O.; El Moutaouakil, A. A Review on MoS2 Energy Applications: Recent Developments and Challenges. *Energies* **2021**, *14*, 4586. [CrossRef]
200. Tiouitchi, G.; Ali, M.A.; Benyoussef, A.; Hamedoun, M.; Lachgar, A.; Kara, A.; Ennaoui, A.; Mahmoud, A.; Boschini, F.; Oughaddou, H.; et al. Efficient Production of Few-Layer Black Phosphorus by Liquid-Phase Exfoliation. *R. Soc. Open Sci.* **2020**, *7*, 201210. [CrossRef]
201. Abed, J.; Rajput, N.S.; Moutaouakil, A.E.; Jouiad, M. Recent Advances in the Design of Plasmonic Au/TiO2 Nanostructures for Enhanced Photocatalytic Water Splitting. *Nanomaterials* **2020**, *10*, 2260. [CrossRef]
202. Moutaouakil, A.E.; Kang, H.-C.; Handa, H.; Fukidome, H.; Suemitsu, T.; Sano, E.; Suemitsu, M.; Otsuji, T. Room Temperature Logic Inverter on Epitaxial Graphene-on-Silicon Device. *Jpn. J. Appl. Phys.* **2011**, *50*, 070113. [CrossRef]
203. Moutaouakil, A.E. Two-Dimensional Electronic Materials for Terahertz Applications: Linking the Physical Properties with Engineering Expertise. In Proceedings of the 2018 6th International Renewable and Sustainable Energy Conference (IRSEC), Rabat, Morocco, 5–8 December 2018; pp. 1–4.
204. Moutaouakil, A.E.; Suemitsu, T.; Otsuji, T.; Coquillat, D.; Knap, W. Nonresonant Detection of Terahertz Radiation in High-Electron-Mobility Transistor Structure Using InAlAs/InGaAs/InP Material Systems at Room Temperature. *J. Nanosci. Nanotechnol.* **2012**, *12*, 6737–6740. [CrossRef]
205. Moutaouakil, A.E.; Komori, T.; Horiike, K.; Suemitsu, T.; Otsuji, T. Room Temperature Intense Terahertz Emission from a Dual Grating Gate Plasmon-Resonant Emitter Using InAlAs/InGaAs/InP Material Systems. *IEICE Trans. Electron.* **2010**, *93*, 1286–1289. [CrossRef]
206. El Moutaouakil, A.; Suemitsu, T.; Otsuji, T.; Videlier, H.; Boubanga-Tombet, S.-A.; Coquillat, D.; Knap, W. Device Loading Effect on Nonresonant Detection of Terahertz Radiation in Dual Grating Gate Plasmon-Resonant Structure Using InGaP/InGaAs/GaAs Material Systems. *Phys. Status Solidi C* **2011**, *8*, 346–348. [CrossRef]
207. Hijazi, A.; Moutaouakil, A.E. Graphene and MoS2 Structures for THz Applications. In Proceedings of the 2019 44th International Conference on Infrared, Millimeter, and Terahertz Waves (IRMMW-THz), Paris, France, 1–6 September 2019; pp. 1–2.
208. Moutaouakil, A.E.; Fukidome, H.; Otsuji, T. Investigation of Terahertz Properties in Graphene Ribbons. In Proceedings of the 2020 45th International Conference on Infrared, Millimeter, and Terahertz Waves (IRMMW-THz), Buffalo, NY, USA, 8–13 November 2020; pp. 1–2.
209. El Moutaouakil, A.; Al Ahmad, M.; Soopy, A.K.K.; Najar, A. Porous Silicon NWs with FiTC-Doped Silica Nanoparticles. In Proceedings of the 2021 6th International Conference on Renewable Energy: Generation and Applications (ICREGA), Al Ain, United Arab Emirates, 2–4 February 2021; pp. 6–8.

210. Moutaouakil, A.E.; Watanabe, T.; Haibo, C.; Komori, T.; Nishimura, T.; Suemitsu, T.; Otsuji, T. Spectral Narrowing of Terahertz Emission from Super-Grating Dual-Gate Plasmon-Resonant High-Electron Mobility Transistors. *J. Phys. Conf. Ser.* **2009**, *193*, 012068. [CrossRef]
211. Moutaouakil, A.E.; Suemitsu, T.; Otsuji, T.; Coquillat, D.; Knap, W. Room Temperature Terahertz Detection in High-Electron-Mobility Transistor Structure Using InAlAs/InGaAs/InP Material Systems. In Proceedings of the 35th International Conference on Infrared, Millimeter, and Terahertz Waves, Rome, Italy, 5–10 September 2010; pp. 1–2.
212. Meziani, Y.M.; Garcia, E.; Velazquez, E.; Diez, E.; El Moutaouakil, A.; Otsuji, T.; Fobelets, K. Strained Silicon Modulation Field-Effect Transistor as a New Sensor of Terahertz Radiation. *Semicond. Sci. Technol.* **2011**, *26*, 105006. [CrossRef]

Article

Investigations of Fused Deposition Modeling for Perovskite Active Solar Cells

Leland Weiss *[ ] and Tyler Sonsalla

Institute for Micromanufacturing, College of Engineering and Science, Louisiana Tech University Ruston, Ruston, LA 71272, USA; tjs041@latech.edu
* Correspondence: lweiss@latech.edu

Abstract: The advent of Fused Deposition Modeling (FDM; or 3D printing) has significantly changed the way many products are designed and built. It has even opened opportunities to fabricate new products on-site and on-demand. In addition, parallel efforts that introduce new materials into the FDM process have seen great advances as well. New additives have been demonstrably utilized to achieve thermal, electrical, and structural property improvements. This combination of fabrication flexibility and material additives make FDM an ideal candidate for investigation of perovskite materials in new solar cell efforts. In this work, we fabricate and characterize a perovskite-based solar cell polymer designed for the FDM fabrication processes. Perovskite solar cells have garnered major research interest since their discovery in 2009. Perovskites, specifically methylammonium lead iodide, offer beneficial properties to solar cell fabrication such as long minority charge carrier distance, high light absorption, and simple fabrication methods. Despite the great potential of these materials, however, stability remains an issue in solar cell utilization as the material degrades under ultraviolet light, exposure to oxygen and water, as well as increased temperatures. To mitigate degradation, different fabrication methods have been utilized. Additionally, multiple groups have utilized encapsulation methods post-fabrication and in situ solution processed integration of polymer materials into the solar cell to prevent degradation. In this paper, we leverage the unique ability of FDM to encapsulate perovskite materials and yield a $MAPbI_3$-PCL solar material as the active layer for solar cell use. In this manner, increased ability to resist UV light degradation and material stability from other environmental factors can be achieved. This study provides characterization of the material via multiple techniques like SEM (Scanning Electron Microscopy) and XRD (X-ray Diffraction) as well as absorbance, transmittance, and photocurrent response. Investigations of processing on perovskite degradation as well as initial solar simulated response are recorded. Unique aspects of the resulting material and process are noted including improved performance with increased operating temperature. Increased electron–hole pair generation is observed for 200 μm FDM-printed PCL film, achieving a 45% reduction in resistance under peak incident flux of 590 W/m^2 with the addition of $MAPbI_3$. This work establishes insight into the use of FDM for full solar cell fabrication and points to the next steps of research and development in this growing field.

Keywords: micro; perovskite; solar cell; 3D printing; fused deposition modeling

Citation: Weiss, L.; Sonsalla, T. Investigations of Fused Deposition Modeling for Perovskite Active Solar Cells. *Polymers* **2022**, *14*, 317. https://doi.org/10.3390/polym14020317

Academic Editor: Bożena Jarząbek

Received: 12 November 2021
Accepted: 5 January 2022
Published: 13 January 2022

Publisher's Note: MDPI stays neutral with regard to jurisdictional claims in published maps and institutional affiliations.

Copyright: © 2022 by the authors. Licensee MDPI, Basel, Switzerland. This article is an open access article distributed under the terms and conditions of the Creative Commons Attribution (CC BY) license (https://creativecommons.org/licenses/by/4.0/).

1. Introduction

Over the course of the past decade, significant changes have occurred in the field of electricity production [1]. Specifically, while the use of traditional fuel sources like coal have diminished, there continues to be a rise in the use of so-called "green" energy sources like solar and wind. It is projected that a continued move from traditional, fossil fuel-based energy sources to renewable sources will continue in the decades ahead. Solar energy power production is poised to be increasingly important among the renewable energy sources between now and 2050 [1]. To support new growth in solar energy production, new methods and materials are needed. Fortunately, there have also been significant advances in unique materials and new methods of manufacture. Fused Deposition Modeling (FDM), or 3D printing, is one such advance.

Given this growing interest in solar-based energy and expanding capabilities of FDM, we investigate new FDM processes and materials that allow the fabrication of 3D printed solar cell active layers. There is inherent promise in the approach given that FDM-based devices are no longer constrained to planar type structures, but may be designed and fabricated for unique spaces, shapes and applications well beyond traditional manufacturing constraints. As an example, a curved-radius active layer will provide improved incidence angles with solar input as the day progresses and the sun moves across the sky. FDM also allows future devices to be fabricated on-site and on-demand in situations where traditional shipping of completed devices is more difficult. This could have significant application in non-terrestrial settings [2,3].

Perovskite solar cells, first discovered in 2009, have quickly achieved efficiencies of over 20% utilizing simple solution-based fabrication methods [4], potentially providing a cost-effective alternative to traditional solar cell technologies. However, perovskite solar cells have also shown stability issues when exposed to ambient atmosphere. It has been shown that UV light, continuous light, oxygen, water, and temperature can affect the stability of perovskite solar cells [5–7]. For these reasons, various methods have been utilized during fabrication and post-fabrication to prolong stability. For example, the use of various epoxies and desiccant have been popular methods in post-fabrication [8–10]. However, these methods increase the fabrication complexity which ultimately effects scalability to commercial application. Other groups have utilized different chemical pathways (i.e., anti-solvents, etc.) to extend the stability and fabricate perovskite solar cells in high humidity environments [11–13]. Other efforts have used different electron and hole transport layer combinations to extend the stability of perovskite solar cells in ambient conditions [4,14]. However, commonly utilized electron and hole transport materials can be cost prohibitive for larger scale application (>$300 per gram).

In this current effort, the general approach employed to achieve these unique devices encapsulates perovskite materials for extrusion as polymer-composite filaments through a heated nozzle in a layer-by-layer FDM process. This builds on our prior work and allows the incorporation and utilization of various polymer-based materials with unique, entrained additives [15].

Over the past several years, multiple investigations have undertaken the general challenge of three-dimensionally printed perovskite materials. Beyond the general advantages of FDM noted previously, there are several perovskite-specific advantages to this approach. First, FDM processes utilizing composite filaments have been previously demonstrated to align filler particles upon printing [15]. This improves thermal and electrical properties, which in the context of a solar cell can improve charge extraction and performance. Second, the FDM process integrates a polymer barrier around filler particles, which can protect the particles from ambient conditions (i.e., oxygen and water). In this manner, the long-term stability issues related to perovskite solar cells can be naturally addressed via the fabrication process itself.

Many of the FDM-oriented approaches have focused on uses like photoelectric sensing [16] or light production [17–19]. Zhao et al. advanced the general formation of perovskite crystals from disordered to ordered via electric field, producing a more orderly crystal deposition for use in a photosensor with high sensitivity, high stability, and good response [16]. Tai et al. investigated several thermoplastic polymers as protective coatings for perovskite materials and incorporated the effort into light-emitting diode (LED) devices with various emission wavelengths. It was further noted that there was fluorescent behavior of some of the printed structures resulting from this effort [17]. Others have focused on photo-luminescent materials as structural materials [18]. Qaid et al. have investigated the opportunity for light amplification via the alternate $CsPbBr_3$ perovskite material while encapsulating in PMMA polymers [20].

In this work, we fabricate and evaluate MAPbI$_3$ perovskite material, suitable for FDM printing application and end use as the photo-active layer in solar cell constructions. Fabrication that produces the polymer-encapsulated printable material and full characterization is presented, followed by initial testing of photocurrent operational results in a 3D-printed test piece. This effort sets the foundation for efforts that incorporate the material into a fully realized solar cell.

2. Experimental Methods

To achieve a prototype FDM perovskite active layer, processing, characterization, and printing of the required materials was needed. Material was characterized via several processes including SEM (Scanning Electron Microscopy), XRD (X-ray Diffraction), and photocurrent analysis. This section reviews the fabrication processing as well as the test and characterization steps for the fabricated structures and test pieces.

2.1. Materials and Fabrication

2.1.1. Perovskite Material Fabrication

This section reviews the material fabrication steps necessary for the creation of the encapsulated perovskite solar material, ready for application as an FDM material. The fabrication of perovskite microcrystals was accomplished utilizing a previously published procedure by Johansson et al. [12]. Methylammonium iodide (MAI) was mixed in isopropyl alcohol until the crystals were completely dissolved. After the crystals were dissolved, lead iodide (PbI$_2$) was added in at a 1:1 molar ratio. The resulting solution had a concentration of 0.69 M. After addition of PbI$_2$, the solution transparency changed from clear to black signifying the formation of MAPbI$_3$.

After formation of the perovskite, 1 wt.% of 3-(2-aminoethylamino) propy-ldimethoxy-methylsilane (i.e., coupling agent (LICA-38)) was added and allowed to stir for 5 min before the addition of polycaprolactone (PCL). PCL was selected as it has a low melting temperature of about 60 °C and performs well as a water/oxygen barrier [21–23]. The coupling agent was utilized in order to create a voidless interface between the PCL and MAPbI$_3$ crystals. The agent did not degrade the perovskite during preliminary lab testing. Additionally, the printability of filaments was increased with the addition of coupling agent.

PCL was added to the mixture to create a 50 wt.% MAPbI$_3$-PCL mixture. A 50 wt.% concentration of MAPbI$_3$ was selected as it allowed the filament to contain the maximum amount of perovskite particles, while still providing the filament with the ability to be extruded/printed at low temperatures (<100 °C) which prevented degradation. The components were stirred for another 5 min before the solution was added to a recovery flask and IPA was removed from the mixture via rotary evaporator. Figure 1 graphically shows the process.

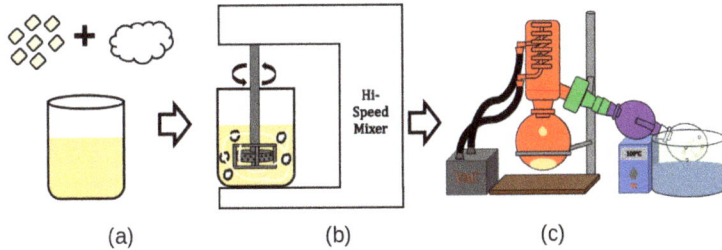

Figure 1. Perovskite composite fabrication process: (**a**) Combining PCL and MAPbI$_3$-PCL mixture. (**b**) High-speed mixing. (**c**) Rotary evaporator use to remove IPA.

Following evaporation, the MAPbI$_3$-PCL mixture was placed in a Heraeus Instruments Vacutherm oven equipped with a Fisher Scientific Maxima C vacuum pump (delivering pressure less than 1×10^{-4} mbar) and dried overnight under light vacuum at 19.1 °C. The resulting MAPbI$_3$-PCL powder was extruded into a filament using a Filabot EX2. The temperature of the extruder was set between 70 and 80 °C. The final resulting filament was optically black in color with a slight shine. Temperatures above this range would result in a yellow filament indicating degradation of the perovskite crystals.

2.1.2. FDM Perovskite Thin Film Fabrication

A fused deposition modeling system (Lulzbot Mini, Lulzbot Taz 5, or Ultimaker 3) controlled by Cura was used to fabricate all plain and composite filament materials characterized in this work. The nozzle diameter utilized was 2.0 mm. The software was programmed to print a thin-film with thickness measured via digital micrometer after printing. The test pieces were designed in Solidworks and imported into the Cura software as .STL files. The thicknesses tested were 25 µm, 50 µm, 100 µm and 200 µm. Overall length and width of these thin-film samples was 25 mm by 25 mm. A Nikon Digimicro was used to determine the thickness of the FDM printed thin films.

2.1.3. FDM Perovskite Photoconductivity Test Piece

Following thin film preparation and evaluation, the MAPbI$_3$-PCL composite was printed via FDM into a disk to allow initial investigation of photoconductivity as well as larger-scale 3D printer capability with the new material. The photoconductivity test piece had a diameter of 25 mm and thickness of 0.9 mm. Two circular holes with a diameter of 3 mm, separated by 12.5 mm were printed into the structure as electrode attachment points. Printing was accomplished via commercially available FDM printers. Twin through-bolts were utilized as electrodes and attached to the photoconductivity test piece via MG Chemicals Silver Conductive Epoxy to ensure electrical connection. Figure 2 shows the fabricated test piece and design.

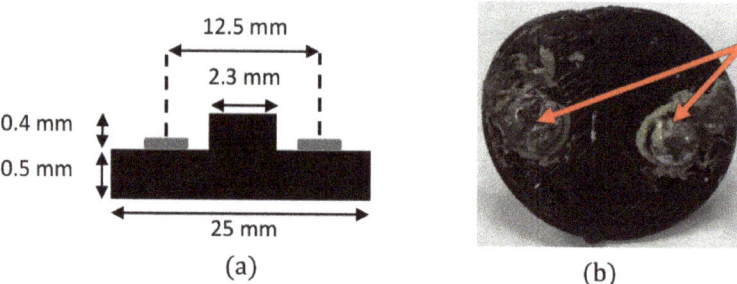

Figure 2. (**a**) Layout of photoconductivity sample. (**b**) Photoconductivity sample and electrode attachment points (red arrow).

2.2. *Test Setup*

Tests were conducted that included characterization of the fabricated perovskite materials as well as photoconductivity and other operationally-oriented tests. Test setups and characterization procedures are discussed in this section.

2.2.1. Material Characterization Testing

Material characterization of the FDM perovskite material included the use of SEM, UV–Vis (Ultraviolet–visible Spectroscopy), and XRD prior to any material operational characterization. A Hitachi S4800 FE-SEM (Hitachi City: White Plains, NY, USA) was utilized to image the MAPbI$_3$ crystals, MAPbI$_3$ crystals after coupling agent treatment, and MAPbI$_3$-PCL composite film after FDM printing. Specific attention was paid to potential damage caused by the FDM print process as well as action of the coupling agent in the final material.

To more completely establish the successful formation of MAPbI$_3$ microcrystals, a Bruker D8 X-ray diffractometer XRD) (Bruker City: Madison, WI, USA) was utilized. This allowed insight into the crystal structure of the perovskite crystals and the printed photoconductivity test pieces (thin films) of MAPbI$_3$-PCL. For clarity, XRD spectrum of PbI$_2$ powder, MAI powder, MAPbI$_3$ powder and PCL-MAPbI$_3$ thin film was examined and compared against characteristic peaks associated with MAPbI$_3$, and MAPBI$_3$-PCL to validate integration and processing into the desired perovskite filament.

2.2.2. Absorbance and Transmittance/Reflectance Testing of PCL Thin Films

A Jasco V-530 UV–Vis spectrophotometer (Jasco Inc., Easton, MD, USA) was utilized to characterize the absorbance spectrum of PCL and MAPbI$_3$-PCL thin-films. For PCL and MAPBI$_3$-PCL the thin films, absorbance was monitored across a range of about 400–1050 nm. The range was selected based on prior work in the field that indicated MAPBI$_3$ absorbance peaking in this range [24–26]. This was done for each thin film sample, with particular attention to the effects of perovskite and its addition into the test materials.

Transmittance spectrum was also investigated. A Filmetrics RT-10 refractometer (Filmetrics, Roselle, CA USA) was utilized to determine the transmittance and reflectance characteristics of the various FDM printed thin-films. The fabricated thin-film pieces (25 µm, 50 µm, 100 µm, and 200 µm thick) were investigated to assess transmittance through polycaprolactone as well as transmittance vs. thickness levels and light allowance into active layers.

Based on those tests, 200 µm film was tested further including tests at room temperature and after heating (40 °C, 5 min) to monitor performance changes. In addition, tests were also conducted on PCL, MAPbI$_3$-PCL, and samples of MAPbI$_3$-PCL-LICA 38 (coupling agent).

2.2.3. I-V Sweep and Photocurrent Testing

Using the FDM perovskite test pieces constructed for the purpose, electrodes were attached via wire to a Keithley 2400 Sourcemeter(Keithley/Tektronix, Beaverton, OR, USA). Current–voltage (I-V) sweeps from −10 V to 10 V by 0.1 V were performed to determine current increase once the material was exposed to light. Dark conditions were also tested in order to provide conclusive results indicating the material reacted to light and not I-V sweep.

All photoconductivity tests were conducted at room temperature. The outcome of these tests was compared to validate perovskite absorption ranges based on prior work in the field [24–26]. The light was applied with a Spectra-Physics 66900 solar simulator (Newport/Spectra-Physics, Milpitas, CA, USA). Input power was varied from 50–80 W, which corresponded to an incident flux of 252–590 W/m^2. Flux was measured with a Newport 91150V Reference Cell and Meter (Newport, Milpitas, CA, USA). This product followed ISO-17025 standards and the output reading was in sun units (where 1 sun is equal to 1000 W/m^2 at 25 °C and Air Mass 1.5 Global Reference). This corresponded to approximately half of what could be expected on a sunny, summer day (approximately 1000 W/m^2). This provided an adequate but conservative simulated solar input for these tests and investigated how light intensity affected photoconductivity. Incident fluxes at each input power level are summarized in Table 1.

Table 1. Incident flux versus input power.

Input Power (W)	Incident Flux (W/m^2)
50 W	252 W/m^2
60 W	364 W/m^2
70 W	459 W/m^2
80 W	590 W/m^2

3. Results and Discussion

This section presents the specific results of the various tests that were conducted on both material and full solar cell constructs. Specific material test results are presented first, followed by the solar photo-conductivity test piece performance test results.

3.1. Material Characterization Test Results

Figure 3 shows SEM images of perovskite, perovskite treated with coupling agent, FDM printed MAPbI$_3$-PCL thin-film, and PCL/perovskite interface. Figure 3a shows the perovskite crystals, which had a cubic structure before coupling agent introduction. The different size of crystals remained following fabrication as well. Figure 3b shows the perovskite crystals after coupling agent introduction. As can be noted in the image, the cubic structure remained, however it was coated with the silane coupling agent used in the study. After printing the MAPbI$_3$-PCL thin-film, the surface morphology was rough and had the appearance of embedded cubes as depicted in Figure 3c. The coupling agent aided in creating a voidless interface between the perovskite and PCL shown in Figure 3d.

This investigation via SEM images showed that coupling agent introduction, filament extrusion, and FDM printing process did not damage the structure of the MAPbI$_3$ crystals. Thus, the FDM filament manufacture and printing process was found to be a viable option to produce MAPbI$_3$ composites.

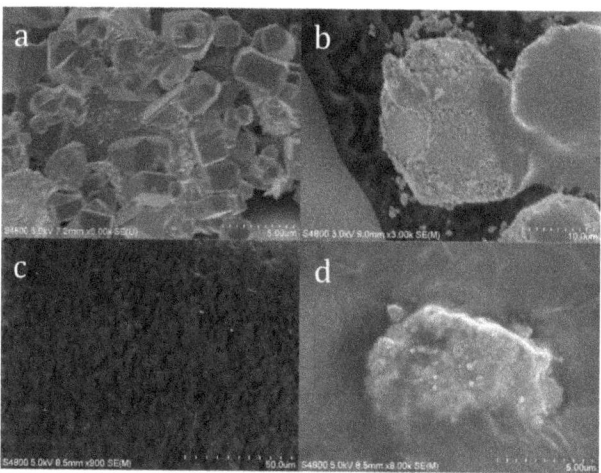

Figure 3. SEM images of (**a**) perovskite (**b**) coupling agent treated perovskite (**c**) PCL-50 wt.% Perovskite (**d**) PCL-50 wt.% Perovskite.

XRD characterization was performed and is presented in Figures 4 and 5. Table 2 summarizes the findings. In general, these findings indicated the successful formation of MAPbI$_3$ microcrystals, integration of the MAPbI$_3$ microcrystals into the PCL matrix, and good stability with lack of degradation during the FDM printing process.

Figure 4 shows XRD spectrum of PbI$_2$ powder, MAI powder, MAPbI$_3$ powder, and PCL-MAPBI$_3$ thin film. The diffraction peaks at $2\Theta = 26°, 34°, 39°, 46°$ (denoted with Δ) correspond to the (101), (102), (110), and (103) lattice planes of PbI$_2$ [27]. Peaks at $2\Theta = 20°$ and $30°$ (denoted with □) correspond to the (002) and (003) lattice planes of MAI [28]. MAPbI$_3$ contained peaks at $2\Theta = 14°, 23°, 28.3°$ and $28.6°$ (denoted with ○), which corresponded to the (002), (121), (004), and (220) lattice planes [8]. The characteristic peaks associated with MAPbI$_3$ remained in the PCL-MAPbI$_3$ composite.

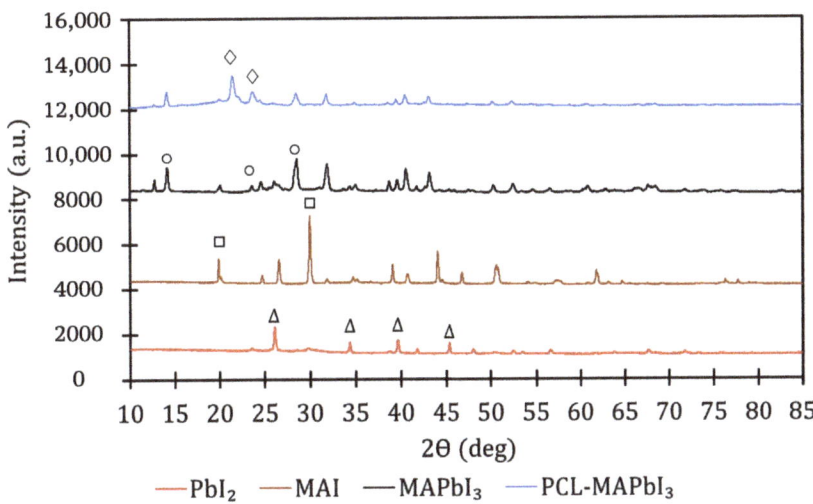

Figure 4. XRD of PbI$_2$, MAI, MAPbI$_3$, and MAPbI$_3$-PCL.

Figure 5 shows an XRD spectrum of perovskite overlaid on the spectrum for PCL-MAPbI$_3$ composite. Although the peak intensities were reduced in the PCL-MAPbI$_3$ composite, the MAPbI$_3$ peaks show the integration and processing of the MAPbI$_3$ into PCL-MAPbI$_3$ filament. The peaks at 2Θ = 21° and 23° (denoted with ∗ in Figure 4) peaks corresponded to the (110) and (200) lattice planes of PCL and agree with literature values [29–32].

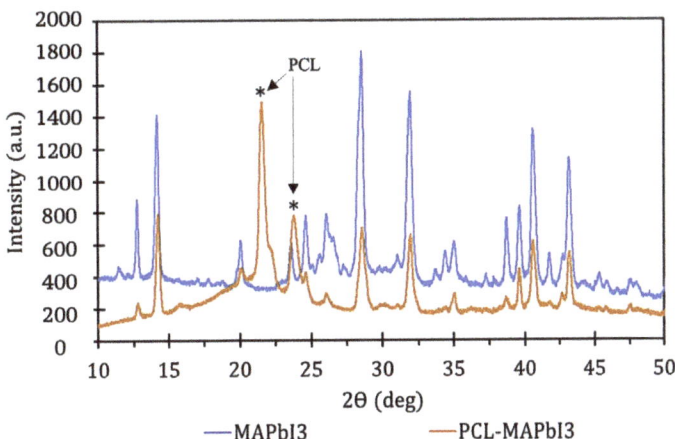

Figure 5. XRD of MAPbI3 and PCL-MAPbI$_3$. Note: ∗ indicates PCL corresponding lattice plane peaks.

One key goal of these essential material evaluations was to confirm that the perovskite material itself did not exhibit significant degradation either in the preliminary mixing, or subsequent processing as it was made suitable for FDM utilization as a 'printable' polymer. This was shown through both SEM and XRD analysis where expected perovskite properties were evident throughout. This allowed continued investigation of the polymer in more optically-oriented testing.

Table 2. Summary of diffraction peaks and lattice planes for PbI$_2$, MAI, MAPbI$_3$, and PCL-MAPbI$_3$.

Material	Diffraction Peak 2Θ	Lattice Plane (hkl)	Reference
Lead Iodide (PbI$_2$)	26°	(101)	[27]
	34°	(102)	
	39°	(110)	
	46°	(103)	
Methylammonium Iodide (MAI)	20°	(002)	[28]
	30°	(003)	
Methylammonium Lead Triiodide (MAPbI$_3$)	14°	(002)	[12]
	23°	(121)	
	28.3°/28.6°	(004)/(220)	
Polycaprolactone Methylammonium Lead Triiodide (PCL-MAPbI$_3$)	21°	(110)	[33,34]
	23°	(200)	

3.2. Absorbance and Transmittance Spectrum Results

With material properties verified, evaluation turned to optical performance and properties. Figure 6 shows the transmittance spectrum for PCL thin films from 400–1050 nm. The thicknesses tested were 25 µm, 50 µm, 100 µm, and 200 µm. The 200 µm thicknesses were also tested with one or two coatings of Smooth-On XTC-3D. The maximum transmittance occurred in the 25 µm sample, which had a transmittance of 29.6% at 1050 nm. Samples had decreasing transmittance with increasing thickness, the lowest transmittance occurred in the 200 µm sample with two coatings of the Smooth-On XTC-3D at 0.9% and 393 nm. Maximum transmittance did increase from a single coating of Smooth-On XTC-3D on a 200 µm PCL sample to 4.1% when compared to the 200 µm plain PCL sample, which attained a maximum transmittance of 1.4%. Transmittance increased for all samples with increasing wavelength except the 200 µm sample with a double coating of Smooth-On XTC 3D.

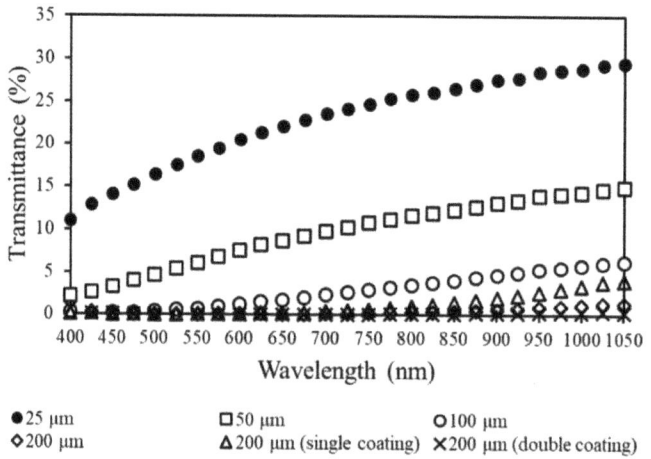

Figure 6. Transmittance spectrums for PCL thin films.

This demonstrated the effect of thickness on optical transparency with thinner FDM-printed samples allowing more light into the active layer in a solar cell structure. However, even at a thickness of 200 µm light still entered the active layer. This also confirmed that PCL allowed transmission of optical wavelengths absorbed by MAPbI$_3$ to transmit through the layer.

The use and performance of the XTC-3D coating is also worth noting. The primary purpose of the coating was to reduce surface roughness, which reduces light scattering effects of the 3D printed film. However, while one layer improved performance, a second layer had negative effects. One obvious reason is that increased layering effectively reduced transmittance simply via thickness, despite the initial benefit of reduced light scatter. Further, the use of a second layer may cause negative light-transmittance reflecting effects. For instance, the second layer of the XTC-3D may produce an XTC–air–XTC interface to form causing light to bounce back and forth. This is an outcome of the tendency of XTC-3D to bond to itself in layering processes. Consequently, this would lead to the longer wavelength range being absent in the transmittance spectrum of the double-coated samples.

Figure 7 shows the transmittance spectrum for 200 µm samples of PCL, PCL-MAPbI$_3$, and samples of PCL-MAPbI$_3$-LICA38 at room temperature and after heating at 40 °C for 5 min. Transmittance generally increased for all samples as wavelength increased. Additionally, transmittance increased once the samples were heated except in the case of the PCL-MAPbI$_3$-LICA38, which decreased in transmittance after heating. The maximum transmittance occurred in the PCL sample after heating at 30.4%. The addition of MAPbI$_3$ into PCL reduced the maximum transmittance to 4.9% after heating the sample. While initially seemingly a negative outcome, this reduction in transmittance could be a good indicator of absorbance due to the presence of MAPbI$_3$ micro/nanocrystals.

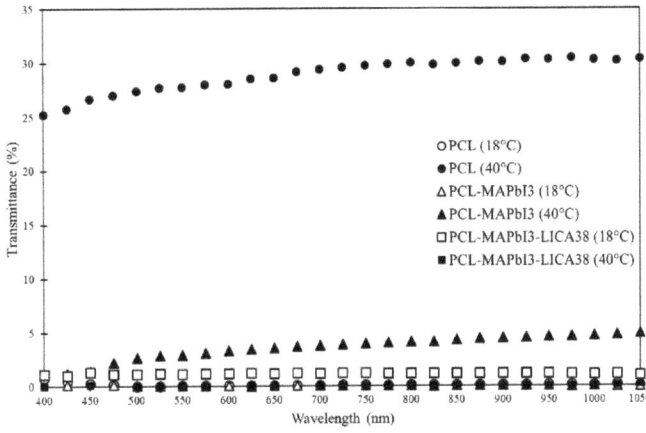

Figure 7. Transmittance spectrum of PCL, PCL-MAPbI$_3$, and PCL-MAPbI$_3$-LICA38.

Figure 8 shows the UV–Vis absorbance spectrum of a plain PCL thin-film and a MAPbI$_3$-PCL thin-film from 350–1000 nm. After the addition of perovskite, the absorbance across all wavelengths was increased. The most prominent increases occurring from 400–800 nm. Given this known absorption range from prior work in the field, this confirmed an active and receptive perovskite material [24–26].

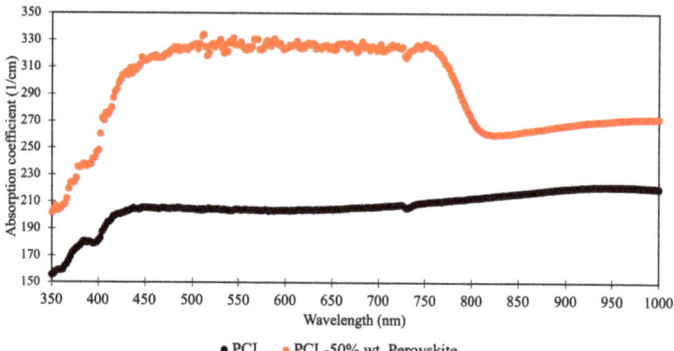

Figure 8. UV–Vis of PCL thin-film and PCL-MAPbI$_3$ thin-film.

3.3. Photocurrent Characterization Results

Following setup and test procedure described in Section 2, photocurrent characterization tests were completed on the fabricated test piece with attached electrodes. Figure 9 shows I-V curve test result of the photoconductivity of the MAPbI$_3$-PCL composite at different power input solar simulator levels. These ranged from 50–80 W. The linear current–voltage curve indicated ohmic contacts were formed at the composite/electrode interfaces. In dark conditions, the resistance was 9.79×10^9 ohms. At 80 W, the resistance was 5.46×10^9 ohms. It is interesting to note that resistance was decreased as lighting intensity was increased (at the same voltage levels). This signified an increase in photocurrent at higher light intensities, which indicated increased electron–hole pair generation. Table 3 further summarizes the results.

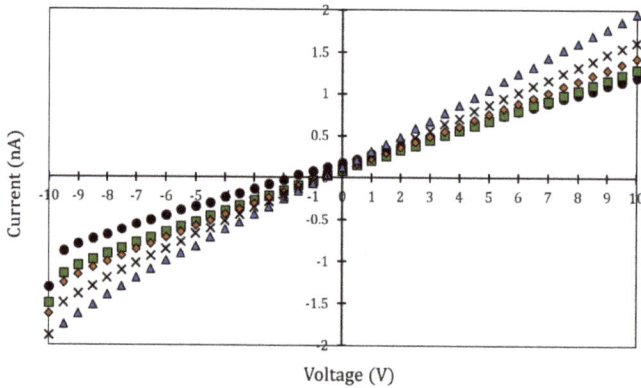

Figure 9. I-V curve of MAPbI$_3$-PCL composite in dark, 50 W, 60 W, 70 W, and 80 W.

Table 3. Resistances of photoconductivity test piece at different power levels.

Input Power (W)	Resistance (M-ohms)
Dark, 0 W	9.79×10^3
50 W	8.29×10^3
60 W	7.58×10^3
70 W	6.62×10^3
80 W	5.46×10^3

In summary, reaction of the new polymer FDM materials showed the promise of an active solar layer even given the fabrication steps that included FDM extrusion. Photo-reaction performance of the material in the thin layer FDM pieces and 3D fabricated photo conductivity test pieces both showed solar absorption capability and electron–hole pair generation. Further, heating the material increased transmission and performance which holds promise for real devices that operate at more elevated temperatures.

4. Conclusions

Perovskite solar cells have generated major research interest since their discovery and offer beneficial properties to solar cell fabrication like long minority charge carrier distance, high light absorption, and simple fabrication methods. Despite the potential for these solar cells, however, stability remains an issue due to material degradation in UV light and other common atmospheric elements. This work investigates the potential for fused deposition modeling (FDM), or 3D printing, to produce perovskite solar cells that can address some of these challenges. Many of these are achieved specifically through the encapsulation and printing process of FDM.

Specific steps have been demonstrated to produce the printable polymer material including $MAPbI_3$ microcrystal fabrication in ambient atmosphere, integration of $MAPbI_3$ into a polymer matrix without degradation, and usage of that $MAPbI_3$-PCL composite to fabricate a solar active material. Outcome of testing included a noted increasing transparency with increasing temperature and hence, improving conductivity when exposed to simulated solar light. Further, increased electron-hole pair generation was observed for 200 µm FDM-printed PCL film, achieving a 45% reduction in resistance under peak incident flux of 590 W/m^2 with the addition of $MAPbI_3$.

While these tests were conducted at relatively conservative simulated solar loadings, additional testing at higher solar loading is planned for future work alongside construction of a full solar stack. Fused deposition modeling establishes opportunity to integrate aligned conductive fillers into this composite to improve charge extraction and power output via integrated electrodes. This demonstrates the feasibility and potential of FDM in solar cell fabrication.

Author Contributions: Conceptualization, L.W. and T.S.; methodology, L.W. and T.S.; formal analysis, L.W. and T.S.; investigation, T.S.; resources, L.W.; data curation, T.S.; writing—original draft preparation, L.W. and T.S.; writing—review and editing, L.W. and T.S.; supervision, L.W.; project administration, L.W. All authors have read and agreed to the published version of the manuscript.

Funding: This research was funded by NSF EPSCoR CIMM project under cooperative agreement #OIA-1541079.

Institutional Review Board Statement: Not applicable.

Informed Consent Statement: Not applicable.

Data Availability Statement: Data archived and maintained by authors.

Conflicts of Interest: The authors declare no conflict of interest.

References

1. EIA. *Annual Energy Outlook 2021 with Projections to 2050, Narrative*; Technical Report; Energy Information Administration: Washington, DC, USA, 2021.
2. Werkheiser, N. *Overview of NASA Initiatives in 3D Printing and Additive Manufacturing*; Technical Report M15-4252; NASA: Washington, DC, USA, 2014.
3. Prater, T.; Werkheiser, N.; Ledbetter, F.; Timucin, D.; Wheeler, K.; Snyder, M. 3D Printing in Zero G Technology Demonstration Mission: Complete experimental results and summary of related material modeling efforts. *Int. J. Adv. Manuf. Technol.* **2019**, *101*, 391–417. [CrossRef]
4. Arora, N.; Dar, M.I.; Hinderhofer, A.; Pellet, N.; Schreiber, F.; Zakeeruddin, S.M.; Grätzel, M. Perovskite solar cells with CuSCN hole extraction layers yield stabilized efficiencies greater than 20%. *Science* **2017**, *358*, 768–771. [CrossRef]
5. Domanski, K.; Alharbi, E.A.; Hagfeldt, A.; Grätzel, M.; Tress, W. Systematic investigation of the impact of operation conditions on the degradation behaviour of perovskite solar cells. *Nat. Energy* **2018**, *3*, 61–67. [CrossRef]

6. Bryant, D.; Aristidou, N.; Pont, S.; Sanchez-Molina, I.; Chotchunangatchaval, T.; Wheeler, S.; Durrant, J.R.; Haque, S.A. Light and oxygen induced degradation limits the operational stability of methylammonium lead triiodide perovskite solar cells. *Energy Environ. Sci.* **2016**, *9*, 1655–1660. [CrossRef]
7. Moiz, S.A.; Alahmadi, A.N.M. Design of Dopant and Lead-Free Novel Perovskite Solar Cell for 16.85 Efficiency. *Polymers* **2021**, *13*, 2110. [CrossRef]
8. Dong, Q.; Liu, F.; Wong, M.K.; Tam, H.W.; Djurišić, A.B.; Ng, A.; Surya, C.; Chan, W.K.; Ng, A.M.C. Encapsulation of perovskite solar cells for high humidity conditions. *ChemSusChem* **2016**, *9*, 2597–2603. [CrossRef]
9. Han, Y.; Meyer, S.; Dkhissi, Y.; Weber, K.; Pringle, J.M.; Bach, U.; Spiccia, L.; Cheng, Y.B. Degradation observations of encapsulated planar $CH_3NH_3PbI_3$ perovskite solar cells at high temperatures and humidity. *J. Mater. Chem. A* **2015**, *3*, 8139–8147. [CrossRef]
10. Bush, K.A.; Palmstrom, A.F.; Zhengshan, J.Y.; Boccard, M.; Cheacharoen, R.; Mailoa, J.P.; McMeekin, D.P.; Hoye, R.L.; Bailie, C.D.; Leijtens, T.; et al. 23.6%-efficient monolithic perovskite/silicon tandem solar cells with improved stability. *Nat. Energy* **2017**, *2*, 1–7. [CrossRef]
11. Yang, F.; Kapil, G.; Zhang, P.; Hu, Z.; Kamarudin, M.A.; Ma, T.; Hayase, S. Dependence of acetate-based antisolvents for high humidity fabrication of $CH_3NH_3PbI_3$ perovskite devices in ambient atmosphere. *ACS Appl. Mater. Interfaces* **2018**, *10*, 16482–16489. [CrossRef]
12. Johansson, M.B.; Edvinsson, T.; Bitter, S.; Eriksson, A.I.; Johansson, E.M.; Göthelid, M.; Boschloo, G. From quantum dots to micro crystals: Organolead triiodide perovskite crystal growth from isopropanol solution. *ECS J. Solid State Sci. Technol.* **2016**, *5*, P614. [CrossRef]
13. Li, M.; Yan, X.; Kang, Z.; Liao, X.; Li, Y.; Zheng, X.; Lin, P.; Meng, J.; Zhang, Y. Enhanced efficiency and stability of perovskite solar cells via anti-solvent treatment in two-step deposition method. *ACS Appl. Mater. Interfaces* **2017**, *9*, 7224–7231. [CrossRef]
14. You, J.; Meng, L.; Song, T.B.; Guo, T.F.; Yang, Y.M.; Chang, W.H.; Hong, Z.; Chen, H.; Zhou, H.; Chen, Q.; et al. Improved air stability of perovskite solar cells via solution-processed metal oxide transport layers. *Nat. Nanotechnol.* **2016**, *11*, 75–81. [CrossRef]
15. Sonsalla, T.; Moore, A.; Radadia, A.; Weiss, L. Printer orientation effects and performance of novel 3-D printable acrylonitrile butadiene styrene (ABS) composite filaments for thermal enhancement. *Polym. Test.* **2019**, *80*, 106125. [CrossRef]
16. Zhao, Z.; Li, Y.; Du, Y.; Zhang, L.; Wei, J.; Lin, F. Preparation and testing of anisotropic $MAPbI_3$ perovskite photoelectric sensors. *ACS Appl. Mater. Interfaces* **2020**, *12*, 44248–44255. [CrossRef]
17. Tai, C.L.; Hong, W.L.; Kuo, Y.T.; Chang, C.Y.; Niu, M.C.; Ochathevar, M.K.P.; Hsu, C.L.; Horng, S.F.; Chao, Y.C. Ultrastable, Deformable, and Stretchable Luminescent Organic–Inorganic Perovskite Nanocrystal–Polymer Composites for 3D Printing and White Light-Emitting Diodes. *ACS Appl. Mater. Interfaces* **2019**, *11*, 30176–30184. [CrossRef] [PubMed]
18. Zhang, Y.; Zhao, Y.; Wu, D.; Xue, J.; Qiu, Y.; Liao, M.; Pei, Q.; Goorsky, M.S.; He, X. Homogeneous freestanding luminescent perovskite organogel with superior water stability. *Adv. Mater.* **2019**, *31*, 1902928. [CrossRef] [PubMed]
19. Chen, M.; Yang, J.; Wang, Z.; Xu, Z.; Lee, H.; Lee, H.; Zhou, Z.; Feng, S.P.; Lee, S.; Pyo, J.; et al. 3D nanoprinting of perovskites. *Adv. Mater.* **2019**, *31*, 1904073. [CrossRef]
20. Qaid, S.M.H.; Ghaithan, H.M.; AlHarbi, K.K.; Al-Asbahi, B.A.; Aldwayyan, A.S. Enhancement of Light Amplification of $CsPbBr_3$ Perovskite Quantum Dot Films via Surface Encapsulation by PMMA Polymer. *Polymers* **2021**, *13*, 2574. [CrossRef]
21. Fiedler, T.; Belova, I.; Murch, G.; Poologasundarampillai, G.; Jones, J.; Roether, J.; Boccaccini, A. A comparative study of oxygen diffusion in tissue engineering scaffolds. *J. Mater. Sci. Mater. Med.* **2014**, *25*, 2573–2578. [CrossRef] [PubMed]
22. Cava, D.; Giménez, E.; Gavara, R.; Lagaron, J. Comparative performance and barrier properties of biodegradable thermoplastics and nanobiocomposites versus PET for food packaging applications. *J. Plast. Film Sheeting* **2006**, *22*, 265–274. [CrossRef]
23. Sanchez-Garcia, M.; Ocio, M.; Gimenez, E.; Lagaron, J. Novel polycaprolactone nanocomposites containing thymol of interest in antimicrobial film and coating applications. *J. Plast. Film Sheeting* **2008**, *24*, 239–251. [CrossRef]
24. Tang, X.; Brandl, M.; May, B.; Levchuk, I.; Hou, Y.; Richter, M.; Chen, H.; Chen, S.; Kahmann, S.; Osvet, A.; et al. Photoinduced degradation of methylammonium lead triiodide perovskite semiconductors. *J. Mater. Chem. A* **2016**, *4*, 15896–15903. [CrossRef]
25. Roldán-Carmona, C.; Gratia, P.; Zimmermann, I.; Grancini, G.; Gao, P.; Graetzel, M.; Nazeeruddin, M.K. High efficiency methylammonium lead triiodide perovskite solar cells: The relevance of non-stoichiometric precursors. *Energy Environ. Sci.* **2015**, *8*, 3550–3556. [CrossRef]
26. Abdelmageed, G.; Mackeen, C.; Hellier, K.; Jewell, L.; Seymour, L.; Tingwald, M.; Bridges, F.; Zhang, J.Z.; Carter, S. Effect of temperature on light induced degradation in methylammonium lead iodide perovskite thin films and solar cells. *Sol. Energy Mater. Sol. Cells* **2018**, *174*, 566–571. [CrossRef]
27. Acuna, D.; Krishnan, B.; Shaji, S.; Sepulveda, S.; Menchaca, J.L. Growth and properties of lead iodide thin films by spin coating. *Bull. Mater. Sci.* **2016**, *39*, 1453–1460. [CrossRef]
28. Song, Z.; Watthage, S.C.; Phillips, A.B.; Tompkins, B.L.; Ellingson, R.J.; Heben, M.J. Impact of processing temperature and composition on the formation of methylammonium lead iodide perovskites. *Chem. Mater.* **2015**, *27*, 4612–4619. [CrossRef]
29. Diba, M.; Fathi, M.; Kharaziha, M. Novel forsterite/polycaprolactone nanocomposite scaffold for tissue engineering applications. *Mater. Lett.* **2011**, *65*, 1931–1934. [CrossRef]
30. Augustine, R.; Malik, H.N.; Singhal, D.K.; Mukherjee, A.; Malakar, D.; Kalarikkal, N.; Thomas, S. Electrospun polycaprolactone/ZnO nanocomposite membranes as biomaterials with antibacterial and cell adhesion properties. *J. Polym. Res.* **2014**, *21*, 1–17. [CrossRef]

31. Barud, H.S.; Ribeiro, S.J.; Carone, C.; Ligabue, R.; Einloft, S.; Queiroz, P.; Borges, A.P.; Jahno, V. Optically transparent membrane based on bacterial cellulose/polycaprolactone. *Polímeros* **2013**, *23*, 135–142. [CrossRef]
32. Christensen, P.; Egerton, T.; Martins-Franchetti, S.; Jin, C.; White, J. Photodegradation of polycaprolactone/poly (vinyl chloride) blend. *Polym. Degrad. Stab.* **2008**, *93*, 305–309. [CrossRef]
33. Balu, R.; Kumar, T.; Ramalingam, M.; Ramakrishna, S. Electrospun Polycaprolactone/Poly (1,4-butylene adipate-co-polycaprolactam) blends: Potential biodegradable scaffold for bone tissue regeneration. *J. Biomater. Tissue Eng.* **2011**, *1*, 30–39. [CrossRef]
34. Yahiaoui, F.; Benhacine, F.; Ferfera-Harrar, H.; Habi, A.; Hadj-Hamou, A.S.; Grohens, Y. Development of antimicrobial PCL/nanoclay nanocomposite films with enhanced mechanical and water vapor barrier properties for packaging applications. *Polym. Bull.* **2015**, *72*, 235–254. [CrossRef]

Communication

Synthesis of a Low-Cost Thiophene-Indoloquinoxaline Polymer Donor and Its Application to Polymer Solar Cells

Yiping Guo [1], Zeyang Li [1], Mengzhen Sha [2], Ping Deng [1,3,*], Xinyu Lin [1], Jun Li [1], Liang Zhang [1], Hang Yin [2,*] and Hongbing Zhan [1]

1. College of Materials Science and Engineering, Fuzhou University, Fuzhou 350108, China; gyp010601@163.com (Y.G.); n191820032@fzu.edu.cn (Z.L.); lxy153694@163.com (X.L.); lj03040010@163.com (J.L.); 18379452130@139.com (L.Z.); hbzhan@fzu.edu.cn (H.Z.)
2. State Key Laboratory of Crystal Materials, School of Physics, Shandong University, Jinan 250100, China; 202112136@mail.sdu.edu.cn
3. Key Laboratory of Eco-materials Advanced Technology Fuzhou University, Fuzhou 350108, China
* Correspondence: pingdeng@fzu.edu.cn (P.D.); hyin@sdu.edu.cn (H.Y.)

Abstract: A simple wide-bandgap conjugated polymer based on indoloquinoxaline unit (**PIQ**) has been newly designed and synthesized via cheap and commercially available starting materials. The basic physicochemical properties of the **PIQ** have been investigated. **PIQ** possesses a broad and strong absorption band in the wavelength range of 400~660 nm with a bandgap of 1.80 eV and lower-lying highest occupied molecular orbital energy level of −5.58 eV. Polymer solar cells based on **PIQ** and popular acceptor Y6 blend display a preliminarily optimized power conversion efficiency of 6.4%. The results demonstrate indoloquinoxaline is a promising building unit for designing polymer donor materials for polymer solar cells.

Keywords: indoloquinoxaline; low-cost polymer donor; wide-bandgap polymer; polymer solar cells

1. Introduction

Polymer solar cells (PSCs) are attractive as a promising new energy device for solar-to-electric conversion [1]. In a typical device, the active-layer blending film consists of a donor material and an acceptor material [2,3]. One of the most successful blends in recent years contains a p-type polymer as the donor and an n-type non-fullerene molecule as the acceptor [4]. Thanks to rational molecular design and device optimization [5], the power conversion efficiencies (PCEs) of PSCs have consistently improved [6]. However, one issue that must be critically considered is the cost of the active-layer materials [7,8]. Because of the complicated molecular structures, tedious multi-step organic synthesis, and laborious purifications, the costs of the efficient active-layer materials reported so far were too high to meet commercial application of PSCs [9]. Therefore, developing low-cost and efficient active-layer materials is one of the key challenges for the application of PSCs [10,11].

Recently, a low-cost and high-performance polymer donor, PTQ10 [12], has been demonstrated as a promising polymer donor for commercial application of PSCs. Compared to the classical benzo[1,2-b:4,5-b']dithiophene (BDT)-based polymers [13–15] (see Figure 1a) synthesized via multi-step synthesis, PTQ10 has a very simple molecular structure (see Figure 1b), and it can be synthesized via simple two-step reactions with cheap raw materials. Low-cost and efficient polymer donors have gained relative less attention in recent years, and only a few polymers with these features have been developed until now [9]. We had reviewed and summarized the representative low-cost and efficient polymer donors (see Table S1 and Figure S7). It was shown that these types of polymers are promising donor materials for high-performance PSCs [16–22]. Therefore, we expected to develop new low-cost and efficient polymer donors.

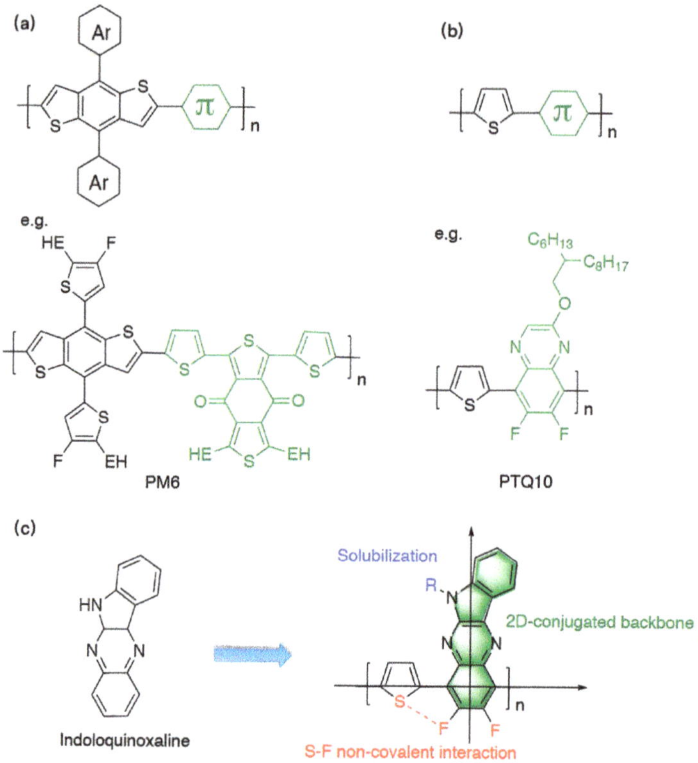

Figure 1. (a) BDT-based and (b) thiophene-based polymer donors; (c) molecular design strategy of the indoloquinoxaline-based polymer.

Indolo[2,3-*b*] quinoxaline (IQ) is a unique planar built-in donor–acceptor heterocyclic unit that can be considered as the fusion of electron-deficient quinoxaline and electron-rich indole. Some IQ small molecular derivatives have been applied as promising multifunctional anti-Alzheimer agents [23], photosensitizers [24–26], hole injection-layer materials [27] and non-fullerene acceptors [28]. In this work, we designed an IQ-based polymeric p-type semiconductor material (**PIQ**) for polymer solar cells. The molecular design strategy is shown in Figure 1c. This polymer contains simple thiophene and difluorine-substituted IQ units with two-dimensional (2D) conjugated backbone. The 2D conjugated structure is favorable for intermolecular carrier transporting [29–31]. The fluorination of the IQ unit is to improve molecular planarity via S···F non-covalent interactions and further enhance carrier transporting [32–35]. The alkyl side chain on the IQ unit is to ensure good solubility.

2. Materials and Methods
2.1. Materials

9-(Iodomethyl)nonadecane (97%, Lyntech), 3,6-Dibromo-4,5-difluorobenzene-1,2-diamine (98%, Zhengzhou Ruke Biological, Zhengzhou, China), indoline-2,3-dione (97%, Rhawn), and 2,5-bis(trimethylstannyl)thiophene (99%, bidepharm), potassium carbonate (K_2CO_3, 98%, Aladdin), *N*,*N*-Dimethylformamide (DMF, AR, 99.5%, Aladdin), toluene (99.5%, Aladdin), acetic acid (CH_3COOH, 99.7%, Aladdin), Tris(dibenzylideneacetone)dipalladium(0) ($Pd_2(dba)_3$, 97%, Aladdin), Tri(*o*-tolyl)phosphine (P(*o*-tolyl)$_3$, 97%, Aladdin), calcium hydride (95%, Aladdin), molecular sieves (3Å, Aladdin), 2,2'-((2Z,2'Z)-((12,13-

bis(2-ethylhexyl)-3,9-diundecyl-12,13-dihydro-[1,2,5]thiadiazolo-[3,4-e]thieno[2″,3″:4′,5′]-thieno[2′,3′:4,5]pyrrolo[3,2-g]thieno[2′,3′:4,5]thieno[3,2-b]indole-2,10-diyl)bis(methanylylidene))bis(5,6-difluoro-3-oxo-2,3-dihydro-1H-inde (Y6, 98%, Zhengzhou Alfachem Co., Ltd., Zhengzhou, China) were used as received. Toluene was distilled over calcium hydride under an argon atmosphere and was then dried with 3Å molecular sieves. The detailed synthesis routes are shown in Scheme 1.

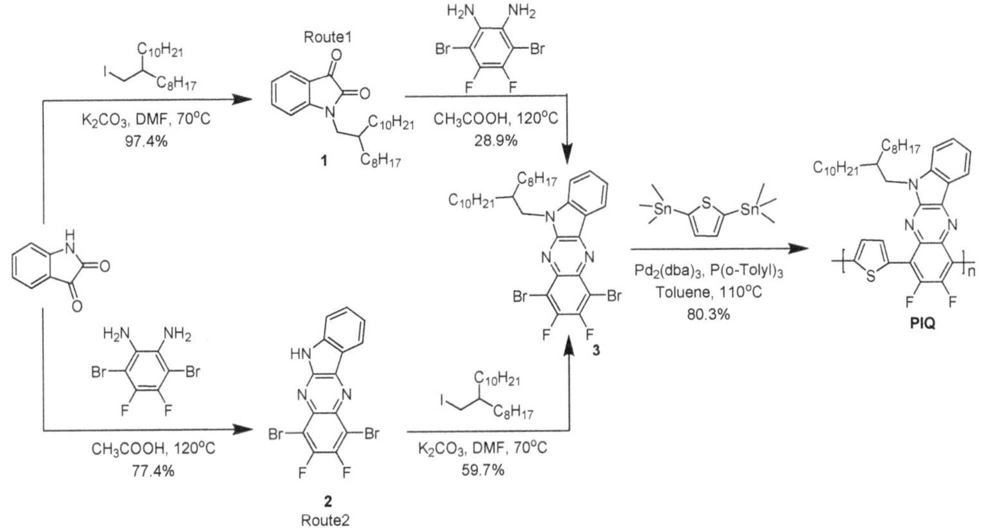

Scheme 1. Synthetic routes of the **PIQ**.

2.2. Synthesis of 1,4-Dibromo-2,3-difluoro-6-(2-octyldodecyl)-6H-indolo[2,3-b]quinoxaline

[Route 1] 1-(2-octyldodecyl) indoline-2,3-dione (**1**) [35] was synthesized according to literature procedures. Quantities of 3,6-dibromo-4,5-difluorobenzene-1,2-diamine (0.66 mmol, 0.2033 g) and 1-(2-octyldodecyl)indoline-2,3-dione (0.54 mmol, 0.2309 g) were added to a Schlenk reaction flask (38 mL) under an argon atmosphere, followed by the addition of deoxygenated acetic acid (3.5 mL), and the reaction was carried out at 120 °C for 16 h. After the reaction was cooled to room temperature, the mixture was poured into cold water (100 mL). It was then extracted with dichloromethane (50 mL × 3). The combined organic layers were washed with water and brine then dried over anhydrous magnesium sulfate. After removing the solvent, the crude product was purified by flash column chromatography (silica gel, dichloromethane: petroleum ether = 1:2, v/v) to afford the titled compound (0.1083 g, 28.9%) as a yellow solid.

[Route 2] Indoline-2,3-dione (0.66 mmol, 0.0991 g) and 3,6-dibromo-4,5-difluorobenzene-1,2-diamine (0.55 mmol, 0.1694 g) were added to a Schlenk reaction flask (38 mL) under argon atmosphere, followed by the addition of deoxygenated acetic acid (1.8 mL), and the reaction was carried out at 120 °C for 16 h. After the reaction was cooled to room temperature, the mixture was poured into water (100 mL). The precipitate was filtered and then washed with methanol (5 mL × 4) and dried under vacuum to obtain the 1,4-dibromo-2,3-difluoro-6H-indolo[2,3-b]quinoxaline (0.1759 g, 77.4%) as a yellow solid. Next, 1,4-dibromo-2,3-difluoro-6H-indolo[2,3-b]- quinoxaline was transferred to a double-necked flask (250 mL) under argon atmosphere. K_2CO_3 (0.86 mmol, 0.1189 g), and deoxygenated DMF (2 mL) were added. Subsequently, 9-(iodomethyl)nonadecane (0.65 mmol, 0.2709 g) was added slowly dropwise. The reaction was carried out at 70 °C for 21 h. After the reaction was cooled to room temperature, the product was poured into water (100 mL). It was then extracted with dichloromethane (50 mL × 3). The combined organic layers were washed with water and brine and dried over anhydrous magnesium sulfate. After removing the solvent, the crude product was purified by flash column chromatography (silica gel, dichloromethane: petroleum ether = 1:4, v/v) to afford the titled compound (0.1779 g, 59.7%) as a yellow solid.

^{1}**H NMR** (400 MHz, CDCl$_3$, ppm): δ 8.55 (d, J = 7.7 Hz, 1H), 7.75 (t, J = 7.7 Hz, 1H), 7.50 (d, J = 8.2 Hz, 1H), 7.43 (t, J = 7.6 Hz, 1H), 4.41 (d, J = 7.4 Hz, 2H), 2.31–2.22 (m, 1H), 1.51–1.47 (m, 2H), 1.28–1.18 (m, 30H), 0.86 (q, J = 7.1 Hz, 6H).

^{13}**C NMR** (125 MHz, CDCl$_3$, ppm): δ 150.60, 148.80, 147.12, 146.09, 145.22, 140.65, 135.75, 133.58, 132.02, 123.74, 121.56, 118.81, 110.24, 108.14, 46.30, 37.26, 32.05, 31.76, 30.08, 29.85, 29.75, 29.68, 29.63, 29.46, 26.36, 22.83, 14.27.

2.3. Synthesis of the Polymer PIQ

Quantities of 1,4-dibromo-2,3-difluoro-6-(2-octyldodecyl)-6H-indolo[2,3-b]quinoxaline (0.2 mmol, 0.1387 g), 2,5-bis(trimethylstannyl)thiophene (0.2 mmol, 0.0828 g), and toluene (6 mL) were added to a oven-dried Schlenk tube (100 mL) under argon atmosphere. The mixture was degassed with argon for 30 min. Next, Pd$_2$(dba)$_3$ (0.004 mmol, 0.0037 g) and P(o-tolyl)$_3$ (0.016 mmol, 0.009 g) were added. After being degassed with argon for another 10 min, the tube was sealed. The tube was placed in a 110 °C oil bath. After 48 h, it was cooled down to room temperature. The reaction mixture was poured into stirring methanol to precipitate the crude product. The precipitate was collected by filtration and was further purified by sequential Soxhlet extractions with methanol and petroleum ether. The residue after Soxhlet extractions was then extracted with chloroform. The chloroform solution was re-precipitated with methanol. The resulting solid was collected and then dried to obtain the title polymer (0.0989 g, 80.3%) as a purple-black solid.

^{1}**H NMR** (600 MHz, CDCl$_3$, ppm): δ 8.43(br, 2H), 7.55–7.08(br, 4H), 4.27 (br, 2H), 2.02–0.88 (br, 38H).

GPC (THF): M_n = 7.1 kDa, Đ = 1.98.

T_d (5% loss) = 464 °C.

2.4. Device Fabrication and Characterization

The OPV device structure was set to ITO/PEDOT:PSS/**PIQ**:Y6/PDINN/Ag. The ITO glass substrates were ultrasonicated in deionized water with various reagents (acetone and 1,2-propanol), and dried in ambient atmosphere for 15 h. The dried glass substrates were treated with UV ozone for 20 min, and the PEDOT:PSS layers were spin-coated onto substrates at 7000 rpm for 60 s. The PEDOT:PSS layers had a thickness of 30 nm. Next, the film underwent an annealing process in the air at 150 °C for 15 min. The substrates were transferred into an Ar-filled glove box to spin-coat the active layers. The active layer materials **PIQ** and Y6 were dissolved in chloroform with a 1:1 weight ratio at a total concentration of 16.8 mg/mL. The solution of the **PIQ** and Y6 was subsequently spin-coated onto the hole transport layer (PEDOT: PSS), and the spin speed was 2000 rpm for 50s, to form ca. 80 nm uniform active layers. After that, the active layer needed to anneal for 12 min at 100 °C in the vacuum glove box. Finally, a thin PDINN layer (ca. 1 nm) and Ag (ca. 120 nm) were evaporated in a high vacuum chamber (ca. 4×10^{-6} torr). After this step, the device can be used for corresponding characterizations. Under AM1.5 solar illumination, J-V curves were measured by PV Test Solutions solar simulator. The external quantum efficiency (EQE) of the solar cells was tested using Zolix SolarCellScan 100.

3. Results and Discussion

PIQ can be synthesized with low cost via a three-step reaction from cheap raw materials. Two synthetic routes were explored to obtain monomer **3** with indoline-2,3-dione as the cheap raw material. N-alkylation reaction between 1-(2-octyldodecyl)indoline-2,3-dione and 9-(iodomethyl)nonadecane was used to synthesize compound **1** in a high yield of 97.4%. The acetic acid-catalyzed condensation reaction between compound **1** and 3,6-dibromo-4,5- difluorobenzene-1,2-diamine was conducted to synthesize monomer **3**

in a low yield of 28.9%. Thus, monomer **3** was obtained with a low overall yield of 28% through this synthetic route. An improved route is to conduct the acetic acid-catalyzed condensation reaction followed by the N-alkylation reaction, as monomer **3** could be obtained with a reasonable overall yield of 46%. The Stille cross-coupling polycondensation of 2,5-bis(trimethylstannyl) thiophene and monomer **3** was performed to gain the target polymer **PIQ** as a purple-black solid (80.3% yield). The number average molecular weight and polydispersity index for **PIQ** were 7.1 kDa and 1.98, respectively, and **PIQ** had good solubility in common organic solvents. We performed synthesis cost calculations of the polymer **PIQ** using the model developed by Li et al. [36], which can be used as a rough indication of synthetic complexity. The results were displayed in Supporting Information (see Table S2). The cost of **PIQ** synthesized via route 1 is approximately 414.3 ¥/g, whereas the cost of **PIQ** synthesized via route 2 is approximately 241.1 ¥/g. The latter is significantly lower than the former, indicating that route 2 is the preferable route. As shown in Table S3, the synthesis cost of **PIQ** is compared to those of some famous polymer donors (e.g., PTQ10, PBDB-T, and PM6) [37–39].

3.1. Optical Properties

To study the optical properties of polymer **PIQ**, the UV-vis absorption spectra of monomer **3** and **PIQ** were tested. The photograph and absorption spectra of monomer **3** and polymer **PIQ** in dilute chlorobenzene solutions are shown in Figure 2a. **PIQ** solution exhibited absorption edge at 663 nm, which red-shifted over 188 nm relative to the monomer **3** solution. Introduction of the electron-donating thiophene to conjugated backbone can significantly enhance electronic delocalization along the chain axis via intramolecular charge transfer [40–42]. The monomer **3** solution has a strong absorption at 400–470 nm with a maximum absorption coefficient (ε) of 1.25×10^5 M^{-1} cm^{-1}, whereas the polymer **PIQ** solution has a much larger range of absorption, showing strong absorption in the 400–660 nm range with a slightly higher maximum absorption coefficient of 1.37×10^5 M^{-1} cm^{-1} (Figure 2b). The absorption spectra of **PIQ** as a thin film is also displayed in Figure 2a. Compared to that of its solution, a distinct red shift by 22 nm was observed due to stronger aggregation in a solid state. The bandgap of **PIQ** as a thin film was estimated to be 1.80 eV, which could be comparable to that of PTQ10 and matched well with the typical low-bandgap acceptor of Y6 (see Figure S9 for its molecular structure) to show a complementary absorption [12,43]. The optical properties of **PIQ** are summarized in Table 1.

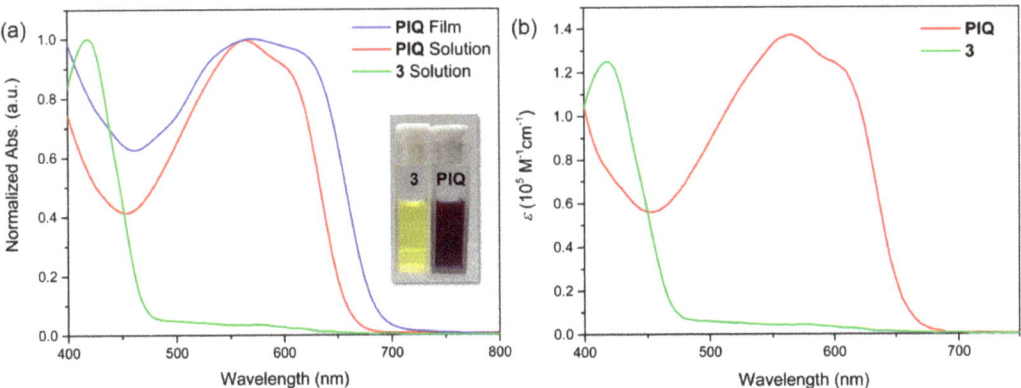

Figure 2. (**a**) Normalized UV-vis absorption spectra of monomer **3** and polymer **PIQ** in chlorobenzene solutions and **PIQ** as thin film (insert, photograph); (**b**) UV-vis absorption spectra of monomer **3** and **PIQ** in chlorobenzene solutions.

Table 1. The optical and electrochemical properties of **PIQ**.

Polymer	λ_{max}^{sol} (nm)	λ_{max}^{film} (nm)	λ_{onset}^{film} (nm)	E_g^{opt} [1] (eV)	E_{red}/E_{LUMO} [2] (V/eV)	E_{ox}/E_{HOMO} [3] (V/eV)
PIQ	564	570	687	1.80	−1.20/−3.51	0.87/−5.58

[1] Calculated by the equation: $E_g^{opt} = \frac{1240}{\lambda_{onset}^{film}}\ eV$; [2] $E_{LUMO} = -e(E_{red} + 4.71)\ eV$; [3] $E_{HOMO} = -e(E_{ox} + 4.71)\ eV$.

3.2. Electrochemical Properties

The electronic energy levels of **PIQ** were measured by electrochemical cyclic voltammetry (Figure 3). The highest occupied molecular orbital (HOMO) and lowest unoccupied molecular (LUMO) levels of **PIQ** were estimated to be −5.58/−3.51 eV from the first onset oxidation and onset reduction potentials, respectively. The electrochemical properties of **PIQ** are also summarized in Table 1. The value of the electrochemical band gap for **PIQ** thin film was found to be 2.07 eV, which was larger than that of its optical band gap (1.8 eV). This may be due to the exciton binding energy for conjugated polymers [44].

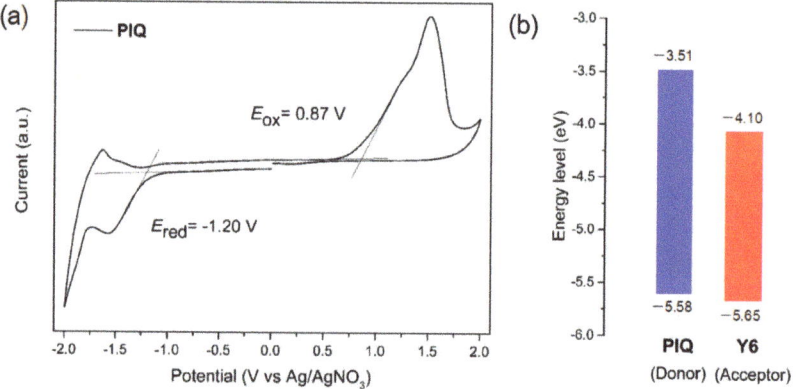

Figure 3. (a) Cyclic voltammetry curve of the **PIQ** thin film; (b) energy level diagram of the **PIQ** donor and Y6 acceptor.

3.3. Thermal Properties and X-ray Diffraction Characterization

The thermal stability of the **PIQ** polymer was tested by thermogravimetric analysis by taking approximately 6 mg of sample and placing it in an alumina ceramic crucible under nitrogen protection at a temperature increase rate of 20 °C/min up to 600 °C. The mass change of the sample at different temperatures was observed by heating. Organic polymer semiconductor materials can be considered to have good thermal stability when the temperature of 5% thermal weight loss is above 300 °C, which fully meets the requirements of optoelectronic device construction and testing. The temperature of 5% thermal weight loss of **PIQ** was 464 °C (see Figure S6), indicating that **PIQ** has good thermal stability.

To investigate the crystallinity of **PIQ** film, the X-ray diffraction (XRD) measurement was performed on a drop-cast film of **PIQ** (Figure S6). The sample showed distinct 100 peak at 5.17°, corresponding to a lamellar distance of 17.08 Å.

3.4. Photovoltaic Properties and Photoluminescence Characterization

To study the photovoltaic properties of PIQ, we fabricated BHJ polymer solar cells with a device structure of ITO/PEDOT:PSS/**PIQ**:Y6/PDINN/Ag (Figure 4c). The corresponding energy level diagram of the related materials is shown in Figure 4d. The polymer **PIQ** and Y6 were dissolved in chloroform. Devices with a donor/acceptor (D/A) ratio of 1:1 were fabricated. As illustrated in Figure 4a, a power conversion efficiency (PCE) of 6.41% was achieved with the fill factor (FF) of 46.6%, combined with the J_{SC} of 18.65 mA/cm^2)

and V_{OC} of 0.737 V. The J_{SC} value of polymer solar cells can be confirmed by the external quantum efficiency (EQE) measurement, and the result is shown in Figure 4b. Thin-film photoluminescence (PL) spectra of **PIQ**, **PIQ**:Y6 blend were measured (Figure 5). Blending PIQ with Y6 results in strong fluorescence quenching, indicating efficient photo-induced charge transfer [2,45] between **PIQ** and Y6 in blend.

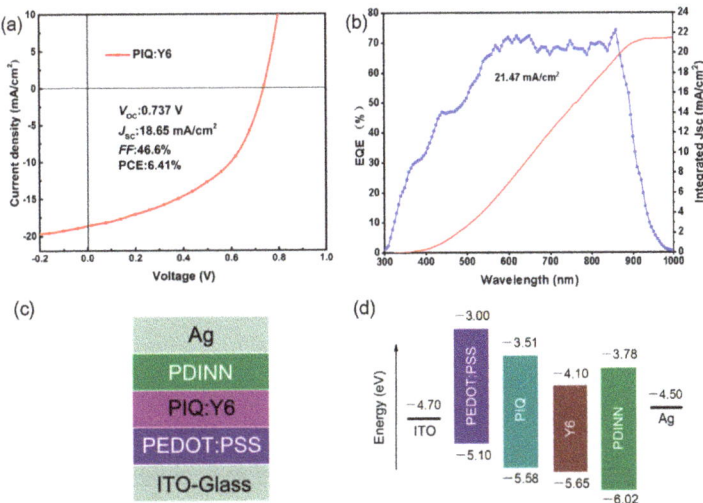

Figure 4. (**a**) *J-V* curve of the **PIQ**: Y6 blend-based polymer solar cells; (**b**) EQE spectrum of the PIQ: Y6 blend-based polymer solar cells; (**c**) schematic diagram of the device structure; (**d**) energy level diagram of the related materials.

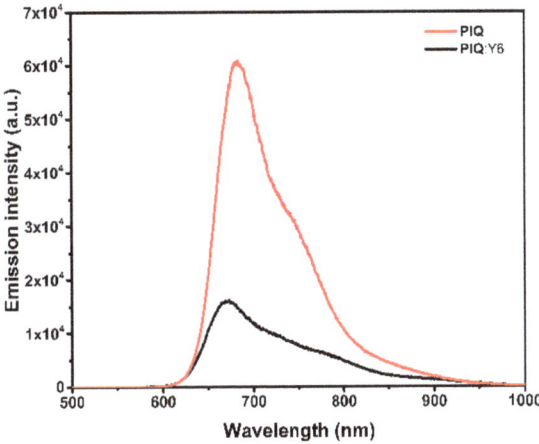

Figure 5. Thin-film photoluminescence spectra of **PIQ** and **PIQ**: Y6 blend.

4. Conclusions

In summary, a new polymer donor, **PIQ**, has been developed. **PIQ** can be easily gained via a simple three-step reaction from cheap raw materials with reasonable overall yield. **PIQ** has a medium bandgap of 1.80 eV, a broad and strong absorption feature in the wavelength range of 400~650 nm, and a low-lying HOMO energy level. The PSCs based on binary blend with **PIQ** as donor and Y6 as acceptor displayed a reasonable PCE of 6.41%. We believe that the tuning of physicochemical properties of the **PIQ** via optimization of conjugated backbones and side chains and its polymerization reaction may bring about further improvement in photovoltaic performance. We have developed the indoloquinoxaline-based polymer as the donor material for organic solar cells, and we also believe that indoloquinoxaline-based polymers can be promising low-cost and efficient polymer donor photovoltaic materials.

Supplementary Materials: The following supporting information can be downloaded at: https://www.mdpi.com/article/10.3390/polym14081554/s1, Figures S1–S3: NMR spectrum; Figure S4: GPC test result of the polymer **PIQ**; Figure S5: TGA curve of the polymer **PIQ**; Figure S6: XRD pattern of the polymer **PIQ** thin film; Figure S7: Chemical structures of polymer donors involved in Table S1; Figure S8: Chemical structures of PBDB-T and PBDB-T-2F; Figure S9: Chemical structures of Y6; Table S1: Survey of polymer solar cells based on some representative low-cost and efficient donor polymers materials; Table S2: Survey of calculated chemical synthesis costs for **PIQ**; Table S3: Comparison of the synthetic steps and synthesis costs for polymer donor materials.

Author Contributions: Y.G.—methodology, investigation, writing—original draft; Z.L.—methodology, investigation, writing—original draft; M.S.—methodology, investigation; P.D.—project administration, funding acquisition, conceptualization, supervision, writing—review and editing; X.L.—methodology, investigation; J.L.—methodology, investigation; L.Z.—methodology, investigation; H.Y.—project administration, funding acquisition, supervision, writing—review and editing. H.Z.—project administration, resources, supervision. All authors have read and agreed to the published version of the manuscript.

Funding: This work was supported by (1) the Natural Science Foundation of Fujian (2021J01596); (2) the National Undergraduate Training Program for Innovation and Entrepreneurship (202110386033); (3) the Shandong Provincial Natural Science Foundation (ZR2021QF016).

Institutional Review Board Statement: Not applicable.

Informed Consent Statement: Not applicable.

Data Availability Statement: The data presented in this study are available on request from the corresponding author.

Conflicts of Interest: The authors declare no conflict of interest.

References

1. Gopalakrishnan, V.; Balaji, D.; Dangate, M.S. Review-Conjugated Polymer Photovoltaic Materials: Performance and Applications of Organic Semiconductors in Photovoltaics. *ECS J. Solid State Sci. Technol.* **2022**, *11*, 035001. [CrossRef]
2. Lim, I.; Hoa Thi, B.; Shrestha, N.K.; Lee, J.K.; Han, S.-H. Interfacial Engineering for Enhanced Light Absorption and Charge Transfer of a Solution-Processed Bulk Heterojunction Based on Heptazole as a Small Molecule Type of Donor. *ACS Appl. Mater. Inter.* **2016**, *8*, 8637–8643. [CrossRef] [PubMed]
3. Nitschke, P.; Jarząbek, B.; Vasylieva, M.; Godzierz, M.; Janeczek, H.; Musioł, M.; Domiński, A. The Effect of Alkyl Substitution of Novel Imines on Their Supramolecular Organization, Towards Photovoltaic Applications. *Polymers* **2021**, *13*, 1043. [CrossRef] [PubMed]
4. Luo, D.; Jang, W.; Babu, D.D.; Kim, M.S.; Wang, D.H.; Kyaw, A.K.K. Recent Progress in Organic Solar Cells Based on Non-Fullerene Acceptors: Materials to Devices. *J. Mater. Chem. A* **2022**, *10*, 3255–3295. [CrossRef]
5. Li, Y.; Huang, W.; Zhao, D.; Wang, L.; Jiao, Z.; Huang, Q.; Wang, P.; Sun, M.; Yuan, G. Recent Progress in Organic Solar Cells: A Review on Materials from Acceptor to Donor. *Molecules* **2022**, *27*, 1800. [CrossRef]
6. Zheng, Z.; Wang, J.; Bi, P.; Ren, J.; Wang, Y.; Yang, Y.; Liu, X.; Zhang, S.; Hou, J. Tandem Organic Solar Cell with 20.2% Efficiency. *Joule* **2022**, *6*, 171–184. [CrossRef]
7. Xue, R.; Zhang, J.; Li, Y.; Li, Y. Organic Solar Cell Materials toward Commercialization. *Small* **2018**, *14*, 1801793. [CrossRef]
8. Li, S.; Yuan, X.; Zhang, Q.; Li, B.; Li, Y.; Sun, J.; Feng, Y.; Zhang, X.; Wu, Z.; Wei, H.; et al. Narrow-Bandgap Single-Component Polymer Solar Cells with Approaching 9% Efficiency. *Adv. Mater.* **2021**, *33*, 2101295. [CrossRef]
9. Zhao, F.; Zhou, J.; He, D.; Wang, C.; Lin, Y. Low-Cost Materials for Organic Solar Cells. *J. Mater. Chem. C* **2021**, *9*, 15395–15406. [CrossRef]
10. Jessop, I.A.; Chong, A.; Graffo, L.; Camarada, M.B.; Espinoza, C.; Angel, F.A.; Saldías, C.; Tundidor-Camba, A.; Terraza, C.A. Synthesis and Characterization of a 2,3-Dialkoxynaphthalene-Based Conjugated Copolymer Via Direct Arylation Polymerization for Organic Electronics. *Polymers* **2020**, *12*, 1377. [CrossRef]

11. Ramoroka, M.E.; Mdluli, S.B.; John-Denk, V.S.; Modibane, K.D.; Arendse, C.J.; Iwuoha, E.I. Synthesis and Photovoltaics of Novel 2,3,4,5-Tetrathienylthiophene-Co-Poly(3-Hexylthiophene-2,5-Diyl) Donor Polymer for Organic Solar Cell. *Polymers* **2020**, *13*, 2. [CrossRef] [PubMed]
12. Sun, C.; Pan, F.; Bin, H.; Zhang, J.; Xue, L.; Qiu, B.; Wei, Z.; Zhang, Z.; Li, Y. A Low Cost and High Performance Polymer Donor Material for Polymer Solar Cells. *Nat. Commun.* **2018**, *9*, 743. [CrossRef] [PubMed]
13. Ye, L.; Zhang, S.; Huo, L.; Zhang, M.; Hou, J. Molecular Design toward Highly Efficient Photovoltaic Polymers Based on Two-Dimensional Conjugated Benzodithiophene. *Acc. Chem. Res.* **2014**, *47*, 1595–1603. [CrossRef] [PubMed]
14. Zhao, J.; Li, Q.; Liu, S.; Cao, Z.; Jiao, X.; Cai, Y.-P.; Huang, F. Bithieno [3,4-c]Pyrrole-4,6-Dione-Mediated Crystallinity in Large-Bandgap Polymer Donors Directs Charge Transportation and Recombination in Efficient Nonfullerene Polymer Solar Cells. *ACS Energy Lett.* **2020**, *5*, 367–375. [CrossRef]
15. Li, H.; Yang, W.; Wang, W.; Wu, Y.; Wang, T.; Min, J. Wide Bandgap Donor Polymers Containing Carbonyl Groups for Efficient Non-Fullerene Polymer Solar Cells. *Dye. Pigment.* **2021**, *186*, 108987. [CrossRef]
16. Sun, C.; Pan, F.; Chen, S.; Wang, R.; Sun, R.; Shang, Z.; Qiu, B.; Min, J.; Lv, M.; Meng, L.; et al. Achieving Fast Charge Separation and Low Nonradiative Recombination Loss by Rational Fluorination for High-Efficiency Polymer Solar Cells. *Adv. Mater.* **2019**, *31*, 1905480. [CrossRef]
17. Qiu, B.; Chen, S.; Sun, C.; Yuan, J.; Zhang, X.; Zhu, C.; Qin, S.; Meng, L.; Zhang, Y.; Yang, C.; et al. Understanding the Effect of the Third Component PC$_{71}$BM on Nanoscale Morphology and Photovoltaic Properties of Ternary Organic Solar Cells. *Sol. RRL* **2020**, *4*, 1900540. [CrossRef]
18. Yang, J.; Geng, Y.F.; Li, J.F.; Zhao, B.M.; Guo, Q.; Zhou, E.J. A-DA' D-A-Type Non-fullerene Acceptors Containing a Fused Heptacyclic Ring for Poly(3-hexylthiophene)-Based Polymer Solar Cells. *J. Phys. Chem. C* **2020**, *124*, 24616–24623. [CrossRef]
19. He, K.; Kumar, P.; Abd-Ellah, M.; Liu, H.; Li, X.; Zhang, Z.; Wang, J.; Li, Y. Alkyloxime Side Chain Enabled Polythiophene Donors for Efficient Organic Solar Cells. *Macromolecules* **2020**, *53*, 8796–8808. [CrossRef]
20. Ren, J.; Bi, P.; Zhang, J.; Liu, J.; Wang, J.; Xu, Y.; Wei, Z.; Zhang, S.; Hou, J. Molecular Design Revitalizes the Low-cost PTV-polymer for Highly Efficient Organic Solar Cells. *Natl. Sci. Rev.* **2021**, *8*, nwab031. [CrossRef]
21. Xiao, J.; Jia, X.; Duan, C.; Huang, F.; Yip, H.-L.; Cao, Y. Surpassing 13% Efficiency for Polythiophene Organic Solar Cells Processed from Nonhalogenated Solvent. *Adv. Mater.* **2021**, *33*, 2008158. [CrossRef] [PubMed]
22. Yang, C.; Zhang, S.; Ren, J.; Bi, P.; Yuan, X.; Hou, J. Fluorination Strategy Enables Greatly Improved Performance for Organic Solar Cells Based on Polythiophene Derivatives. *Chin. Chem. Lett.* **2021**, *32*, 2274–2278. [CrossRef]
23. Manna, K.; Agrawal, Y. Microwave Assisted Synthesis of New Indophenazine 1,3,5-Trisubstruted Pyrazoline Derivatives of Benzofuran and Their Antimicrobial Activity. *Bioorg. Med. Chem. Lett.* **2009**, *19*, 2688–2692. [CrossRef] [PubMed]
24. Qian, X.; Wang, X.; Shao, L.; Li, H.; Yan, R.; Hou, L. Molecular Engineering of D-D-pi A Type Organic Dyes Incorporating Indoloquinoxaline and Phenothiazine for Highly Efficient Dye Sensitized Solar Cells. *J. Power Sour.* **2016**, *326*, 129–136. [CrossRef]
25. Su, R.; Ashraf, S.; Lyu, L.; El-Shafei, A. Tailoring Dual-Channel Anchorable Organic Sensitizers with Indolo [2,3-b]quinoxaline Moieties: Correlation Between Structure and DSSC Performance. *Sol. Energy.* **2020**, *206*, 443–454. [CrossRef]
26. Venkateswararao, A.; Tyagi, P.; Thomas, K.; Chen, P.; Ho, K. Organic Dyes Containing Indolo [2,3-b]quinoxaline as a Donor: Synthesis, Optical and Photovoltaic Properties. *Tetrahedron* **2014**, *70*, 6318–6327. [CrossRef]
27. Dong, D.; Fang, D.; Li, H.; Zhu, C.; Zhao, X.; Li, J.; Jin, L.; Xie, L.; Chen, L.; Zhao, J.; et al. Direct Arylated 6H-Indolo [2,3-b]quinoxaline Derivative as a Thickness-Dependent Hole-Injection Layer. *Chem. Asian J.* **2017**, *12*, 920–926. [CrossRef]
28. Payne, A.; Li, S.; Dayneko, S.; Risko, C.; Welch, G. An Unsymmetrical Non-Fullerene Acceptor: Synthesis via Direct Heteroarylation, Self-Assembly, and Utility as A Low Energy Absorber in Organic Photovoltaic Cells. *Chem. Commun.* **2017**, *53*, 10168–10171. [CrossRef]
29. Chao, P.; Wang, H.; Qu, S.; Mo, D.; Meng, H.; Chen, W.; He, F. From Semi- to Full-Two-Dimensional Conjugated Side-Chain Design: A Way toward Comprehensive Solar Energy Absorption. *Macromolecules* **2017**, *50*, 9617–9625. [CrossRef]
30. Zhang, Z.; Bai, Y.; Li, Y. Benzotriazole Based 2D-conjugated Polymer Donors for High Performance Polymer Solar Cells. *Chinese J. Polym. Sci.* **2021**, *39*, 1–13. [CrossRef]
31. Kim, T.; Lee, J.; Heo, J.; Lim, B.; Kim, J. Highly Efficient Polymer Solar Cells with a Thienopyrroledione and Benzodithiophene Containing Planar Random Copolymer. *Polym. Chem.* **2018**, *9*, 1216–1222. [CrossRef]
32. Lim, B.; Long, D.; Han, S.; Nah, Y.; Noh, Y. Well-defined Alternative Polymer Semiconductor Using Large Size Regioregular Building Blocks as Monomers: Electrical and Electrochemical Properties. *J. Mater. Chem. C* **2018**, *6*, 5662–5670. [CrossRef]
33. Choi, E.; Eom, S.; Song, C.; Nam, S.; Lee, J.; Woo, H.; Jung, I.; Yoon, S.; Lee, C. Synthesis and Characterization of a Wide Bandgap Polymer Based on a Weak Donor-Weak Acceptor Structure for Dual Applications in Organic Solar Cells and Organic Photodetectors. *Org. Electron.* **2017**, *46*, 173–182. [CrossRef]
34. Yang, J.; Uddin, M.; Tang, Y.; Wang, Y.; Wang, Y.; Su, H.; Gao, R.; Chen, Z.; Dai, J.; Woo, H.; et al. Quinoxaline-Based Wide Band Gap Polymers for Efficient Nonfullerene Organic Solar Cells with Large Open-Circuit Voltages. *ACS Appl. Mater. Interfaces* **2018**, *10*, 23235–23246. [CrossRef]
35. He, K.; Li, X.; Liu, H.; Zhang, Z.; Kumar, P.; Ngai, J.; Wang, J.; Li, Y. D-A Polymer with a Donor Backbone-Acceptor-side-chain Structure for Organic Solar Cells. *Asian J. Org. Chem.* **2020**, *9*, 1301–1308. [CrossRef]
36. Li, X.; Pan, F.; Sun, C.; Zhang, M.; Wang, Z.; Du, J.; Wang, J.; Xiao, M.; Xue, L.; Zhang, Z.-G.; et al. Simplified Synthetic Routes for Low Cost and High Photovoltaic Performance n-type Organic Semiconductor Acceptors. *Nat. Commun.* **2019**, *10*, 519. [CrossRef]

37. Loewe, R.S.; Khersonsky, S.M.; McCullough, R.D. A Simple Method to Prepare Head-to-tail Coupled, Regioregular Poly(3-alkylthiophenes) Using Grignard Metathesis. *Adv. Mater.* **1999**, *11*, 250–253. [CrossRef]
38. Qian, D.; Ye, L.; Zhang, M.; Liang, Y.; Li, L.; Huang, Y.; Guo, X.; Zhang, S.; Tan, Z.; Hou, J. Design, Application, and Morphology Study of a New Photovoltaic Polymer with Strong Aggregation in Solution State. *Macromolecules* **2012**, *45*, 9611–9617. [CrossRef]
39. Zhang, M.; Guo, X.; Ma, W.; Ade, H.; Hou, J. A Large-Bandgap Conjugated Polymer for Versatile Photovoltaic Applications with High Performance. *Adv. Mater.* **2015**, *27*, 4655–4660. [CrossRef]
40. Xie, R.; Song, L.; Zhao, Z. Comparing Benzodithiophene Unit with Alkylthionaphthyl and Alkylthiobiphenyl Side-Chains in Constructing High-Performance Nonfullerene Solar Cells. *Polymers* **2020**, *12*, 1673. [CrossRef]
41. Murad, A.R.; Iraqi, A.; Aziz, S.B.; Abdullah, S.N.; Brza, M.A.; Saeed, S.R.; Abdulwahid, R.T. Fabrication of Alternating Copolymers Based on Cyclopentadithiophene-Benzothiadiazole Dicarboxylic Imide with Reduced Optical Band Gap: Synthesis, Optical, Electrochemical, Thermal, and Structural Properties. *Polymers* **2020**, *13*, 63. [CrossRef] [PubMed]
42. Osw, P.; Nitti, A.; Abdullah, M.N.; Etkind, S.I.; Mwaura, J.; Galbiati, A.; Pasini, D. Synthesis and Evaluation of Scalable D-A-D π-Extended Oligomers as P-Type Organic Materials for Bulk-Heterojunction Solar Cells. *Polymers* **2020**, *12*, 720. [CrossRef] [PubMed]
43. He, K.; Kumar, P.; Yuan, Y.; Zhang, Z.; Li, X.; Liu, H.; Wang, J.; Li, Y. A Wide Bandgap Polymer Donor Composed of Benzodithiophene and Oxime-Substituted Thiophene for High-Performance Organic Solar Cells. *ACS Appl. Mater. Interfaces* **2021**, *13*, 26441–26450. [CrossRef] [PubMed]
44. Leenaers, P.J.; Maufort, A.J.L.A.; Wienk, M.M.; Janssen, R.A.J. Impact of π-Conjugated Linkers on the Effective Exciton Binding Energy of Diketopyrrolopyrrole–Dithienopyrrole Copolymers. *J. Phys. Chem. C* **2020**, *124*, 27403–27412. [CrossRef]
45. Zhang, Z.; Miao, J.; Ding, Z.; Kan, B.; Lin, B.; Wan, X.; Ma, W.; Chen, Y.; Long, X.; Dou, C.; et al. Efficient and thermally stable organic solar cells based on small molecule donor and polymer acceptor. *Nat. Commun.* **2019**, *10*, 3271. [CrossRef]

Article

Design of an Efficient PTB7:PC70BM-Based Polymer Solar Cell for 8% Efficiency

Ahmed N. M. Alahmadi

Device Simulation Lab, Department of Electrical Engineering, Umm Al-Qura University, Makkah 21955, Saudi Arabia; anmahmadi@uqu.edu.sa

Abstract: Polymer semiconductors may have the potential to fully replace silicon in next-generation solar cells because of their advantages such as cheap cost, lightweight, flexibility, and the ability to be processed for very large area applications. Despite these advantages, polymer solar cells are still facing a certain lack of power-conversion efficiency (PCE), which is essentially required for commercialization. Recently, bulk heterojunction of PTB7:PC70BM as an active layer showed remarkable performance for polymer solar cells in terms of PCE. Thus, in this paper, we developed and optimized a novel design using PEDOT:PSS and PFN-Br as electron and hole transport layers (ETL and HTL) for ITO/PEDOT:PSS/PT7B:PC70BM/PFN-Br/Ag as a polymer solar cell, with the help of simulation. The optimized solar cell has a short-circuit current (Isc) of 16.434 mA.cm^{-2}, an open-circuit voltage (Voc) of 0.731 volts, and a fill-factor of 68.055%, resulting in a maximum PCE of slightly above 8%. The findings of this work may contribute to the advancement of efficient bulk-heterojunction-based polymer solar cells.

Keywords: polymer; solar cell; bulk heterojunction; PEDOT:PSS; PTB7:PC70BM; PFN-Br; SCAPS 1D

Citation: Alahmadi, A.N.M. Design of an Efficient PTB7:PC70BM-Based Polymer Solar Cell for 8% Efficiency. *Polymers* **2022**, *14*, 889. https://doi.org/10.3390/polym14050889

Academic Editor: Bożena Jarząbek

Received: 6 February 2022
Accepted: 21 February 2022
Published: 23 February 2022

Publisher's Note: MDPI stays neutral with regard to jurisdictional claims in published maps and institutional affiliations.

Copyright: © 2022 by the author. Licensee MDPI, Basel, Switzerland. This article is an open access article distributed under the terms and conditions of the Creative Commons Attribution (CC BY) license (https://creativecommons.org/licenses/by/4.0/).

1. Introduction

Organic semiconductor-based solar cells have gained considerable popularity over the last few years, and some scientists believe they have the potential to completely replace silicon-based solar cells in the near future [1–5]. Organic semiconductors offer many advantages for solar cell applications such as lightweight, low cost, fabrication on various substrates, wide-area applications, and flexible and tunable processing at room temperature [6]. Despite these well-reported advantages, organic solar cell efficiency is far behind Si solar cells. It is generally believed that some combination of a proper absorber layer with a hole and electron may yield a high-efficiency device for next-generation solar cells [7–10].

Researchers are exploiting a variety of techniques to enhance the power-conversion efficiency (PCE) of organic solar cells. Some schools of thought still believe that the combination of the most suited hole, electron transport, and buffer layer with a highly efficient bulk-heterojunction as an absorber layer may yield an excellent photovoltaic response [11–13]. Bulk heterojunction has attracted great interest due to various advantages such as low cost, tunable bandgap and electron affinity, lightweight, and most importantly excellent power conversion efficiency compared to other organic/polymer materials. The bulk-heterojunction layer consists of a blend of acceptor and donor materials (organic/polymer) at the nanoscale and broadly speaking donor materials are usually polymer/organic while fullerene derivatives (PCBM) are used as acceptor materials for bulk heterojunctions layer such as P3HT:PCBM, MEH-PPV:PCBM, PCPDTBT:PCBM, and PTB7:PC70BM [14,15].

Suitable electron and hole transport layers (ETL and HTL) for PTB7:PC70BM create challenges, as PTB7:PC70BM has strong binding (low dielectric constant) energy for exciton with low diffusion length, and despite its heterogeneous nature most of the excitons are lost in recombination [11,12]. If a very thin PTB7:PC70BM layer is used, then these issues can be improved, but the issue of inefficient optical absorption will arise. On the other hand, the optimum thickness of the PTB7:PC70BM layer emphasizes the importance of an efficient hole and electron transport layer, which attract the required free carriers and also block the injection of opposite free carriers. The optical absorption spectra of PTB7:PCBM bulk-heterojunction polymer can be found in the reference [16].

For the hole transport layer, poly(3,4-ethenedioxythiophene):poly(styrenesulfonate) (PEDOT:PSS) is accepted as one of the best polymers for hole transport materials and especially for inverted polymer solar cells. It has many advantages such as lightweight, high conductivity, low cost, and thin-film processing even at room temperature [17,18]. However, the most important reason for its success as a hole transport layer is that PEDOT:PSS offers not only a well-coordinated work function for HOMO (Highest Occupied Molecular Orbital) level of the donor semiconducting polymer but also offers highly matched work function with ITO (tin-doped indium oxide) over a glass substrate [19]. As well as proper work function, PEDOTT:PSS also offers excellent visible transparency as well as good air stability essentially required for photovoltaic applications [20]. As a result, PEDOT:PSS can remove holes efficiently from the semiconducting polymer layer and forward them towards the cathode. Hence, in this work, we employed PEDOT:PSS as a HTL.

Similarly, for an electron transport layer, [6,6]-phenyl C60 butyric acid methyl ester (PC60BM) is another common material for inverted (p-i-n) polymer solar cells. It facilitates the electron-transport process and has very high electron-affinity which helps to extract the electron efficiently [21]. However, it has some limitations which cause degradation to the PCE of polymer solar cells. Some of these limitations are low electron mobility, high leakage current, and recombination at interfaces [5]. On the other hand, a polyfluorene derivative such as PFN-Br is reported to show excellent electron extraction and transport behavior [22]. Figure 1 shows the overall architecture of the novel ITO/PEDOT:PSS/PTB7:PC70BM/PFN-Br/Ag photovoltaic device proposed for this study. The photovoltaic response of the solar cell described above was numerically simulated in order to identify the optimal doping density and thickness of ETL, HTL, and the absorber layer.

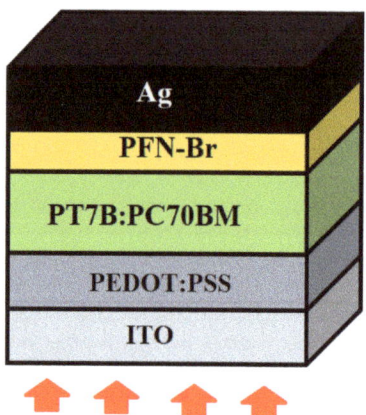

Figure 1. Shows the schematic view of the proposed ITO/PEDOT:PSS/PTB7:PC70BM/PFN-Br/Ag photovoltaic device for simulation.

2. Simulation Methods and Physical Parameters

2.1. Simulation Software

Simulation of a photovoltaic response for an organic solar cell is a highly mature field and has already played a vital to overall improving the PCE of the solar cell. In industry, various types of software are available for the simulation of photovoltaic response. Among simulation software, SCAPS-1D is very attractive as open-source, simple, highly reliable, and provides comprehensive tools for simulations. Similarly, SCAPS-1D software also offers high consistency between simulation and experimental results [23–25]. On the other hand, various simulation results for organic/polymer materials as absorbers or transport layers for different solar cells have already been reported in the literature [26–28]. Therefore, SCAPS 1D software (SCAPS 3.8, ELIS-University of Gent, Gent, Belgium) was chosen for the simulation study of the proposed solar cell.

2.2. Simulation Method

SCAPS 1D simultaneously solves many fundamental semiconductor photovoltaic equations for both electron and hole separately such as (i) continuity equation, (ii) Poisson equations, (iii) charge transport equations, (iv) diffusivity equations, and (v) optical absorption equations. The following reference [29,30] contains in-depth information on these fundamental equations These equations are driven by physical and geometrical parameters associated with each layer and lead to the overall photovoltaic response of the given solar cell. These equations are listed as

$$\frac{d^2\varnothing(x)}{dx^2} = \frac{q}{\epsilon_o \epsilon_r}\left(p(x) - n(x) + N_D - N_A + \rho_p - \rho_n\right) \quad (1)$$

$$\frac{dJ_n}{dx} = G - R \quad (2)$$

$$\frac{dJ_p}{dx} = G - R \quad (3)$$

$$J = J_n + J_p \quad (4)$$

$$J_n = D_n \frac{dn}{dx} + \mu_n\, n \frac{d\varnothing}{dx} \quad (5)$$

$$J_p = D_p \frac{dp}{dx} + \mu_p\, p \frac{d\varnothing}{dx} \quad (6)$$

$$\alpha(\lambda) = \left(A + \frac{B}{h\nu}\right)\sqrt{h\nu - E_g} \quad (7)$$

Here $\varnothing(x)$, q, ϵ_o, ϵ_r, ρ_P, ρ_N, N_A, N_D, $p(x)$, $n(x)$, G, R, J_p, J_n, and J, are the electrostatic potential, electrical charge, absolute permittivity of vacuum, relative permittivity of a semiconductor, hole defect density, electron defect density, shallow acceptor doping density, shallow donor doping density, hole carrier density as a function of the thickness (x), electron carrier density as a function of the thickness (x), carrier generation rate of free carriers, total carrier recombination rate, hole current density, and electron current density, total current density, respectively. Similarly, D_p, D_n, μ_p, and μ_n are the free hole diffusion coefficient, free electron diffusion coefficient, free hole carrier mobility, and free-electron carrier mobility, respectively. Finally, h, $\alpha(\lambda)$, E_g, and ν are the plank constant, absorption coefficient, energy bandgap, optical frequency, and few arbitrary constant, respectively.

The simulation of the proposed solar cell is divided into six-well defined steps, these simulation steps are summarized as a flowchart in Figure 2. Firstly, the hole transport layers' thickness and doping density are optimized and then the electron transport layers' thickness and doping density are optimized. Similarly, in the next step, the absorber layer thickness is optimized, while in the second last and last step the final photovoltaic response of the optimized device is determined.

2.3. Physical Parameters

The physical parameters for each transport and absorber layer required by the software are the backbone of the simulation, special attention was paid to the selection of these parameters. These parameters are selected from the published results and are listed in Table 1. As organic semiconductor is considered disordered material, it inherently offers a high density of traps [31–34]. The photovoltaic performances of solar cells are seriously affected by the existence of both shallow and deep traps. Therefore, high traps density (10^{15} cm^{-3}) is introduced in both bulk and layer interface for the hole/electron transport layer and absorber layer, as shown in Table 1. Similarly, all calculations were performed at an ambient temperature environment of 300 K with 100 mW/cm² of power spectral density as a 1.5 AM solar radiation light source.

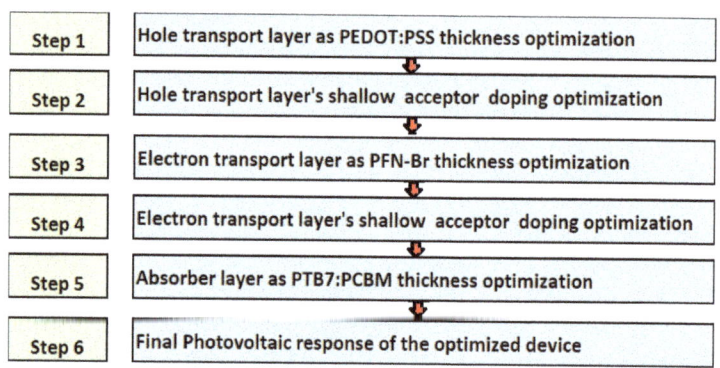

Figure 2. The steps of the methods followed in this work to optimize the proposed ITO/PEDOT:PSS/PTB7:PCBM/PFN-Br/Ag solar cell.

Table 1. The parameters of the photovoltaic device utilized in these simulations, including the initial estimation of the doping concentrations and thicknesses of each layer, which will be improved in the subsequent stages.

Physical Parameters	Symbol	Unit	PEDOT:PSS	PTB7:PC$_{70}$BM	PFN-Br
Thickness	Th	Nm	-	250	250
Energy Band Gap	E_g	eV	1.6	0.9	2.98
Electron Affinity	X	eV	3.5	3.7	4
Dielectric Permittivity (Relative)	E	-	3	3.9	5
Effective Density of States at Valence Band	N_V	cm^{-3}	1×10^{22}	1×10^{18}	1×10^{19}
Effective Density of States at Conduction Band	N_C	cm^{-3}	1×10^{22}	1×10^{18}	1×10^{19}
Hole Thermal Velocity	V_e	cm/s	1×10^7	1×10^7	1×10^7
electron Thermal Velocity	V_h	cm/s	1×10^7	1×10^7	1×10^7
Electron Mobility	μ_e	cm²/V.s	0.01	5.00×10^{-4}	1.00×10^{-4}
Hole Mobility	μ_h	cm²/V.s	$9.9 \times 10^{-0.5}$	5.00×10^{-4}	2.00×10^{-6}
Uniform Shallow Donor Doping	N_d	cm^{-3}	0.00	1×10^{19}	-
Uniform Shallow Acceptor Doping	N_a	cm^{-3}	-	1×10^{19}	0
Defect Density	N_t	cm^{-3}	1×10^{15}	1×10^{15}	1×10^{15}
References			[35]	[36,37]	[38]

3. Results and Discussion

3.1. Thickness Optimization of PEDOT:PSS

Thickness optimization of PEDOT:PSS as a hole transport layer is very crucial for the proposed solar cell because at one side PEDOT:PSS interacts with semitransparent ITO and on the other side it interacts with PT7B:PC70BM absorber layer. As a result, optical transmission, hole extraction and blocking of the electron from the absorber, hole transportation, and collection to the respective ITO anode depend critically on the PEDOT:PSS layer thickness [39]. The thickness optimization of PEDOT:PSS was performed by determining the photovoltaic characteristics such as PCE, short-circuit current (Isc), open-circuit voltage (Voc), and fill-factor, as functions of the thickness of PEDOT:PSS, shown in Figure 3.

Among these photovoltaic parameters, fill-factor is unique and defined as the percentage ratio between the actual and maximum possible power.

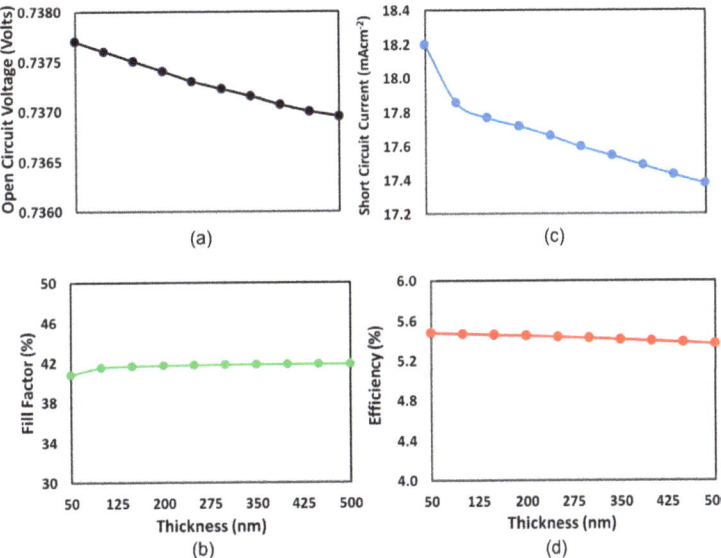

Figure 3. Performance characteristics such as (**a**) Voc, (**b**) fill factor, (**c**) Isc, and (**d**) PCE of the proposed ITO/PEDOT:PSS/PTb7:PC70BM/PFN-Br/Ag solar cell as a function of the PEDOT:PSS (HTL) thickness.

The thickness range of PEDOT:PSS is selected from 50 nm to 500 nm according to their efficiency with the high repeatability for the photovoltaic response [40]. Figure 3 demonstrates that both open-circuit voltage and fill-factor, as well as short-circuit current and efficiency, follow different trends. At almost 125 nm, the fill factor of the cell hits a maximum and then nearly remains constant as the thickness of PEDOT:PSS increases, while Voc is sharply declined with the increase in PEDOT:PSS thickness. On the other hand, PCE and short-circuit current is dropped from 50 nm thickness of PEDOT:PSS. Because PCE is the decisive factor, the optimal thickness of PEDOT:PSS as an HTL for the current solar cell is 50 nm.

3.2. Shallow Doping Density Optimization of PEDOT:PSS

Another significant parameter to consider when optimizing a solar cell for efficiency is the doping density for PEDOT:PSS as the HTL. Doping of PEDOT:PSS as the hole transport layer significantly improves both charge extraction and charge transport process by reducing the series resistance and the establishment of ohmic contacts to the ITO electrodes, which overall enhances the solar cell's photovoltaic parameters [41]. However, the higher dopant concentration may cause the creation of traps, which in turn behave as electron–hole recombination centers for PEDOT:PSS, thus we selected the range of doping density from 10^{12} to 10^{20} cm^{-3} based on published results [42]. PEDOT:PSS doping is critical for the proposed solar cell to have an efficient photovoltaic response. Before beginning the doping simulation, the optimized PEDOT:PSS thickness was updated in the software, and then photovoltaic parameters, such as Voc, fill factor, Isc, and PCE, as functions of shallow acceptor doping of PEDOT:PSS. The layer was simulated as shown in Figure 4. The figure depicts similar trends for all photovoltaic parameters except open-circuit voltage, which increases sharply and reaches a maximum at 10^{16} cm^{-3} and then starts to decrease. While other photovoltaic parameters also increase at early doping density with a slow rate, sharply rise to 10^{18} cm^{-3}, and then slightly increase up to 10^{20} cm^{-3}, which is the typical behavior of trapped space charge, limited current behavior was also observed for many organic/polymer semiconductors [43–45]. Consequently, the optimal doping density for the PEDOTPSS (HTL) in the proposed solar cell is inferred to be 10^{20} cm^{-3}.

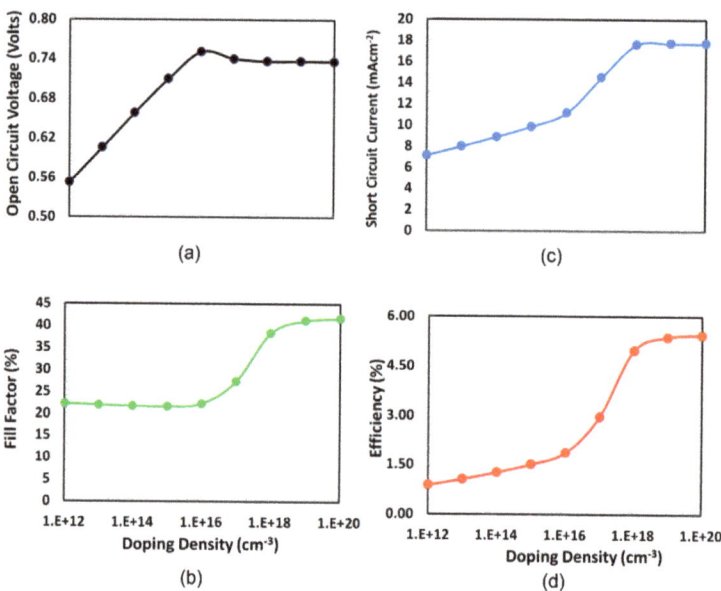

Figure 4. Performance characteristics such as (**a**) Voc, (**b**) fill factor, (**c**) Isc, and (**d**) PCE of the proposed ITO/PEDOT:PSS/PTb7:PC70BM/PFN-Br/Ag solar cell as a function of the PEDOT:PSS (HTL) doping density.

3.3. Electron Transport Layer Thickness Optimization

The optimal thickness of PFN-Br as an ETL is obtained in the third step of the simulation. Just like PEDOT:PSS as the HTL, the thickness of PFN-Br as ETL is also very important for electron extraction from PT7B-PC70BM, electron transport, and collection of electrons at the Ag cathode. The thickness optimization of PFN-Br was performed by determining the photovoltaic characteristics such as Voc, fill factor, Isc, and PCE as functions of PFN-Br thickness, shown in Figure 5. The Voc and Isc response of the current solar cell are degraded when the thickness of PFN-Br increases, while fill-factor is slightly increased up to 125 nm and then remains nearly constant. The efficiency is also degraded but at a very slow rate. The figures clearly show that the optimal thickness of PFN-Br as an ETL is 50 nm. Therefore, it can be inferred that the 50 nm thickness of PFN-Br provides the balance trade-off between electron–hole recombination, electron extraction, and blocking of the hole from the absorber, electron transportation, and hence collection to the respective Ag cathode.

3.4. Shallow Doping Density Optimization of the PFN-Br

In the fourth step of the simulation, the optimum doping density of PFN-Br as an electron transport layer is determined. The optimized donor doping of PFN-Br can be attributed to the efficient electron extraction and good ohmic contact between Ag cathode and the active PTB7:PC70BM layer. The optimized donor doping PFN-Br was estimated by determining the photovoltaic parameters, such as Voc, fill factor, Isc, and PCE, by altering the shallow donor doping of PFN-Br from 10^{12} to 10^{20} cm^{-3}, as shown in Figure 6. According to the Figure, it can be seen that higher doping of PFN-Br causes the open-circuit voltage response to degrade, which may be due to the creation of extra traps density at higher doping and the relaxation of the free carriers at these traps may cause to reduce the open-circuit voltage [46]. While PCE, fill-factor, and short-circuit current are increased with doping, PCE performed well, reaching the maximum at 10^{18} cm^{-3} doping and then starting to degrade. Thus, on the basis of these results, it can be concluded that the most optimal doping for PFN-Br as an ETL is 10^{18} cm^{-3}.

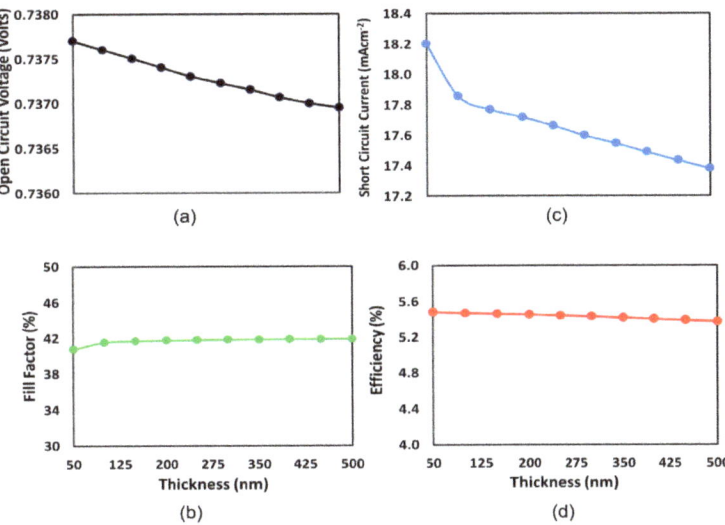

Figure 5. Performance characteristics such as (**a**) Voc, (**b**) fill factor, (**c**) Isc, and (**d**) PCE of the proposed ITO/PEDOT:PSS/PTb7:PC70BM/PFN-Br/Ag solar cell as a function of PFN-Br (ETL) thickness.

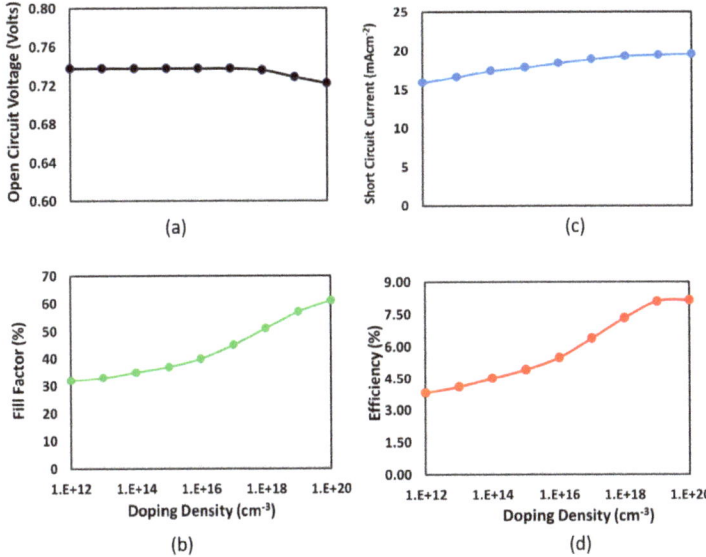

Figure 6. Performance characteristics such as (**a**) Voc, (**b**) fill factor, (**c**) Isc, and (**d**) PCE of the proposed ITO/PEDOT:PSS/PTb7:PC70BM/PFN-Br/Ag solar cell as a function of PFN-Br (electron-transport layer) doping density.

3.5. Thickness Optimization of PTB7-PC70BM

Thickness optimization of bulk-heterojunction polymer absorber layer (e.g., PTB7-PC70BM) is one of the main challenging tasks because it depends on many inter-related processes such as strong optical absorption, generation of electron–hole pairs, conversion of bounded electron–hole pairs into free carriers, reducing carrier recombination losses, efficient charge transportation to the respective transport layers, mechanical and environmental stability. All these factors required different thicknesses of the absorber layer for their efficient individual response and a compromise between these processes is required for an efficient photovoltaic response [47–49]. In literature, various thicknesses of bulk heterojunction absorber layer for organic/polymer solar cells are reported [50–52]. Therefore, we varied the thickness of PTB7:PC70BM from 50 to 500 nm for simulation. Consequently, the thickness optimization of bulk heterojunction PTB7:PC70BM absorber layer was performed by simulating the photovoltaic characteristics such as Voc, fill factor, Isc, and PCE by altering the thickness of absorber layer and the results are shown in Figure 7. Both Voc and fill-factor decrease with thickness, while PCE and short-circuit current, initially, slightly increase up and reached the maximum at nearly 100 nm thickness, then they gradually decrease. Hence, based on the simulation results, it can justify that the 100 nm thickness of the PTB7:PC70BM is the optimum thickness of the current solar cell.

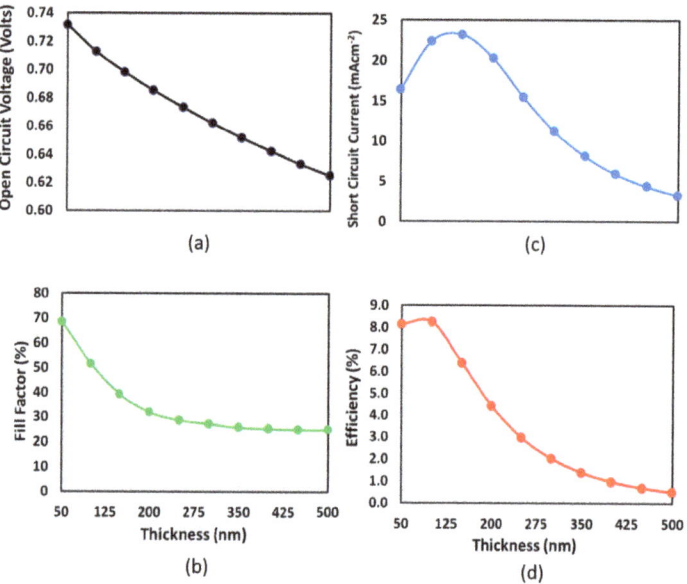

Figure 7. Performance characteristics such as (**a**) Voc, (**b**) fill factor, (**c**) Isc and (**d**) PCE of the proposed ITO/PEDOT:PSS/PTb7:PC70BM/PFN-Br/Ag solar cell as a function of PTB7:PC70BM (absorber layer) thickness.

3.6. Photo Current–Voltage Response of Proposed Solar Cell

The final phase of the simulation was to combine all of the optimum doping density and thickness for the PEDOT:PSS, PFB-Br, and PTB7:PC70BM layers and determine the current solar cell's overall photocurrent–voltage response, as shown in Figure 8.

The proposed solar cell's photovoltaic parameters are shown in Figure 8. The optimized ITO/PEDOT:PSS/PTB7:PC70BM/PFN-Br/Ag solar cell has an Isc of 16.434 mA.cm^{-2}, Voc of 0.731 volts, a fill-factor of 68.055%, and a PCE of 8.18%. The higher value of short-circuit current may be due to the commutative effects of wider optical absorption, exciton generation, efficient exciton dissociation leads to the free carrier generation, and then transportation at their respective transport layer before collection at electrodes [53]. The proposed solar cell's open-circuit voltage still has space for future improvement.

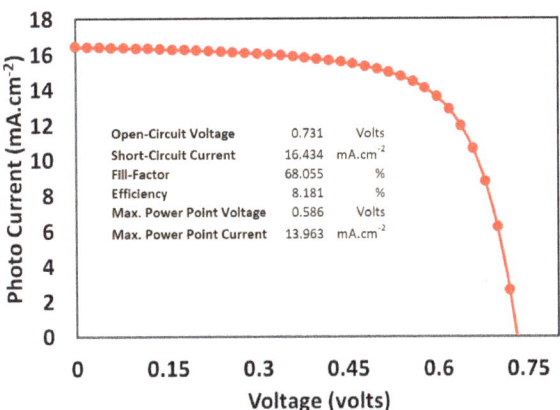

Figure 8. The simulated photocurrent–voltage response at AM 1.5 for the proposed ITO/PEDOT:PSS/PT7:PC70BM/PFN-Br/Ag solar cell.

On the other hand, lower PCE compared to the other reported simulation of bulk-hybrid solar cells is maybe due to the incorporation of a higher density of traps [38]. It is experimentally evident that polymers are full of traps, and these traps may be presented due to many factors such as humidity, structural defects, distortion, impurity, and/or any other known or unknown reasons. However, these traps act as the recombination centers and cause severely degrade the overall photovoltaic response. Therefore, a high density of traps in each layer is introduced in order to make the simulation more realistic and comparable to the experimental results, which in turn show the lower PCE.

4. Conclusions

In conclusion, we have efficiently designed and optimized a polymer-based novel bulk heterojunction solar cell as ITO/PEDOT:PSS/PTB7:PC70BM/PFN-Br/Ag through SCAPS 1D simulations. For this purpose PEDOT:PSS, PFN-Br, and PT7B:PC70BM layers were selected as a HTL, ETL, and bulk-heterojunction absorber layers, respectively, and sandwiched between transparent ITO and Ag electrodes. Doping density and thickness of both PEDOT:PSS and PFN-Br were optimized and then PT7B:PC70BM is investigated for an efficient photovoltaic response. The proposed ITO/PEDOT:PSS/PTB7:PC70BM/PFN-Br/Ag solar cells yield an Isc of 16.434 mA.cm-2, a Voc of 0.731 volts, and a fill factor of 68.055%, resulting in a PCE of just over 8 %. Similarly, it is also indicated that all photovoltaic parameters are considerably affected by the doping density as well as the layer thickness of both ETL and HTL, and the bulk-heterojunction absorber layer. The higher short circuit current may the result of efficient optical absorption, exciton generation, exciton dissociation, free carrier generation, and then transportation at their respective transport layers before collection at the electrodes. As it is accepted that polymers are full of traps, we introduced a high density of traps in each layer in order to make the simulation more realistic, which in turn shows the lower PCE. Additionally, the proposed solar cell's open-circuit voltage still has room for improvement.

Funding: This research received no external funding.

Institutional Review Board Statement: Not applicable.

Informed Consent Statement: Not applicable.

Data Availability Statement: Not applicable.

Acknowledgments: We acknowledge that all simulations in this work were performed using the SCAPS-1D program developed by Marc Burgelman et al. (University of Gent).

Conflicts of Interest: The author declares no conflict of interest.

References

1. Peng, W.; Lin, Y.; Jeong, S.Y.; Genene, Z.; Magomedov, A.; Woo, H.Y.; Chen, C.; Wahyudi, W.; Tao, Q.; Deng, J.; et al. Over 18% Ternary Polymer Solar Cells Enabled by a Terpolymer as the Third Component. *Nano Energy* **2022**, *92*, 106681. [CrossRef]
2. Murad, A.R.; Iraqi, A.; Aziz, S.B.; Abdullah, S.N.; Brza, M.A. Conducting Polymers for Optoelectronic Devices and Organic Solar Cells: A Review. *Polymers* **2020**, *12*, 2627. [CrossRef] [PubMed]
3. Ghosekar, I.C.; Patil, G.C. Review on Performance Analysis of P3HT:PCBM-Based Bulk Heterojunction Organic Solar Cells. *Semicond. Sci. Technol.* **2021**, *36*, 045005. [CrossRef]
4. Moiz, S.A.; Nahhas, A.M.; Um, H.-D.; Jee, S.-W.; Cho, H.K.; Kim, S.-W.; Lee, J.-H. A Stamped PEDOT:PSS–Silicon Nanowire Hybrid Solar Cell. *Nanotechnology* **2012**, *23*, 145401. [CrossRef]
5. Moiz, S.A.; Alahmadi, A.N.M.; Karimov, K.S. Improved Organic Solar Cell by Incorporating Silver Nanoparticles Embedded Polyaniline as Buffer Layer. *Solid State Electron.* **2020**, *163*, 107658. [CrossRef]
6. Li, W.; Ye, L.; Li, S.; Yao, H.; Ade, H.; Hou, J. A High-Efficiency Organic Solar Cell Enabled by the Strong Intramolecular Electron Push–Pull Effect of the Nonfullerene Acceptor. *Adv. Mater.* **2018**, *30*, e1707170. [CrossRef]
7. Helgesen, M.; Søndergaard, R.; Krebs, F.C. Advanced Materials and Processes for Polymer Solar Cell Devices. *J. Mater. Chem.* **2010**, *20*, 36–60. [CrossRef]
8. Hau, S.K.; Yip, H.L.; Jen, A.K.Y. A Review on the Development of the Inverted Polymer Solar Cell Architecture. *Polym. Rev.* **2010**, *50*, 474–510. [CrossRef]
9. Chen, C.C.; Chang, W.H.; Yoshimura, K.; Ohya, K.; You, J.; Gao, J.; Hong, Z.; Yang, Y. An Efficient Triple-Junction Polymer Solar Cell Having a Power Conversion Efficiency Exceeding 11%. *Adv. Mater.* **2014**, *26*, 5670–5677. [CrossRef]
10. Malik, A.; Hameed, S.; Siddiqui, M.J.; Haque, M.M.; Umar, K.; Khan, A.; Muneer, M. Electrical and Optical Properties of Nickel- and Molybdenum-Doped Titanium Dioxide Nanoparticle: Improved Performance in Dye-Sensitized Solar Cells. *J. Mater. Eng. Perform.* **2014**, *23*, 3184–3192. [CrossRef]
11. Sharma, N.; Gupta, S.K.; Negi, C.M. Influence of Active Layer Thickness on Photovoltaic Performance of PTB7:PC$_{70}$BM Bulk Heterojunction Solar Cell. *Superlattices Microstruct.* **2019**, *135*, 106278. [CrossRef]
12. Ramírez-Como, M.; Balderrama, V.S.; Sacramento, A.; Marsal, L.F.; Lastra, G.; Estrada, M. Fabrication and Characterization of Inverted Organic PTB7:PC$_{70}$BM Solar Cells Using Hf-In-ZnO as Electron Transport Layer. *Sol. Energy* **2019**, *181*, 386–395. [CrossRef]
13. Dridi, C.; Touafek, N.; Mahamdi, R. Inverted PTB7:PC$_{70}$BM Bulk Heterojunction Solar Cell Device Simulations for Various Inorganic Hole Transport Materials. *Optik* **2022**, *252*, 168447. [CrossRef]
14. Zhang, Z.; Qin, Y. Cross-Conjugated Poly(Selenylene Vinylene)s. *Polym. Chem.* **2019**, *10*, 1018–1025. [CrossRef]
15. Meng, L.; Fan, H.; Lane, J.M.D.; Qin, Y. Bottom-up Approaches for Precisely Nanostructuring Hybrid Organic/Inorganic Multi-Component Composites for Organic Photovoltaics. *MRS Adv.* **2020**, *5*, 2055–2065. [CrossRef]
16. Mola, G.T.; Dlamini, W.E.; Oseni, S.O. Improving Optical Absorption Bandwidth Using Bi-Layer Bulkheterojunction Organic Photoactive Medium. *J. Mater. Sci. Mater. Electron.* **2016**, *27*, 11628–11633. [CrossRef]
17. Louwet, F.; Groenendaal, L.; D'haen, J.; Manca, J.; van Luppen, J.; Verdonck, E.; Leenders, L. PEDOT/PSS: Synthesis, Characterization, Properties and Applications. *Synth. Met.* **2003**, *135–136*, 115–117. [CrossRef]
18. Moiz, S.A.; Alahmadi, A.N.M.; Aljohani, A.J. Design of Silicon Nanowire Array for PEDOT:PSS-Silicon Nanowire-Based Hybrid Solar Cell. *Energies* **2020**, *13*, 3797. [CrossRef]
19. Han, W.; Ren, G.; Liu, J.; Li, Z.; Bao, H.; Liu, C.; Guo, W. Recent Progress of Inverted Perovskite Solar Cells with a Modified PEDOT:PSS Hole Transport Layer. *ACS Appl. Mater. Interfaces* **2020**, *12*, 49297–49322. [CrossRef]
20. Rivnay, J.; Inal, S.; Collins, B.A.; Sessolo, M.; Stavrinidou, E.; Strakosas, X.; Tassone, C.; Delongchamp, D.M.; Malliaras, G.G. Structural Control of Mixed Ionic and Electronic Transport in Conducting Polymers. *Nat. Commun.* **2016**, *7*, 11287. [CrossRef]
21. Bai, Y.; Yu, H.; Zhu, Z.; Jiang, K.; Zhang, T.; Zhao, N.; Yang, S.; Yan, H. High Performance Inverted Structure Perovskite Solar Cells Based on a PCBM:Polystyrene Blend Electron Transport Layer. *J. Mater. Chem. A* **2015**, *3*, 9098–9102. [CrossRef]
22. Yin, H.; Chiu, K.L.; Bi, P.; Li, G.; Yan, C.; Tang, H.; Zhang, C.; Xiao, Y.; Zhang, H.; Yu, W.; et al. Enhanced Electron Transport and Heat Transfer Boost Light Stability of Ternary Organic Photovoltaic Cells Incorporating Non-Fullerene Small Molecule and Polymer Acceptors. *Adv. Electron. Mater.* **2019**, *5*, 1900497. [CrossRef]
23. Niemegeers, A.; Gillis, S.; Burgelman, M. A User Program for Realistic Simulation of Polycrystalline Heterojunction Solar Cells: SCAPS-1D. In *Proceedings of the 2nd World Conference on Photovoltaic Energy Conversion, Vienna, Austria, 6–10 July 1998*; Joint Research Centre, European Commission: Ispra, Italy, 1998; pp. 672–675.
24. Burgelman, M.; Nollet, P.; Degrave, S. Modelling Polycrystalline Semiconductor Solar Cells. *Thin Solid Films* **2000**, *361–362*, 527–532. [CrossRef]
25. Verschraegen, J.; Burgelman, M. Numerical Modeling of Intra-Band Tunneling for Heterojunction Solar Cells in Scaps. *Thin Solid Films* **2007**, *515*, 6276–6279. [CrossRef]
26. Nithya, K.S.; Sudheer, K.S. Device Modelling and Optimization Studies on Novel ITIC-OE Based Non-Fullerene Organic Solar Cell with Diverse Hole and Electron Transport Layers. *Opt. Mater.* **2022**, *123*, 111912. [CrossRef]

27. Abdelaziz, W.; Zekry, A.; Shaker, A.; Abouelatta, M. Numerical Study of Organic Graded Bulk Heterojunction Solar Cell Using SCAPS Simulation. *Sol. Energy* **2020**, *211*, 375–382. [CrossRef]
28. Moiz, S.A. Optimization of Hole and Electron Transport Layer for Highly Efficient Lead-Free Cs_2TiBr_6-Based Perovskite Solar Cell. *Photonics* **2022**, *9*, 23. [CrossRef]
29. Decock, K.; Zabierowski, P.; Burgelman, M. Modeling Metastabilities in Chalcopyrite-Based Thin Film Solar Cells. *J. Appl. Phys.* **2012**, *111*, 43703. [CrossRef]
30. Moiz, S.A.; Alahmadi, A.N.M. Design of Dopant and Lead-Free Novel Perovskite Solar Cell for 16.85% Efficiency. *Polymers* **2021**, *13*, 2110. [CrossRef]
31. Ndiaye, N.S.; Simonetti, O.; Nguyen, T.P.; Giraudet, L. Generation-Recombination in Disordered Organic Semiconductor: Application to the Characterization of Traps. *Org. Electron.* **2021**, *99*, 106350. [CrossRef]
32. McMahon, D.P.; Troisi, A. Organic Semiconductors: Impact of Disorder at Different Timescales. *ChemPhysChem* **2010**, *11*, 2067–2074. [CrossRef] [PubMed]
33. Karimov, K.; Ahmed, M.M.; Moiz, S.A.; Babadzhanov, P.; Marupov, R.; Turaeva, M.A. Electrical Properties of Organic Semiconductor Orange Nitrogen Dye Thin Films Deposited from Solution at High Gravity. *Eurasian Chem. Technol. J.* **2007**, *5*, 109. [CrossRef]
34. Moiz, S.A.; Karimov, K.; Gohar, N.D. Orange Dye Thin Film Resistive Hygrometers. *Eurasian Chem. Technol. J.* **2017**, *6*, 179. [CrossRef]
35. Moiz, S.A.; Alahmadi, A.N.M.; Aljohani, A.J. Design of a Novel Lead-Free Perovskite Solar Cell for 17.83% Efficiency. *IEEE Access* **2021**, *9*, 54254–54263. [CrossRef]
36. Hedley, G.J.; Ward, A.J.; Alekseev, A.; Howells, C.T.; Martins, E.R.; Serrano, L.A.; Cooke, G.; Ruseckas, A.; Samuel, I.D.W. Determining the Optimum Morphology in High-Performance Polymer-Fullerene Organic Photovoltaic Cells. *Nat. Commun.* **2013**, *4*, 2867. [CrossRef]
37. Khelifi, S.; Voroshazi, E.; Spoltore, D.; Piersimoni, F.; Bertho, S.; Aernouts, T.; Manca, J.; Lauwaert, J.; Vrielinck, H.; Burgelman, M. Effect of Light Induced Degradation on Electrical Transport and Charge Extraction in Polythiophene:Fullerene (P3HT:PCBM) Solar Cells. *Sol. Energy Mater. Sol. Cells* **2014**, *120*, 244–252. [CrossRef]
38. Sharma, B.; Mathur, A.S.; Rajput, V.K.; Singh, I.K.; Singh, B.P. Device Modeling of Non-Fullerene Organic Solar Cell by Incorporating CuSCN as a Hole Transport Layer Using SCAPS. *Optik* **2022**, *251*, 168457. [CrossRef]
39. Lattante, S. Electron and Hole Transport Layers: Their Use in Inverted Bulk Heterojunction Polymer Solar Cells. *Electronics* **2014**, *3*, 132–164. [CrossRef]
40. Kim, G.W.; Shinde, D.V.; Park, T. Thickness of the Hole Transport Layer in Perovskite Solar Cells: Performance versus Reproducibility. *RSC Adv.* **2015**, *5*, 99356–99360. [CrossRef]
41. Zhao, Z.; Chen, X.; Liu, Q.; Wu, Q.; Zhu, J.; Dai, S.; Yang, S. Efficiency Enhancement of Polymer Solar Cells via Zwitterion Doping in PEDOT:PSS Hole Transport Layer. *Org. Electron.* **2015**, *27*, 232–239. [CrossRef]
42. Skompska, M.; Mieczkowski, J.; Holze, R.; Heinze, J. In Situ Conductance Studies of P- and n-Doping of Poly(3,4-Dialkoxythiophenes). *J. Electroanal. Chem.* **2005**, *577*, 9–17. [CrossRef]
43. Tanase, C.; Blom, P.W.M.; de Leeuw, D.M. Origin of the Enhanced Space-Charge-Limited Current in Poly(p-Phenylene Vinylene). *Phys. Rev. B* **2004**, *70*, 193202. [CrossRef]
44. Chung, D.S.; Lee, N.H.; Yang, C.; Hong, K.; Park, C.E.; Park, J.W.; Kwon, S.K. Origin of High Mobility within an Amorphous Polymeric Semiconductor: Space-Charge-Limited Current and Trap Distribution. *Appl. Phys. Lett.* **2008**, *93*, 033303. [CrossRef]
45. Karimov, K.S.; Qazi, I.; Moiz, S.A.; Murtaza, I. Electrical Properties of Organic Semiconductor Copper Phthalocyanine Thin Films Deposited from Solution at High Gravity. *Optoelectron. Adv. Mater. Rapid Commun.* **2008**, *2*, 219–223.
46. Qi, B.; Wang, J. Open-Circuit Voltage in Organic Solar Cells. *J. Mater. Chem.* **2012**, *22*, 24315–24325. [CrossRef]
47. Ye, L.; Zhang, S.; Zhao, W.; Yao, H.; Hou, J. Highly Efficient 2D-Conjugated Benzodithiophene-Based Photovoltaic Polymer with Linear Alkylthio Side Chain. *Chem. Mater.* **2014**, *26*, 3603–3605. [CrossRef]
48. Heeger, A.J. 25th Anniversary Article: Bulk Heterojunction Solar Cells: Understanding the Mechanism of Operation. *Adv. Mater.* **2014**, *26*, 10–28. [CrossRef]
49. Nam, S.; Seo, J.; Woo, S.; Kim, W.H.; Kim, H.; Bradley, D.D.C.; Kim, Y. Inverted Polymer Fullerene Solar Cells Exceeding 10% Efficiency with Poly(2-Ethyl-2-Oxazoline) Nanodots on Electron-Collecting Buffer Layers. *Nat. Commun.* **2015**, *6*, 8929. [CrossRef]
50. Nam, S.; Song, M.; Kim, H.; Bradley, D.D.C.; Kim, Y. Thickness Effect of Bulk Heterojunction Layers on the Performance and Stability of Polymer:Fullerene Solar Cells with Alkylthiothiophene-Containing Polymer. *ACS Sustain. Chem. Eng.* **2017**, *5*, 9263–9270. [CrossRef]
51. Lenes, M.; Koster, L.J.A.; Mihailetchi, V.D.; Blom, P.W.M. Thickness Dependence of the Efficiency of Polymer: Fullerene Bulk Heterojunction Solar Cells. *Appl. Phys. Lett.* **2006**, *88*, 243502. [CrossRef]
52. Moulé, A.J.; Bonekamp, J.B.; Meerholz, K. The Effect of Active Layer Thickness and Composition on the Performance of Bulk-Heterojunction Solar Cells. *J. Appl. Phys.* **2006**, *100*, 094503. [CrossRef]
53. Park, S.; Jeong, J.; Hyun, G.; Kim, M.; Lee, H.; Yi, Y. The Origin of High PCE in PTB7 Based Photovoltaics: Proper Charge Neutrality Level and Free Energy of Charge Separation at PTB7/PC71BM Interface. *Sci. Rep.* **2016**, *6*, 35262. [CrossRef]

Article

Thermo-Optical and Structural Studies of Iodine-Doped Polymer: Fullerene Blend Films, Used in Photovoltaic Structures

Bożena Jarząbek [1,*], Paweł Nitschke [1,*], Marcin Godzierz [1], Marcin Palewicz [2], Tomasz Piasecki [2] and Teodor Paweł Gotszalk [2]

1 Centre of Polymer and Carbon Materials, Polish Academy of Sciences, 34 M. Curie-Skłodowska Str., 41-819 Zabrze, Poland; mgodzierz@cmpw-pan.edu.pl
2 Department of Nanometrology, Faculty of Electronics, Photonics and Microsystem, Wroclaw University of Science and Technology, 50-372 Wroclaw, Poland; marcin.palewicz@pwr.edu.pl (M.P.); tomasz.piasecki@pwr.edu.pl (T.P.); teodor.gotszalk@pwr.edu.pl (T.P.G.)
* Correspondence: bozena.jarzabek@cmpw-pan.edu.pl (B.J.); pnitschke@cmpw-pan.edu.pl (P.N.)

Citation: Jarząbek, B.; Nitschke, P.; Godzierz, M.; Palewicz, M.; Piasecki, T.; Gotszalk, T.P. Thermo-Optical and Structural Studies of Iodine-Doped Polymer: Fullerene Blend Films, Used in Photovoltaic Structures. *Polymers* 2022, 14, 858. https://doi.org/10.3390/polym14050858

Academic Editors: Gianmarco Griffini and Rong-Ho Lee

Received: 3 January 2022
Accepted: 17 February 2022
Published: 22 February 2022

Publisher's Note: MDPI stays neutral with regard to jurisdictional claims in published maps and institutional affiliations.

Copyright: © 2022 by the authors. Licensee MDPI, Basel, Switzerland. This article is an open access article distributed under the terms and conditions of the Creative Commons Attribution (CC BY) license (https://creativecommons.org/licenses/by/4.0/).

Abstract: Optical and structural properties of a blend thin film of (1:1 wt.) of poly(3-hexylthiophene) (P3HT) and [6,6]-phenyl-C61-butyric acid methyl ester (PCBM) doped with iodine (I_2) and then exposed to a stepwise heating were reported and compared with the properties of doped P3HT films. The UV-Vis(T) absorption measurements were performed in situ during annealing runs, at the precisely defined temperatures, in a range of 20–210 °C. It was demonstrated that this new method allows one to observe the changes of absorption spectra, connected with the iodine release and other structural processes upon annealing. In addition, the thermally-induced changes of the exciton bandwidth (W) and the absorption edge parameters, i.e., the energy gap (E_G) and the Urbach energy (E_U) were discussed in the context of different length of conjugation and the structural disorder in polymers and blends films. During annealing, several stages were distinguished and related to the following processes as: the iodine escape and an increase in P3HT crystallinity, the orderly stacking of polymer chains, the thermally inducted structural defects and the phase separation caused by an aggregation of PCBM in the polymer matrix. Moreover, the detailed X-ray diffraction studies, performed for P3HT and P3HT:PCBM films, before and after doping and then after their thermal treatment, allowed us to consider the structural changes of polymer and blend films. The effect of iodine content and the annealing process on the bulk heterojunction (BHJ) solar cells parameters was checked, by the impedance spectroscopy (IS) measurements and the J-V characteristics registration. All of the investigated P3HT:PCBM blend films showed the photovoltaic effect; the increase in power conversion efficiency (PCE) upon iodine doping was demonstrated.

Keywords: polymer:fullerene blend films; iodine doping; annealing effect; absorption edge parameters; exciton bandwidth; structural changes; BHJ solar cells

1. Introduction

Organic photovoltaic (OPV) systems have attracted much attention and have been intensively investigated in recent decades due to their advantages, such as: low production cost, light weight and mechanical flexibility [1]. Bulk-heterojunction (BHJ) PV cells, whose photoactive layers are composed of a blend of electron donating (donor-D) and electron accepting (acceptor-A) materials can maximize the interfacial D-A area, which allows for higher power conversion efficiencies of polymer OPV systems. A classic polymer donor, such as poly(3-hexylthiophene) (P3HT) blended with a fullerene derivative [6,6]-phenyl-C_{61}-butyric acid methyl ester (PCBM) acting as an acceptor were widely investigated due to the broad absorption spectra and suitable energy gap, together with the good photo-generation of mobile charge carriers [2]. Nevertheless, the power conversion efficiency of this type of BHJ structure reaches only 5–5.5% [3,4] (while the state-of-the-art devices [5] currently reach over 18%) the different aspects of P3HT:PCBM blends are still presented in an enormous amount of publications. For many years, this type of polymer:fullerene structure has been recognized as a model system for the organic photovoltaic application, despite its rather low performance as an active layer [6]. In addition, other conjugated polymers based on polythiophenes (PTs) have been investigated in BHJ photovoltaic structures [7,8].

To improve the efficiency of these BHJ solar cells, many various post-depositions strategies were developed, such as: annealing (in the air or argon) films or substrates [4,9–15] and annealing solvents [2,15–17], with the controlled drying rate of obtained films [18,19]. Another way to improve photovoltaic properties (used both for non-organic semiconductors, as for conjugated polymers) is suitable doping. Among various dopants, halogens (Br, I, Cl) are one of the most often used doping agents. Iodine-doped-polymer films used in the light emitting diodes and/or solar cells were presented in [11–13], while for BHJ, structures this type of doping is also often described [20]. The iodine doping process may be realized from the gas phase, especially for the insoluble polymer films, obtained, e.g., by the CVD process [21,22], or in the case of soluble polymers, when thin films can be obtained using "wet" methods (spin-on, spray-on, printing); the doping process can be realized in solution, where the amount of doping factor can be precisely controlled [20].

In this work, we present the effect of stepwise, controlled annealing (up to 210 °C) of the iodine (I_2)-doped (0, 1, 5, 10 mol.%) P3HT:PCBM (1:1) blend films, on the basis of in situ thermo-optical investigations. The same thermal treatment process was also used for the iodine-doped polymer (P3HT) films. This method of UV-Vis-NIR(T) measurements turned out to also be a very useful tool for investigations of the mesomorphic behavior of compounds [23] and to evaluate thermal stability of polymer thin films [24].

In our previous work [25], we investigated the behavior of neat P3HT and P3HT:PCBM blend films during annealing/cooling runs; several stages were distinguished and related to thermally inducted structural changes, using the similar thermo-optical studies.

Now, the main idea of this work was to check how the presence of iodine, as a doping factor, changes the properties of both pure polymer and blended polymer with fullerene thin films, also at the higher temperatures. It was demonstrated that the presence of PCBM affected the polymers behavior, both after doping and during annealing. These changes were discussed on the basis of absorption edge parameters (E_G, E_U) and the exciton bandwidth (W), obtained as a function of temperature, for a different % of the iodine content. Moreover, the detailed X-ray diffraction studies of polymer and blends films, before and after doping and annealing, allowed us to describe the changes of structural order and to confirm our explanation of thermo-optical results. To check the influence of doping and thermal treatment on BHJ solar cells with the P3HT:PCBM active layer, the impedance spectroscopy (IS) measurements and the current density-voltage (J-V) characteristics, together with solar cells parameters, were presented.

2. Materials and Methods

2.1. Materials

Poly(3-hexylthiophene) (P3HT, M102, M_n = 66 225 g/mol), [6,6]-Phenyl-C61-butyric acid methyl ester (PCBM, M111, >99% wt.), poly(3,4-ethylenedioxythiophene) polystyrene sulfonate dispersion in water (PEDOT:PSS, M124) were purchased from Osilla (Sheffield, UK) and used as received. Iodine crystals (p.a.) were purchased in POCH (currently Avantor Performance Materials, Gliwice, Poland) and used without further purification. Chlorobenzene was purchased from Avantor Performance Materials (Gliwice, Poland), and used as received.

2.2. Thin Films-Deposition and Thickness Measurements

The iodine doping was conducted in chlorobenzene solutions of P3HT or P3HT:PCBM (1:1 wt.) blend of 10 mg/mL concentration. The iodine was introduced in various contents

(0, 1, 5 or 10 % mol.) towards P3HT and, subsequently, prepared solutions were spin-coated on the quartz or glass substrates, at 1500 rpm, which resulted in the formation of thin films.

Thicknesses of thin films and the roughness of their surfaces were measured by the atomic force microscopy (AFM) technique, using AFM Topo-Metrix Explorer microscopy, working in a contact mode in the air, in the constant force regime. All obtained thicknesses of thin films, together with their root mean squares (RMS) of surface's roughness are gathered in Table S1 in the Supplementary Information part.

2.3. Measurements Techniques

2.3.1. UV-Vis-NIR Optical Investigations

Optical measurements were carried out using a two-beam UV-Vis-NIR, JASCO V-570 spectrophotometer, working with the Spectra Manager Program. Transmission ($T\%$) and reflectivity ($R\%$) spectra of thin films on quartz substrates were registered at room temperature, within the spectral range of 200–2500 nm. During the reflectivity measurements, a special two-beam reflectance arrangement was used, with an Al mirror in the reference beam, as a reflectance standard. Due to the small level of films reflectivity (5–8%) within the whole spectral range, the absorption coefficient (α) was calculated neglecting the reflectivity, using the simple equation [26]:

$$\alpha = \left(\frac{1}{d}\right) \ln\left(\frac{1}{T\%}\right) \quad (1)$$

where d is films' thickness

Moreover, the temperature (T) dependence of absorption coefficient, i.e., $\alpha(T)$ was obtained on the basis of transmission measurements at higher temperatures. All investigated thin films were subjected to a stepwise annealing in a special auto-controlled equipment of the JASCO spectrophotometer, which enabled the registration of transmission spectra, at precisely defined temperatures (± 0.5 °C). The special in situ computer program was used to control the heating protocol and the temperature of investigated samples. Transmission spectra of thin films were measured within the range of temperature from 20 °C up to 210 °C, every 20 °C. Between steps, the temperature was gradually increased, with a rate of 2 °C/min; the short isothermal phase was used to stabilize the target temperature. After the last step, during annealing (the measurement at 210 °C), the samples were left in the spectrometer and then transmission spectra were registered at room temperature, once more.

2.3.2. X-Ray Diffraction Studies

X-ray diffraction studies were performed using the D8 Advance diffractometer (Bruker, Karlsruhe, Germany) with Cu-Kα cathode (λ = 1.54 Å). The critical angle for conjugated polymers using copper radiation is ~0.17° (2) [27,28] and layer thickness of sample (~100 nm); for the 2D-GIWAXS setup, the 0.18° incidence angle was applied, which is just above the critical angle for polymer layer and below the critical angle for SiO$_2$ support material. The scan rate was 1.2°/min with a scanning step of 0.02° in the range of 2.5° to 60° 2Θ (dwell time 1 s). Measurements were performed in 7 variations, using different φ (Phi) angle, which corresponded to the sample rotation. As a φ = 0°, a longer edge was set as parallel to the X-Ray beam direction. The esulting φ rotation (15, 30, 45, 60, 75 and 90°) was programmed with a resolution of 0.1° φ. Obtained 2D patterns (with width of 3° 2θ) for different φ angles were integrated to 1D patterns. Background subtraction, occurring from air scattering, was performed using DIFFRAC.EVA program.

2.4. Photovoltaic Cells—Preparation and Characterization

Devices with the bulk-heterojunction structure were prepared on ITO-coated glass substrates (6 pixels, each with area of 4.5 mm^2). After cleaning the substrate with isopropanol in the ultrasonic bath, a thin film of PEDOT:PSS was deposited by spin coating. Solutions in chlorobenzene of the active layer were prepared by dissolving blends of each individual

P3HT:PCBM (1:1 wt) with 5% or 10% mol. of iodine. Such prepared solutions were spin coated on the PEDOT:PSS layer, and, subsequently, an aluminum counter electrode was evaporated on the top of the blend thin film.

The impedance spectroscopy (IS) measurements were performed using the technique both in dark and illuminated conditions, with the precise RLC meter Agilent HP E4980A in the frequency range from 20 Hz to 1 MHz with the small signal voltage excitation of 20 mVrms. In order to identify the phenomenon of photo-generation of charges, experiments in the dark and under illumination (white cold LED COB with the electrical power of 10 W, viewing the angle 140°, color temperature 6500 K and luminosity 850 mL) were completed.

The J-V curves of obtained photovoltaic devices were measured by the PV Test Solutions Solar Simulator under the AM1.5 solar illumination and using the Keithley 2400 Source Meter SMU Instrument.

3. Results and Discussion

3.1. Optical Properties

3.1.1. Iodine-Doped P3HT Thin Films

Absorption coefficient (α) spectra, obtained at the room temperature, according to the Equation (1), for the neat and iodine (I_2) doped P3HT films are presented in Figure 1.

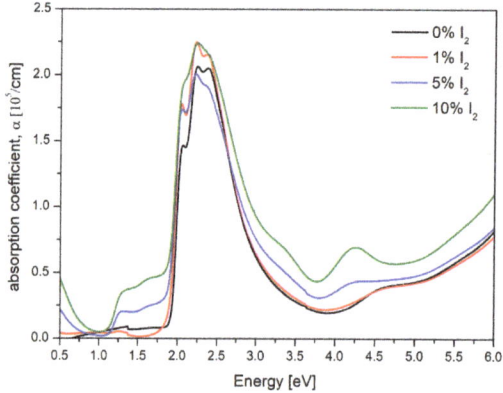

Figure 1. Absorption coefficient spectra, within the whole UV-Vis-NIR spectral range, of P3HT thin films with 0% (black), 1% (red), 5% (blue) and 10% (green) I_2 mol. concentration.

The characteristic changes of absorption spectra of P3HT film at various iodine doping level (see Figure 1) were observed: (i) at about 1.5 eV, where the absorption was connected with the polaron states, (ii) within the range 2–3 eV where the strong peak was characteristic for $\pi \rightarrow \pi^*$ electronic transitions and (iii) the absorption at about 4.2 eV, connected with the J-type aggregation (red shift of absorption peak) and intra-chain interactions [29]. The vibronic progression was clearly seen at the $\pi \rightarrow \pi^*$ absorption band but changed with the iodine doping level. Some of these vibronic bands were very distinct, particularly for the neat P3HT film; however, in the case of the film with content of 10% iodine, these features were not evident. Thus, to find precisely the position of individual peaks, the second derivative method was used (i.e., minimum of the second derivative of absorption corresponds to the absorption maximum). Then, the vibronic progression of bands was deconvoluted, with the modified Fourier self-deconvolution and finite response operator (FIRO) methods [30]. Positions of all vibronic peaks: λ_A^{0-2}, λ_A^{0-1}, λ_A^{0-0} in [nm] and [eV] and their intensities: I_A^{0-2}, I_A^{0-1}, I_A^{0-0}, obtained for all the spectra from Figure 1 are gathered in the Supplementary Information part in Table S2. Then, the exciton bandwidth

(W) parameter was estimated (assuming a Huang-Rhys factor of unity) from the ratio of (0-0) and (0-1) absorbance peaks' intensities, according to the formula [31–33]:

$$\frac{I_A^{0-0}}{I_A^{0-1}} \approx \left(\frac{1 - 024W/E_p}{1 + 0.73W/E_p}\right)^2 \quad (2)$$

using the I_A^{0-0}, I_A^{0-1} values from Table S2 and where the phonon energy E_p was involved with the main oscillator coupled to the electronic transition (a symmetric ring-stretching mode with energy 0.18 eV) [33]. The exciton bandwidth was connected with: the intra- and intermolecular excitonic coupling, electron-vibrational coupling and correlated energetic disorder, which led to the aggregate behavior in polymeric semiconductors. As shown in [31], the polymer P3HT can behave as both an H-type aggregate and a J-type aggregate, depending on the morphology (preparation method).

The edge of absorption, being the low-energy wing of the first low-energy band (the $\pi\to\pi^*$ transition band of investigated thin films) was subjected to a more detailed analysis, which is the designation of absorption edge parameters, i.e., the energy gap width (E_G) and the Urbach energy (E_U). Overall, the value of energy gap of conjugated polymers depended on the length of conjugation in the polymer chain, while the Urbach energy was connected with the localized defect states within the energy gap. The absorption edges of investigated thin films exhibited an exponential region, which could be described by the Urbach relation [34]:

$$\alpha \propto \exp\left(\frac{E}{E_U}\right) \quad (3)$$

So, the E_U values of thin films were obtained based on the slope of the exponential edge, as it is seen in Figure 2a. The Urbach energy, as a "width of the band tail" occurring due to localized states within the energy gap, is caused by possible structural defects, such as a break, torsion or aberration of the polymer chains or molecules [35]; hence the "Urbach–like" behavior of absorption edges of investigated films was observed.

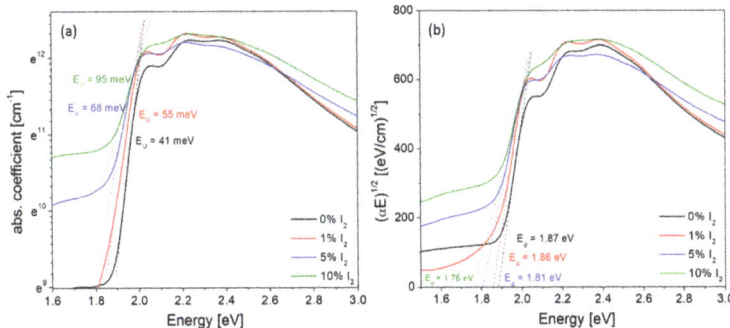

Figure 2. Absorption edges of iodine-doped P3HT thin films, used to obtain (**a**) the Urbach energy (**b**) the energy gap.

The values of energy gaps of neat and iodine-doped P3HT films were obtained based on the linear approximation to the energy axis, of the following relation [36]:

$$\alpha \propto (E - E_G)^2, \quad (4)$$

true for the energy $E > E_G$.

This dependence, known as the Tauc relation, is typical for amorphous semiconductors, and is often used for polymers thin films and freestanding foils [37–39]. Since the X-ray diffraction studies for all investigated films before and after annealing demonstrated that the crystallinity was below 50% and that absorption edges were well fitted (as seen in

Figure 2a), this dependence was used. The methods used to determine E_U and E_G is depicted in Figure 2a,b, respectively

All obtained absorption edge parameters (E_U, E_G) and exciton bandwidths' (W) values are gathered in Table S3 in the Supplementary Information part and are presented, as a function of the iodine % content, in Figure 3.

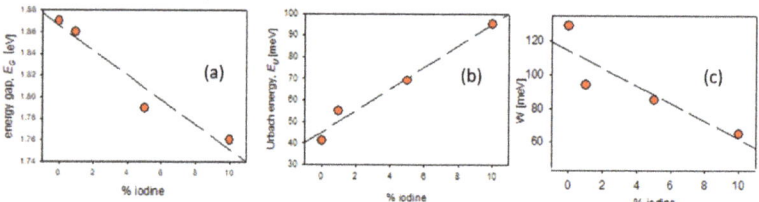

Figure 3. Optical parameters of neat and iodine-doped P3HT films (**a**) energy gap (**b**) the Urbach energy (**c**) the exciton bandwidth, as a function of % mol. content of iodine.

As it is seen in Figure 3a, the value of the energy gap decreased with the content of iodine, which could confirm the better conjugation after doping; simultaneously, the amount of defects increased, seen as an increase in the Urbach energy (Figure 3b), due to the localized defect states within the energy gap. Both the exciton bandwidth and energy gap decreased with an increase in iodine content (Figure 3c) which suggests the extension of the π-conjugation area in polymer chains.

Thermo-optical properties for the neat P3HT film were presented in [25], where this film was exposed to a stepwise heating and cooling and the changes of absorption edge parameters were discussed; the heat-inducted movement of elastic hexyl side chains and formation of defects at higher temperatures increased the free volume and decreased the order between the polymer chains [6]. Moreover, these changes turned out to be reversible and recurrent during annealing/cooling runs, while the energy gap was almost constant, which means that annealing up to 210 °C did not influence the conjugation in the main, rigid chain of P3HT film [23]. The same behavior during annealing was also observed in [38] for polymers with flexible octyloxy side chains.

In this work, thermo-optical properties of iodine-doped P3HT films were investigated and transmission spectra of P3HT thin films with 5% and 10% of iodine (I_2) mol. concentration were recorded in situ, every 20 °C, in the temperature range 20–210 °C. The absorption coefficient spectra, obtained for each temperature, are presented in Figure 4. These two concentrations of iodine were chosen for further experiments due to the best effect of power efficiency of such doped solar cells, as reported in [20].

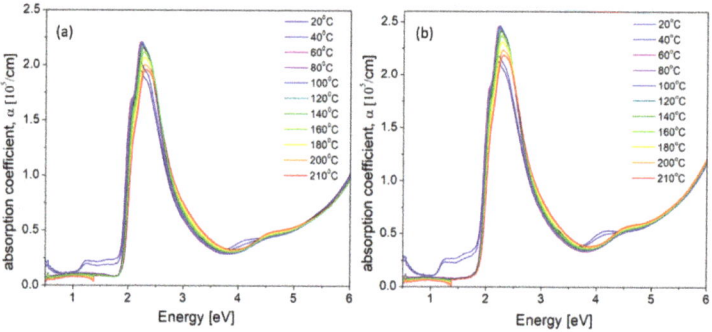

Figure 4. Absorption coefficient spectra, measured at different temperatures, within the whole UV-Vis-NIR spectral range, of iodine-doped P3HT thin films with (**a**) 5% and (**b**) 10% mol. concentration.

Then, using the Equations (3) and (4) and the same procedure, as it is seen in Figure 2, the Urbach energy and energy gap for each temperature were obtained. The method of determining absorption edge parameters, at representative temperatures, is shown in the Supplementary Information part, in Figure S1, together with all obtained optical parameters, gathered in Table S4. The temperature dependence of all obtained absorption edge parameters (E_U, E_G) and the exciton bandwidth (W) for 5% and 10% mol. concentration of iodine-doped P3HT films are presented in Figure 5. As it is seen in this figure, the content of iodine did not influence these temperature dependences of any calculated optical parameters.

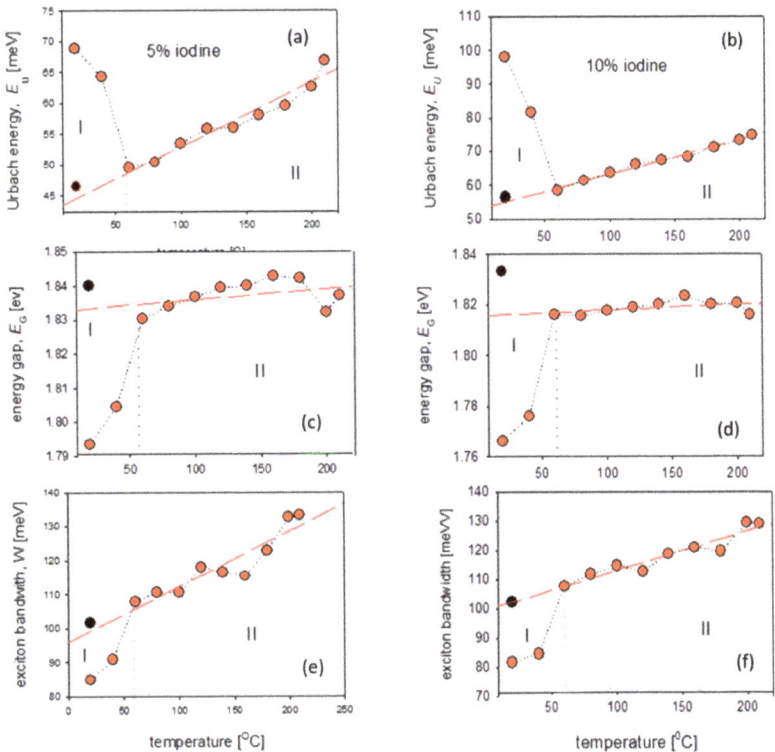

Figure 5. Temperature dependences of absorption edge parameters (E_U, E_G) and the exciton bandwidth (W) for 5% (**a,c,e**) and 10% (**b,d,f**) iodine-doped P3HT films; (black points in figures denote the values at 20 °C, after thermal treatment).

The most characteristic changes of absorption coefficient spectra (Figure S1) and optical parameters (Figure 5) connected with the iodine escape process were seen at the temperature of 60 °C. The analysis of dependences of the absorption edge parameters (E_U, E_G) and the exciton bandwidth (W) on temperature, allowed to divide these runs into two stages:

(I) In the range 20–60 °C, the Urbach energy decreased and, simultaneously, the values of energy gap and exciton bandwidth increased. Annealing of iodine-doped P3HT films led to the releasing process of dopant atoms, connected with the disappearance of localized defect states within the energy gap (the lower Urbach energy) and the extinction of polaron bands and vibronic structure (see Figure 4), which results in the worse conjugation and higher values of energy gap and exciton bandwidth.

(II) Above 60 °C, during annealing up to 210 °C, the energy gap turned out to be almost constant, both for 5% ($E_G \cong 1.84$ eV) and for 10% ($E_G \cong 1.82$ eV) iodine-doped P3HT films, while the Urbach energy and exciton bandwidth slightly increased (Figure 5). Within this range of temperature, such behavior was similar to that for neat P3HT films [25], where the linear dependence of E_U on temperature was connected with the presence of flexible, hexyl side chains, while the almost constant value of E_G during heating confirmed unchanging conjugation of polymer main chains.

Then, after cooling to the room temperature, absorption spectra and optical parameters of investigated films were obtained once more. As it is seen in Figure 5 (black points) these values differed both from initials as from these parameters at 210 °C. Due to the relaxation of structural defects during the cooling process, the Urbach energies turned out to be smaller, while energy gaps and exciton bandwidths were larger than the values obtained at 60 °C, (when P3HT films were already without iodine atoms).

3.1.2. Iodine-Doped P3HT:PCBM Blends Thin Films

Absorption coefficient spectra obtained during annealing process of 5% and 10% iodine-doped blend thin films are presented in Figure 6a,b, respectively.

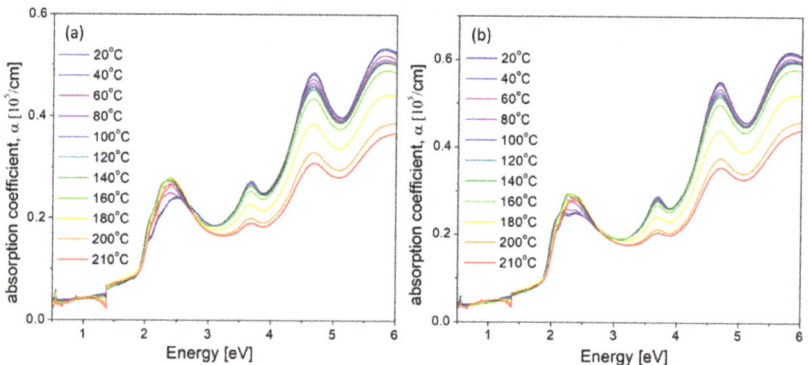

Figure 6. Absorption coefficient spectra, measured at different temperatures, within the whole UV-Vis-NIR spectral range, of iodine-doped P3HT:PCBM (1:1) blend thin films with (**a**) 5% and (**b**) 10% mol. concentration.

Changes of absorption coefficient spectra (seen in Figure 6) under the influence of higher temperatures were clearly seen for all absorption bands. The band at about 2.5 eV was connected with the electron transitions between $\pi \rightarrow \pi^*$ molecular orbitals of P3HT polymer, while three subsequent bands, seen in Figure 6, positioned at 3.70, 4.69 and 5.80 eV originated from the electron transitions in PCBM fullerene, as is described in [40]. The intensity of all bands connected with PCBM decreased during annealing, but their positions were unchanged. Decrease in the PCBM absorption bands during annealing can be explained by the formation of PCBM clusters in the P3HT matrix [23], while the bathochromic shift and increase in the P3HT band intensity with increasing temperature (seen in Figure 6 and Figure S2) are caused by an increase in P3HT crystallinity and the orderly stacking of polymer chains, respectively [23]. More information about the doped polymer:fullerene blend films behavior upon annealing may be obtained by analyzing the changes of absorption edge parameters and exciton bandwidth in higher temperatures. Similarly as for P3HT films, the way of determining absorption edge parameters, at representative temperatures, is shown in the Supplementary Information part, as Figure S2, together with all obtained optical parameters, gathered in Table S5. The temperature dependences of all obtained absorption edge parameters (E_U, E_G) and the exciton bandwidth (W) for 5% and 10% iodine-doped P3HT films are presented in Figure 7.

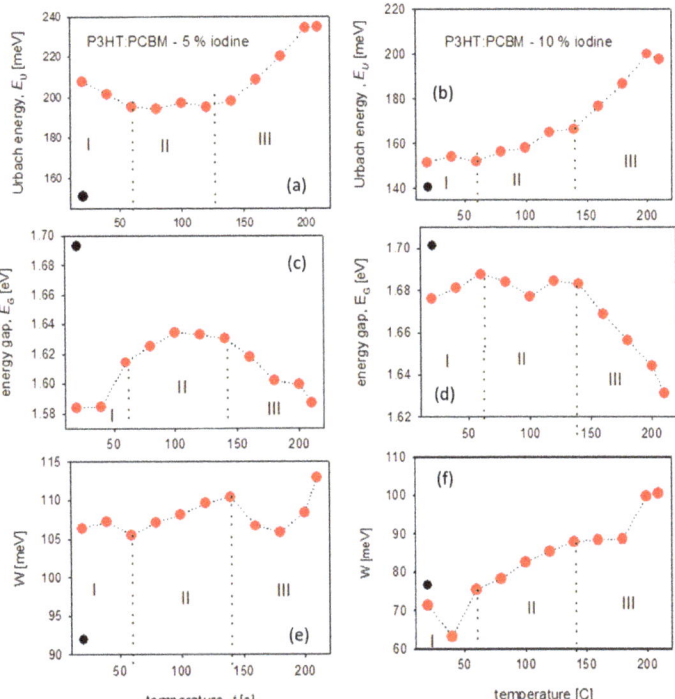

Figure 7. Temperature dependences of absorption edge parameters (E_U, E_G) and the exciton bandwidth (W) for 5% (**a,c,e**) and 10% (**b,d,f**) iodine-doped P3HT:PCBM blend films. (black points in figures mean the values at 20 °C, after thermal treatment).

Contrary to the annealing process of iodine-doped P3HT films, where two main stages were determined (see Figure 5) the behavior of P3HT:PCBM blend film during thermal treatment was more complicated and connected not only with the release of iodine (Figure 7). Moreover, we could observe unexpected differences between these runs for 5% and 10% iodine-doped blends films. Generally, three stages during annealing process were obtained:

(I) The first region, from 20 to 60 °C, was connected with the gradual iodine escape process and, simultaneously, with the increase in P3HT order. For 5% iodine, the blend predominated the polymer ordering (decrease in the Urbach energy) while, for 10% content of iodine, these two processes were seen to be in equilibrium (almost constant E_U). Since the polymer ordering is connected with an increase in P3HT crystallinity, such differences might be explained by the higher crystallinity of 10% mol. doped thin film at the beginning of annealing.

(II) The stage within the temperature range of 60–140 °C was related to the thermally-induced movements of flexible side chains of P3HT, while the conjugation in the main polymer chain was almost unchanged because the values of energy gaps within this stage were approximately on the same level. We could observe the slight increase in the Urbach energy and exciton bandwidth, together with the energy gap of about 1.63 eV for 5% and 1.68 eV for 10% iodine-doped blends films.

(III) Above 140 °C, the rapid increase in the Urbach energy and decrease in the energy gap was due to the phase separation process (which probably could have started above the PCBM glass transition temperature T_g = 124 °C [41]) and the formation of PCBM clusters, which have introduced the permanent defects.

Due to the degradation process of P3HT:PCBM blends under the influence of such high temperatures, all parameters obtained after cooling to the room temperature were difficult to interpret (see black points in Figure 7). This behavior is different from that of such thermal treatment undoped blend films [25], wich confirmed that the presence of iodine may introduce permanent structural changes.

3.2. X-Ray Diffraction Investigations

To closely investigate the morphology of active blends, X-ray diffraction measurements were performed for both neat and doped P3HT and P3HT:PCBM thin films. For the structural analysis, the unit cell parameter a is related to the short oligomer axis and c corresponds to the long axis of molecule, while b is related to the π-stacking period [28]. The calculated d-spacing for all thin films are gathered in Table S6.

3.2.1. Iodine-Doped P3HT Thin Films

Registered diffractograms of neat and iodine-doped thin films before and after thermal treatment are presented in Figure 8.

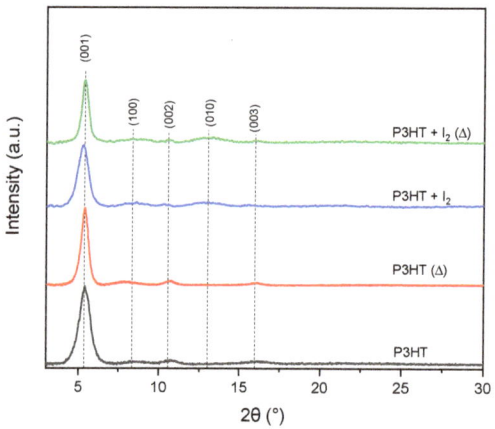

Figure 8. XRD patterns of neat P3HT (P3HT), 10% iodine-doped P3HT (P3HT + I_2) before and after (Δ) thermal treatment (in 210 °C).

For neat P3HT, the crystallinity of thin film was ~34% and slightly increased, after thermal treatment, to ~39%. Doped P3HT (10% mol. I_2) thin films revealed higher crystallinity, 43%, which increased after thermal treatment to ~46%. Observed peak for the P3HT sample did not allow us to determine lattice parameters, due to lack of a (010) peak. However, a comparison of d-spacing of neat and iodine-doped samples showed that an introduction of 10% I_2 slightly changed P3HT orthorhombic lattice. The c parameter (001 Miller index, long oligomer axis) of non-treated doped P3HT increased its dimension by 3.9%, while the enlargement for thermally treated film was only 1.5%. Annealing also slightly enlarged the chain axis (by 0.4%). In case of the a parameter (100 Miller index, short oligomer axis), introduction of I_2 provoked a slight reduction in the a axis (by approx. 1%), while thermal treatment provoked its enlargement (9.4%) in comparison to the non-treated P3HT.

Application of both, iodine and annealing, also provoked an enlargement of a parameter (3.1%). The b parameter (010 Miller index) was not visible in any neat P3HT thin film, in contrary to iodine-doped layers, suggesting random orientation of lamellas. The b parameter for I_2-doped P3HT decreased after annealing from 6.95 Å to 6.66 Å. The value decreased after thermal treatment, suggesting that this treatment allows material to obtain the higher arrangement. However, π-stacking peaks, which should be present at ~25.8° 2θ [42], were not visible, suggesting random or near-random orientation of lamellas in all samples. The π-stacking analysis of conjugated polymer systems is usually particularly inaccessible, because only the first-order peak is measurable (010 peak) [43]. Moreover, d-spacing calculated for peaks in the P3HT samples was gathered in Table S6 in the Supplementary Information part.

3.2.2. Iodine-Doped P3HT:PCBM Blends Thin Films

Introduction of PCBM into P3HT resulted in the presence of only primary peaks of orthorhombic lattice (001 and 100), with much lower lattice parameters than in the case of neat P3HT (Figure 9).

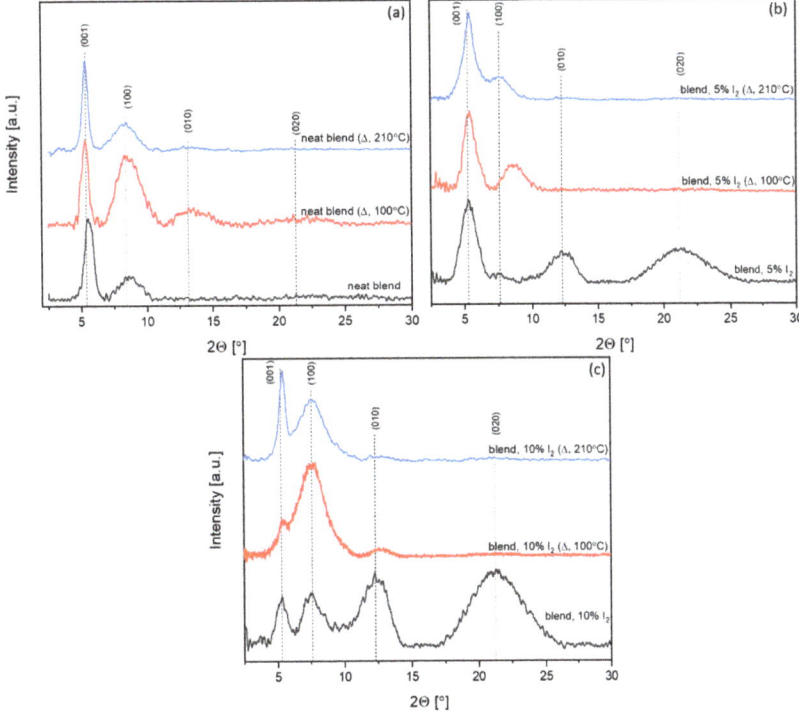

Figure 9. XRD patterns of (**a**) neat, (**b**) 5%, (**c**) 10% I_2 doped P3HT:PCBM blend before and after thermal treatment.

Registered diffractograms revealed the high contraction of chain (long) axis (3.9%), while, for the short molecule axis, it was much lower (1%). For P3HT:PCBM, the enlargement of unit cells occurred, even in comparison to the neat P3HT after thermal treatment. Compared to the non-annealed neat blend, in the sample annealed at 210 °C, the chain axis enlarged by about 5.6%, while the short oligomer axis enlarged by 6.1%. Moreover, the presence of the (010) peak was detected, which suggests that introduction of PCBM combined with thermal treatment allowed us to obtain a higher order of the P3HT structure, with random orientation of lamellas (due to absence of π-stacking peak). In the neat active blend, annealed at 100 °C, enlargement of the P3HT lattice occurred in comparison to the non-treated blend. Moreover, the higher order of the P3HT structure was detected, due to the presence of (010) and (020) peaks. Intensity of the (010) peak was higher for the blend annealed at 100 °C than for the one at 210 °C, which may suggest disordering of the P3HT structure during long-term thermal exposition, which is consistent with the optical results presented above, where a degradation of blend was observed.

In non-annealed doped P3HT:PCBM blends (5% and 10% mol. I_2 content), a high and broad (020) peak was visible, which corresponded to π-stacking. Peaks that corresponded to the *a* and *c* axis were smaller than peaks that corresponded to the *b* axis, while, in the sample, after thermal treatment, enlargement of (001) and (100) peaks occurred, with a simultaneous decrease in the intensity of (010) and (020) peaks. That might suggest the positive effect of iodine introduction on crystallization of P3HT. Crystallinity was higher for the neat blend thin film treated with 100 °C than for non-treated or treated with 210 °C and was 48%, 36% and 46%, respectively (see Table 1).

Table 1. Crystallinity of investigated thin films after heat treatment.

Heat-Treatment	Neat P3HT:PCBM	P3HT:PCBM + 5% I_2	P3HT:PCBM + 10% I_2
non-annealed	36%	37%	39%
annealed at 100 °C	48%	49%	51%
annealed at 210 °C	46%	46%	49%

In all doped P3HT:PCBM samples, the crystallinity increased along with the iodine content, for each investigated heat treatment variant. An increase in the blend order was observed after annealing at 100 °C, which slightly decreased after treatment with 210 °C. This is consistent with the results, presented above, from in situ UV-Vis measurements, where ordering of P3HT and further blend degradation were observed, respectively.

3.3. Photovoltaic Response of BHJ Devices

3.3.1. Photo-Active Impedance Spectroscopy (IS) Investigations

IS experiments of reference devices (ITO/PEDOT:PSS/P3HT:PCBM/Al) and organic solar cells with modified active layer (by adding iodine into the P3HT:PCBM solution) were conducted to define the influence of incorporated iodine on the photovoltaic phenomena and electrical parameters. Obtained spectra from IS measurements allowed for fitting the experimental data by an electrical equivalent circuit (EEC) and also on the estimate resistances and relaxation times (see Table S7) and capacitance behavior of organic solar cells. The proposed approach to the fitting of the obtained data from IS experiments was also mentioned in articles [44–47]; the same equivalent electrical circuit of the described phenomena that occurred in organic solar cells based on polymers from the polythiophene family was also used in [44–47].

In Figure 10, the Nyquist plots of all working photovoltaic devises are presented. For all devices measured in dark conditions, one semicircle in the Nyquist plots (Figure 10a,c,e) was observed. On the other hand, for samples measured under illumination, two semicircles in the Nyquist chart (Figure 10b,d,f) could be easily detected. Furthermore, for all devices, the intensive reduction in the real part of impedance after irradiation in the relation to dark measurements was noticed. This is a confirmation of the intense photo-generation of charges in the active layer of devices. Obtained impedance spectra were analysed using electric equivalent circuit (EEC) modelling. The proposed structure of EEC was shown in Figure 11.

The reduction in all resistances after the illumination was clearly noticeable in the EEC modelling results, gathered in Table S7 in the Supplementary Information part. The most instant photo-generation effect, in the case of modified samples, was observed for the sample with 10% of iodine in active layer. Furthermore, a decrease in time constants (τ) in devices with modified active layer (with 5% and 10% iodine) vs. reference undoped sample, for measurements conducted under dark conditions, was observed (from 185.70 µs to 21.25 µs, and from 748.3 µs to 193.8 µs for τ_1 and τ_2, respectively). Moreover, reduction in time constants (τ) for all devices after illumination was noticed. Such behaviour of the τ parameter confirms the fact of the improvement of photo-generation of chargers and charge transfer phenomena in modified devices, especially for the sample with 10% of iodine content.

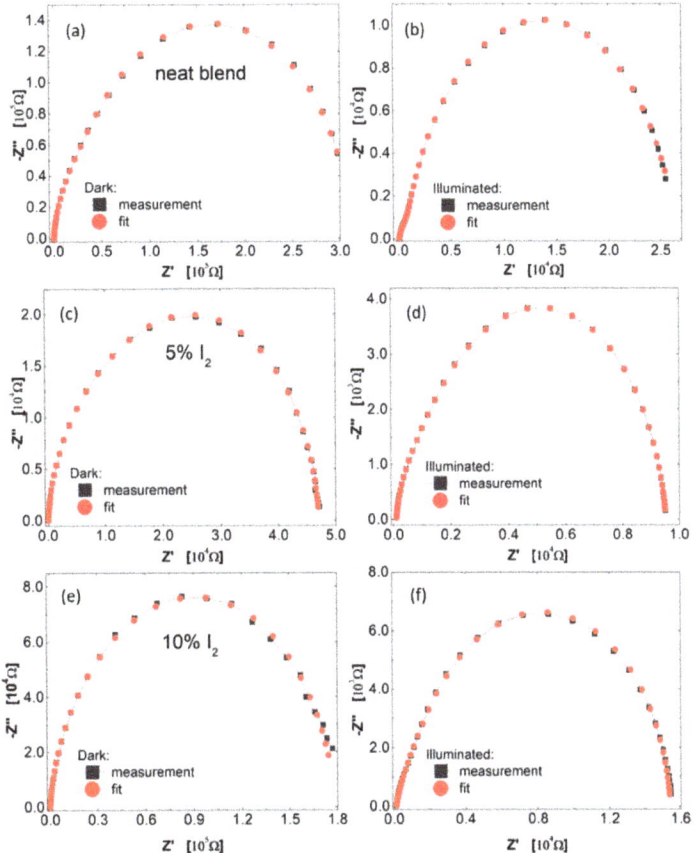

Figure 10. Electrochemical impedance spectra of standard organic solar cells (left side—in dark; right side -illuminated) with the active layer (**a,b**) undoped; (**c,d**) 5% (**e,f**) 10% iodine.

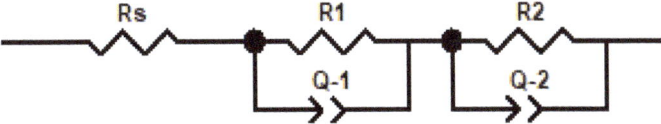

Figure 11. Electric equivalent circuit (EEC) used to model the photovoltaic devices impedance spectra in the dark and under illumination.

3.3.2. J-V Characteristics

- *Influence of iodine content*

To correlate the results from optical, structural and IS measurements, presented above, the bulk-heterojunction (BHJ) photovoltaic cells were prepared and their current density–voltage (*J-V*) characteristics were registered. Firstly, the iodine content in the P3HT:PCBM blend films was considered. Since the most pronounced effects of iodine doping on optical spectra were visible when 5% and 10% mol. iodine was introduced, such active layers were used in BHJ devices (Figure 12) and compared with a neat reference. All active layers were annealed at temperature of 100 °C for 10 min.

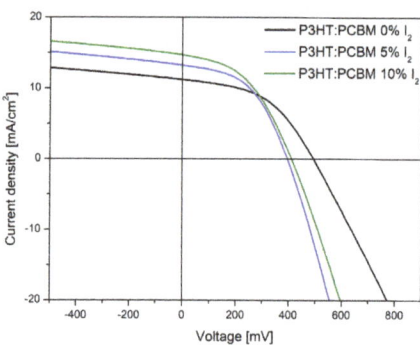

Figure 12. J-V characteristics of studied BHJ devices.

Designated parameters (Table 2) showed an increase in device performance, along with an increase in iodine content. Such an enhancement was caused by a lower active layer series resistance (R_s), which increased the short-circuit current density. This was most likely caused by a more favorable morphology of doped thin films, connected with their higher crystallinity, and is in agreement with IS results, where the improvement of photo-generation of chargers and charge transfer phenomena upon doping was observed. Introduction of 5% mol. of iodine caused a decrease in both series resistance (R_s) and shunt resistance (R_{sh}); however, the decrease in R_s was more pronounced; thus, an increase in power conversion efficiency from 2.24% to 2.41% was observed. Further, an increase in iodine content did not affect the series resistance; however, it increased the shunt resistance, lowering the current losses, and causing a further increase in PCE.

Table 2. Parameters of BHJ devices under illumination.

Layer	V_{OC} [mV]	J_{SC} [mA/cm^2]	FF	PCE [%]	R_s [Ω]	R_{sh} [kΩ]
neat	465.2 ± 21.1	9.94 ± 0.91	0.48 ± 0.01	2.24 ± 0.29	354.89 ± 39.55	5.18 ± 0.75
5% I$_2$	394.2 ± 4.4	12.95 ± 0.59	0.46 ± 0.02	2.41 ± 0.19	240.10 ± 33.59	3.87 ± 0.78
10% I$_2$	405.1 ± 6.1	14.18 ± 0.59	0.44 ± 0.01	2.61 ± 0.13	246.69 ± 26.41	4.24 ± 0.88

V_{OC}—open circuit voltage, J_{SC}—short circuit current density, FF—fill factor, PCE—power. conversion efficiency, R_s—series resistance, R_{sh}—shunt resistance.

- *Influence of the thermal treatment*

Subsequently, the thermal treatment effect of BHJ structure was considered. The 10% mol. iodine-doped active blends were tested in a non-treated form, and annealed at 40 °C, 100 °C and 180 °C. All registered current density–voltage characteristics (Figure 13) were compared with the results from the device with neat active blend (annealed at 100 °C).

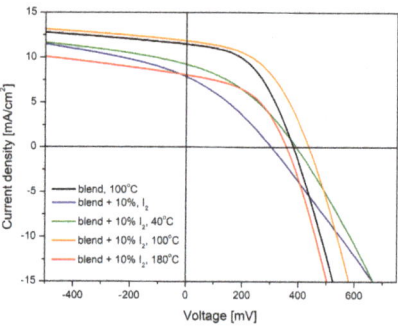

Figure 13. Current density-voltage characteristics of studied devices.

Based on the registered J-V curves, all characteristic parameters of prepared photovoltaic cells were designated and gathered in Table 3.

Table 3. Parameters of BHJ devices under illumination.

Layer (Blend)	V_{OC} [mV]	J_{SC} [mA/cm^2]	FF	PCE [%]	R_s [Ω]	R_{sh} [kΩ]
net 100 °C	376.26 ± 3.7	11.14 ± 0.32	0.48 ± 0.01	2.06 ± 0.06	248.18 ± 6.59	6.49 ± 0.82
+10% I$_2$	281.1 ± 12.3	7.68 ± 0.43	0.30 ± 0.01	0.66 ± 0.07	591.33 ± 24.91	2.62 ± 0.50
+10% I$_2$-40 °C	380.0 ± 8.0	8.80 ± 0.56	0.36 ± 0.01	1.24 ± 0.09	509.81 ± 26.60	2.61 ± 0.42
+10% I$_2$-100 °C	430.7 ± 7.0	11.10 ± 0.53	0.48 ± 0.01	2.35 ± 0.13	271.03 ± 6.95	5.89 ± 0.25
+10% I$_2$-180 °C	348.1 ± 9.2	7.83 ± 0.43	0.46 ± 0.01	1.28 ± 0.11	304.95 ± 15.76	3.97 ± 0.32

V_{OC}—open circuit voltage, J_{SC}—short circuit current density, FF—fill factor, PCE—power conversion efficiency, R_s—series resistance, R_{sh}—shunt resistance.

Analysis of registered parameters revealed that the non-annealed, doped, P3HT:PCBM active blend provided much lower efficiency than a reference device with the neat active layer. This is associated with much smaller conductivity of such thin film (higher series resistance), resulting from low crystallinity of the conducting polymer (P3HT) [48]. Slight increase in the J_{SC} after annealing at 40 °C probably resulted from an increase in the polymer crystallinity, induced by heat-induced ordering of P3HT chains [4]. The device with doped active layer, annealed at 100 °C, showed further improvement of registered parameters, achieving higher efficiency (2.35%) than a reference (2.06%). Such enhancement was most likely caused by a lower donor–acceptor interface area, suppressing the non-radiative recombination voltage losses [49]. Apart from this, the orientation of P3HT chains took place, decreasing the series resistance, and increasing R_{sh}. Further elevation of annealing temperature (180 °C), due to blend degradation, significantly lowered the device performance [41].

4. Summary and Conclusions

In conclusion, this work reported on the iodine (I$_2$)-doped polymer (P3HT) and blended with the fullerene (PCBM) thin films and their thermo-optical and structural properties, towards photovoltaic applications in BHJ structures. Results of absorption and X-ray diffraction studies showed a positive effect of iodine dopant on the crystallinity of polymer and its blend, with fullerene, inducing a bathochromic shift of the low-energy absorption band and the decrease in energy gap. However, thermo-stability of these doped blend films can be a problem, particularly in the case of their photovoltaic applications, since the solar cells are exposed to the effect of higher temperatures.

Herein, presented results provide new information about the changes in optical and structural properties of doped polymer and polymer:fullerene blend films upon annealing. The novel method of in situ optical measurements of thin films, during their annealing, allows one to observe the changes of absorption spectra, connected with the iodine release and other thermal-induced structural changes. Several processes that took place in the doped P3HT or its blend with PCBM thin films were distinguished. Temperature dependencies of the exciton bandwidth (W) and absorption edge parameters (E_U, E_G) were used to obtain:

- for the iodine-doped P3HT films, the temperature range of iodine escape (up to 60 °C). Above this temperature, the energy gap turned out to be almost constant, both for 5% ($E_G \cong 1.84$ eV) and for 10% ($E_G \cong 1.82$ eV) of iodine content, while the Urbach energy and exciton bandwidth slightly increased. These changes were connected with the presence of flexible side chains, while the almost constant value of E_G confirmed unchanging conjugation in polymer main chains.
- for iodine-doped blends films, three different stages: (i) up to 60 °C: the gradual iodine escape process and simultaneously the increase in P3HTcrystallinity; (ii) 60–140 °C: changes related to the thermally induced movements of the flexible side chains of

P3HT and the initiation of phase separation; (iii) above 140 °C: the gradual blend degradation, due to the formation of PCBM clusters, which introduced permanent defects.

These thermo-optical investigations confirmed that the presence of fullerene in blend with P3HT influenced the properties of polymer. Moreover, the content of iodine changed the structural order of blend films and their behavior during annealing.

The structural changes of neat and doped polymer and blend films were considered by X-ray diffraction studies, performed for samples annealed at 100 °C and 210 °C. In all doped P3HT:PCBM films, the crystallinity increased along with the iodine content, for each of the investigated heat treatment variants. An increase in the blend order was observed after annealing at 100 °C for all samples; the highest crystallinity (51%) was obtained for 10% iodine content blend film. Crystallinity decreased after treatment with 210 °C due to the blend thermal degradation processes.

Solar cells with these iodine-doped P3HT:PCBM active layers were investigated by the impedance spectroscopy (IS) and current density-voltage (J-V) characteristics. The highest power conversation efficiency (2.61 ± 0.13%) and the most effective photo-generation effect were detected for devices with the modified active layer by 10% of iodine. Thermal treatment of this photovoltaic device confirmed the fact that annealing at 100 °C is the best post-deposition strategy to improve solar cells parameters.

Supplementary Materials: The following supporting information can be downloaded at: https://www.mdpi.com/article/10.3390/polym14050858/s1, Table S1. References [28,42,43,45,46,50–53] are cited in the Supplementary Materials. Thicknesses and *RMS* of surfaces' roughness of polymer and blend films, Table S2. Positions and intensities of absorption bands of P3HT thin films neat and doped with various I_2 mole ratios, Table S3. Optical parameters of iodine-doped P3HT films, Figure S1. Absorption coefficient spectra within the whole spectral range of P3HT iodine-doped films, Table S4. Temperature dependence of optical parameters of P3HT iodine-doped 5% and 10% films, Figure S2. Absorption coefficient spectra within the whole spectral range of the P3HT:PCBM blend iodine-doped films, Table S5. Temperature dependence of optical parameters of thin films of P3HT:PCBM blends iodine-doped 5% and 10% films, Table S6. d-spacing calculated for peaks in P3HT samples, Table S7. The results of modeling procedure (d-dark; i-illuminated).

Author Contributions: B.J.—conceptualization, supervision, writing—review and editing; P.N.—conceptualization, methodology, investigation, writing—original draft; M.G.—methodology, investigation; M.P.—methodology, investigation, T.P.—methodology, investigation; T.P.G.—supervision. All authors have read and agreed to the published version of the manuscript.

Funding: This research received no external funding.

Institutional Review Board Statement: Not applicable.

Informed Consent Statement: Not applicable.

Data Availability Statement: The data presented in this study are available on request from the corresponding authors.

Conflicts of Interest: The authors declare no conflict of interest.

References

1. Klimov, E.; Li, W.; Yang, X.; Hoffmann, G.G.; Loos, J. Scanning Near-Field and Confocal Raman Microscopic Investigation of P3HT–PCBM Systems for Solar Cell Applications. *Macromolecules* **2006**, *39*, 4493–4496. [CrossRef]
2. Liu, Y.X.; Lü, L.F.; Ning, Y.; Lu, Y.Z.; Lu, Q.P.; Zhang, C.M.; Fang, Y.; Tang, A.W.; Hu, Y.F.; Lou, Z.D.; et al. Effects of acetone-soaking treatment on the performance of polymer solar cells based on P3HT/PCBM bulk heterojunction. *Chinese Phys. B* **2014**, *23*, 118802. [CrossRef]
3. Street, R.A.; Hawks, S.; Khlyabich, P.P.; Li, G.; Schwartz, B.; Thompson, B.C.; Yang, Y. Electronic Structure and Transition Energies in Polymer–Fullerene Bulk Heterojunctions. *J. Phys. Chem. C* **2014**, *118*, 21873–21883. [CrossRef]
4. Kadem, B.; Hassan, A.; Cranton, W. Efficient P3HT:PCBM bulk heterojunction organic solar cells; effect of post deposition thermal treatment. *J. Mater. Sci. Mater. Electron.* **2016**, *27*, 7038–7048. [CrossRef]

5. Lin, Y.; Firdaus, Y.; Isikgor, F.H.; Nugraha, M.I.; Yengel, E.; Harrison, G.T.; Hallani, R.; El-Labban, A.; Faber, H.; Ma, C.; et al. Self-assembled monolayer enables hole transport layer-free organic solar cells with 18% efficiency and improved operational stability. *ACS Energy Lett.* **2020**, *5*, 2935–2944. [CrossRef]
6. Wang, T.; Pearson, A.J.; Lidzey, D.G.; Jones, R.A.L. Evolution of structure, optoelectronic properties, and device performance of polythiophene:fullerene solar cells during thermal annealing. *Adv. Funct. Mater.* **2011**, *21*, 1383–1390. [CrossRef]
7. Lee, R.H.; Yang, L.C.; Wu, J.Y.; Jeng, R.J. Synthesis of di(ethylene glycol)-functionalized diketopyrrolopyrrole derivative-based side chain-conjugated polymers for bulk heterojunction solar cells. *RSC Adv.* **2017**, *7*, 1016–1025. [CrossRef]
8. Lee, W.H.; Liu, B.T.; Lee, R.H. Difluorobenzothiadiazole based two-dimensional conjugated polymers with triphenylamine substituted moieties as pendants for bulk heterojunction solar cells. *Express Polym. Lett.* **2017**, *11*, 910–923. [CrossRef]
9. Motaung, D.E.; Malgas, G.F.; Nkosi, S.S.; Mhlongo, G.H.; Mwakikunga, B.W.; Malwela, T.; Arendse, C.J.; Muller, T.F.G.; Cummings, F.R. Comparative study: The effect of annealing conditions on the properties of P3HT:PCBM blends. *J. Mater. Sci.* **2013**, *48*, 1763–1778. [CrossRef]
10. Kim, Y.; Choulis, S.A.; Nelson, J.; Bradley, D.D.C.; Cook, S.; Durrant, J.R. Composition and annealing effects in polythiophene/fullerene solar cells. *J. Mater. Sci.* **2005**, *40*, 1371–1376. [CrossRef]
11. Kim, H.; So, W.W.; Moon, S.J. The importance of post-annealing process in the device performance of poly(3-hexylthiophene): Methanofullerene polymer solar cell. *Sol. Energy Mater. Sol. Cells* **2007**, *91*, 581–587. [CrossRef]
12. Zhokhavets, U.; Erb, T.; Gobsch, G.; Al-Ibrahim, M.; Ambacher, O. Relation between absorption and crystallinity of poly(3-hexylthiophene)/fullerene films for plastic solar cells. *Chem. Phys. Lett.* **2006**, *418*, 347–350. [CrossRef]
13. Lee, W.H.; Chuang, S.Y.; Chen, H.L.; Su, W.F.; Lin, C.H. Exploiting optical properties of P3HT:PCBM films for organic solar cells with semitransparent anode. In *Thin Solid Films*; Elsevier: Amsterdam, The Netherlands, 2010; Volume 518, pp. 7450–7454.
14. Watts, B.; Belcher, W.J.; Thomsen, L.; Ade, H.; Dastoor, P.C. A Quantitative Study of PCBM Diffusion during Annealing of P3HT:PCBM Blend Films. *Macromolecules* **2009**, *42*, 8392–8397. [CrossRef]
15. Campoy-Quiles, M.; Ferenczi, T.; Agostinelli, T.; Etchegoin, P.G.; Kim, Y.; Anthopoulos, T.D.; Stavrinou, P.N.; Bradley, D.D.C.; Nelson, J. Morphology evolution via self-organization and lateral and vertical diffusion in polymer:fullerene solar cell blends. *Nat. Mater.* **2008**, *7*, 158–164. [CrossRef]
16. Li, G.; Yao, Y.; Yang, H.; Shrotriya, V.; Yang, G.; Yang, Y. "Solvent annealing" effect in polymer solar cells based on poly(3-hexylthiophene) and methanofullerenes. *Adv. Funct. Mater.* **2007**, *17*, 1636–1644. [CrossRef]
17. Chen, F.C.; Ko, C.J.; Wu, J.L.; Chen, W.C. Morphological study of P3HT:PCBM blend films prepared through solvent annealing for solar cell applications. *Sol. Energy Mater. Sol. Cells* **2010**, *94*, 2426–2430. [CrossRef]
18. Li, G.; Shrotriya, V.; Huang, J.; Yao, Y.; Moriarty, T.; Emery, K.; Yang, Y. High-efficiency solution processable polymer photovoltaic cells by self-organization of polymer blends. *Nat. Mater.* **2005**, *4*, 864–868. [CrossRef]
19. Sasaki, K.; Yamanari, T.; Ohashi, N.; Ogo, H.; Yoshida, Y.; Ueda, Y. Elucidation of formation mechanism of bulk heterojunction active layer by real-time uv-visible absorption and grazing-incidence wide-angle X-ray scattering. *Appl. Phys. Express* **2013**, *6*, 4–7. [CrossRef]
20. Zhuo, Z.; Zhang, F.; Wang, J.; Wang, J.; Xu, X.; Xu, Z.; Wang, Y.; Tang, W. Efficiency improvement of polymer solar cells by iodine doping. *Solid. State. Electron.* **2011**, *63*, 83–88. [CrossRef]
21. Jarząbek, B.; Hajduk, B.; Jurusik, J.; Domański, M. In situ optical studies of thermal stability of iodine-doped polyazomethine thin films. *Polym. Test.* **2017**, *59*, 230–236. [CrossRef]
22. Jarząbek, B.; Weszka, J.; Hajduk, B.; Jurusik, J.; Domanski, M.; Cisowski, J. A study of optical properties and annealing effect on the absorption edge of pristine- and iodine-doped polyazomethine thin films. *Synth. Met.* **2011**, *161*, 969–975. [CrossRef]
23. Iwan, A.; Janeczek, H.; Jarzabek, B.; Domanski, M.; Rannou, P. Characterization, optical and thermal properties of new azomethines based on heptadecafluoroundecyloxy benzaldehyde. *Liq. Cryst.* **2009**, *36*, 873–883. [CrossRef]
24. Schab-Balcerzak, E.; Siwy, M.; Jarzabek, B.; Kozanecka-Szmigiel, A.; Switkowski, K.; Pura, B. Post and prepolymerization strategies to develop novel photochromic poly(esterimide)s. *J. Appl. Polym. Sci.* **2011**, *120*, 631–643. [CrossRef]
25. Jarząbek, B.; Nitschke, P.; Hajduk, B.; Domański, M.; Bednarski, H. In situ thermo-optical studies of polymer:fullerene blend films. *Polym. Test.* **2020**, *88*, 106573. [CrossRef]
26. Jarzabek, B.; Schab-Balcerzak, E.; Chamenko, T.; Sek, D.; Cisowski, J.; Volozhin, A. Optical properties of new aliphatic-aromatic co-polyimides. *J. Non. Cryst. Solids* **2002**, *299–302*, 1057–1061. [CrossRef]
27. Schuettfort, T.; Thomsen, L.; McNeill, C.R. Observation of a Distinct Surface Molecular Orientation in Films of a High Mobility Conjugated Polymer. *J. Am. Chem. Soc.* **2013**, *135*, 1092–1101. [CrossRef]
28. Zajaczkowski, W.; Nanajunda, S.K.; Eichen, Y.; Pisula, W. Influence of alkyl substitution on the supramolecular organization of thiophene- and dioxine-based oligomers. *RSC Adv.* **2017**, *7*, 1664–1670. [CrossRef]
29. Baghgar, M.; Labastide, J.; Bokel, F.; Dujovne, I.; McKenna, A.; Barnes, A.M.; Pentzer, E.; Emrick, T.; Hayward, R.; Barnes, M.D. Probing Inter- and Intrachain Exciton Coupling in Isolated Poly(3-hexylthiophene) Nanofibers: Effect of Solvation and Regioregularity. *J. Phys. Chem. Lett.* **2012**, *3*, 1674–1679. [CrossRef]
30. Jones, R.N.; Shimokoshi, K. Some Observations on the Resolution Enhancement of Spectral Data by the Method of Self-Deconvolution. *Appl. Spectrosc.* **1983**, *37*, 59–67. [CrossRef]
31. Spano, F.C.; Silva, C. H- and J-aggregate behavior in polymeric semiconductors. *Annu. Rev. Phys. Chem.* **2014**, *65*, 477–500. [CrossRef]

32. Clark, J.; Chang, J.F.; Spano, F.C.; Friend, R.H.; Silva, C. Determining exciton bandwidth and film microstructure in polythiophene films using linear absorption spectroscopy. *Appl. Phys. Lett.* **2009**, *94*, 2007–2010. [CrossRef]
33. Clark, J.; Silva, C.; Friend, R.H.; Spano, F.C. Role of intermolecular coupling in the photophysics of disordered organic semiconductors: Aggregate emission in regioregular polythiophene. *Phys. Rev. Lett.* **2007**, *98*, 206406. [CrossRef] [PubMed]
34. Cody, G.D. Chapter 2 The Optical Absorption Edge of a-Si: H. *Semicond. Semimet.* **1984**, *21*, 11–82. [CrossRef]
35. Jarzabek, B.; Weszka, J.; Domański, M.; Jurusik, J.; Cisowski, J. Optical studies of aromatic polyazomethine thin films. *J. Non. Cryst. Solids* **2008**, *354*, 856–862. [CrossRef]
36. Tauc, J.; Menth, A. States in the gap. *J. Non. Cryst. Solids* **1972**, *8–10*, 569–585. [CrossRef]
37. Jarząbek, B.; Wójtowicz, M.; Wolińska-Grabczyk, A. Optical Studies of Poly(hydroxy imide) to Polybenzoxazole Thermal Rearrangement. *Macromol. Chem. Phys.* **2015**, *216*, 2377–2385. [CrossRef]
38. Jarząbek, B.; Kaczmarczyk, B.; Jurusik, J.; Siwy, M.; Weszka, J. Optical properties of thin films of polyazomethine with flexible side chains. *J. Non. Cryst. Solids* **2013**, *375*, 13–18. [CrossRef]
39. Nitschke, P.; Jarząbek, B.; Wanic, A.; Domański, M.; Hajduk, B.; Janeczek, H.; Kaczmarczyk, B.; Musioł, M.; Kawalec, M. Effect of chemical structure and deposition method on optical properties of polyazomethines with alkyloxy side groups. *Synth. Met.* **2017**, *232*, 171–180. [CrossRef]
40. Karagiannidis, P.G.; Georgiou, D.; Pitsalidis, C.; Laskarakis, A.; Logothetidis, S. Evolution of vertical phase separation in P3HT:PCBM thin films induced by thermal annealing. *Mater. Chem. Phys.* **2011**, *129*, 1207–1213. [CrossRef]
41. Agostinelli, T.; Lilliu, S.; Labram, J.G.; Campoy-Quiles, M.; Hampton, M.; Pires, E.; Rawle, J.; Bikondoa, O.; Bradley, D.D.C.; Anthopoulos, T.D.; et al. Real-time investigation of crystallization and phase-segregation dynamics in P3HT:PCBM solar cells during thermal annealing. *Adv. Funct. Mater.* **2011**, *21*, 1701–1708. [CrossRef]
42. Tang, M.; Zhu, S.; Liu, Z.; Jiang, C.; Wu, Y.; Li, H.; Wang, B.; Wang, E.; Ma, J.; Wang, C. Tailoring π-Conjugated Systems: From π-π Stacking to High-Rate-Performance Organic Cathodes. *Chem* **2018**, *4*, 2600–2614. [CrossRef]
43. Rivnay, J.; Mannsfeld, S.C.B.; Miller, C.E.; Salleo, A.; Toney, M.F. Quantitative Determination of Organic Semiconductor Microstructure from the Molecular to Device Scale. *Chem. Rev.* **2012**, *112*, 5488–5519. [CrossRef] [PubMed]
44. Romero, B.; Del Pozo, G.; Arredondo, B.; Reinhardt, J.P.; Sessler, M.; Wurfel, U. Circuital model validation for s-shaped organic solar cells by means of impedance spectroscopy. *IEEE J. Photovolt.* **2015**, *5*, 234–237. [CrossRef]
45. Perrier, G.; De Bettignies, R.; Berson, S.; Lemaître, N.; Guillerez, S. Impedance spectrometry of optimized standard and inverted P3HT-PCBM organic solar cells. *Sol. Energy Mater. Sol. Cells* **2012**, *101*, 210–216. [CrossRef]
46. Arredondo, B.; Romero, B.; Del Pozo, G.; Sessler, M.; Veit, C.; Würfel, U. Impedance spectroscopy analysis of small molecule solution processed organic solar cell. *Sol. Energy Mater. Sol. Cells* **2014**, *128*, 351–356. [CrossRef]
47. Knipper, M.; Parisi, J.; Coakley, K.; Waldauf, C.; Brabec, C.J.; Dyakonov, V. Impedance spectroscopy on polymer-fullerene solar cells. *Zeitschrift fur Naturforsch. Sect. A J. Phys. Sci.* **2007**, *62*, 490–494. [CrossRef]
48. Reisdorffer, F.; Haas, O.; Le Rendu, P.; Nguyen, T.P. Co-solvent effects on the morphology of P3HT:PCBM thin films. *Synth. Met.* **2012**, *161*, 2544–2548. [CrossRef]
49. Tang, Z.; Wang, J.; Melianas, A.; Wu, Y.; Kroon, R.; Li, W.; Ma, W.; Andersson, M.R.; Ma, Z.; Cai, W.; et al. Relating open-circuit voltage losses to the active layer morphology and contact selectivity in organic solar cells. *J. Mater. Chem. A* **2018**, *6*, 12574–12581. [CrossRef]
50. Kronemeijer, A.J.; Gili, E.; Shahid, M.; Rivnay, J.; Salleo, A.; Heeney, M.; Sirringhaus, H. A selenophene-based low-bandgap donor-acceptor polymer leading to fast ambipolar logic. *Adv. Mater.* **2012**, *24*, 1558–1565. [CrossRef]
51. Xiao, X.; Wang, Z.; Hu, Z.; He, T. Single Crystals of Polythiophene with Different Molecular Conformations Obtained by Tetrahydrofuran Vapor Annealing and Controlling Solvent Evaporation. *J. Phys. Chem. B* **2010**, *114*, 7452–7460. [CrossRef]
52. Privitera, A.; Righetto, M.; De Bastiani, M.; Carraro, F.; Rancan, M.; Armelao, L.; Granozzi, G.; Bozio, R.; Franco, L.; Carraro,F. Hybrid Organic/Inorganic Perovskite−Polymer Nanocomposites: Toward the Enhancement of Structural and Electrical Properties. *J. Phys. Chem. Lett.* **2017**, *8*, 5981–5986. [CrossRef] [PubMed]
53. Macdonald, J.R. *Impedance Spectroscopy*; Pergamon Press Ltd.: London, UK, 1992; Volume 20, ISBN 9780471831228.

Article

Investigation of Dye Dopant Influence on Electrooptical and Morphology Properties of Polymeric Acceptor Matrix Dedicated for Ternary Organic Solar Cells

Gabriela Lewińska [1,*], Piotr Jeleń [2], Jarosław Kanak [1], Łukasz Walczak [3], Robert Socha [4], Maciej Sitarz [2], Jerzy Sanetra [†] and Konstanty Waldemar Marszałek [1]

1 Faculty of Computer Science, Electronics and Telecommunication, AGH University of Science and Technology, 30 Mickiewicza Ave., 30059 Krakow, Poland; kanak@agh.edu.pl (J.K.); marszale@agh.edu.pl (K.W.M.)
2 Department of Silicate Chemistry and Macromolecular Compounds, Faculty of Materials Science and Ceramics, AGH University of Science and Technology, 30059 Krakow, Poland; pjelen@agh.edu.pl (P.J.); msitarz@agh.edu.pl (M.S.)
3 Science & Research Division, PREVAC sp. z o.o., Raciborska 61, 44362 Rogow, Poland; lukasz.walczak@prevac.pl
4 Jerzy Haber Institute of Catalysis and Surface Chemistry, Polish Academy of Sciences, Niezapominajek 8, 30239 Krakow, Poland; robert.socha@ikifp.edu.pl
* Correspondence: glewinska@agh.edu.pl; Tel.: +48-692-376-639
† The author Jerzy Sanetra is retired; jsanetra@agh.edu.pl.

Abstract: The publication presents the results of investigations of the influence of dye dopant on the electrooptical and morphology properties of a polymeric donor:acceptor mixture. Ternary thin films (polymer:dye:fullerene) were investigated for potential application as an active layer in organic solar cells. The aim of the research is to determine the effect of selected dye materials (dye D131, dye D149, dye D205, dye D358) on the three-component layer and their potential usefulness as an additional donor in ternary cells, based on P3HT donor and PC71BM acceptor. UV–vis spectroscopy studies were performed, and absorption and luminescence spectra were determined. Ellipsometry parameters for single dye and ternary layers have been measured. The analyses were performed using the Raman spectroscopy method, and the Raman spectra of the mixtures and single components have been determined. Organic layers were prepared and studied using scanning electron microscope and atomic force microscope. For dyes, ultraviolet photoelectron spectroscopy and X-ray photoelectron spectroscopy studies were carried out and the ternary system was presented and analyzed in terms of energy bands.

Keywords: dye; organic solar cells; ellipsometry; ternary organic films; morphology examination

1. Introduction

Organic photovoltaics (OPV) has been developing in recent years due to its cost, solution processibility, semi-transparency, mechanical flexibility, light weight, and has become increasingly interesting for technology [1–3]. The possibility of application in wearable systems, connection to IOT (Internet of Things) and powering personal devices [4–6] as well as the hope for the use of organic cells in space [7], causes substantial popularity of the organic photovoltaic. It should also be mentioned the possibility of using OPV in indoor applications, which have excellent potential for use in flexible and wearable electronics [8–10]. One of the rising trends in organic photovoltaics are ternary solar cells.

In ternary solar cells [11], a suitably selected third component (additional donor or additional acceptor) is introduced into the active layer to increase cell efficiency. The mixture (donor 1:donor 2:acceptor blend) forms a bulk heterostructure and in the photovoltaic process, and photons are absorbed. Materials that broaden the absorption spectrum and facilitate the transport of holes and electrons are selected as the third element [12].

A beneficial effect of the extra component on the morphology of the thin film has also been observed.

From the first mention of two-component active layer additives and three-component systems appeared around the mid-2000s. Since then, quite a few ternary systems have been synthesized and tested, with efficiency increasing from 2% to about 17% [13,14].

In ternary systems, different chemical compounds were used because of their molecular structure (polymers or small molecules). The output efficiency of the binary system was increased by 12% (in the ratio of P3HT: PCBM PCPDTBT 0.8:1:0.2) using the addition of the polymer poly[2,6-(4,4-bis-(2-ethylhexyl)-4H-cyclopenta[2,1-b;3,4-b′]-dithiophene)-alt-4,7-(2,1,3-benzothiadiazoles)] (PCPDTBT) [15]. In 2015, a efficiency over 5% (a 20% improvement, relative to the binary system) was achieved on the doping of naphthalene azomethylenediimides [16]. Another cells operating using poly[4-(5-(4,8-bis(dodecyloxy)-4,8-dihydrobenzo[1,2-b:4,5-b′]dithiophen-2-yl)-alt-5,8-bis-(thiophen-2-yl)-6,7-bis(3,4-bis(dodecyloxy)phenyl)-2-dodecyl-2H-[1,2,3]triazolo[4,5-g]quinoxaline) (PBDT- BTzQx-C12) with BDT as the donor building block and TzQ$_x$ as the acceptor building block. These systems (depending on the percentage amount from 1 to 4%) achieved efficiency from 2.73 to 3.54% [17]. Poly[[4,8-bis[(2-ethylhexyl)oxy]benzo[1,2-b:4,5-b′]dithiophen-2,6-diyl][3-fluoro-2-[(2-ethylhexyl)carbonyl]thieno[3,4-b]thiophenediyl]] (PTB7)-based cells present some of the highest reported efficiencies for polymer fullerene solar cells due to enhanced near-infrared absorption and lower HOMO. Polymer-doped cells in the donor-acceptor configuration of PTB7: PCBM showed efficiencies above 8.6% [18] and 10.4 % [19].

Dyes are also a type of additive being developed and studied in ternary systems. Their main advantages are (usually) high values of absorption coefficients. Many dyes are characterized by absorption spectra in the sunlight range. For P3HT:PCBM systems doped with dyes bis(trihexylsilyl oxide) silicon phthalocyine (SiPc) and bis(trihexylsilyl oxide) silicon naphthalocyanine (SiNc). The addition of both compounds increased the energy conversion efficiency to 4.3% compared to single dye ternary solar cells when illuminated with the AM1.5G spectrum SiPC:P3HT:PCBM and SiNC:P3HT:PCBM systems achieved 4.1% and 3.7%, respectively [20]. The SiPC derivative silicon bis(6-azidohexanoate)phthalocyanine ((HxN3)2-SiPC) with crosslinking groups was also used to obtain cells with an efficiency of 3.4% [13]. High efficiencies were achieved by ternary systems with the addition of small-molecule hinges efficiencies of 7–10% [21–23].

This paper presents a study of ternary mixtures of polymer:dye:fullerene for potential applications as an active layer in ternary organic cells [24,25]. The introduction of an additional substance (donor-dye) is aimed at reducing the roughness and improving the quality of the layer. The aim of the study was to check the influence of the investigated compounds on the morphology of layers in ternary systems.

The dye materials selected were chosen for their requested properties. Surface sensitization dyes are standard process for DSSCs, the aim of this research is to investigate them in the context of applications in organic electronics, more specifically in organic photovoltaic cells. It is also required to consider their potential for use as colors controller in a polymer matrix [26], in organic transistors [27] or as organic memories [28].

The typical active layer in organic cells is a mixture of donor and acceptor materials (bulk heterojunction, BHJ). BHJ cells are much cheaper and simpler to construct, in comparison to bilayer devices as they can be applied from wet phase. In the case of ternary solar cells, we are dealing with three chemical compounds. It is therefore important to select a common solvent. In an ideal case, the acceptor is homogeneously dispersed in the matrix formed by the donors, thus forming a three-dimensional network. The mixture allows a large increase in the contact area between these materials, leading to an increase in the number of excitons produced, as well as facilitating their dissociation into free charge carriers.

2. Materials and Methods

Thiophene-based compounds, polyacetylene derivatives, or polyanilines are used as donor materials, among others. Compounds based on fullerene or perylene diamides are commonly used as acceptor materials. In this case, poly(3-hexylthiophene-2,5-diyl) (P3HT) was used as a donor material, phenyl-C71-butyric acid methyl ester (PC71BM) [29] was chosen as an acceptor (the chemical formulas are shown in Figure 1a,b). Series of four dye materials were tested as dopants (potential second donors) in the study: dye D131 (2-cyano-3-[4-[4-(2,2-diphenylethenyl)phenyl]-1,2,3,3a,4,8b-hexahydrocyclopent[b]indol-7-yl]-2-propenoic acid), dye D149 (5-[[4-[4-(2,2-diphenylethenyl)phenyl]-1,2,3-3a,4,8b-hexahydrocyclopent[b]indol-7-yl]methylene]-2-(3-ethyl-4-oxo-2-thioxo-5-thiazolidinylidene)-4-oxo-3-thiazolidineacetic acid), dye D205 (5-[[4-[4-(2,2-diphenylethenyl)phenyl]-1,2,3,3a,4,8b-hexahydrocyclopent[b]indol-7-yl]methylene]-2-(3-octyl-4-oxo-2-thioxo-5-thiazolidinylidene)-4-oxo-3-thiazolidineacetic acid) and D358 (5-[3-(carboxymethyl)-5-[[4-[4-(2,2-diphenylethenyl)phenyl]-1,2,3,3a,4,8b-hexahydrocyclopent[b]indol-7-yl]methylene]-4-oxo-2-thiazolidinylidene]-4-oxo-2-thioxo-3-thiazolidinedodecanoic acid). They were provided by Merck KGaA (Darmstadt, Germany), along with other reagents. The information about dyes were collected in Table S1 in Supplementary Materials. Chemical formulas of the tested dye materials are shown in Figure 1c–f [30]. The materials were provided in solid form (colored powder). The solutions were made at concentrations of 10 mg/mL. In this study, the materials were dissolved in spectroscopically grade chloroform.

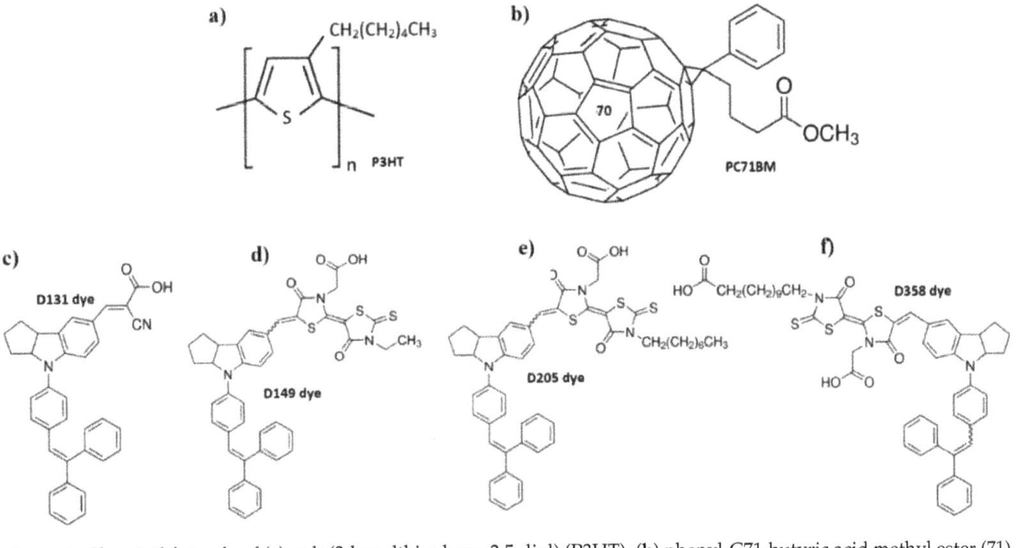

Figure 1. Chemical formula of (**a**) poly(3-hexylthiophene-2,5-diyl) (P3HT), (**b**) phenyl-C71-butyric acid methyl ester (71) and dye materials under investigation: (**c**) dye D131, (**d**) dye D149, (**e**) dye D205, (**f**) dye D358.

Spectroscopic studies were performed using Avantes Sensline Ava-Spec ULS-RS-TEC fiber optic spectrophotometer (Avantes, Appelsdorn, The Netherlands) with Avantes AvaLight DH-S-BAL-Hal lamp. Absorption and luminescence spectra of solid were investigated as a thin film (quartz was used as a reference) within the range 250–1100 nm. Photoluminescence tests were carried out using laser excitation at wavelengths λ_{ex} = 405 nm and λ_{ex} = 532 nm. Raman measurements were performed using Witec Alpha 300 M+ spectrometer (WITec, Ulm, Germany) equipped with the 488 nm laser, 600 groove grating, and a 100× ZEISS objective (Oberkochen, Germany). Laser power was adjusted to prevent sample degradation. The samples were deposited on a glass substrate.

For spectroscopic ellipsometry, analysis was carried out using Woolam M-2000 ellipsometer (Lincoln, NE, USA), with a wavelength range from UV to near-infrared. Atomic force microscope pictures were taken using the NTMDT Ntegra Aura (Apeldoorn, The Netherlands) system in SemiContact mode. Scanning electron microscope (SEM) analyses were done using ultra-high resolution scanning electron microscopy with field emission (FEG—Schottky emitter) NOVA NANO SEM 200 (Hitachi, Japan) cooperating with an EDAX EDS analyzer.

The X-ray photoelectron spectroscopy/ultraviolet photoelectron spectroscopy (XPS/UPS) experiment (PREVAC sp. z o.o., Rogow, Poland) was performed using the XPS/UPS/ARPES PREVAC setup, in an ultra-high vacuum (UHV) chamber with a base pressure around 8×10^{-10} mbar and at room temperature. The analysis chamber was equipped with a Ea15 PREVAC hemispherical analyzer and UVS 40B source PREVAC (UV power U = 0.56 kV, PUV = 55 W, He I). Binding energy (BE) scale was calibrated at the Fermi level 16.87 eV.

3. Results and Discussions

3.1. Optical Properties

The absorption spectra for all compounds considered are presented in Figure 2. Each of the compounds showed a broad absorption spectrum, in the case of D131 with one maximum, the others show the presence of two maxima. D139 has a distinct absorption maximum at 464 nm. The other three compounds have dual-maximum spectra, located for dye D149:400 nm, 560 nm, dye D205:395 nm, 545 nm, dye D358:398 nm, 545 nm, respectively. Examining the absorption edges of the compounds, the energy gaps were determined as follows: for D131 Eg = 2.2 eV, for D149 Eg = 1.8 eV, for D205 Eg = 1.9eV, and for D385 Eg = 1.9 eV.

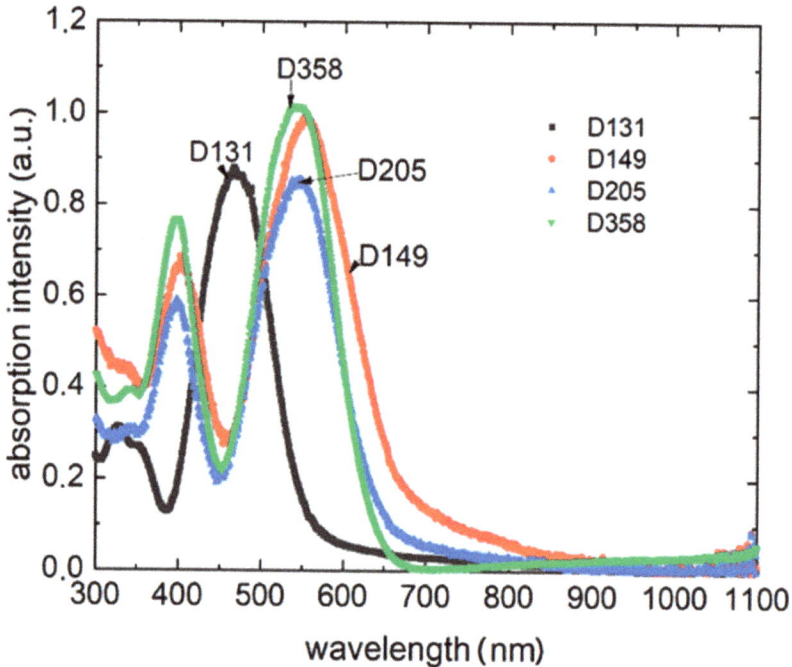

Figure 2. Absorption spectra of the investigated dyes.

All dyes exhibit clear photoluminescence (Figure 3) for blue and green laser excitation wavelength excitation ($\lambda_{ex} = 405$ nm and $\lambda_{ex} = 532$ nm).

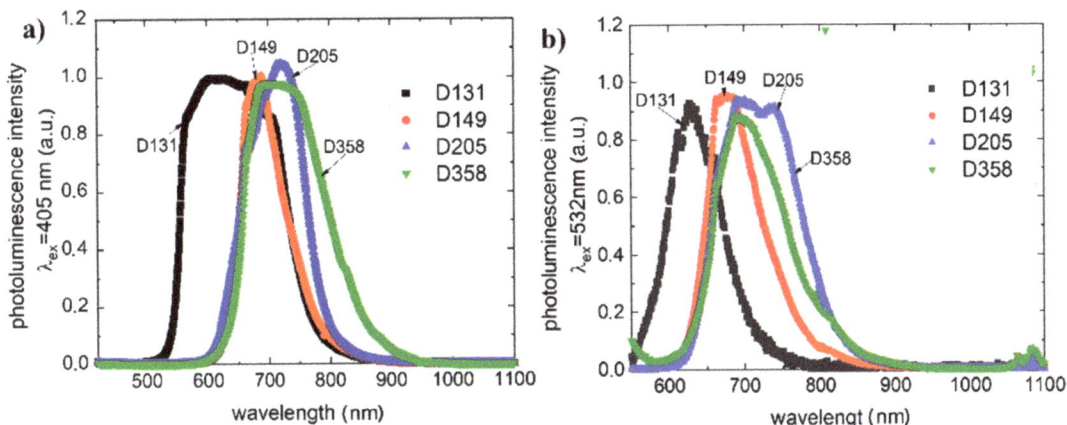

Figure 3. Photoluminescence spectra of the studied dyes (**a**) $\lambda_{ex} = 405$ nm, (**b**) $\lambda_{ex} = 532$ nm.

For compound D131 in the case of excitation (405 nm) we obtained a maximum indicating (photoluminescence) state at 559, 607, 664 and 708 nm. The green laser-induced (532 nm) spectrum does not show the first maximum. The other excitations taper slightly at 611, 623, and 708 nm, respectively. For the D149 dye, the absorption maxima shown for both excitations are: 661, 673, and 688 nm. The luminescence spectrum for dye D205 excited by blue laser shows two minor maxima at 662 and 685 nm and expressed at 720 nm. Treated with green excitation, the spectrum of D205 splits into two distinct luminescence bands with maxima at 694 and 739 nm. The maximum observed for 405 nm excitation is almost invisible. The luminescence spectrum of D358 has small maxima around 660 nm and main band with a maximum of 689 nm. It is observed for both investigated excitations, however, in the case of green laser excitation the luminescence band narrows.

Figure 4 shows the absorption and photoluminescence ($\lambda_{ex} = 405$ nm) spectra for example dyes for different weight ratios of dye to donor (P3HT). The weight ratio of donors to acceptor (PCBM) was 1:1.

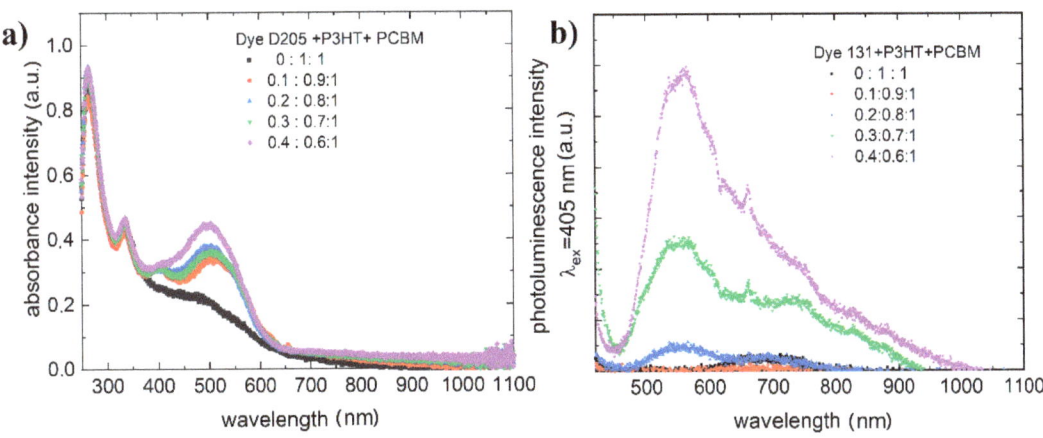

Figure 4. Absorption (**a**) and photoluminescence (**b**) ($\lambda_{ex} = 405$ nm) spectra for mixtures dye:P3HT:PCBM.

The photoluminescence coming from the P3HT:PCBM mixture is quenched by absorption of D131 (Figure 4) and only emission from D131 is visible (Figure 3).

The results of the dispersion relations for the refractive indices (n) of the studied compounds and for the extraction coefficients (k) are presented in Figure 5. The mode used to model the active layer was general oscillators (Gen-Osc) composed of Cody–Lorentz and Tauc–Lorentz type oscillators after initial spline modeling.

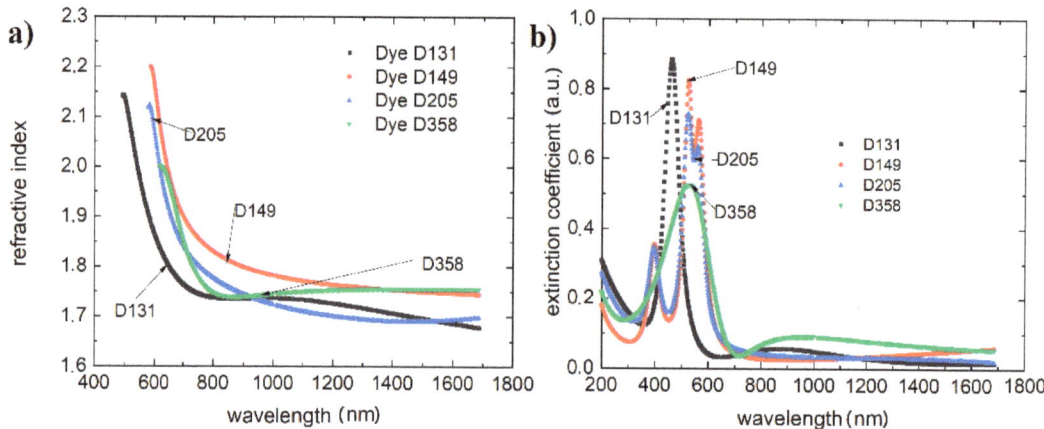

Figure 5. (a) Refractive indices versus wavelength; (b) extinction coefficients versus wavelength for the investigated materials.

The Lorentz-type oscillator and derived oscillator model were used during fitting. The classical harmonic oscillator model is very similar to the Lorentz oscillator, but is derived from the quantum theory of mechanical perturbations.

$$\varepsilon(E) = \varepsilon_1(E) - i\varepsilon_2(E) = \varepsilon_\infty + \frac{2AE_0}{E_0^2 - E^2 + i\Gamma E + \frac{1}{4}\Gamma^2} \quad (1)$$

In Equation (1), A is approximately the peak of ε_2 at the resonant energy and Γ is the full width at half the maximum value found with the Lorentz oscillator. Tauc–Lorentz [31] (Equation (2)) and Cody–Lorentz [32] (Equation (3)), models provide a more realistic representation of real materials and are widely used to describe many amorphous dielectrics and semiconductors (a detailed description can be found in Supplementary Materials).

The Tauc–Lorentz absorption formula is as follows:

$$\varepsilon_2 \propto \frac{(E - E_g)^2}{E_g^2} \quad (2)$$

and the Cody–Lorentz model is shown by the equation:

$$\varepsilon_2 \propto (E - E_g)^2 \quad (3)$$

All dyes exhibit complex multi-oscillator spectra.

The simplest single-maximal dispersion spectra of extinction coefficients have been recorded for dyes D131 and D358, which corresponds to the absorption spectra. The spectra of D149 and D205 show a multipeak character. With respect to the dispersion dependence of the refractive index, they show differentiated maxima for D131 at 496 nm reaching 2.14, for D149 at 587 nm reaching 2.19, D205 at 585 nm having a refractive index of 2.12 and compound D358 at 625 nm having a refractive index of 2.00. The dispersion dependence of the refractive indices after passing through the maximum is stabilized at the level of about 1.7–1.8. The refractive index and extinction coefficient values for the 633 nm wavelength are summarized in Table 1. The thicknesses of the layers tested in can be found in the Supplementary Materials (Table S2).

Table 1. Refractive indices and extinction coefficients values for wavelength $\lambda = 633$ nm investigated materials.

Compound	Refractive Index for 633 nm	Extinction Coefficient for 633 nm
D131	1.81	0.171
D149	2.04	0.112
D205	2.03	0.101
D358	1.94	0.0385

Due to the application potential, we found the effect of dye in ternary systems (with PCBM and P3HT) on the optical properties of the active layer (Figure 6).

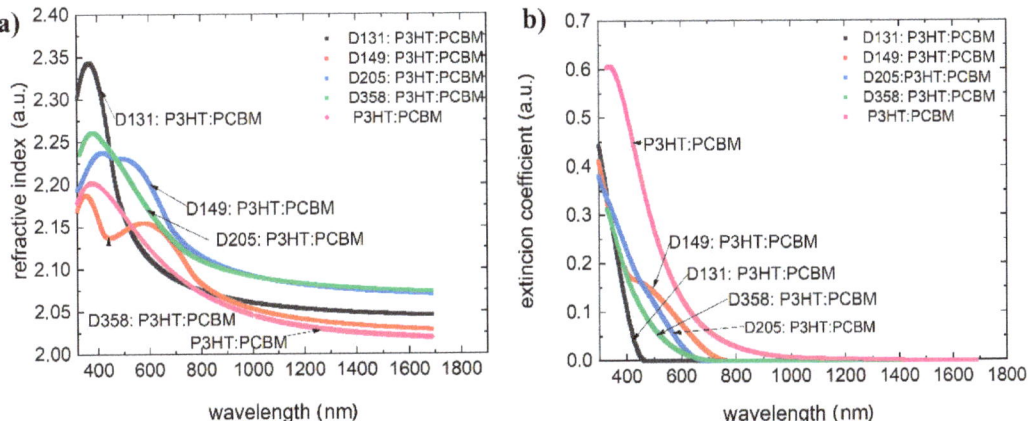

Figure 6. (a) Refractive indices versus wavelength, and (b) extinction coefficients versus wavelength for mixtures dye:P3HT:PCBM.

3.2. Surface Morphology

To evaluate the surface morphology, atomic force microscopy (AFM) and scanning electron microscopy (SEM) were performed. The study of ternary systems was carried out with reference to the donor-acceptor system. Images of the layers under consideration are shown in the AFM: Figure 7 and SEM: Figure 8. AFM profiles are included in Figure S1 (Supplementary Materials).

The results at 10,000× magnification confirm that the obtained layers are smooth and well distributed (apart from negligible surface defects). At 100,000× magnification, the structure of the thin film is a mixture of elongated, slightly twisted elements (similar to fingerprint texture). The mixture in which clear structures are observed is a mixture with the dye D205:P3HT:PCBM. Strong inhomogeneities and emerging spherical formation can be seen. It can be concluded that in the case of D131, D149 and D358 substances we obtain a mixture in which the second donor is embedded in the donor, both donors are distributed in the acceptor [33–35].

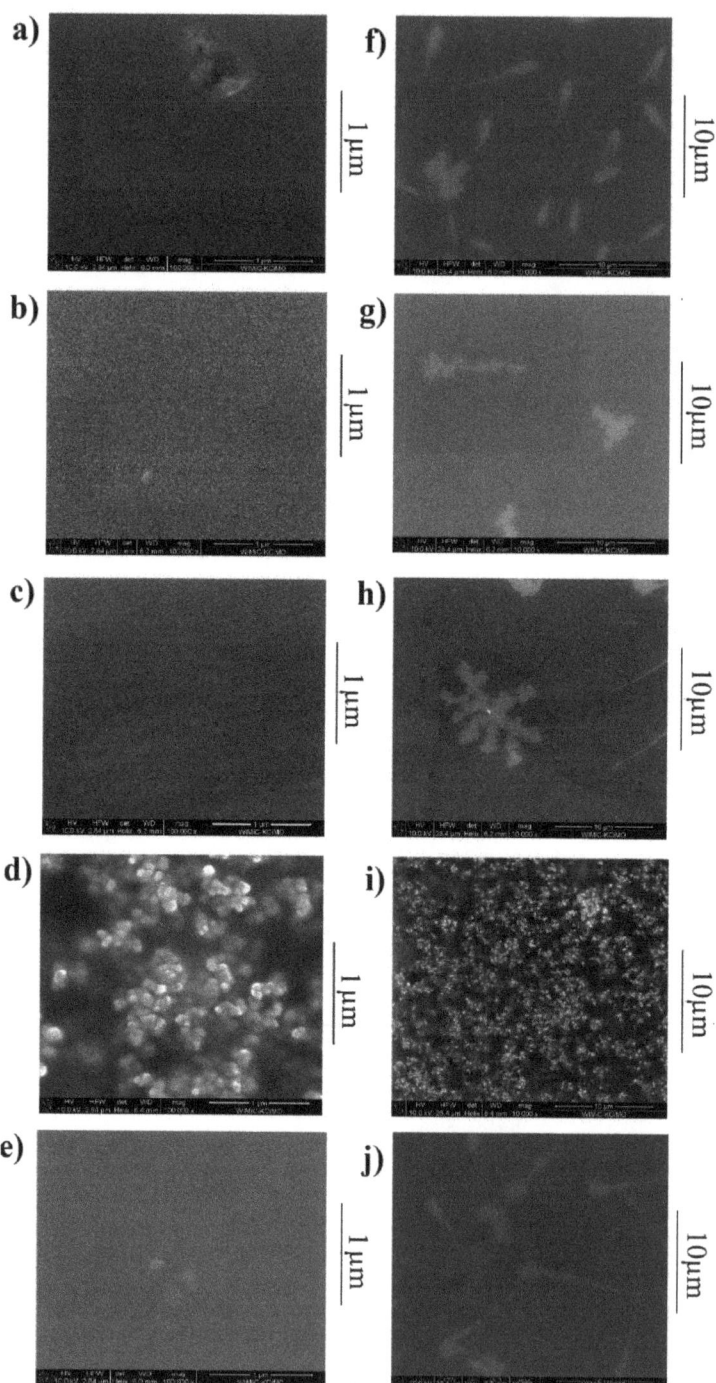

Figure 7. SEM images of the tested mixtures: (**a,f**) P3HT:PCBM; (**b,g**) dye D131:P3HT:PCBM; (**c,h**) dye D149:P3HT:PCBM; (**d,i**) dye D205:P3HT:PCBM; (**e,j**) dye D358:P3HT:PCBM. Left row shows images taken at a magnification of 100,000× times, the right row at a magnification of 10,000× times.

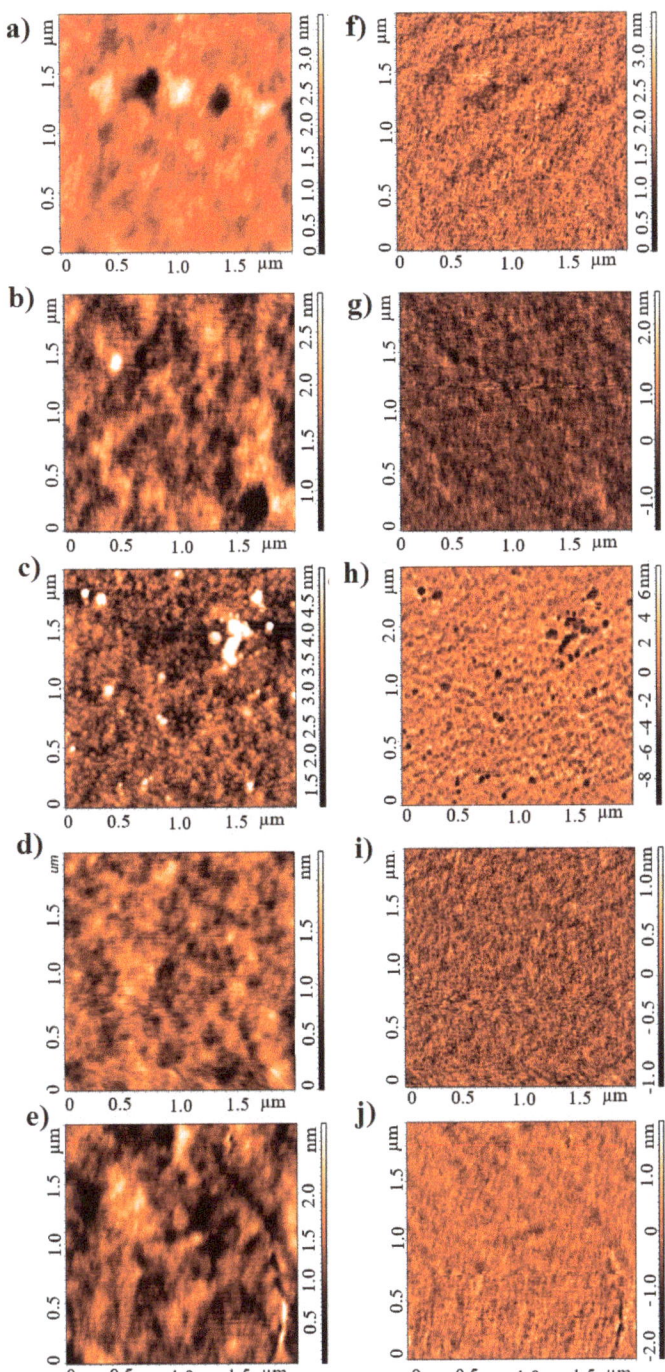

Figure 8. AFM images of the tested mixtures (**a**,**f**) P3HT:PCBM; (**b**,**g**) dye 201 D131:P3HT:PCBM; (**c**,**h**) dye D149:P3HT:PCBM; (**d**,**i**) dye D205:P3HT:PCBM; (**e**,**j**) dye 202 D358:P3HT:PCBM. Left row shows the morphology, the right row presents phase contrast.

The layer roughness's obtained from the AFM measurements are summarized in Table 2. The addition of a third dye component (with the exception of D205), reduces the layer roughness.

Table 2. Roughnes Rq parameter values for the investigated materials.

Layer Composition	Roughness Rq Parameter (nm)
P3HT:PCBM 1:1	0.640
D131:P3HT:PCBM	0.379
D149:P3HT:PCBM	0.323
D205:P3HT:PCBM	0.528
D358:P3HT:PCBM	0.248

3.3. Raman Spectroscopy

Due to the low robustness of material D131, it was not possible to obtain a Raman spectrum. Material D205 also showed slight degradation. The degradation of dye 131 is likely related to the fact that there is no sulfur in the compound, which affects the thermal properties, unlike the other investigated dyes [36]. The resulting spectra are shown in Figure 9.

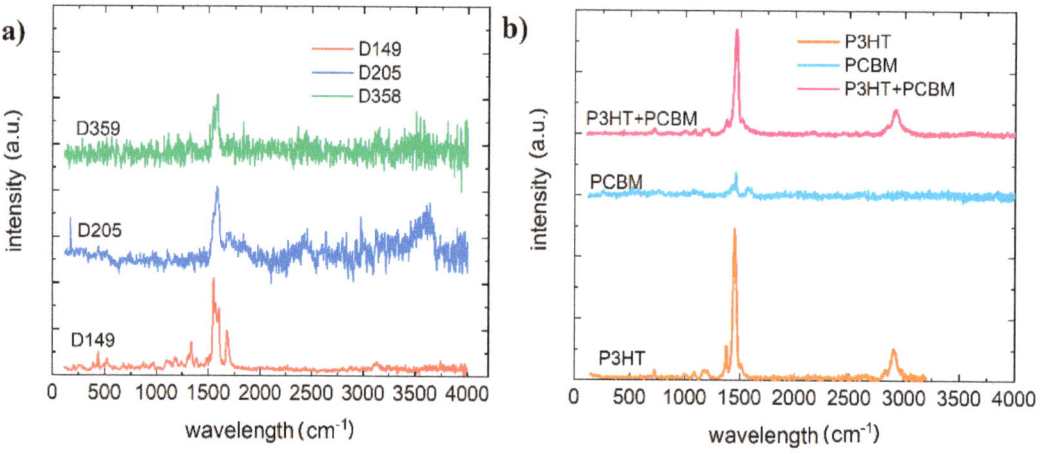

Figure 9. Raman plot (**a**) for dyes D149, D205, and D358 (**b**) P3HT, PCBM, and mixture P3HT and PCBM.

Pure PCBM and P3HT test was carried out by Yadov et al. [37,38]. For P3HT C–S–C ring deformation (725 cm^{-1}), C–C intra-ring stretching 1379 cm^{-1} and symmetric C=C stretching vibrations (1447 cm^{-1}) were identified. PCBM Raman spectrum shows four peaks located at 658, 1038, 1127 and 1570 cm^{-1}. The intensity of the P3HT peaks is much higher than that of the PCBM, so that mainly the peaks of P3HT are visible in the mixture. Due to the extended range of measurements, a peak was also recorded at 2920 cm^{-1} corresponding to sp3 C–H stretching mode. The quantitative effects of the dye material and the effects of the investigated materials on the ternary system are shown in the Figure 10.

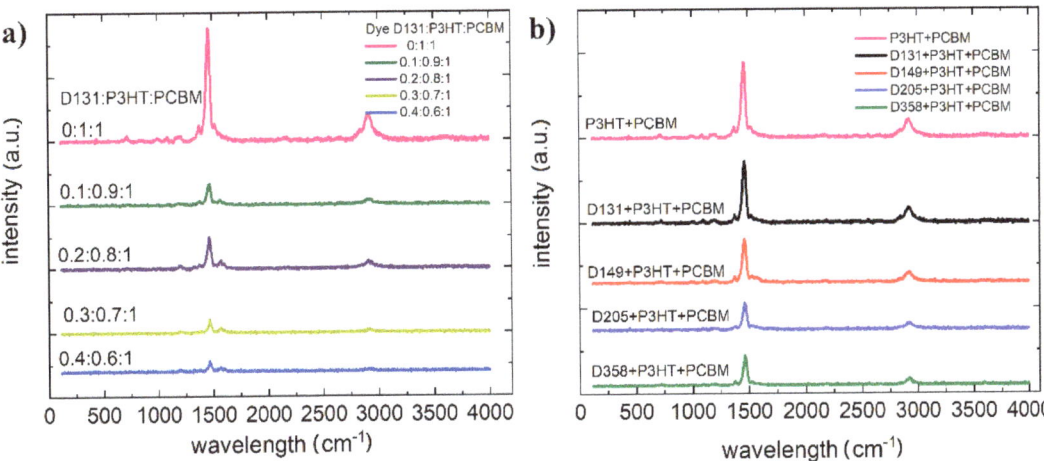

Figure 10. Raman plot (**a**) of dye D139:P3HT:PCBM ternary mixtures with different concentrations (**b**) of dye:P3HT:PCBM ternary mixtures for the investigated dyes. The intensity of the peaks has been normalized.

All dye compounds have intense vibrations of strong double bonds (C=C) around 1453 cm^{-1}. Dye D149 shows peaks at 1590, 1619, and 1713 cm^{-1} and D205 at 1650 cm^{-1} and D358 at 1515 cm^{-1} and 1605 cm^{-1}. This is identified with the vibrations of two carbon atoms linked by strong double bonds (C=C). In three-component mixtures this is quenched by the 1465 cm^{-1} vibration from P3HT.

3.4. X-ray Photoelectron Spectroscopy/Ultraviolet Photoelectron Spectroscopy

The He I UPS spectra of thin films investigated materials deposited on Au (1 1 1) were presented in Figure 11. We obtained a survey spectrum for investigated materials (shown in Figure 11a). A comparison of the cut-offs for all samples is also provided (Figure 11b). The obtained valence spectrum in different ranges is shown in Figure 11c,d. On the diagrams the postulated subsequent quantum states (P1–P4) are marked.

The S 2p spectra (Figure 11a) show main electronic state of sulfur at electron binding energy of approximately 164.5 eV, which is assigned to sulfur in thiazolidine ring. The lower maximum at 162.3 eV is ascribed to thioketone group (S=C). Additionally, some oxidation of sulfur to sulfate species is indicated by slight increase of background above 168 eV. The latter is observed well for D358 compound but it is insignificant for the other ones.

The C 1s (Figure 12b) spectra are very similar for D131, D205 and D358 compounds but different for D149. The difference is related to a shoulder at approximately 287 eV, which indicates presence of carbonyl/ketone groups (C=O) well related to dye structure. On the other hand, D205 has similar structure but the C 1s spectrum differs significantly of D149. Such difference can be an effect of much longer aliphatic chain (8 carbons) of D205 than in case of D149 (1 carbon).

The N 1s (Figure 12c) spectra show similar shape of all spectrum envelopes with some negative shift of the intensity maximum for D131. The shift is app. 1 eV and it is correlated with lower electronegativity of nitrogen in the indole structure than in thiazolidine ring. Additionally, these spectra suggest that indole structure is rather screened by thiazolidine ring that can indicate external location of the latter one at the D149, D205 and D358 samples surfaces.

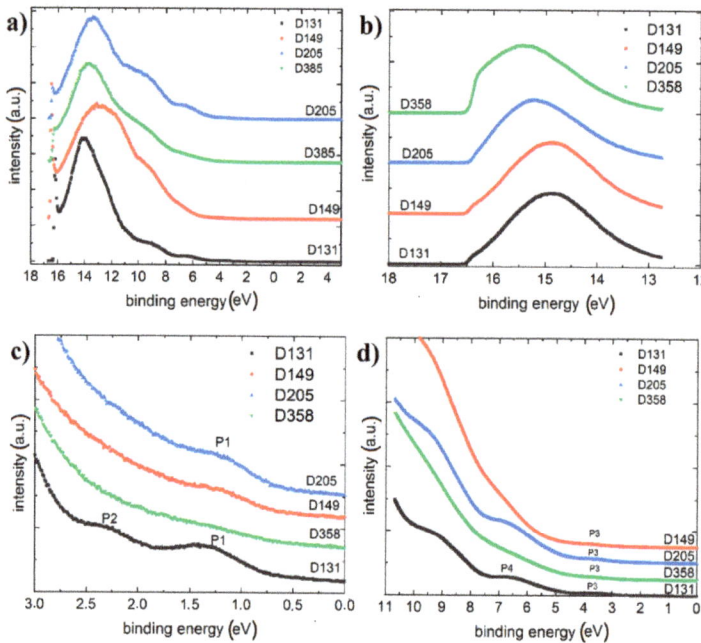

Figure 11. UPS results for investigated materials (**a**) survey spectrum for investigated materials (**b**) comparison of the cut-offs for all samples (**c**), UPS spectrum of the valence bands for all samples (**d**).

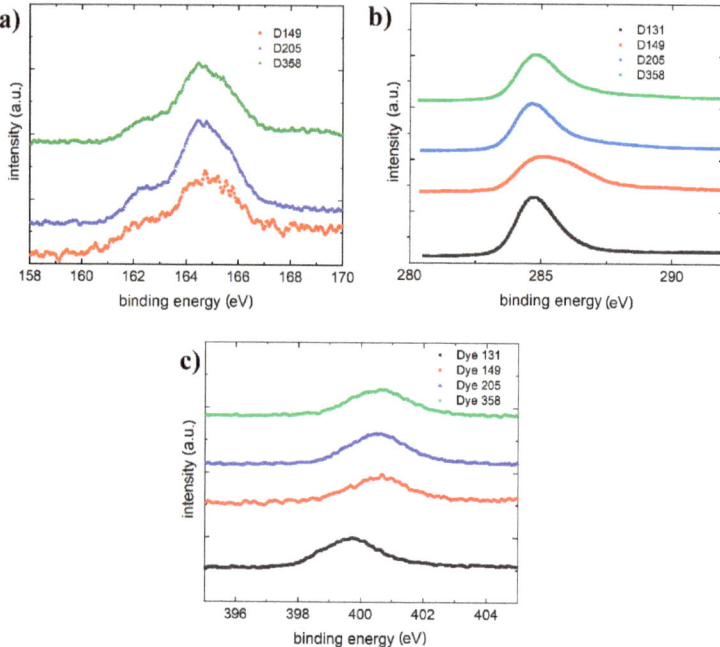

Figure 12. (**a**) The S 2p core excitations of D149, D205 and D358 thin films surface (**b**) the C 1s core excitations (**c**) the N 1s core excitations of D131, D149, D205 and D358 thin films surfaces.

4. Discussion

Investigated dyes have broad absorption spectra, overlapping with the spectrum of sunlight, which makes it suitable for use as a donor element. Dyes were previously simulated using the Cauchy model [39] and EMA-layer [40], however, for the compounds under consideration, the oscillator model [41,42] was chosen because of the broad absorption band.

The law of additivity of absorbance [43] is used in the spectrophotometric analysis of multicomponent systems. If there are more substances in a solution that absorb radiation at a selected wavelength, then the absorbance of this solution is equal to the sum of the absorbances of its individual components. From this point of view, the inserting of additional components is beneficial to produce the most absorbent material.

It can be seen that the photoluminescence increases with the concentration of the dye. This type of effect is observed up to a certain concentration [44]. Absorption of the dye in the region of 400 to 600 nm further absorbs the emission of the P3HT:PCBM mixture [45].

According to the energy levels, the acceptor picks up all the excited electrons causing all the luminescence coming from dyes and P3HT to be quenched. Subsequent studies have indicated however, that this relationship deviates from the linear function, even for non-interacting compounds [46]. For multicomponent systems, an analysis that considers the Kramers–Kronig relation is warranted, linking concentration to oscillator strength. PCBM and P3HT mixtures are quite well established. Their spectra of extinction coefficient and refractive index were studied by Ng et al. [47]. Furthermore, the addition of a third component shows that the reflection coefficient of the active layer is also reduced. The antireflective property is also very favorable.

Surface tests on P3HT: PCBM have already been carried out by Hajduk et al. [48]. Increasing the donor proportion, caused an increase in roughness. AFM images showed a smoother surface for binary mixtures with a lower amount of donor (P3HT) [49]. We therefore focused on observing the effect of third component on the donor:acceptor system. Improved quality (reduced roughness) is the expected result [50]. Only the additive dye D205 is not promising, due to the structure formation. The other layers present a blended structure. The homogeneity of the layer has an additional advantage: exciton in the mixture is generally short-lived, so the size of the individual domains is therefore critical. It is also essential that the domains are interconnected so that there are uninterrupted pathways for both electrons and holes to be transported to the electrodes. Despite the beneficial effect of an increased amount of donor on the optical properties, in view of the electrical properties, the amount of donor material cannot be increased continuously. The active layer ingredients should have a homogeneous structure to ensure an adequate transport of the carriers under the energy levels. Chemically, Raman spectroscopy results showed a domination of molecular dynamics by vibrations originating from P3HT, both over the PCBM acceptor and additional donors (dyes).

Based on the HOMO levels obtained from the UPS spectra and the energy gap obtained from the absorption spectrum, we can determine the energy diagram for potential ternary solar cells (Figure 13). The HOMO level was determined by determination from Figure 11b (comparison of the cut-offs for all samples) Fermi level position and Fermi level position relative to the HOMO level (VB). These quantities were subtracted from the He-I radiation energy radiation of 21.2 eV [51,52]. It should be further considered here that the number of P3HT molecules decreases, which molecule receives the holes of the excited molecules quenching the possible photoluminescence from the dyes. From the data shown in Figure 13, it is further evident that the acceptor molecules receive excited electrons via photoluminescence.

Due to the similar energy levels of both donors, it can be concluded that the transport of charge carriers will be supported by an additional donor. The application of the materials and the fabrication of the cells require further specified research.

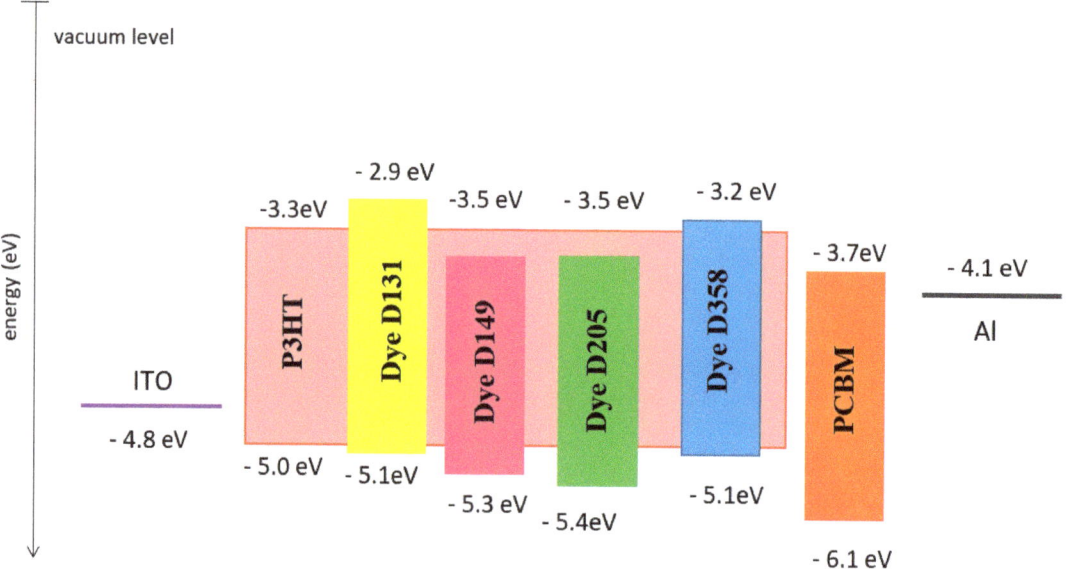

Figure 13. Energy diagram for the investigated materials in the context of the use of their ternary organic cells.

5. Summary

The optical properties, morphology, and molecular states of the selected dyes as well as their ternary mixtures with a donor (P3HT) and an acceptor (PCBM) have been studied in this publication. The aim was to determine their potential for use in the active layer of ternary organic cells. Considering the absorption spectra, all the considered dyes are excellent materials for active layer enrichment, as they have a wide absorption spectra in the region of maximum solar radiation. All dyes also exhibit luminescence, so they are able to produce a stable excited state. Analysis of thin film morphology indicates that D205 dye is not necessarily suitable for the application due to the lack of homogeneity of the thin film at nanometer level. The other dyes with donor (P3HT) and acceptor (PCBM) form very consistent layers and in terms of surface morphology are most optimistic for three components active layer. The resulting energy levels structure also places itself in the range supporting charge transport. The presented properties of the considered compounds indicate the high applicability for implementation in organic ternary solar cells. Tests have been conducted in air but research in vacuum is planned as polymer photovoltaics are likely to be used in space.

Supplementary Materials: The following are available online at https://www.mdpi.com/article/10.3390/polym13234099/s1, Figure S1: AFM profiles for the tested mixtures (a) P3HT:PCBM (b) dye D131:P3HT:PCBM (c) dye D149:P3HT:PCBM (d) dye D205:P3HT:PCBM (e) dye D358:P3HT:PCBM, Table S1: Information about materials, Table S2: Thicknesses of investigated layers determined by spectroscopic ellipsometry.

Author Contributions: Conceptualization, G.L.; investigation, G.L., J.K., P.J., M.S., Ł.W., J.S. and R.S.; writing—original draft preparation, G.L. and R.S.; writing—review and editing, G.L. and P.J.; visualization, G.L.; supervision, K.W.M. All authors have read and agreed to the published version of the manuscript.

Funding: This research was funded by National Science Center, grant number 2020/04/X/ST5/00792.

Institutional Review Board Statement: Not applicable.

Informed Consent Statement: Not applicable.

Data Availability Statement: The data presented in this study are available on request from the corresponding author.

Conflicts of Interest: The authors declare no conflict of interest.

References

1. Choi, J.Y.; Park, I.P.; Heo, S.W. Ultra-Flexible Organic Solar Cell Based on Indium-Zinc-Tin Oxide Transparent Electrode for Power Source of Wearable Devices. *Nanomaterials* **2021**, *11*, 2633. [CrossRef]
2. Martinez, F.; Neculqueo, G.; Vasquez, S.O.; Lemmetyinen, H.; Efimov, A.; Vivo, P. Branched Thiophene Oligomer/Polymer Bulk Heterojunction Organic Solar Cell. *MRS Proc.* **2015**, *1737*. [CrossRef]
3. Liu, C.; Xiao, C.; Xie, C.; Li, W. Flexible organic solar cells: Materials, large-area fabrication techniques and potential applications. *Nano Energy* **2021**, *89*, 106399. [CrossRef]
4. Choi, J.; Kwon, D.; Kim, B.; Kang, K.; Gu, J.; Jo, J.; Na, K.; Ahn, J.; Del Orbe, D.; Kim, K.; et al. Wearable self-powered pressure sensor by integration of piezo-transmittance microporous elastomer with organic solar cell. *Nano Energy* **2020**, *74*, 104749. [CrossRef]
5. Get, R.; Islam, S.M.; Singh, S.; Mahala, P. Organic polymer bilayer structures for applications in flexible solar cell devices. *Microelectron. Eng.* **2020**, *222*, 111200. [CrossRef]
6. Prakash, J.; Choudhary, S.; Raliya, R.; Chadha, T.S.; Fang, J.; Biswas, P. Real-time source apportionment of fine particle inorganic and organic constituents at an urban site in Delhi city: An IoT-based approach. *Atmos. Pollut. Res.* **2021**, *12*, 101206. [CrossRef]
7. Cardinaletti, I.; Vangerven, T.; Nagels, S.; Cornelissen, R.; Schreurs, D.; Hruby, J.; Vodnik, J.; Devisscher, D.; Kesters, J.; D'Haen, J.; et al. Organic and perovskite solar cells for space applications. *Sol. Energy Mater. Sol. Cells* **2018**, *182*, 121–127. [CrossRef]
8. Kim, S.H.; Saeed, M.A.; Lee, S.Y.; Shim, J.W. Investigating the Indoor Performance of Planar Heterojunction Based Organic Photovoltaics. *IEEE J. Photovolt.* **2021**, *11*, 997–1003. [CrossRef]
9. You, Y.-J.; Saeed, M.A.; Shafian, S.; Kim, J.; Kim, S.H.; Kim, S.H.; Kim, K.; Shim, J.W. Energy recycling under ambient illumination for internet-of-things using metal/oxide/metal-based colorful organic photovoltaics. *Nanotechnology* **2021**, *32*, 465401. [CrossRef] [PubMed]
10. Ahsan Saeed, M.; Hyeon Kim, S.; Baek, K.; Hyun, J.K.; Youn Lee, S.; Won Shim, J. PEDOT:PSS: CuNW-based transparent composite electrodes for high-performance and flexible organic photovoltaics under indoor lighting. *Appl. Surf. Sci.* **2021**, *567*, 150852. [CrossRef]
11. Privado, M.; de la Cruz, P.; Gupta, G.; Singhal, R.; Sharma, G.D.; Langa, F. Highly efficient ternary polymer solar cell with two non-fullerene acceptors. *Sol. Energy* **2020**, *199*, 530–537. [CrossRef]
12. Cowart, J.S.; Liman, C.; Garnica, A.; Page, Z.; Lim, E.; Zope, R.R.; Baruah, T.; Hawker, C.J.; Chabinyc, M.L. Donor-fullerene dyads for energy cascade organic solar cells. *Inorganica Chim. Acta* **2017**, *468*, 192–202. [CrossRef]
13. Ma, Q.; Jia, Z.; Meng, L.; Zhang, J.; Zhang, H.; Huang, W.; Yuan, J.; Gao, F.; Wan, Y.; Zhang, Z.; et al. Promoting charge separation resulting in ternary organic solar cells efficiency over 17.5%. *Nano Energy* **2020**, *78*, 105272. [CrossRef]
14. Wang, S.; Zhang, H.; Zhang, B.; Xie, Z.; Wong, W.-Y. Towards high-power-efficiency solution-processed OLEDs: Material and device perspectives. *Mater. Sci. Eng. R Rep.* **2020**, *140*, 100547. [CrossRef]
15. Koppe, M.; Egelhaaf, H.-J.; Dennler, G.; Scharber, M.C.; Brabec, C.J.; Schilinsky, P.; Hoth, C.N. Near IR Sensitization of Organic Bulk Heterojunction Solar Cells: Towards Optimization of the Spectral Response of Organic Solar Cells. *Adv. Funct. Mater.* **2010**, *20*, 338–346. [CrossRef]
16. Nowak, E.M.; Sanetra, J.; Grucela, M.; Schab-Balcerzak, E. Azomethine naphthalene diimides as component of active layers in bulk heterojunction solar cells. *Mater. Lett.* **2015**, *157*, 93–98. [CrossRef]
17. Sun, Q.; Zhang, F.; Hai, J.; Yu, J.; Huang, H.; Teng, F.; Tang, W. Doping a D-A structural polymer based on benzodithiophene and triazoloquinoxaline for efficiency improvement of ternary solar cells. *Electron. Mater. Lett.* **2015**, *11*, 236–240. [CrossRef]
18. Zhang, S.; Zuo, L.; Chen, J.; Zhang, Z.; Mai, J.; Lau, T.-K.; Lu, X.; Shi, M.; Chen, H. Improved photon-to-electron response of ternary blend organic solar cells with a low band gap polymer sensitizer and interfacial modification. *J. Mater. Chem. A* **2016**, *4*, 1702–1707. [CrossRef]
19. Li, D.; Song, L.; Chen, Y.; Huang, W. Modeling Thin Film Solar Cells: From Organic to Perovskite. *Adv. Sci.* **2020**, *7*, 1901397. [CrossRef] [PubMed]
20. Honda, S.; Ohkita, H.; Benten, H.; Ito, S. Multi-colored dye sensitization of polymer/fullerene bulk heterojunction solar cells. *Chem. Commun.* **2010**, *46*, 6596–6598. [CrossRef]
21. Lin, Y.-C.; Cheng, H.-W.; Su, Y.-W.; Lin, B.-H.; Lu, Y.-J.; Chen, C.-H.; Chen, H.-C.; Yang, Y.; Wei, K.-H. Molecular engineering of side chain architecture of conjugated polymers enhances performance of photovoltaics by tuning ternary blend structures. *Nano Energy* **2018**, *43*, 138–148. [CrossRef]
22. Kumari, T.; Lee, S.M.; Kang, S.-H.; Chen, S.; Yang, C. Ternary solar cells with a mixed face-on and edge-on orientation enable an unprecedented efficiency of 12.1%. *Energy Environ. Sci.* **2017**, *10*, 258–265. [CrossRef]
23. Xiao, L.; Gao, K.; Zhang, Y.; Chen, X.; Hou, L.; Cao, Y.; Peng, X. A complementary absorption small molecule for efficient ternary organic solar cells. *J. Mater. Chem. A* **2016**, *4*, 5288–5293. [CrossRef]
24. Huang, J.-S.; Hsiao, C.-Y.; Syu, S.-J.; Chao, J.-J.; Lin, C.-F. Well-aligned single-crystalline silicon nanowire hybrid solar cells on glass. *Sol. Energy Mater. Sol. Cells* **2009**, *93*, 621–624. [CrossRef]

25. Li, W.; Liu, W.; Zhang, X.; Yan, D.; Liu, F.; Zhan, C. Quaternary Solar Cells with 12.5% Efficiency Enabled with Non-Fullerene and Fullerene Acceptor Guests to Improve Open Circuit Voltage and Film Morphology. *Macromol. Rapid Commun.* **2019**, *40*, 1900353. [CrossRef] [PubMed]
26. Chen, D.; Li, W.; Gan, L.; Wang, Z.; Li, M.; Su, S.-J. Non-noble-metal-based organic emitters for OLED applications. *Mater. Sci. Eng. R Rep.* **2020**, *142*, 100581. [CrossRef]
27. Huseynova, G.; Xu, Y.; Yawson, B.N.; Shin, E.-Y.; Lee, M.J.; Noh, Y.-Y. P-type doped ambipolar polymer transistors by direct charge transfer from a cationic organic dye Pyronin B ferric chloride. *Org. Electron.* **2016**, *39*, 229–235. [CrossRef]
28. Han, J.; Lian, H.; Cheng, X.; Dong, Q.; Qu, Y.; Wong, W. Study of Electronic and Steric Effects of Different Substituents in Donor–Acceptor Molecules on Multilevel Organic Memory Data Storage Performance. *Adv. Electron. Mater.* **2021**, *7*, 2001097. [CrossRef]
29. Bao, X.; Wang, T.; Yang, A.; Yang, C.; Dou, X.; Chen, W.; Wang, N.; Yang, R. Annealing-free P3HT:PCBM-based organic solar cells via two halohydrocarbons additives with similar boiling points. *Mater. Sci. Eng. B* **2014**, *180*, 7–11. [CrossRef]
30. Selvaraj, A.R.K.; Hayase, S. Molecular dynamics simulations on the aggregation behavior of indole type organic dye molecules in dye-sensitized solar cells. *J. Mol. Model.* **2012**, *18*, 2099–2104. [CrossRef]
31. Jellison, G.E.; Modine, F.A. Parameterization of the optical functions of amorphous materials in the interband region. *Appl. Phys. Lett.* **1996**, *69*, 371–373. [CrossRef]
32. Ferlauto, A.; Ferreira, G.M.; Pearce, J.; Wronski, C.R.; Collins, R.W.; Deng, X.; Ganguly, G. Analytical model for the optical functions of amorphous semiconductors from the near-infrared to ultraviolet: Applications in thin film photovoltaics. *J. Appl. Phys.* **2002**, *92*, 2424–2436. [CrossRef]
33. Ware, W.; Wright, T.; Mao, Y.; Han, S.; Guffie, J.; Danilov, E.O.; Rech, J.; You, W.; Luo, Z.; Gautam, B. Aggregation Controlled Charge Generation in Fullerene Based Bulk Heterojunction Polymer Solar Cells: Effect of Additive. *Polymers* **2021**, *13*, 115. [CrossRef] [PubMed]
34. Nunzi, J.; Lebel, O. Revisiting the Optimal Nano-Morphology: Towards Amorphous Organic Photovoltaics. *Chem. Rec.* **2019**, *19*, 1028–1038. [CrossRef] [PubMed]
35. Yin, W.; Dadmun, M. A New Model for the Morphology of P3HT/PCBM Organic Photovoltaics from Small-Angle Neutron Scattering: Rivers and Streams. *ACS Nano* **2011**, *5*, 4756–4768. [CrossRef] [PubMed]
36. Thiounn, T.; Lauer, M.K.; Bedford, M.S.; Smith, R.C.; Tennyson, A.G. Thermally-healable network solids of sulfur-crosslinked poly(4-allyloxystyrene). *RSC Adv.* **2018**, *8*, 39074–39082. [CrossRef]
37. Yadav, A.; Upadhyaya, A.; Gupta, S.K.; Verma, A.S.; Negi, C.M.S. Solution processed graphene as electron transport layer for bulk heterojunction based devices. *Superlattices Microstruct.* **2018**, *120*, 788–795. [CrossRef]
38. Yadav, A.; Negi, C.M.; Verma, A.; Gupta, S. Electrical, optical and photoresponse characteristics of P3HT:PCBM bulk heterojunction device. In Proceedings of the 2018 3rd International Conference on Microwave and Photonics, ICMAP 2018, Dhanbad, India, 9–11 February 2018; Volume 2018, pp. 1–2.
39. Wang, C.; Ma, S.; Zeng, H.; Li, J.; Chen, L.; Wang, W.; Tian, H. Spectroscopic ellipsometry on a novel cyanine dyes in Langmuir–Blodgett multilayers. *Colloids Surf. A Physicochem. Eng. Asp.* **2006**, *284-285*, 414–418. [CrossRef]
40. Åkerlind, C.; Arwin, H.; Jakobsson, F.; Kariis, H.; Järrendahl, K. Optical properties and switching of a Rose Bengal derivative: A spectroscopic ellipsometry study. *Thin Solid Films* **2011**, *519*, 3582–3586. [CrossRef]
41. Pandey, S.K.; Awasthi, V.; Verma, S.; Gupta, M.; Mukherjee, S. Spectroscopic ellipsometry study on electrical and elemental properties of Sb-doped ZnO thin films. *Curr. Appl. Phys.* **2015**, *15*, 479–485. [CrossRef]
42. Shahrokhabadi, H.; Bananej, A.; Vaezzadeh, M. Investigation of Cody–Lorentz and Tauc–Lorentz Models in Characterizing Dielectric Function of $(HfO_2)_x (ZrO_2)_{1-x}$ Mixed Thin Film. *J. Appl. Spectrosc.* **2017**, *84*, 1–8. [CrossRef]
43. Beer, P. Bestimmung der Absorption des rothen Lichts in farbigen Flüssigkeiten. *Ann. Phys.* **1852**, *162*, 78–88. [CrossRef]
44. Gasparini, N.; Lucera, L.; Salvador, M.; Prosa, M.; Spyropoulos, G.D.; Kubis, P.; Egelhaaf, H.-J.; Brabec, C.J.; Ameri, T. High-performance ternary organic solar cells with thick active layer exceeding 11% efficiency. *Energy Environ. Sci.* **2017**, *10*, 885–892. [CrossRef]
45. Zhu, K.; Wang, X.; He, Y.; Zhai, F.; Gao, C.; Wang, Q.; Jing, X.; Yu, L.; Sun, M. Ester-substituted copolymer-based ternary semitransparent polymer solar cells with enhanced FF and PCE. *Polymer* **2021**, *229*, 123973. [CrossRef]
46. Mayerhöfer, T.G.; Pipa, A.V.; Popp, J. Beer's Law-Why Integrated Absorbance Depends Linearly on Concentration. *ChemPhysChem* **2019**, *20*, 2748–2753. [CrossRef]
47. Ng, A.; Liu, X.; To, C.H.; Djurišić, A.B.; Zapien, J.A.; Chan, W.K. Annealing of P3HT:PCBM Blend Film—The Effect on Its Optical Properties. *ACS Appl. Mater. Interfaces* **2013**, *5*, 4247–4259. [CrossRef]
48. Hajduk, B.; Bednarski, H.; Jarząbek, B.; Nitschke, P.; Janeczek, H. Phase diagram of P3HT:PC70BM thin films based on variable-temperature spectroscopic ellipsometry. *Polym. Test.* **2020**, *84*, 106383. [CrossRef]
49. Kadem, B.; Hassan, A. The Effect of Fullerene Derivatives Ratio on P3HT-based Organic Solar Cells. *Energy Procedia* **2015**, *74*, 439–445. [CrossRef]
50. Lewińska, G.; Danel, K.S.; Łukaszewska, I.; Lewiński, G.; Niemiec, W.; Sanetra, J. Ternary organic solar cells doped methoxyphenyl indenopyrazoloquinoline derivatives. *J. Mater. Sci. Mater. Electron.* **2018**, *29*, 17809–17817. [CrossRef]

51. McLeod, J.A.; Pitman, A.L.; Kurmaev, E.Z.; Finkelstein, L.D.; Zhidkov, I.S.; Savva, A.; Moewes, A. Linking the HOMO-LUMO gap to torsional disorder in P3HT/PCBM blends. *J. Chem. Phys.* **2015**, *143*, 224704. [CrossRef] [PubMed]
52. Schlaf, R.; Parkinson, B.A.; Lee, P.A.; Nebesny, A.K.W.; Armstrong, N.R. HOMO/LUMO Alignment at PTCDA/ZnPc and PTCDA/ClInPc Heterointerfaces Determined by Combined UPS and XPS Measurements. *J. Phys. Chem. B* **1999**, *103*, 2984–2992. [CrossRef]

Article

Electrical Transport, Structural, Optical and Thermal Properties of [(1−x)Succinonitrile: xPEO]-LiTFSI-Co(bpy)$_3$(TFSI)$_2$-Co(bpy)$_3$(TFSI)$_3$ Solid Redox Mediators

Ravindra Kumar Gupta [1,*], Hamid Shaikh [2], Ahamad Imran [1], Idriss Bedja [3], Abrar Fahad Ajaj [4] and Abdullah Saleh Aldwayyan [4,5]

1. King Abdullah Institute for Nanotechnology, King Saud University, Riyadh 11451, Saudi Arabia; aimran@ksu.edu.sa
2. SABIC Polymer Research Center, College of Engineering, King Saud University, Riyadh 11421, Saudi Arabia; hamshaikh@ksu.edu.sa
3. Cornea Research Chair, Department of Optometry, College of Applied Medical Sciences, King Saud University, Riyadh 11433, Saudi Arabia; bedja@ksu.edu.sa
4. Department of Physics and Astronomy, College of Science, King Saud University, Riyadh 11451, Saudi Arabia; abrar.fa.ksa@gmail.com (A.F.A.); dwayyan@ksu.edu.sa (A.S.A.)
5. K.A. CARE Energy Research and Innovation Center, King Saud University, Riyadh 11451, Saudi Arabia
* Correspondence: rgupta@ksu.edu.sa

Abstract: The solar cell has been considered one of the safest modes for electricity generation. In a dye-sensitized solar cell, a commonly used iodide/triiodide redox mediator inhibits back-electron transfer reactions, regenerates dyes, and reduces triiodide into iodide. The use of iodide/triiodide redox, however, imposes several problems and hence needs to be replaced by alternative redox. This paper reports the first Co^{2+}/Co^{3+} solid redox mediators, prepared using [(1−x)succinonitrile: xPEO] as a matrix and LiTFSI, Co(bpy)$_3$(TFSI)$_2$, and Co(bpy)$_3$(TFSI)$_3$ as sources of ions. The electrolytes are referred to as SN_E ($x = 0$), Blend 1_E ($x = 0.5$ with the ethereal oxygen of the PEO-to-lithium ion molar ratio (EO/Li$^+$) of 113), Blend 2_E ($x = 0.5$; EO/Li$^+$ = 226), and PEO_E ($x = 1$; EO/Li$^+$ = 226), which achieved electrical conductivity of 2.1×10^{-3}, 4.3×10^{-4}, 7.2×10^{-4}, and 9.7×10^{-7} S cm^{-1}, respectively at 25 °C. Only the blend-based polymer electrolytes exhibited the Vogel-Tamman-Fulcher-type behavior (vitreous nature) with a required low pseudo-activation energy (0.05 eV), thermal stability up to 125 °C, and transparency in UV-A, visible, and near-infrared regions. FT-IR spectroscopy demonstrated the interaction between salt and matrix in the following order: SN_E < Blend 2_E < Blend 1_E << PEO_E. The results were compared with those of acetonitrile-based liquid electrolyte, ACN_E.

Keywords: dye-sensitized solar cells; redox mediator; solid polymer electrolytes; succinonitrile; electrical conductivity

1. Introduction

Efficient utilization of fossil-fuel-based energy sources is one of the key factors of human social and economic development. However, this has led to an increase in the levels of greenhouse gases and pollution. The nuclear energy source is also not safe because of the hazards associated with it. Owing to the abundance of sunlight, the photovoltaic cell has emerged as an energy source, especially in regions near to and between the Tropics of Cancer and Capricorn (sunlight irradiance ~2 MWh m^{-2}) [1].

One of the third generation photovoltaic cells, the dye-sensitized solar cell (DSSC), is highly attractive due to several advantages of DSSCs over other solar cells [2–7]. Some of the advantages are the simple cell structure, they are flexible and lightweight, the absence of toxic and less-available elements; energy payback time is less than a year, their all-direction-capturing of incident light; and their performance under real indoor and outdoor conditions. The first DSSC was reported by O'Regan and Gratzel in 1991 [8] with a power conversion efficiency (η) of 7.1% at 75 mW cm^{-2}. They used a liquid electrolyte: tetrapropylammonium iodide, KI, and I$_2$ in ethylene carbonate and acetonitrile (ACN). This electrolyte provided fast transport of the I$^-$/I$_3^-$ redox couple, (i) to regenerate dyes via the oxidation of I$^-$ into I$_3^-$ at the mesoporous and nanostructured TiO$_2$ working electrode, (ii) to reduce I$_3^-$ into I$^-$ at the platinum counter electrode, and (iii) to inhibit back-electron transfer reactions. Since then, several I$^-$/I$_3^-$ redox mediators in a form of liquid, gel, or solid have been synthesized [9–21]. The DSSC has also been commercialized with the highest η-value of 11.9% at 100 mW cm^{-2} (1 sun), utilizing a liquid electrolyte: dimethyl-propyl imidazolium iodide (an ionic liquid), I$_2$, LiI, and 4-*tert*-butylpyridine (TBP) in ACN [7,22]. The ionic liquid helped to reduce the organic solvent-related problems, thereby improving the stability of the device, though this required a hermetic sealing.

Owing to their corrosive nature, dissolving of many of the commonly used sealants and metal interconnects, sublimation, and partial absorption of visible light around 430 nm of the iodine-based I$^-$/I$_3^-$ redox-couple, the researchers started to think of replacing this by one using a molecular species of similar type such as Br$^-$/Br$_2$ redox (e.g., Br$_2$ and LiBr in ACN), one using metal complex-based redox such as Co^{2+}/Co^{3+}, and one using organic radicals, such as TEMPO [2,6,9,10,23–29]. The Co^{2+}/Co^{3+} redox electrolytes, in general, with ACN as an organic solvent showed η of more than 10% at 1 sun for several dyes. For example, 11.9% for YD2-o-C8 dye [23], 12.3% for YD2-o-C8+Y123 dyes [23], 10.3% for JF419 dye [30], 13% for SM315 dye [24], 10.6% for Y123 dye [31], 11.4% for YD2-o-C8 dye [32], 10.2% for C101 dye [33], 10.5% for LEG4+D35+Dyenamo Blue dyes [34], 10.42% for FW1+WS5 dyes [35], 12.8% for SM342+Y123 dyes [36], 11% for AQ310 dye [37], 13.6% for ZL003 dye [38], 10.3% for H2 dye [39], and 11.2% for YD2-o-C8 dye [40].

The researchers used ionic liquid to suppress the organic solvent-related problems. Xu et al. [41] synthesized a compound [Co{3,3'-(2,2'-bipyridine-4,4'-diyl-bis(methylene)) bis(1-methyl-1H-imidazol-3-ium) hexafluorophosphate}$_3$]$^{2+/3+}$ and mixed with 1-propyl-3-methylimidazolium iodine, 1-ethyl-3-methyl imidazolium thiocyanate, guanidinium thiocyanate (GuSCN), and TBP. They reported η~7.37% at 1 sun for N719 dye. Kakiage et al. reported η~12.5% at 1 sun for ADEKA-1 dye [42] and η~14.3% at 1 sun for ADEKA-1+LEG4 dyes [43], utilizing an electrolyte solution: [Co^{2+}(phen)$_3$](PF$_6$)$_2$, [Co^{3+}(phen)$_3$](PF$_6$)$_3$, LiClO$_4$, NaClO$_4$, tetrabutyl ammonium hexafluorophosphate, tetrabutylphosphonium hexafluorophosphate, 1-hexyl-3-methylimidazolium hexa fluorophosphate, TBP, 4-trimethylsilylpyridine, 4-methylpyridine, 4-cyano-4'-propyl biphenyl, 4-cyano-4'-pentylbiphenyl, 4-cyano-4'-octylbiphenyl in ACN. Wang et al. [44] reported η~8.1% at 1 sun for D205 dye with a redox mediator, bis(3-butyl-1-methylimidazolium) tetraisothiocyanato cobalt, 1-propyl-3-methyl-imidazolium iodine, nitrosyl tetrafluoroborate, LiClO$_4$, and TBP in methoxy propionitrile.

The researchers synthesized the Co^{2+}/Co^{3+} redox mediators in a gel (quasi-solid) form as well. The gel was prepared by incorporating a large amount of organic solvent mixed with a redox couple into an inorganic or organic frame. So far nanoparticles of SiO$_2$ (η~2.58% at 1 sun for D35 dye) [45], TiO$_2$ (η~5.1% at 1 sun for N719 dye) [46], and TiC (η~6.29% at 1 sun for N719 dye) [46] were used to form an inorganic frame. An organic frame was prepared using poly(ethylene glycol) with gelatin (η~4.1% at 1 sun for MK2 dye) [47], bisphenol A ethoxylate dimethacrylate with poly(ethylene glycol) methyl ether methacrylate (η~6.4% at 1 sun for LEG4 dye) [48], poly(ethylene glycol)/poly(methyl methacrylate) (η~1.9% at 1 sun for N719 dye) [49], poly(vinylidene fluoride-co-hexafluoropropylene) (η~8.7% at 1 sun for MK2 dye [50]; η~4.34% at 1 sun for Z907 dye [51], η~7.1% at 1 sun for MK2 dye [52]), poly(ethylene oxide-co-2-(2-methoxyethoxy) ethyl glycidyl ether-co-allyl glycidyl ether) (η~3.59% at 1 sun for MK2 dye and η~1.74% at 1 sun for Z907 dye) [53], poly(ethylene oxide-co-2-(2-methoxyethoxy) ethyl glycidyl ether (η < 0.1% at 1 sun for L0 dye) [54], poly(ethylene oxide) (η~21.1% at 200 lx for Y123 dye) [55], poly(ethylene oxide)-poly(methyl methacrylate) blend (η~18.7% at 200 lx for Y123 dye) [55], hydroxypropyl cellulose (η~9.1% at 0.7 sun for N719 dye) [56], and hydroxyethyl cellulose (η~4.5% at 1 sun for N3 dye) [57].

Unfortunately, the liquid nature of electrolytes creates internal pressure in the DSSCs at the ambient temperature range (50–80 °C), resulting in a leakage of solvent, thereby requiring hermetic sealing [9–17]. This also makes the manufacturing of DSSCs non-scale-up. A gel electrolyte exhibits problems similar to those of a liquid electrolyte, hence, it needs to be replaced by a solid one to sustain it in the hot weather of Gulf countries. However, until now no Co^{2+}/Co^{3+} redox mediator in solid form has been reported.

Earlier, a high-molecular-weight poly(ethylene oxide) (PEO) was used as a polymer matrix of the I^-/I_3^- redox-based solid polymer electrolytes, PEO-PQ-MI-I_2, where M represents an alkali metal cation [9–17]. The PEO offered its self-standing film-forming, thermal stability up to 200 °C, it was eco- and bio-benign, was of relatively low material cost, the dissociation/complexation of salt due to its moderate dielectric constant ($\varepsilon_{25°C}$)-value (5–8), Gutmann donor number of 22, had just the right spacing between coordinating ethereal oxygens for maximum solvation of the Li^+ ions, and the segmental motion of polymeric chains for the ion transport through ethereal oxygen [58–61]. Ionic liquids, low molecular weight polymers, and copolymerization were used as plasticizers (PQs) to reduce PEO crystallinity (χ), thereby increasing the $\sigma_{25°C}$- and η-values.

Gupta et al. [62–68] showed that the equal weight proportion of succinonitrile (SN) can be used as a plasticizer without hampering the thin-film forming property of the PEO. This blending resulted in several beneficial properties, such as a higher $\sigma_{25°C}$-value, $\sim 10^{-8}$ S cm^{-1} than the PEO ($\sigma_{25°C}$ $\sim 10^{-10}$ S cm^{-1}), a lower χ-value, $\sim 25\%$ than the PEO ($\sim 82\%$), and higher thermal stability up to $\sim$125 °C than the SN ($\sim$75 °C) [62]. The PEO-SN-MI-I_2 solid polymer electrolytes achieved $\sigma_{25°C}$-value $3–7 \times 10^{-4}$ S cm^{-1}, transparency more than 95% in visible and IR regions, $\chi \sim 0\%$, thermal stability up to $\sim$125 °C, and η-value between 2 and 3.7% at 1 sun with Ru-based N719 dye. The solid solvent/ plasticizing property of the plastic crystal, SN is due to its low molecular weight, high molecular diffusivity at the plastic phase between -35 °C (crystal-to-plastic-crystal phase transition temperature, T_{pc}) and 58 °C (melting temperature, T_m), low T_m-value, high ε-value $\sim$55 at 25 °C and 62.6 at 58 °C, nitrile group for ion transport, and waxy nature [69–73].

In this work, we have extended the concept of blending for achieving the high electrical conductivity of the Co^{2+}/Co^{3+} solid redox mediators. We reported electrical, structural, optical, and thermal properties of new [(1−x)SN: xPEO]-LiTFSI-Co(bpy)$_3$(TFSI)$_2$-Co(bpy)$_3$(TFSI)$_3$ solid redox mediators. The composition, x is 0, 0.5, and 1 in weight fraction. Other notations, bpy and TFSI stand for tris-(2,2'-bipyridine) and bis(trifluoromethyl) sulfonylimide, respectively. These solid redox mediators are based on a liquid electrolyte (0.1-M LiTFSI, 0.25-M Co(bpy)$_3$(TFSI)$_2$, and 0.06-M Co(bpy)$_3$(TFSI)$_3$ in acetonitrile), which resulted in η of 13% with SM315 dye [24]. This liquid electrolyte is hereafter referred to as ACN_E. We just replaced acetonitrile with succinonitrile for synthesizing SN_E (x = 0). Succinonitrile was then replaced by PEO for synthesizing PEO_E (x = 1 in weight fraction). This had the ethereal oxygen of the PEO-to-lithium ion mole ratio, abbreviated as EO/Li$^+$ of 226. We also used a blend containing SN and PEO in an equal weight fraction for retaining the beneficial properties of SN and PEO, as discussed earlier. The value of EO/Li$^+$ was kept at either 113 (Blend 1_E) or 226 (Blend 2_E) for understanding its effect on the electrical transport properties [65]. Figure 1a shows the chemical structure of the ingredients. The solid nature of SN_E, PEO_E, Blend 1_E, and Blend 2_E is shown in Figure 1b. Ionic salts with TFSI$^-$ anion were used because TFSI$^-$ offers a low value of lattice energy with delocalized electrons, making the salt highly dissociable in the solvent with a less anionic contribution to the total conductivity [58–61]. The lithium salt is thermally and electrochemically stable as well [58–61,74]. Owing to the small size, the Li$^+$ ions get intercalated on the TiO$_2$ nanoparticles of the DSSC, leading to faster electron injection from the excited dye molecules to the conduction band of the TiO$_2$, thereby the higher photocurrent [67,75]. Contrary to this, the cobalt ions adsorb on the skirt of the TiO$_2$ nanoparticles, resulting in a negative shift of the Fermi level of the TiO$_2$ nanoparticles, thereby resulting in the higher open-circuit voltage. It is also known that an ion with a large size acts as a plasticizer in a polymer electrolyte, resulting in higher electrical conductivity [65,67,75].

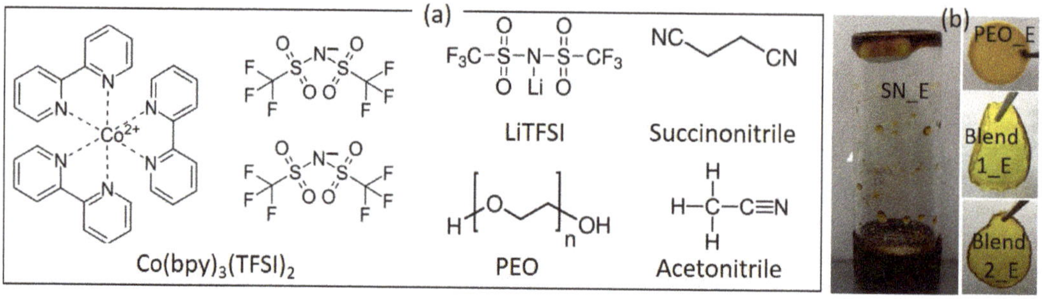

Figure 1. (a) Chemical structure of ingredients. (b) Optical image of solid redox mediators.

We determined σ-value at different temperatures for knowing the nature of the electrolyte and determining the activation energy. The electrical transport properties were elucidated by X-ray diffractometry (XRD), Fourier-transform infrared (FT-IR) spectroscopy, UV-visible spectroscopy, polarized optical microscopy (POM), and differential scanning calorimetry (DSC). We used thermogravimetric analysis (TGA) for the thermal stability study.

2. Materials and Methods

2.1. Materials

Succinonitrile, PEO (1-M g mol^{-1}), and LiTFSI were procured from Sigma Aldrich, Inc., St. Louis, MO, USA. Cobalt salts, Co(bpy)$_3$(TFSI)$_2$ (DN-C13) and Co(bpy)$_3$(TFSI)$_3$ (DN-C14) were procured from Dyenamo AB, Stockholm, Sweden. These chemicals were used without purification.

2.2. Synthesis

Table 1 shows the composition of ACN_E, SN_E, PEO_E, Blend 1_E, and Blend 2_E. As per the procedure of Mathew et al. [24], ACN_E was synthesized using LiTFSI (0.1 M), DN-C13 (0.25 M), and DN-C14 (0.06 M) in acetonitrile under stirring at 65 °C for 24 h. The SN_E was prepared identically by dissolving the salts in succinonitrile. The PEO_E and Blends 1_E & 2_E were prepared using the solution cast method. The ingredients were dissolved in 20-mL acetonitrile by vigorous stirring at 65 °C for 48 h. This resulted in a homogeneous polymeric solution which was poured on a Teflon Petri dish followed by drying at room temperature in a nitrogen gas atmosphere for two weeks and in a vacuum desiccator for a day. This produced a self-standing film of the solid polymer electrolyte.

Table 1. Composition of liquid and solid redox mediators.

x (Electrolyte)	EO/Li$^+$	PEO (g)	ACN/ SN (g)	LiTFSI (g)	DN-C13 (g)	DN-C14 (g)
(ACN_E)	-	-	0.4425	0.0162	0.1531	0.0462
0 (SN_E)	-	-	0.5600	0.0162	0.1531	0.0462
0.5 (Blend 1_E)	113	0.2800	0.2800	0.0162	0.1531	0.0462
0.5 (Blend 2_E)	226	0.5600	0.5600	0.0162	0.1531	0.0462
1 (PEO_E)	226	0.5600	-	0.0162	0.1531	0.0462

2.3. Characterizations

A specific sample holder [72] was used to measure the electrical conductivity of the ACN_E and SN_E electrolytes. The liquid electrolyte was poured on a space (area, $A \sim 0.16$ cm^{-2} and thickness, $l \sim 0.05$ cm) created by a Teflon spacer between platinum plates (blocking electrode). For determining the electrical conductivity of the PEO_E and Blends 1_E & 2_E solid polymer electrolytes, another sample holder [64], having stainless steel plate as a blocking electrode, was used. The sandwiched electrolyte was subjected to 20 mV ac voltage and monitoring of real and imaginary impedances from 100 kHz to 1 Hz by a

Palmsens4 impedance analyzer (PalmSens BV, Houten, the Netherlands). This resulted in a Nyquist curve, thereby a bulk resistance (R_b) and then the electrical conductivity (σ) using the formula, $\sigma = l/(A\,R_b)$ [76].

For the XRD pattern of the solid electrolyte film, a D2 Phaser Bruker x-ray diffractometer (Karlsruhe, Germany) was used. The pattern was collected using the CuKα radiation (1.54184 Å) in a range of 10–40° with a step of 0.06°. The FT-IR spectrum of the electrolyte film on a potassium bromide pellet was recorded in a range of 400–4000 cm^{-1} and a resolution of 1 cm^{-1} using a Spectrum 100 Perkin Elmer FT-IR spectrometer (Waltham, MA, USA). The spectrum was analyzed using EZ-OMNIC software, ver. 7.2a (Thermo Scientific Inc., Waltham, MA, USA).

Transmittance spectrum of the electrolyte film (thickness 2–3 µm) was collected using an Agilent UV-visible spectrometer (model 8453, Santa Clara, CA, USA). The POM image with a magnification of 100× for the polymer electrolyte film (thickness 2–3 µm) was obtained using a computer interfaced ZZCAT polarized optical microscope (Zhuzhou, Hunan, China).

The DSC curve of the electrolyte was measured using a DSC-60A differential scanning calorimeter (Shimadzu, Kyoto, Japan) under the purging of nitrogen gas with 10 °C min^{-1} heating rate and in the range of −50 to 90 °C. For the TGA curve, the weight loss of the electrolyte was monitored using a Shimadzu DTG-60H unit in the temperature range of room temperature to 550 °C with a heating rate of 10 °C min^{-1} under the purging of nitrogen gas.

3. Results

3.1. Electrical Transport Properties

Figure 2 shows the Nyquist curves for the liquid (ACN_E) and solid (SN_E, PEO_E, Blend 1_E, and Blend 2_E) redox mediators at 25 °C. These curves portrayed (i) the blocking electrode effect in the low-frequency domain, and (ii) the ionic diffusion effect in the high-frequency domain [76]. Being a liquid electrolyte, ACN_E depicted a perfect semi-circle in the high-frequency domain. SN_E and PEO_E also had a semi-circle; however, the semi-circle was slightly and largely depressed for the former and latter, respectively. This is most probably due to the existence of the plastic crystalline phase of succinonitrile and the semi-crystalline phase of PEO, respectively [72,76]. Contrary to this, Blends 1_E and 2_E had no semi-circle, indicating the existence of amorphous domains, the semi-random motion of short polymer chains, and the segmental motion, demonstrating the plasticizing effect of the succinonitrile [58–61,64,65,77,78]. The bulk resistance is marked by an arrow in the Nyquist curve and is used to calculate the $\sigma_{25°C}$-value of the electrolyte.

Figure 3a shows electrical conductivity $\sigma_{25°C}$) of solid electrolytes, SN_E, Blend 1_E, Blend 2_E, and PEO_E along with that of the liquid electrolyte, ACN_E. The ACN_E exhibited $\sigma_{25°C} \sim 1.7 \times 10^{-2}$ S cm^{-1}, which is similar to those reported earlier for liquid electrolytes [60]. The high electrical conductivity is due to $\varepsilon_{25°C}$ of 36.6, donor number of 14.1 kcal mol^{-1}, molar enthalpy of 40.6 kJ mol^{-1}, and acceptor number of 18.9, helping to dissolve the salt completely and solvate the ions easily [79,80]. The replacement of ACN by SN resulted in SN_E with the $\sigma_{25°C}$-value less than an order of magnitude to $\sim 2.1 \times 10^{-3}$ S cm^{-1}. This conductivity value is similar to those obtained earlier for the SN-LiTFSI [69] and SN-LiI-I$_2$ [72] electrolytes. As discussed earlier [69,72], this is due to the solid solvent property of the succinonitrile. The replacement of SN by PEO resulted in PEO_E with the $\sigma_{25°C}$-value of $\sim 9.7 \times 10^{-7}$ S cm^{-1}, which is 3-orders of magnitude less. This is legitimate too. The pure PEO-based solid polymer electrolytes are known to have high PEO crystallinity, hindering ion transport [58–61,64,65]. The blend-based solid polymer electrolytes, however, showed $\sigma_{25°C}$-value less than that of SN_E and higher than that of PEO_E. Blend 1_E and Blend 2_E exhibited $\sigma_{25°C}$ of $\sim 4.3 \times 10^{-4}$ and $\sim 7.2 \times 10^{-4}$ S cm^{-1}, respectively. As observed earlier [62–68], this is due to the plasticizing property of the succinonitrile. Also, a competition between the nitrile group of succinonitrile and the ethereal oxygen of PEO to bind metal ions leads to more free ions for transport [64,65].

Besides, the availability of a huge number of large-sized TFSI$^-$ ions is helpful to produce more amorphous regions in the Blends 1_E and 2_E for easy ion transport [74]. One can expect a similar scenario for Co^{2+}/Co^{3+} ions too [65,75]. It is also notable that Blend 2_E had higher electrical conductivity than Blend 1_E. This is due to more amorphous regions for ion transport in the Blend 2_E as demonstrated by the FT-IR spectroscopy, UV-visible spectroscopy, and DSC studies, which will be discussed later.

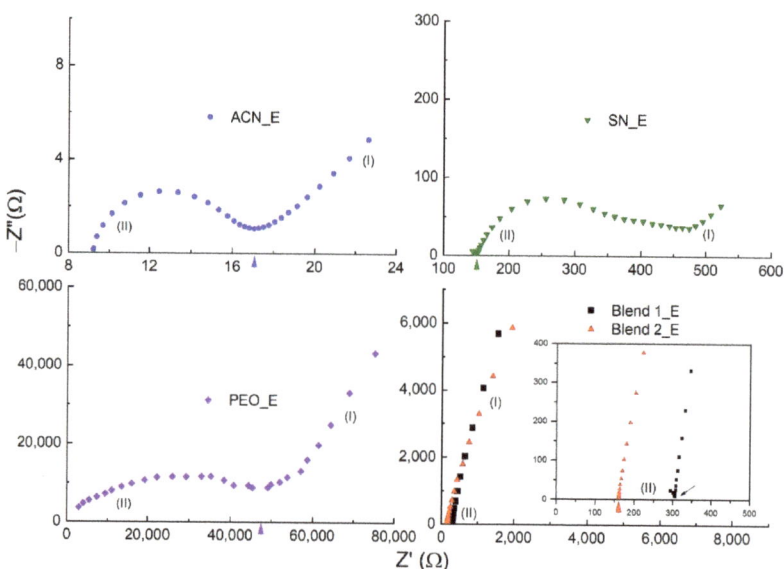

Figure 2. Nyquist curves of the solid (SN_E, PEO_E, Blend 1_E, and Blend 2_E) and liquid (ACN_E) redox mediators at 25 °C. (I) and (II) represent low- and high-frequency domains, respectively. The inset shows the high-frequency domain of the Blends 1_E and 2_E.

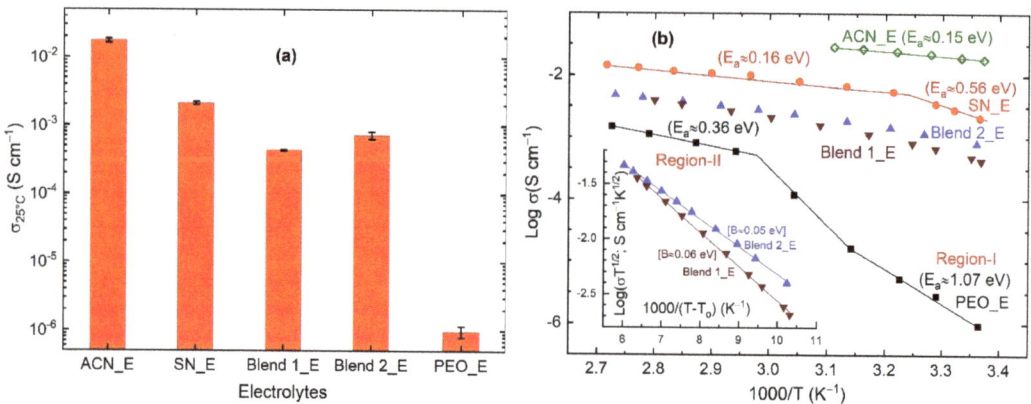

Figure 3. (a) Electrical conductivity ($\sigma_{25°C}$) and (b) log σ vs. T^{-1} plots of the solid redox mediators, SN_E, PEO_E, Blend 1_E, and Blend 2_E. Inset in (b) is VTF plots of Blends 1_E and 2_E. ACN_E, liquid electrolyte.

Figure 3b shows log σ vs. T^{-1} plots of the solid (SN_E, Blend 1_E, Blend 2_E, and PEO_E) and liquid (ACN_E) redox mediators. The ACN_E, SN_E, and PEO_E portrayed a linear curve, revealing the thermally activated Arrhenius-type behavior of molecules/

polymeric chains. Blends 1_E and 2_E depicted a slightly downward curve, indicating the existence of an amorphous phase, which follows the Vogel-Tamman-Fulcher (VTF)-type behavior. We have observed these trends for several I^-/I_3^- redox mediators [64,65,68,72,73]. The Arrhenius behavior is expressed by the equation, $\sigma = \sigma_o \exp[-E_a/k_B T]$, where σ_o is the pre-exponential factor, E_a is the activation energy, and k_B is the Boltzmann constant. While the VTF behavior is represented by an expression, $\sigma = AT^{-1/2} \exp[-B/k_B(T-T_o)]$, where A is the pre-exponential factor, B is the pseudo-activation energy, and T_o is the temperature at which the free volume vanishes. The E_a-value calculated from the slope of the Arrhenius plot is as follows: 0.56 eV (Region-I) and 0.16 eV (Region-II) for SN_E; and 1.07 eV (Region-I) and 0.36 eV (Region-II) for PEO_E. Region-I represents the solid-state region for SN_E and PEO_E, while Region-II corresponds to the liquid state for SN_E and the amorphous phase for PEO_E. The activation energy for SN_E in Region-II is similar to that observed (0.15 eV) for the liquid electrolyte, ACN_E. The pseudo-activation energy (B) calculated from the slope of the VTF plot is as follows: 0.06 eV and 0.05 eV for Blends 1_E and 2_E, respectively. The low activation energy values for the Blends 1_E and 2_E indicate easy ion transport, which is required for the DSSC application.

3.2. Structural Properties

Figure 4 shows XRD patterns of the solid redox mediators, SN_E, PEO_E, Blend 1_E, and Blend 2_E. The SN_E and PEO_E exhibited characteristic reflection peaks of succinonitrile and poly(ethylene oxide), respectively, though their peaks are broader and weaker than those of pure matrices, succinonitrile, and PEO. These indicate molecular disorder for SN_E and an increase in amorphicity for PEO_E [64,65,72], resulting in significantly enhanced electrical conductivity as compared to the pure matrices. The available cations and anions, having a large size, also acted as plasticizers and contributed to increasing the amorphicity [67,72–75]. Also, these electrolytes did not show any peak corresponding to cobalt and lithium salts, indicating complete salt dissociation/ complexation. The Blends 1_E and 2_E portrayed the absence of reflection peaks of ingredients, revealing the arrest of the glassy phase. As mentioned earlier, succinonitrile is a very good plasticizer to decrease the crystallinity of PEO [63–65]. This is also accompanied by the PEO-SN blend matrix-metal ions interaction, where SN molecules are more active [68,77,78]. These results are also supported by the findings of the FT-IR spectroscopy, which are discussed below.

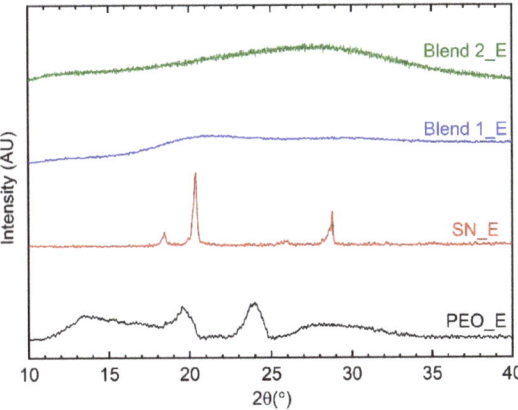

Figure 4. XRD patterns of the solid redox mediators, SN_E, PEO_E, Blend 1_E, and Blend 2_E.

Figure 5 shows FT-IR spectra of the solid redox mediators, SN_E, PEO_E, Blend 1_E, and Blend 2_E. This figure also shows the spectra of liquid electrolyte, ACN_E and Co(bpy)$_3$(TFSI)$_2$ salt for comparison. The spectrum of the Co(bpy)$_3$(TFSI)$_3$ salt (data not shown) was similar to that of the Co(bpy)$_3$(TFSI)$_2$ salt. The observed vibrational frequencies of liquid and solid redox mediators along with those of constituents, ACN [81], SN [82], PEO [83], PEO-SN blend [62,64,65], LiTFSI [74,84,85], and Co(bpy)$_3$(TFSI)$_2$ [84–86], are listed in Table 2. This also shows the corresponding assignments. It is worth mentioning that the PEO-SN blending occurs via the interaction of the ethereal oxygen with the nitrogen of SN [62,64]. Wen et al. [84] asserted that the vibrational peaks at 1057, 1133, 1196, and 1351 cm^{-1} correspond to free and unpaired anions that are strongly solvated. Rey et al. [85] showed that the vibrational peaks at 1229 and 1331 cm^{-1} correspond to the ion-pairing peaks, though these peaks present free ions too if their position did not get changed with increasing salt concentration. The formation of an ion pair reduces the number of free ions. However, being uncharged and having nearly the same size as the cation, it has somewhat higher mobility in the polar polymer/solvent and lower solvent-salt interaction for increased amorphicity [84]. These have been discussed below.

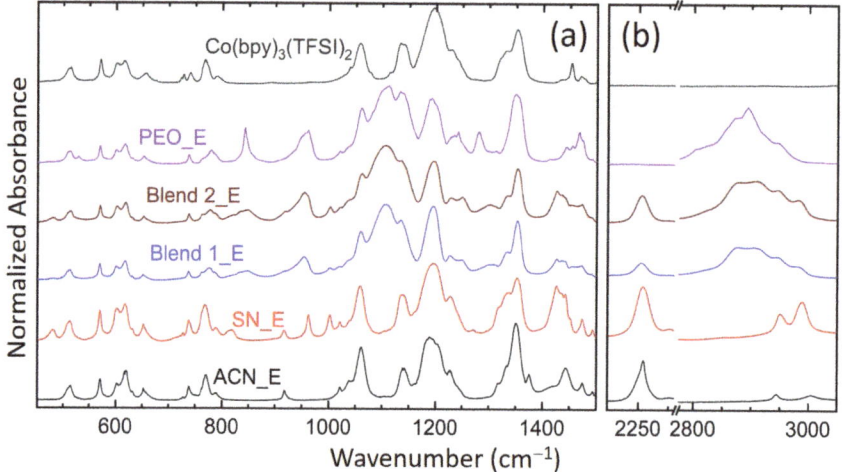

Figure 5. FT-IR spectra of the solid (SN_E, PEO_E, Blend 1_E, and Blend 2_E) and liquid (ACN_E) redox mediators. The spectrum of the Co(II) salt is also included for direct comparison. (**a**) Fingerprint region. (**b**) $\nu_{C\equiv N}$ and ν_{CH} regions.

The solute-solvent interaction in ACN_E is quite low as indicated by the comparatively unaltered position of several modes of ACN and ionic salts. We observed a change only at 739 cm^{-1} ($\nu_{a,C-CN}$) and 2255 cm^{-1} ($\nu_{s,C\equiv N}$) of the ACN_E relative to those of the ACN because of nitrile-metal ions coordination [87]. The ion-pairing peaks were present, however as weak shoulders only. This indicates the availability of a huge number of free ions for migration along with a negligible level of ion-pairing, resulting in a high value of $\sigma_{25°C}$ for the ACN_E. In this electrolyte, the metal cations migrate through the nitrile group of the ACN [69]. The SN_E showed a scenario similar to ACN_E except at 769 (δ_{CH2}, ring), 1228 (t$_{CH2}$, i.p.ring), 1443 ($\delta_{a,CH2}$, ring), and 1474 cm^{-1} ($\delta_{a,CH2}$, ring), revealing the SN-ring interaction. Similar to ACN_E, SN_E portrayed a weak ion-pairing peak at 1228 cm^{-1}. These indicate the availability of a large number of free ions for migration, however, with a higher level of ion-pairing, resulting in a lower $\sigma_{25°C}$-value for SN_E than ACN_E. PEO_E experienced the PEO-salt interaction via shifts to 780 (767 cm^{-1}, ring), 1113 (1109 cm^{-1}, PEO; $\nu_{a,COC}$), 1134 (1149 cm^{-1}, PEO; ν_{CC}, $\nu_{a,COC}$), and 1349 (1342 cm^{-1}, PEO; $\omega_{a,CH2}$, $\nu_{a,SO2}$). These were accompanied by a shift in ν_{CH2} modes of the PEO from 2861 to 2872 cm^{-1} and from 2889 to 2894 cm^{-1}, indicating a decrease in C-H bond length for ion solvation, and thereby increasing the amorphicity of the electrolyte [64,65,74]. However, the increase in amorphicity was inadequate to sufficiently increase the electrical conductivity as suggested by the absence of the ion-paring peaks. The Blend 1_E and Blend 2_E observed no significant change in the position of several modes, except at 777 cm^{-1} (δ_{CH2}, SN) and 1437 cm^{-1}

(δ_{CH2}, ring) corresponding to the SN-bpy ligand interaction, and at 2253 cm^{-1} ($\nu_{s,C\equiv N}$, SN) for the interaction of the nitrile-metal ions [87]. The blend-based electrolytes portrayed ion-pairing peaks at 1227 and 1334 cm^{-1} as well as a blue shift in the stretching C-H modes of PEO in the region, 2800–3050 cm^{-1}, indicating a conformational change to form the amorphous phase. The blue shift was higher for the Blend 2_E. These findings suggest that succinonitrile and cobalt salts are crucial for improving the amorphicity of the blend-based electrolyte, thereby, enhancing the electrical conductivity.

Table 2. Observed vibrational frequencies (in cm^{-1}) of the redox mediators, ACN_E, SN_E, PEO_E, and Blends 1_E & 2_E along with those of solvent/matrices and ionic salts.

ACN [†]	ACN_E	SN	SN_E	Blend	Blend 1_E	Blend 2_E	PEO	PEO_E	Li Salt	Co Salt	Assignments [‡]
			481s	479m		478w					δ_{CCC}
	515m			514m	513m	514m		513m	513m	514m	
	571m			571m	571m	571m		570m	571s	571s	
	602sh	604s	602m	604m	601m	602m		601m	602m	601m	δ_{CCC}
	620s		618s		618m	618m		617m	617m	616m	
	653m		653m		652w	652w		652w	654m	656m	
753m	739m		738m		738w	738w		738m	739m	740m	$\nu_{a,C-CN}$, $\nu_{s,SNS}$
	770s	762s	769s	762m	777m	778m		780m		767s	δ_{CH2}, ring
	789m		789m		786sh	786sh		786sh	789m	789m	$\nu_{a,SNS}$
			819s	818m	819w						ν_{C-CN}
				846m	849m	847m	843s	843m			$\rho_{a,CH2}$, ν_{CO}
919s	918m	918s	918m	918sh	sh	sh					$\nu_{s,C-CN}$
		963s	963s	953s	954m	953s		963s		961s	$\rho_{a,CH2}$, t_{CH2}, ν_{C-CN}
			1002s	1002m	1002m	1003w	1002w				ρ_{CH2}
1039s	1039sh										$\rho_{a,CH3}$
	1061s			1059s	sh	1060sh	1061sh	1061m	1061sh	1059s	1057s	$\nu_{a,COC}$, $\rho_{a,CH2}$, $\nu_{a,SNS}$
				1105s	1107s	1106s	1109s	1112s			$\nu_{s,COC}$
	1138m		1136m		1134sh	1134sh	1149s	1134s	1136s	1133s	ν_{CC}, $\nu_{s,SO2}$
	1189s	1199m	1197s	1196w	1195s	1196s		1193s	1197s	1196s	t_{CH2}, $\nu_{a,CF3}$
	1227sh	1233s	1228sh	sh	1227w	1228w			1228m	1229sh	t_{CH2}, i.p.ring
				1251m	1250w	1250w	1242m	1242m			$t_{a,CH2}$
				1299m	1301w	1301w	1280m	1281m			$t_{a,CH2}$, $t_{s,CH2}$
		1337s	1337sh	sh	1334sh	1334sh			1333sh	1331sh	ω_{CH2}, $\nu_{a,SO2}$
	1350s		1353s	1350m	1353s	1353s	1342s	1349s	1353s	1351s	$\omega_{a,CH2}$, $\nu_{a,SO2}$
1374s	1376m										$\omega_{s,CH2}$
		1426s	1426s	1426s	1427w	1427m		1444w			δ_{CH2}
1443s	1443m		1443m	1453w	1437w	sh	1454m	1454m		1453m	$\delta_{a,CH2}$, ring
	1474m		1474m	1469w	1472w	1472w	1467m	1468m		1470m	$\delta_{a,CH2}$, ring
2252s	2255s	2254s	2254s	2251s	2253m	2253s					$\nu_{s,C\equiv N}$
2293s	2293w			2875s	2876s	2877s	2861sh	2872sh			$\nu_{s,CH2}$
				2899s	2904s	2907s	2889s	2894s			$\nu_{a,CH2}$
2942m	2944w	2952s	2951s	2943sh	2944sh	2946sh					$\nu_{s,CH2}$
3001m	3004w	2989s	2989s	2975sh	2979sh	2981sh					$\nu_{a,CH2}$

[†] Relative intensity notations: w, weak; m, medium; s, strong; and sh, shoulder. [‡] Assignment notations: ν, stretching; δ, bending; ω, wagging; t, twisting; ρ, rocking; s, symmetric; a, asymmetric; i.p., in-plane; and o.p., out-of-plane.

Figure 6 shows relative intensities, ΔI_1 (=I(1105 cm^{-1}; $\nu_{s,COC}$; PEO)/I(1196 cm^{-1}; $\nu_{a,CF3}$; TFSI$^-$) and ΔI_2 (=I(2255 cm^{-1}; $\nu_{s,C\equiv N}$; acetonitrile or succinonitrile)/I(1196 cm^{-1}; $\nu_{a,CF3}$; TFSI$^-$) of the redox mediators. Intensity had the following order: ACN_E = SN_E (=0) << Blend 1_E = Blend 2_E = PEO_E (=1) at 1105 cm^{-1} and ACN_E < SN_E ≈ Blend 1_E > Blend 2_E < PEO_E at 1196 cm^{-1}, resulting in ΔI_1 as Blend 2_E > PEO_E > Blend 1_E >> SN_E = ACN_E (=0). This shows the effect of the PEO-salt interaction on the conformational change of PEO to form the amorphous phase, which is the highest for the Blend 2_E. A similar assertion can be made using ΔI_2, which had the following order: SN_E > ACN_E > Blend 2_E > Blend 1_E >> PEO_E (=0). This demonstrates the effect of nitrile-salt interaction on the conformational change of ACN/SN to form the disordered/amorphous structure, which is higher for SN_E than ACN_E, and more for Blend 2_E than Blend 1_E. These results are also supported by the UV-visible spectroscopy, POM, and DSC studies and are discussed in the later sections.

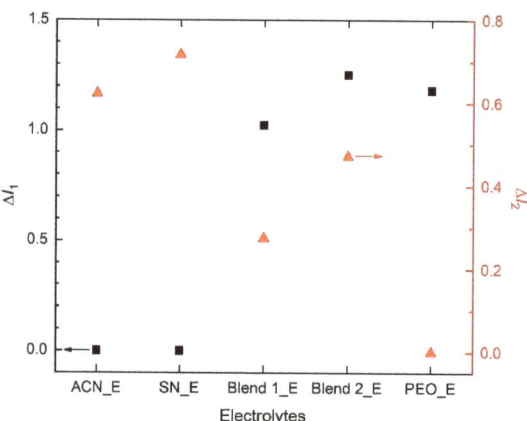

Figure 6. Relative intensities, ΔI_1 and ΔI_2 of the electrolytes. For definition, please see the text.

3.3. Optical Properties

Figure 7a shows the transmittance spectra of the solid redox mediators, SN_E, PEO_E, Blend 1_E, and Blend 2_E as well as liquid electrolytes, ACN_E and SN_E(L). In the UV-A region at 350 nm, the transmittance order was as follows: Blend 2_E (88.1%) > Blend 1_E (81.4%) >> PEO_E (49.6%) >> ACN-E (13.9%) >> SN_E(L) = SN_E(S) (=0%). In the visible region at 555 nm, transparency had the following order: ACN-E (99.9%) ≈ Blend 2_E (99.8%) > SN_E(L) (97%) > Blend 1_E (92.9%) > PEO_E (78.9%) >> SN_E(S) 21.6%. The Blend 2_E portrayed the transparency of all the UV-A, visible, and near-infrared regions; while ACN_E and SN_E(L) showed the transparency in the visible and near-infrared regions only. This shows the supremacy of Blend 2_E over ACN_E and SN_E(L) in terms of transparency. The high value of the transmittance for Blends 1_E and 2_E suggests a low level of the PEO crystallinity [62,65]. The transmittance of Blend 2_E is higher than Blend 1_E, revealing higher amorphicity for the Blend 2_E. These findings are also indicated by the POM study, which has been discussed below.

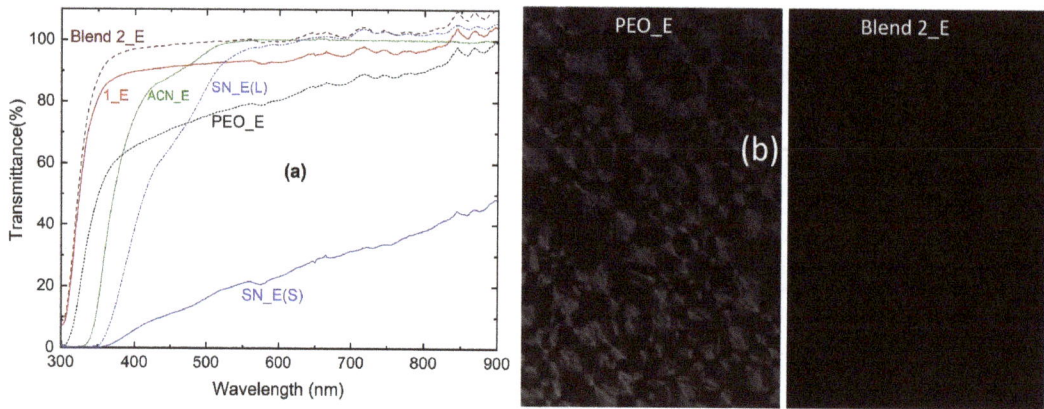

Figure 7. (**a**) Transmittance spectra of the ACN_E, SN_E (L, liquid; S, solid), PEO_E, Blend 1_E, and Blend 2_E. (**b**) polarized optical micrographs of the PEO_E and Blend 2_E.

Figure 7b shows polarized optical micrographs of the PEO_E and Blend 2_E. The Blend 1_E had a micrograph similar to the Blend 2_E. The PEO_E micrograph depicted two parts: (i) several diamond-like spherulites due to the short and randomly oriented PEO chains; and (ii) a little dark region due to the amorphous domain [64,65]. This indicates the presence of a highly crystalline phase of the PEO, resulting in low electrical conductivity. Blend 2_E showed a complete dark region indicating arrest of the amorphous phase, which is responsible for the higher electrical conductivity. These findings are also supported by the DSC study, which has been described below.

3.4. Thermal Properties

Figure 8a shows DSC curves of the solid redox mediators, SN_E, PEO_E, Blend 1_E, and Blend 2_E. The DSC curves showed endothermic peaks corresponding to the melting temperature of the electrolytes, which are as follows: ~63.8 °C for PEO_E, ~47 °C for SN_E, ~6 °C for Blend 1_E, and ~4 °C for Blend 2_E. These values are less than those of pure matrices: ~65.7 °C for PEO [62], ~57.7 °C for SN [72], and ~30.1 °C for PEO-SN blend [62]. This indicates a decrease in the crystallinity of the PEO and SN, which are responsible for the conductivity enhancement of the electrolytes [62–65,72]. It is worth mentioning that the PEO, thereby the related compounds, do not lose the thin film-forming property of the PEO even after the T_m-value [58,59]. In fact, the electrolyte becomes amorphous, which provides highly conducting pathways for easy ion transport [62–65]. The TGA study discussed later showed that Blend-based electrolytes are thermally stable up to 125 °C. The SN_E depicted another endothermic peak at −37.8 °C, which is similar to that of pure SN (−38.4 °C [72]) and corresponds to the T_{pc}. The Blends 1_E and 2_E did not show the T_{pc}-peak. This is due to the matrix-salt interaction phenomenon [62–65]. It is also worth mentioning that the area under the melting point peak corresponds to the heat enthalpy of the electrolyte [62–65]. This area showed the following order for the PEO-based electrolytes: PEO_E ≫ Blend 1_E > Blend 2_E, indicating an extremely low level of crystallinity for the Blends 1_E and 2_E, which is one of the unique properties of the SN-PEO blend-based electrolytes [64,65,77,78].

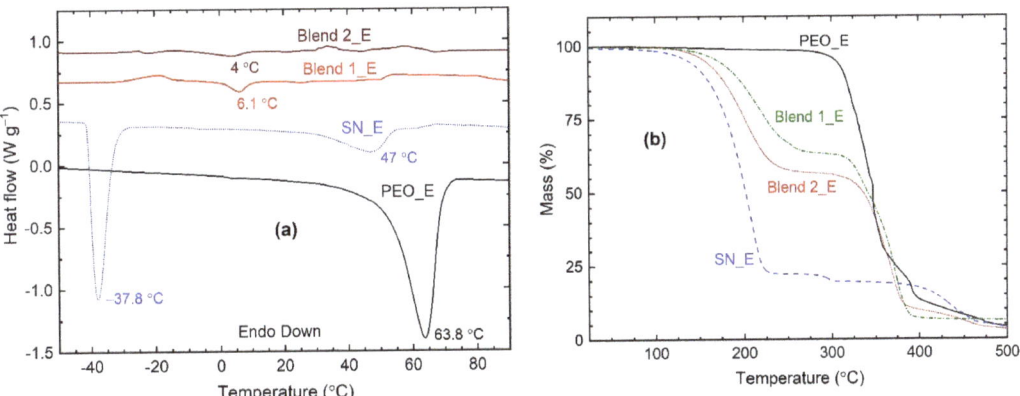

Figure 8. (a) DSC and (b) TGA curves of the solid redox mediators.

Figure 8b shows TGA curves of the solid redox mediators, SN_E, PEO_E, Blend 1_E, and Blend 2_E. The thermal stability of the electrolyte is estimated by the initial plateau region for the mass, which is as follows: ~75 °C for SN_E, ~200 °C for PEO_E, and ~125 °C for Blends 1_E and 2_E. These values are similar to pure matrices reported earlier [62]. The SN_E and PEO_E exhibited a huge drop at ~125 °C and ~300 °C, respectively, due to single-stage decomposition. However, the Blends 1_E and 2_E portrayed two-stage degradation, first at ~125 °C and second at ~300 °C, corresponding to the decomposition of the ingredients, earlier SN, and later PEO.

4. Discussion

We synthesized solid redox mediators using $[(1-x)SN: xPEO]$ as a solid matrix and LiTFSI, Co(bpy)$_3$(TFSI)$_2$, and Co(bpy)$_3$(TFSI)$_3$ ionic solids as sources of ions, following the procedure of Mathew et al. [24], which had acetonitrile as a solvent and 0.1-M LiTFSI, 0.25-M Co(bpy)$_3$(TFSI)$_2$, and 0.06-M Co(bpy)$_3$(TFSI)$_3$ as sources of ions. The acetonitrile-based liquid electrolyte, ACN_E exhibited $\sigma_{25°C}$ of $\sim 1.7 \times 10^{-2}$ S cm^{-1}. The composition, $x = 0$ resulted in a pure plastic crystal-based electrolyte, SN_E with $\sigma_{25°C}$ of $\sim 2.1 \times 10^{-3}$ S cm^{-1}. This value is similar to those of other succinonitrile-based electrolytes [69,72] and is attributed to the solid solvent property of the succinonitrile. The $x = 1$ yielded a pure PEO-based solid polymer electrolyte (EO/Li$^+$ = 226), PEO_E with $\sigma_{25°C}$ of $\sim 9.7 \times 10^{-7}$ S cm^{-1}. The PEO is a well-known polymer matrix for synthesizing a solid polymer electrolyte; however, the highly crystalline nature of the PEO results in poor electrical conductivity [58–61,64,65]. The blend-based solid polymer electrolytes ($x = 0.5$), Blend 1_E (EO/Li$^+$ = 113) and Blend 2_E (EO/Li$^+$ = 226) had $\sigma_{25°C}$ of $\sim 4.3 \times 10^{-4}$ and $\sim 7.2 \times 10^{-4}$ S cm^{-1}, respectively, which were closer to that of the SN_E, disclosing the effect of plasticization property of succinonitrile [64,65].

The investigation of the temperature variation of electrical conductivity resulted in a log $\sigma - T^{-1}$ plot, which was linear for ACN_E, SN_E, and PEO_E, and downward for Blend 1_E and Blend 2_E. The former corresponds to the thermally activated behavior of a homogeneous electrolyte. The latter corresponds to a mixed effect of amorphous domains, the semi-random motion of short polymer chains, and the segmental motion, which was produced by succinonitrile through the interaction with PEO [58–62,64,65,74,77,78]. The Arrhenius-type plot resulted in activation energy of 0.56 eV for SN_E and 1.07 eV for PEO_E in the solid-state region, more than the limiting condition (0.3 eV) for a device application [88]. The ACN_E had an activation energy of 0.15 eV. The VTF-type plot resulted in pseudo-activation energy of 0.06 eV for Blend 1_E and 0.05 eV for Blend 2_E, which are less than the limiting condition.

The XRD patterns of SN_E and PEO_E portrayed weak and broad characteristic peaks of succinonitrile and PEO, respectively, without the ionic salts' peaks [62,64,65]. This indicated the molecular disorderedness of succinonitrile and a decrease in crystallinity of PEO along with a complete dissolution of ionic salts. On the contrary, Blends 1_E and 2_E had no characteristic peaks of ingredients, demonstrating the arrest of the glassy phase because of an interaction between PEO, succinonitrile, and ions [68,77,78]. These assertions can also be made using FT-IR spectroscopy. The FT-IR spectroscopy exhibited no significant change in modes of ionic salts and acetonitrile in ACN_E, revealing the least solvent-solute interaction. SN_E showed a similar scenario, however with an SN-bpy ligand interaction. PEO_E experienced a significant change in modes of ionic salts and PEO, revealing a conformational change of PEO by the large-sized ions. In contrast, the SN-PEO blend-based electrolytes, Blends 1_E and 2_E observed no significant change in modes, except at 777 and 1437 cm^{-1} for the SN-bpy ligand interaction and the ν_{CH2} modes. The matrix/solvent-salts interaction can be put in an order as: ACN_E < SN_E < Blends 1_E & 2_E << PEO_E. It is also worth mentioning that the FT-IR spectra did not show the stability of the electrolytes via the hydrogen interaction with the nitrile group.

The transmittance spectra had the following order: Blend 2_E > Blend 1_E >> PEO_E >> ACN-E >> SN_E (=0%) in the UV-A region and ACN-E $\approx$ Blend 2_E ($\sim$100%) > SN_E(L) > Blend 1_E > PEO_E >> SN_E(S) in the visible region. These electrolytes were transparent in the near-infrared region too. This showed the transparency of Blend 2_E in a wide wavelength range, which makes it superior to ACN_E and I^-/I_3^- redox couple redox mediators. The high level of transparency makes the blend-based electrolyte suitable for various types of solar cells, such as the Gratzel cells, back-illuminated DSSCs, and tandem solar cells [2–6,89]. Also, nearly 100% of transparency for this electrolyte revealed its glassy nature. The same was observed by the polarized optical microscopy too.

The DSC curve showed T_m peak at $\sim$63.8 °C for PEO_E ($\sim$65.7 °C for PEO), $\sim$47 °C for SN_E ($\sim$57.7 °C for SN), $\sim$6 °C for Blend 1_E, and $\sim$4 °C for Blend 2_E ($\sim$30.1 °C for PEO-SN blend). The area under the T_m peak corresponds to the heat of enthalpy of electrolyte, which had the following order: PEO_E >> SN_E >> Blend 1_E ≈ Blend 2_E (≈0). A decrease in T_m-value and/or the area indicated a decrease in crystallinity. Thus, PEO_E had a high level of crystallinity, and Blends 1_E and 2_E had a glassy nature. The TGA curves showed the thermal stability, up to ~75 °C for SN_E, ~200 °C for PEO_E, and ~125 °C for Blends 1_E and 2_E, which are similar to those of pure matrices [62].

5. Conclusions

We synthesized new Co^{2+}/Co^{3+} solid redox mediators, [(1−x)SN: xPEO]-LiTFSI-Co(bpy)$_3$(TFSI)$_2$-Co(bpy)$_3$(TFSI)$_3$ with x equals to 0 (SN_E), 0.5 (Blends 1_E and 2_E), and 1 (PEO_E) in weight fraction. The electrolyte with SN was prepared identically just by replacing the ACN of the liquid redox mediator (ACN_E). The electrolytes exhibited $\sigma_{25°C}$-value in the following order, ACN_E (1.7 × 10^{-2} S cm^{-1}) > SN_E (2.1 × 10^{-3} S cm^{-1}) > Blend 2_E (7.2 × 10^{-4} S cm^{-1}) > Blend 1_E (4.3 × 10^{-4} S cm^{-1}) >> PEO_E (9.7 × 10^{-7} S cm^{-1}). The log $\sigma-T^{-1}$ study showed Arrhenius behavior for SN_E and PEO_E similar to ACN_E, and VTF behavior for Blends 1_E and 2_E. Only Blend-based solid polymer electrolytes showed activation energy of less than 0.3 eV, a high level of transparency in UV-A, visible, and IR regions, and thermal stability up to 125 °C, which are the basic requirements for the DSSC application in the Gulf region. This electrolyte is also suitable for tandem solar cell application.

Author Contributions: Conceptualization, R.K.G. and I.B.; methodology, R.K.G. and I.B.; formal analysis, R.K.G. and H.S.; investigation, R.K.G., H.S., A.I. and A.F.A.; writing—original draft preparation, R.K.G.; writing—review and editing, R.K.G., I.B. and A.S.A.; supervision, R.K.G.; project administration, R.K.G.; funding acquisition, R.K.G. and I.B. All authors have read and agreed to the published version of the manuscript.

Funding: This work was funded by the National Plan for Science, Technology, and Innovation (MAARIFAH), King Abdulaziz City for Science and Technology, Kingdom of Saudi Arabia, Award Number (13-ENE886-02).

Data Availability Statement: Data is available up on request from the corresponding author.

Conflicts of Interest: The authors declare that they have no conflicts of interest.

References

1. Luque, A.; Hegedus, S. *Handbook of Photovoltaic Science and Engineering*, 2nd ed.; John Wiley & Sons, Ltd.: London, UK, 2011.
2. Hagfeldt, A.; Boschloo, G.; Sun, L.C.; Kloo, L.; Pettersson, H. Dye-sensitized solar cells. *Chem. Rev.* **2010**, *110*, 6595–6663. [CrossRef] [PubMed]
3. Ye, M.D.; Wen, X.R.; Wang, M.Y.; Iocozzia, J.; Zhang, N.; Lin, C.J.; Lin, Z.Q. Recent advances in dye-sensitized solar cells: From photoanodes, sensitizers and electrolytes to counter electrodes. *Mater. Today* **2015**, *18*, 155–162. [CrossRef]
4. Gong, J.W.; Sumathy, K.; Qiao, Q.Q.; Zhou, Z.P. Review on dye-sensitized solar cells (DSSCs): Advanced techniques and research trends. *Renew. Sustain. Energy Rev.* **2017**, *68*, 234–246. [CrossRef]
5. Rondan-Gomez, V.; De Los Santos, I.; Seuret-Jimenez, D.; Ayala-Mato, F.; Zamudio-Lara, A.; Robles-Bonilla, T.; Courel, M. Recent advances in dye-sensitized solar cells. *Appl. Phys. A-Mater. Sci. Process.* **2019**, *125*, 836. [CrossRef]
6. Kokkonen, M.; Talebi, P.; Zhou, J.; Asgari, S.; Soomro, S.A.; Elsehrawy, F.; Halme, J.; Ahmad, S.; Hagfeldt, A.; Hashmi, S.G. Advanced research trends in dye-sensitized solar cells. *J. Mater. Chem. A* **2021**, *9*, 10527–10545. [CrossRef] [PubMed]
7. Green, M.A.; Dunlop, E.D.; Hohl-Ebinger, J.; Yoshita, M.; Kopidakis, N.; Hao, X. Solar cell efficiency tables (version 59). *Prog. Photovolt. Res. Appl.* **2022**, *30*, 3–12. [CrossRef]
8. O'Regan, B.; Gratzel, M. A low-cost, high-efficiency solar-cell based on dye-sensitized colloidal TiO$_2$ films. *Nature* **1991**, *353*, 737–740. [CrossRef]
9. Nogueira, A.F.; Longo, C.; De Paoli, M.A. Polymers in dye sensitized solar cells: Overview and perspectives. *Coord. Chem. Rev.* **2004**, *248*, 1455–1468. [CrossRef]
10. Li, B.; Wang, L.D.; Kang, B.N.; Wang, P.; Qiu, Y. Review of recent progress in solid-state dye-sensitized solar cells. *Sol. Energy Mater. Sol. Cells* **2006**, *90*, 549–573. [CrossRef]
11. Singh, P.K.; Nagarale, R.K.; Pandey, S.P.; Rhee, H.W.; Bhattacharya, B. Present status of solid state photoelectrochemical solar cells and dye sensitized solar cells using PEO-based polymer electrolytes. *Adv. Nat. Sci. Nanosci. Nanotechnol.* **2011**, *2*, 023002. [CrossRef]

12. Wu, J.H.; Lan, Z.; Lin, J.M.; Huang, M.L.; Huang, Y.F.; Fan, L.Q.; Luo, G.G. Electrolytes in dye-sensitized solar cells. *Chem. Rev.* **2015**, *115*, 2136–2173. [CrossRef] [PubMed]
13. Su'ait, M.S.; Rahman, M.Y.A.; Ahmad, A. Review on polymer electrolyte in dye-sensitized solar cells (DSSCs). *Sol. Energy* **2015**, *115*, 452–470. [CrossRef]
14. Singh, R.; Polu, A.R.; Bhattacharya, B.; Rhee, H.W.; Varlikli, C.; Singh, P.K. Perspectives for solid biopolymer electrolytes in dye sensitized solar cell and battery application. *Renew. Sustain. Energy Rev.* **2016**, *65*, 1098–1117. [CrossRef]
15. Mehmood, U.; Al-Ahmed, A.; Al-Sulaiman, F.A.; Malik, M.I.; Shehzad, F.; Khan, A.U.H. Effect of temperature on the photovoltaic performance and stability of solid-state dye-sensitized solar cells: A review. *Renew. Sustain. Energy Rev.* **2017**, *79*, 946–959. [CrossRef]
16. Venkatesan, S.; Lee, Y.L. Nanofillers in the electrolytes of dye-sensitized solar cells—A short review. *Coord. Chem. Rev.* **2017**, *353*, 58–112. [CrossRef]
17. Iftikhar, H.; Sonai, G.G.; Hashmi, S.G.; Nogueira, A.F.; Lund, P.D. Progress on electrolytes development in dye-sensitized solar cells. *Materials* **2019**, *12*, 1998. [CrossRef] [PubMed]
18. Hasan, M.M.; Islam, M.D.; Rashid, T.U. Biopolymer-based electrolytes for dye-sensitized solar cells: A critical review. *Energy Fuels* **2020**, *34*, 15634–15671. [CrossRef]
19. Wang, N.; Hu, J.J.; Gao, L.G.; Ma, T.L. Current progress in solid-state electrolytes for dye-sensitized solar cells: A mini-review. *J. Electron. Mater.* **2020**, *49*, 7085–7097. [CrossRef]
20. Abu Talip, R.A.; Yahya, W.Z.N.; Bustam, M.A. Ionic liquids roles and perspectives in electrolyte for dye-sensitized solar cells. *Sustainability* **2020**, *12*, 7598. [CrossRef]
21. Teo, L.P.; Buraidah, M.H.; Arof, A.K. Polyacrylonitrile-based gel polymer electrolytes for dye-sensitized solar cells: A review. *Ionics* **2020**, *26*, 4215–4238. [CrossRef]
22. Chiba, Y.; Islam, A.; Watanabe, Y.; Komiya, R.; Koide, N.; Han, L.Y. Dye-sensitized solar cells with conversion efficiency of 11.1%. *Jpn. J. Appl. Phys. Part 2-Lett. Express Lett.* **2006**, *45*, L638–L640. [CrossRef]
23. Yella, A.; Lee, H.W.; Tsao, H.N.; Yi, C.Y.; Chandiran, A.K.; Nazeeruddin, M.K.; Diau, E.W.G.; Yeh, C.Y.; Zakeeruddin, S.M.; Gratzel, M. Porphyrin-sensitized solar cells with cobalt (II/III)-based redox electrolyte exceed 12 percent efficiency. *Science* **2011**, *334*, 629–634. [CrossRef] [PubMed]
24. Mathew, S.; Yella, A.; Gao, P.; Humphry-Baker, R.; Curchod, B.F.E.; Ashari-Astani, N.; Tavernelli, I.; Rothlisberger, U.; Nazeeruddin, M.K.; Gratzel, M. Dye-sensitized solar cells with 13% efficiency achieved through the molecular engineering of porphyrin sensitizers. *Nat. Chem.* **2014**, *6*, 242–247. [CrossRef] [PubMed]
25. Giribabu, L.; Bolligarla, R.; Panigrahi, M. Recent advances of cobalt(II/III) redox couples for dye-sensitized solar cell applications. *Chem. Rec.* **2015**, *15*, 760–788. [CrossRef]
26. Bella, F.; Galliano, S.; Gerbaldi, C.; Viscardi, G. Cobalt-based electrolytes for dye-sensitized solar cells: Recent advances towards stable devices. *Energies* **2016**, *9*, 384. [CrossRef]
27. Freitag, M.; Teuscher, J.; Saygili, Y.; Zhang, X.; Giordano, F.; Liska, P.; Hua, J.; Zakeeruddin, S.M.; Moser, J.E.; Gratzel, M.; et al. Dye-sensitized solar cells for efficient power generation under ambient lighting. *Nat. Photonics* **2017**, *11*, 372–378. [CrossRef]
28. Vlachopoulos, N.; Hagfeldt, A.; Benesperi, I.; Freitag, M.; Hashmi, G.; Jia, G.B.; Wahyuono, R.A.; Plentz, J.; Dietzek, B. New approaches in component design for dye-sensitized solar cells. *Sustain. Energy Fuels* **2021**, *5*, 367–383. [CrossRef]
29. Srivishnu, K.S.; Prasanthkumar, S.; Giribabu, L. Cu(II/I) redox couples: Potential alternatives to traditional electrolytes for dye-sensitized solar cells. *Mater. Adv.* **2021**, *2*, 1229–1247. [CrossRef]
30. Yella, A.; Humphry-Baker, R.; Curchod, B.F.E.; Astani, N.A.; Teuscher, J.; Polander, L.E.; Mathew, S.; Moser, J.E.; Tavernelli, I.; Rothlisberger, U.; et al. Molecular engineering of a fluorene donor for dye-sensitized solar cells. *Chem. Mater.* **2013**, *25*, 2733–2739. [CrossRef]
31. Yum, J.H.; Moehl, T.; Yoon, J.; Chandiran, A.K.; Kessler, F.; Gratia, P.; Gratzel, M. Toward higher photovoltage: Effect of blocking layer on cobalt bipyridine pyrazole complexes as redox shuttle for dye-sensitized solar cells. *J. Phys. Chem. C* **2014**, *118*, 16799–16805. [CrossRef]
32. Heiniger, L.P.; Giordano, F.; Moehl, T.; Gratzel, M. Mesoporous TiO_2 beads offer improved mass transport for cobalt-based redox couples leading to high efficiency dye-sensitized solar cells. *Adv. Energy Mater.* **2014**, *4*, 1400168. [CrossRef]
33. Liu, F.; Hu, S.L.; Ding, X.L.; Zhu, J.; Wen, J.; Pan, X.; Chen, S.H.; Nazeeruddin, M.K.; Dai, S.Y. Ligand-free nano-grain Cu_2SnS_3 as a potential cathode alternative for both cobalt and iodine redox electrolyte dye-sensitized solar cells. *J. Mater. Chem. A* **2016**, *4*, 14865–14876. [CrossRef]
34. Hao, Y.; Yang, W.X.; Zhang, L.; Jiang, R.; Mijangos, E.; Saygili, Y.; Hammarstrom, L.; Hagfeldt, A.; Boschloo, G. A small electron donor in cobalt complex electrolyte significantly improves efficiency in dye-sensitized solar cells. *Nat. Commun.* **2016**, *7*, 13934. [CrossRef] [PubMed]
35. Xiang, H.D.; Fan, W.; Li, J.H.; Li, T.Y.; Robertson, N.; Song, X.R.; Wu, W.J.; Wang, Z.H.; Zhu, W.H.; Tian, H. High-performance porphyrin-based dye-sensitized solar cells with iodine and cobalt redox shuttles. *Chemsuschem* **2017**, *10*, 938–945. [CrossRef] [PubMed]
36. Yella, A.; Mathew, S.; Aghazada, S.; Comte, P.; Gratzel, M.; Nazeeruddin, M.K. Dye-sensitized solar cells using cobalt electrolytes: The influence of porosity and pore size to achieve high-efficiency. *J. Mater. Chem. C* **2017**, *5*, 2833–2843. [CrossRef]

37. Hao, Y.; Yang, W.X.; Karlsson, M.; Cong, J.Y.; Wang, S.H.; Lo, X.; Xu, B.; Hua, J.L.; Kloo, L.; Boschloo, G. Efficient dye-sensitized solar cells with voltages exceeding 1 v through exploring tris(4-alkoxyphenyl)amine mediators in combination with the tris(bipyridine) cobalt redox system. *ACS Energy Lett.* **2018**, *3*, 1929–1937. [CrossRef]
38. Zhang, L.; Yang, X.C.; Wang, W.H.; Gurzadyan, G.G.; Li, J.J.; Li, X.X.; An, J.C.; Yu, Z.; Wang, H.X.; Cai, B.; et al. 13.6% efficient organic dye-sensitized solar cells by minimizing energy losses of the excited state. *ACS Energy Lett.* **2019**, *4*, 943–951. [CrossRef]
39. Wu, H.; Xie, X.R.; Mei, Y.Y.; Ren, Y.T.; Shen, Z.C.; Li, S.N.; Wang, P. Phenalenothiophene-based organic dye for stable and efficient solar cells with a cobalt redox electrolyte. *ACS Photonics* **2019**, *6*, 1216–1225. [CrossRef]
40. Peng, J.D.; Wu, Y.T.; Yeh, M.H.; Kuo, F.Y.; Vittal, R.; Ho, K.C. Transparent cobalt selenide/graphene counter electrode for efficient dye-sensitized solar cells with $Co^{2+/(3+)}$-based redox couple. *ACS Appl. Mater. Interfaces* **2020**, *12*, 44597–44607. [CrossRef]
41. Xu, D.; Zhang, H.G.; Chen, X.J.; Yan, F. Imidazolium functionalized cobalt tris(bipyridyl) complex redox shuttles for high efficiency ionic liquid electrolyte dye-sensitized solar cells. *J. Mater. Chem. A* **2013**, *1*, 11933–11941. [CrossRef]
42. Kakiage, K.; Aoyama, Y.; Yano, T.; Otsuka, T.; Kyomen, T.; Unno, M.; Hanaya, M. An achievement of over 12 percent efficiency in an organic dye-sensitized solar cell. *Chem. Commun.* **2014**, *50*, 6379–6381. [CrossRef] [PubMed]
43. Kakiage, K.; Aoyama, Y.; Yano, T.; Oya, K.; Fujisawa, J.; Hanaya, M. Highly-efficient dye-sensitized solar cells with collaborative sensitization by silyl-anchor and carboxy-anchor dyes. *Chem. Commun.* **2015**, *51*, 15894–15897. [CrossRef] [PubMed]
44. Wang, Z.Y.; Wang, L.; Zhang, Y.; Guo, J.N.; Li, H.; Yan, F. Dye-sensitized solar cells based on cobalt-containing room temperature ionic liquid redox shuttles. *RSC Adv.* **2017**, *7*, 13689–13695. [CrossRef]
45. Stergiopoulos, T.; Bidikoudi, M.; Likodimos, V.; Falaras, P. Dye-sensitized solar cells incorporating novel Co(II/III) based-redox electrolytes solidified by silica nanoparticles. *J. Mater. Chem.* **2012**, *22*, 24430–24438. [CrossRef]
46. Venkatesan, S.; Liu, I.P.; Chen, L.T.; Hou, Y.C.; Li, C.W.; Lee, Y.L. Effects of tio2 and tic nanofillers on the performance of dye sensitized solar cells based on the polymer gel electrolyte of a cobalt redox system. *ACS Appl. Mater. Interfaces* **2016**, *8*, 24559–24566. [CrossRef] [PubMed]
47. Xiang, W.C.; Chen, D.H.; Caruso, R.A.; Cheng, Y.B.; Bach, U.; Spiccia, L. The effect of the scattering layer in dye-sensitized solar cells employing a cobalt-based aqueous gel electrolyte. *Chemsuschem* **2015**, *8*, 3704–3711. [CrossRef] [PubMed]
48. Bella, F.; Vlachopoulos, N.; Nonomura, K.; Zakeeruddin, S.M.; Gratzel, M.; Gerbaldi, C.; Hagfeldt, A. Direct light-induced polymerization of cobalt-based redox shuttles: An ultrafast way towards stable dye-sensitized solar cells. *Chem. Commun.* **2015**, *51*, 16308–16311. [CrossRef]
49. Bendoni, R.; Barthelemy, A.L.; Sangiorgi, N.; Sangiorgi, A.; Sanson, A. Dye-sensitized solar cells based on N719 and cobalt gel electrolyte obtained through a room temperature process. *J. Photochem. Photobiol. A-Chem.* **2016**, *330*, 8–14. [CrossRef]
50. Xiang, W.C.; Huang, W.C.; Bach, U.; Spiccia, L. Stable high efficiency dye-sensitized solar cells based on a cobalt polymer gel electrolyte. *Chem. Commun.* **2013**, *49*, 8997–8999. [CrossRef] [PubMed]
51. Lee, D.K.; Ahn, K.S.; Thogiti, S.; Kim, J.H. Mass transport effect on the photovoltaic performance of ruthenium-based quasi-solid dye sensitized solar cells using cobalt based redox couples. *Dye. Pigment.* **2015**, *117*, 83–91. [CrossRef]
52. Zhang, X.L.; Huang, W.C.; Gu, A.N.; Xiang, W.C.; Huang, F.Z.; Guo, Z.X.; Cheng, Y.B.; Spiccia, L. High efficiency solid-state dye-sensitized solar cells using a cobalt(II/III) redox mediator. *J. Mater. Chem. C* **2017**, *5*, 4875–4883. [CrossRef]
53. Sonai, G.G.; Tiihonen, A.; Miettunen, K.; Lund, P.D.; Nogueira, A.F. Long-term stability of dye-sensitized solar cells assembled with cobalt polymer gel electrolyte. *J. Phys. Chem. C* **2017**, *121*, 17577–17585. [CrossRef]
54. dos Santos, G.A.; Nogueira, A.F. Thermal and electrochemical characterization of a new poly (ethylene oxide) copolymer-gel electrolyte containing polyvalent ion pair of cobalt (Co-II/III) or iron (Fe-II/III). *J. Solid State Electrochem.* **2018**, *22*, 1591–1605. [CrossRef]
55. Venkatesan, S.; Liu, I.P.; Shan, C.M.T.; Teng, H.S.; Lee, Y.L. Highly efficient indoor light quasi -solid-state dye sensitized solar cells using cobalt polyethylene oxide -based printable electrolytes. *Chem. Eng. J.* **2020**, *394*, 124954. [CrossRef]
56. Karthika, P.; Ganesan, S.; Thomas, A.; Rani, T.M.S.; Prakash, M. Influence of synthesized thiourea derivatives as a prolific additive with tris(1,10-phenanthroline)cobalt(II/III)bis/tris(hexafluorophosphate)/hydroxypropyl cellulose gel polymer electrolytes on dye-sensitized solar cells. *Electrochim. Acta* **2019**, *298*, 237–247. [CrossRef]
57. Balamurugan, S.; Ganesan, S. Novel cobalt redox materials admitted in natrosol polymer with a thiophene based additive as a gel polymer electrolyte to tune up the efficiency of dye sensitized solar cells. *Electrochim. Acta* **2020**, *329*, 135169. [CrossRef]
58. Bruce, P.G.; Gray, F.M. *Polymer electrolytes II: Physical principles In Solid State Electrochemistry*; Bruce, P.G., Ed.; Cambridge University Press: Cambridge, UK, 1995; pp. 119–162.
59. Agrawal, R.C.; Pandey, G.P. Solid polymer electrolytes: Materials designing and all-solid-state battery applications: An overview. *J. Phys. D Appl. Phys.* **2008**, *41*, 223001. [CrossRef]
60. Arya, A.; Sharma, A.L. Electrolyte for energy storage/conversion (Li^+, Na^+, Mg^{2+}) devices based on PVC and their associated polymer: A comprehensive review. *J. Solid State Electrochem.* **2019**, *23*, 997–1059. [CrossRef]
61. Arya, A.; Sharma, A.L. A glimpse on all-solid-state Li-ion battery (ASSLIB) performance based on novel solid polymer electrolytes: A topical review. *J. Mater. Sci.* **2020**, *55*, 6242–6304. [CrossRef]
62. Gupta, R.K.; Kim, H.M.; Rhee, H.W. Poly(ethylene oxide): Succinonitrile—A polymeric matrix for fast-ion conducting redox-couple solid electrolytes. *J. Phys. D-Appl. Phys.* **2011**, *44*, 205106. [CrossRef]
63. Gupta, R.K.; Rhee, H.W. Highly conductive redox-couple solid polymer electrolyte system: Blend-KI-I_2 for dye-sensitized solar cells. *Adv. OptoElectronics* **2011**, *2011*, 102932. [CrossRef]

64. Gupta, R.K.; Rhee, H.W. Effect of succinonitrile on electrical, structural, optical, and thermal properties of poly(ethylene oxide)-succinonitrile /LiI-I$_2$ redox-couple solid polymer electrolyte. *Electrochim. Acta* **2012**, *76*, 159–164. [CrossRef]
65. Gupta, R.K.; Rhee, H.W. Plasticizing effect of k+ ions and succinonitrile on electrical conductivity of poly(ethylene oxide)-succinonitrile /KI-I$_2$ redox-couple solid polymer electrolyte. *J. Phys. Chem. B* **2013**, *117*, 7465–7471. [CrossRef] [PubMed]
66. Gupta, R.K.; Bedja, I.M. Improved cell efficiency of poly(ethylene oxide)-succinonitrile /LiI-I$_2$ solid polymer electrolyte-based dye-sensitized solar cell. *Phys. Status Solidi A Appl. Mater. Sci.* **2014**, *211*, 1601–1604. [CrossRef]
67. Gupta, R.K.; Bedja, I. Cationic effect on dye-sensitized solar cell properties using electrochemical impedance and transient absorption spectroscopy techniques. *J. Phys. D Appl. Phys.* **2017**, *50*, 245501. [CrossRef]
68. Gupta, R.K.; Rhee, H.W.; Bedja, I.; AlHazaa, A.N.; Khan, A. Effect of laponite (R) nanoclay dispersion on electrical, structural, and photovoltaic properties of dispersed poly(ethylene oxide)-succinonitrile -LiI-I$_2$ solid polymer electrolyte. *J. Power Sources* **2021**, *490*, 229509. [CrossRef]
69. Alarco, P.J.; Abu-Lebdeh, Y.; Abouimrane, A.; Armand, M. The plastic-crystalline phase of succinonitrile as a universal matrix for solid-state ionic conductors. *Nat. Mater.* **2004**, *3*, 476–481. [CrossRef]
70. Wang, P.; Dai, Q.; Zakeeruddin, S.M.; Forsyth, M.; MacFarlane, D.R.; Gratzel, M. Ambient temperature plastic crystal electrolyte for efficient, all-solid-state dye-sensitized solar cen. *J. Am. Chem. Soc.* **2004**, *126*, 13590–13591. [CrossRef] [PubMed]
71. Chen, Z.G.; Yang, H.; Li, X.H.; Li, F.Y.; Yi, T.; Huang, C.H. Thermostable succinonitrile-based gel electrolyte for efficient, long-life dye-sensitized solar cells. *J. Mater. Chem.* **2007**, *17*, 1602–1607. [CrossRef]
72. Gupta, R.K.; Bedja, I.; Islam, A.; Shaikh, H. Electrical, structural, and thermal properties of succinonitrile-LiI-I$_2$ redox-mediator. *Solid State Ion.* **2018**, *326*, 166–172. [CrossRef]
73. Gupta, R.K.; Shaikh, H.; Bedja, I. Understanding the electrical transport–structure relationship and photovoltaic properties of a [succinonitrile–ionic liquid]–LiI-I$_2$ redox electrolyte. *ACS Omega* **2020**, *5*, 12346–12354. [CrossRef] [PubMed]
74. Gupta, R.K.; Rhee, H.W. Detailed investigation into the electrical conductivity and structural properties of poly(ethylene oxide)-succinonitrile -Li(CF$_3$SO$_2$)$_2$N solid polymer electrolytes. *Bull. Korean Chem. Soc.* **2017**, *38*, 356–363. [CrossRef]
75. Bhattacharya, B.; Lee, J.Y.; Geng, J.; Jung, H.T.; Park, J.K. Effect of cation size on solid polymer electrolyte based dye-sensitized solar cells. *Langmuir* **2009**, *25*, 3276–3281. [CrossRef] [PubMed]
76. Haq, N.; Shakeel, F.; Alanazi, F.K.; Shaikh, H.; Bedja, I.; Gupta, R.K. Utilization of poly(ethylene terephthalate) waste for preparing disodium terephthalate and its application in a solid polymer electrolyte. *J. Appl. Polym. Sci.* **2019**, *136*, 47612. [CrossRef]
77. Fan, L.Z.; Hu, Y.S.; Bhattacharyya, A.J.; Maier, J. Succinonitrile as a versatile additive for polymer electrolytes. *Adv. Funct. Mater.* **2007**, *17*, 2800–2807. [CrossRef]
78. Patel, M.; Chandrappa, K.G.; Bhattacharyya, A.J. Increasing ionic conductivity and mechanical strength of a plastic electrolyte by inclusion of a polymer. *Electrochim. Acta* **2008**, *54*, 209–215. [CrossRef]
79. Reichardt, C. *Solvents and Solvent Effects in Organic Chemistry*, 3rd ed.; WILEY-VCH Verlag GmbH & Co. KGaA: Weinheim, Germany, 2003.
80. Lide, D.R. *CRC Handbook of Chemistry and Physics*, 89th ed.; CRC Press/Taylor and Francis: Boca Raton, FL, USA, 2009.
81. Pace, E.L.; Noe, L.J. Infrared spectra of acetonitrile and acetonitrile-d$_3$. *J. Chem. Phys.* **1968**, *49*, 5317–5325. [CrossRef]
82. Fengler, O.I.; Ruoff, A. Vibrational spectra of succinonitrile and its 1,4-c-13(2)-, 2,2,3,3-h-2(4)- and 1,4-c-13(2)-2,2,3,3-h-2(4)-isotopomers and a force field of succinonitrile. *Spectrochim. Acta Part A-Mol. Biomol. Spectrosc.* **2001**, *57*, 105–117. [CrossRef]
83. Yoshihara, T.; Tadokoro, H.; Murahashi, S. Normal vibrations of the polymer molecules of helical conformation. IV. Polyethylene oxide and polyethylene-d$_4$ oxide. *J. Chem. Phys.* **1964**, *41*, 2902–2911. [CrossRef]
84. Wen, S.J.; Richardson, T.J.; Ghantous, D.I.; Striebel, K.A.; Ross, P.N.; Cairns, E.J. Ftir characterization of PEO + LiN(CF$_3$SO$_2$)$_2$ electrolytes. *J. Electroanal. Chem.* **1996**, *408*, 113–118. [CrossRef]
85. Rey, I.; Lassègues, J.C.; Grondin, J.; Servant, L. Infrared and Raman study of the PEO-LITFSI polymer electrolyte. *Electrochim. Acta* **1998**, *43*, 1505–1510. [CrossRef]
86. Castellucci, E.; Angeloni, L.; Neto, N.; Sbrana, G. IR and Raman spectra of a 2,2′-bipyridine single crystal: Internal modes. *Chem. Phys.* **1979**, *43*, 365–373. [CrossRef]
87. Colthup, N.B.; Daly, L.H.; Wiberley, S.E. *Introduction to Infrared and Raman Spectroscopy*, 3rd ed.; Academic Press: San Diego, CA, USA, 1990.
88. Agrawal, R.C.; Gupta, R.K. Superionic solids: Composite electrolyte phase—An overview. *J. Mater. Sci.* **1999**, *34*, 1131–1162. [CrossRef]
89. Liska, P.; Thampi, K.R.; Gratzel, M.; Bremaud, D.; Rudmann, D.; Upadhyaya, H.M.; Tiwari, A.N. Nanocrystalline dye-sensitized solar cell/copper indium gallium selenide thin-film tandem showing greater than 15% conversion efficiency. *Appl. Phys. Lett.* **2006**, *88*, 203103. [CrossRef]

Article

Comparison of Crosslinking Kinetics of UV-Transparent Ethylene-Vinyl Acetate Copolymer and Polyolefin Elastomer Encapsulants

Gernot M. Wallner [1,*], Baloji Adothu [2], Robert Pugstaller [1], Francis R. Costa [3] and Sudhanshu Mallick [4]

1. Institute of Polymeric Materials and Testing & Christian Doppler Laboratory for Superimposed Mechanical-Environmental Ageing of Polymeric Hybrid Laminates (CDL-AgePol), University of Linz, Altenbergerstraße 69, 4040 Linz, Austria; robert.pugstaller@jku.at
2. Dubai Electricity and Water Authority (DEWA) Research & Development Center, MBR Solar Park, Dubai 564, United Arab Emirates; baloji.adothu@dewa.gov.ae
3. Borealis Polyolefine GmbH, St.-Peterstraße 625, 4021 Linz, Austria; francis.costa@borealisgroup.com
4. The National Centre for Photovoltaic Research and Education (NCPRE) and Metallurgical Engineering and Materials Science, Indian Institute of Technology Bombay, Mumbai 400076, India; mallick@iitb.ac.in
* Correspondence: gernot.wallner@jku.at; Tel.: +43-732-2468-6614

Citation: Wallner, G.M.; Adothu, B.; Pugstaller, R.; Costa, F.R.; Mallick, S. Comparison of Crosslinking Kinetics of UV-Transparent Ethylene-Vinyl Acetate Copolymer and Polyolefin Elastomer Encapsulants. *Polymers* **2022**, *14*, 1441. https://doi.org/10.3390/polym14071441

Academic Editor: Bożena Jarząbek

Received: 8 March 2022
Accepted: 30 March 2022
Published: 1 April 2022

Publisher's Note: MDPI stays neutral with regard to jurisdictional claims in published maps and institutional affiliations.

Copyright: © 2022 by the authors. Licensee MDPI, Basel, Switzerland. This article is an open access article distributed under the terms and conditions of the Creative Commons Attribution (CC BY) license (https://creativecommons.org/licenses/by/4.0/).

Abstract: Encapsulants based on ethylene-vinyl acetate copolymers (EVA) or polyolefin elastomers (POE) are essential for glass or photovoltaic module laminates. To improve their multi-functional property profile and their durability, the encapsulants are frequently peroxide crosslinked. The crosslinking kinetics are affected by the macromolecular structure and the formulation with stabilizers such as phenolic antioxidants, hindered amine light stabilizers or aromatic ultraviolet (UV) absorbers. The main objective of this study was to implement temperature-rise and isothermal dynamic mechanical analysis (DMA) approaches in torsional mode and to assess and compare the crosslinking kinetics of novel UV-transparent encapsulants based on EVA and POE. The gelation time was evaluated from the crossover of the storage and loss shear modulus. While the investigated EVA and POE encapsulants revealed quite similar activation energy values of 155 kJ/moles, the storage modulus and complex viscosity in the rubbery state were significantly higher for EVA. Moreover, the gelation of the polar EVA grade was about four times faster than for the less polar POE encapsulant. Accordingly, the curing reaction of POE was retarded up to a factor of 1.6 to achieve a progress of crosslinking of 95%. Hence, distinct differences in the crosslinking kinetics of the UV-transparent EVA and POE grades were ascertained, which is highly relevant for the lamination of modules.

Keywords: EVA; POE; crosslinking kinetics; dynamic mechanical analysis; activation energy; photovoltaics

1. Introduction

Crucial components of glass laminates or photovoltaic modules are film adhesives, which are usually based on polar CHO macromolecules such as ethylene-vinyl acetate copolymer (EVA) or poly vinyl butyral (PVB) [1–5]. In recent years, less polar encapsulants based on polyolefin elastomers (POE) have been established [6–10]. While PVB requires an autoclave lamination process, EVA or POE are converted by the more time-efficient vacuum lamination.

Both EVA and POE require organic peroxide crosslinking to attain stable, robust and durable glass or PV module laminates [11]. During the lamination process, the macromolecular structure of the encapsulant changes from the non-crosslinked, entangled thermoplastic state into a widely meshed, three-dimensional network structure. After crosslinking, the encapsulant exhibits better thermal and UV stability, less mechanical creep, less degree of crystallinity and enhanced adherence to glass substrates, silicon solar cells, gridlines or busbars. In the curing process, vinyl silane adhesion promoters are covalently bonded to the macromolecular structure of the encapsulant and the silicate moieties of the glass [12].

A lower degree of the crystallinity of the encapsulants allows for an enhanced optical clarity [13]. However, improperly crosslinked encapsulants impose an adverse effect on the PV module performance. A common problem is the inhomogeneous distribution of the peroxide in the encapsulant films associated with the lateral variation of the crosslink density and an excess of a non-reacted corrosive peroxide [14–16]. As described in [17,18], some additives such as phenolic, nitroxyl and phosphite antioxidants lower the concentration of the peroxide-induced macroradical intermediates that support the polyolefin modifications. In contrast, UV-transparent additives based on hindered amine stabilizers (HAS) confer oxidative stability to the crosslinked EVA or POE materials without compromising the yields of the peroxide-initiated crosslinking [17,18]. Hence, the UV-transparent EVA or POE adhesives and encapsulants based on hindered amine stabilizers would allow for a more reliable lamination process.

To determine the crosslinking state, several methods have been established [19–28]. Soxhlet extraction (the ratio between the mass of the encapsulant film sample after and before extraction) is quite common in the industry and is standardized. However, it is time consuming and requires some material. Differential scanning calorimetry (DSC) and dynamic mechanical analysis (DMA) are much faster and more reliable methods to assess the curing kinetics and the progress of crosslinking [22,23]. In DSC experiments, a sample mass in the mg range is probed. The exothermic crosslinking reaction enthalpy is evaluated. This enthalpy value is affected by the environment (air vs. oxygen vs. nitrogen) [19,21,26]. The reaction enthalpy is amounted to a few J/g, especially in air or oxygen, and the experimental uncertainty is quite high [19].

In contrast, DMA in the molten state is much more sensitive to assess crosslinking kinetics. The real part of the viscosity and modulus undergoes a significant change of a few magnitudes. Moreover, a more representative sample with dimensions in the mm or g range is required for DMA. While temperature-rise DMA is performed at a fixed frequency, isothermal rheometry is based on time or frequency sweeps at a constant temperature. Non-isothermal temperature-rise tests are quite common to characterize thermal transitions, such as glass transitions, melting, the onset of the peroxide decomposition or the gelation of the encapsulant [19–24,26]. Such experiments allow for the assessment of the material changes that occur as they heat up in the lamination process. In contrast, isothermal rheometry better reflects the structural changes at the lamination temperature, which is commonly around 150 °C for peroxide-crosslinking encapsulants. Hence, isothermal DMA is quite relevant for mimicking the main process step of lamination. Interestingly, little research has been performed to assess the curing kinetics of EVA by isothermal DMA [20,22,24,26]. The final lamination quality and degree of crosslinking strongly depend on the peroxide concentration, additive formulation, lamination time, and temperature [23,26]. So far, no specific attention has been given to characterize the crosslinking kinetics of the UV-transparent EVA- or POE-based encapsulants.

Hence, the main objectives of this paper were to implement an isothermal DMA testing method for the assessment of the peroxide-initiated crosslinking kinetics of the encapsulants and to investigate and compare, for the first time, novel UV-transparent EVA and POE film adhesives, which were modified with hindered amine light stabilizers.

2. Materials and Methods

2.1. Encapsulant Materials and Films

Two UV-transparent, fast-cure, commercially available encapsulant films were investigated and compared: an ethylene-vinyl acetate copolymer film (EVA, F406PS® from Hangzhou FIRST Applied Materials, Hangzhou, China), and a polyolefin elastomer film (POE, TF4 also from Hangzhou FIRST Applied Materials, Hangzhou, China). The thickness of the EVA and POE film was 0.45 and 0.54 mm, respectively. According to the data sheets, a gel content of more than 75% for EVA and 60% for POE was stated by the supplier. A qualitative stabilizer analysis of EVA and POE was performed using high-pressure liquid chromatography with UV and mass-spectroscopy detection [29–31]. The EVA and POE films were stabilized with Irgafos 126 (phosphite-based processing antioxidants) and Tinuvin 770 (a hindered amine light stabilizer (HALS)) with secondary amino groups (–NH–), ester ((C=O)–O–) linkage groups and an aliphatic (–C_8H_{16}–) central group.

The encapsulant films were stored in aluminum envelopes and kept in a vacuum box prior to characterization. The surface topology was assessed by laser confocal microscopy. While the investigated EVA film revealed a non-periodic surface topology with a maximum height difference of 130 µm, the pyramid-like surface topology of POE was periodic with a maximum height difference of 500 µm and a diagonal of 1.7 mm of the quadratic base of the pyramid.

2.2. Infrared Spectroscopy

To assess and confirm the chemical structure of the supplied encapsulant films, Fourier Transform Infrared (FTIR) spectrophotometry was performed in direct transmission mode using a PerkinElmer Spectrum 100 (PerkinElmer, Waltham, MA, USA). FTIR spectra were recorded in the range from 650 to 4000 cm^{-1} with 16 scans at a resolution of 4 cm^{-1}.

2.3. Dynamic Mechanical Analysis (DMA)

DMA allows for the investigation of the viscoelastic properties of solids, gels or melts as a function of temperature, time or frequency at a given temperature. The changes in the viscoelastic properties of the EVA and POE encapsulants were measured using a Modular Compact Rheometer (MCR-502, Anton Paar, Graz, Austria). DMA was conducted in torsional mode from 20 to 200 °C at a frequency of 1 Hz at 0.1% strain. Isothermal DMA experiments were performed in torsional mode from 125 °C to 155 °C in 5 °C intervals at a frequency of 1 Hz. The shear stress was kept constant at 7000 Pa.

After 15 min of pre-heating of the oven at a constant temperature, the disc-shaped encapsulant film specimens were placed between the parallel plates with a diameter of 25 mm. The whole process of opening and closing the oven, placing the specimens, and starting the test took 10 s. Viscoelastic properties such as the storage modulus, loss modulus, and complex viscosity were recorded during the dynamic and isothermal tests. The gelation time (t_{gel}) was obtained from the crossover point of the real and imaginary part of the shear modulus. By modelling the gelation time (t_{gel}) using an Arrhenius approach, the activation energy values were deduced for both encapsulants. The complex viscosity data were evaluated as to the progress of the crosslinking reaction.

3. Results and Discussion

In the following, first the results of the IR spectroscopic investigations are described and discussed. Special attention was given to the POE encapsulant, which allowed for the qualitative assessment of the CHO(Si, N)-based comonomers, curing aids, stabilizers and adhesion promoters. In contrast, the EVA-specific peaks of the additives were partly overlaid by the strong absorptions of the vinyl acetate comonomer. In the second and third subchapter, as elucidated by the temperature-rise and isothermal rheological experiments, the similarities and differences of the crosslinking kinetics are described and discussed for the investigated EVA and POE films.

3.1. Structural Features of the Investigated Encapsulants

FTIR absorption spectra of the investigated EVA and POE films in the non-crosslinked reference (ref) and the fully cured (X) state are illustrated in Figure 1. The spectra were measured in transmission mode. Due to a film thickness of about 0.5 mm, the main peaks of the copolymer backbone (i.e., the resonant state of CH_2 and CH_3 stretching vibrations at 2920 and 2850 cm^{-1}, of C=O stretching in EVA at 1730 cm^{-1}, of CH_2 bending vibrations at 1460 cm^{-1} or ester-specific peaks in EVA at 1240, 1160 and 1020 cm^{-1}) were already totally absorbing and could not be resolved. These specific peaks were confirmed and ascertained by the FTIR measurement in ATR mode. For a polar CHO comonomer content of about 10 m%, the C–O related vibrational peaks in the polar ethylene copolymers were totally absorbed in the transmission mode at the film thicknesses above 100 μm [32,33]. In contrast, the polar comonomer content of the investigated 0.5 mm thick EVA film was even higher (~32 m%), which confirmed the totally absorbing peaks in the carbonyl and ester absorption ranges of EVA.

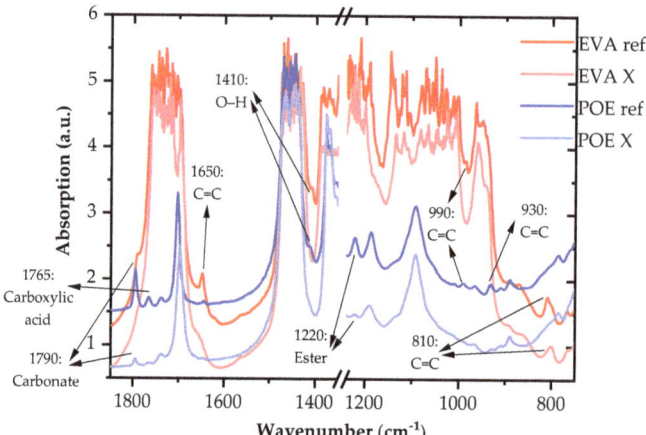

Figure 1. FTIR spectra of UV-transparent, fast-cure EVA and POE grades in the reference (ref) and fully crosslinked (X) state; decaying peaks or shoulders are numbered and highlighted with arrows.

For the investigated EVA film, absorbance peaks at the 1720, 1240, and 1020 cm^{-1} bands, which are also typical for ester and ether groups of cyanurate crosslinking additives, linkage groups of hindered amine light stabilizers or vinyl silane-based adhesion promoters [1,34,35], were totally absorbing. Hence, it was not possible for the investigated EVA film to deduce unambiguous information as to the additives from these peaks. Moreover, in the evaluation of the transmission spectra, special attention was given to the qualitative assessment of the comonomers, crosslinking agents, stabilizers and adhesion promoters in POE. For the POE encapsulant, pronounced peaks at 1790, 1720 and 1090 cm^{-1} were clearly discernible in the transmission spectra. The absorptions are presumably related to the C=O (at 1795, 1720 or 1090 cm^{-1}) or Si–O (at 1090 cm^{-1}) groups of the fast-cure, crosslinking agent tertiary butylperoxy-2-ethylhexylcarbonate (strongest peak at 1790 cm^{-1}), the co-curing agent triallyl isocyanurate (tallest peak at 1700 cm^{-1}), the hindered amine light stabilizer Tinuvin 770 (i.e., Bis(2,2,6,6-tetramethyl-4-piperidyl), sebacate (strongest peak at 1720 cm^{-1}) or the Si–O group of the adhesion promoter 3-(trimethoxysilyl)propyl methacrylate (tallest peak at 1075 cm^{-1}) [34]. Nevertheless, it should be mentioned that the pronounced peaks at 1720 and 1090 cm^{-1} could presumably also be related to low amounts (<10 m%) of the butyl acrylate comonomer in the investigated POE encapsulant [8]. In contrast to the C–H bonds, the C=O and Si–O groups were characterized by high integrated infrared absorption intensity values of more than 10,000 darks [36]. For a given film thickness of 0.5 mm, the measured absorption values, except for the peaks at 1720 and 1090 cm^{-1}, resulted in the content of the additives being around 1 m%. According to [1,11,35], the EVA encapsulants are formulated with 0.1 m% of the HALSs and 1.5 m% of the peroxide curing agents. Due to the superposition of the additives related to the C=O and Si–O peaks with the totally absorbing carbonyl and ester peaks of the vinyl acetate comonomer, just small shoulders were discernible in the EVA spectra.

Peaks which were decreasing significantly upon crosslinking were marked and highlighted by the arrows. In agreement with the data provided in the literature for the peroxide-crosslinking agents or the curing reactions of polyolefins [37–43], the distinct absorption bands for POE or shoulders for EVA at 1790, 1765, 1410 or 1220 cm^{-1} were attributable to the fast-cure, crosslinking agent tertiary butylperoxy-2-ethylhexylcarbonate. Moreover, peaks at 1650 (for EVA), 990 (for both), 930 (for POE) and 810 (for EVA) cm^{-1} were detected in the non-crosslinked reference state. These bands, which were not discernable or just weakly absorbing in the fully cured state, are presumably related to the unsaturated C=C bonds of the co-curing agents (e.g., triallyl isocyanurate) or vinyl silanes. Hence, the investigated EVA and POE films were based on slightly differing curing agent formulations.

3.2. Temperature-Dependent Storage and Loss Modulus and Loss Factor

While the storage modulus (M') is a measure for the elastic or reversible behavior, the loss modulus (M'') describes the viscous response of the encapsulant material. The shear modulus or viscosity of the encapsulant increases as the crosslinking reaction proceeds [23,24]. The thermal transitions of the peroxide-initiated crosslinking reaction such as the decomposition onset (T_{on}), gelation point (T_{gel}) and offset of the crosslinking reaction (T_{off}) temperatures are displayed in Figure 2 and summarized in Table 1. In the temperatures ranging below 120 °C, the decay of the storage and loss modulus could be attributed to the enhanced inner mobility and melting of the crystal lamellae. The melting peak temperatures of EVA and POE were about 55 and 80 °C, respectively [8,19]. The onset of the crosslinking reaction associated with the minimum of the storage modulus was obtained at 125 °C for EVA and at 135 °C for POE. The onset of the crosslinking of EVA was in a similar range from 110 to 125 °C as compared to findings in the literature for the standard and fast-cure EVA encapsulants [8,19,23]. However, in this study, a novel UV-transparent EVA grade with HALSs was used. In contrast, in the literature, focus was given to UV-absorbing EVA grades with phenolic radical scavengers and aromatic UV stabilizers. As well described in [17,18], the peroxide decomposition and crosslinking reaction depends on both the chemical structure of the polymer and the stabilization package.

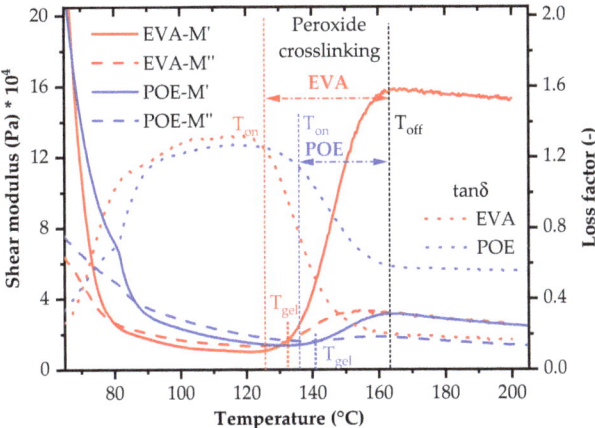

Figure 2. Temperature-rise DMA curves of the investigated peroxide-crosslinking EVA and POE grades (T_{on} ... onset temperature, T_{gel} ... gelation temperature, T_{off} ... offset temperature).

Table 1. Onset, gelation and offset temperatures as well as total curing time of UV-transparent, fast-cure EVA and POE encapsulants.

Encapsulant	T_{on}, °C	T_{gel}, °C	T_{off}, °C	Cure Time, min
EVA	125	130	164	13
POE	135	140	162	9

Due to the curing, an interpenetrating crosslink network is formed that is associated with a crossover of M' and M''. This crossover point is termed the "gelation point" [24]. The gelation temperature and time were indicated by T_{gel} and t_{gel}, respectively. A gelation temperature of 130 °C and 140 °C was obtained for EVA and POE, respectively. After gelation, the difference between M′ and M″ became much more pronounced (Figure 2). A lower gelation temperature might induce more pronounced stresses on solar cells or ribbons during the lamination process. For the investigated EVA grade, a significantly lower flow capability was ascertained in the temperature range from 130 to 150 °C. At the commonly applied lamination temperature of 150 °C, a more than four times higher storage modulus value was deduced for the UV-transparent EVA encapsulant.

Both EVA and POE showed comparable offset temperature values (T_{off}), slightly above 160 °C. Similar values in the range from 155 to 160 °C were reported in the literature for the EVA encapsulants [8,19,23]. At the offset temperature, M′ levelled off due to the completion of the crosslinking reaction. The following slight decrease indicated that there was still a non-crosslinked fraction. The supplier stated a gel content of more than 75% for EVA and 60% for POE. In agreement, the significantly higher storage modulus of the investigated EVA grade at a temperature of more than 165 °C might be an indication for an enhanced crosslinking density of EVA. Furthermore, this conclusion is confirmed by the significant differences of the loss factor values during the peroxide crosslinking. In the rubbery state, EVA revealed a loss factor of about 0.01, whereas it was close to 0.03 for POE.

Interestingly, the absolute, temperature-dependent storage modulus values of the investigated EVA and POE encapsulants differed from the data given in the literature [8,19,23]. While slightly higher values were obtained prior to the onset of the peroxide decomposition and curing, the storage modulus of EVA was lower in the cured state. Most likely, these differences can be attributed to variations in the measurement setup. While the experiments in [8,19,23] were run on a dynamic mechanical analyzer in tensile-shear mode using circular specimens of 9 mm in diameter, in this study, a plate-plate rheometer and well-defined torsional loading was employed. Moreover, more representative disc specimens with a diameter of 25 mm were characterized. For the investigated POE grade, the storage modulus values were markedly different from the data reported in [8]. Especially in the cured state, a factor of five lower M′ values were deduced in this study. Most likely, a quite dissimilar POE grade was investigated in [8]. The POE grades for the PV encapsulation are still under development and are therefore less standardized than the commercially available EVA encapsulants.

In the cured state, a factor of four higher M′ values were ascertained for the investigated EVA as compared to the POE grade. Considering a heating rate of 3 K/min, a total crosslinking reaction time of 13 and 9 min was deduced for EVA and POE, respectively. Hence, the curing of POE was taking place in a narrower processing window (temperature and time), while it was wider and characterized by a lower onset temperature for UV-transparent, fast-cure EVA.

3.3. Curing Kinetics, Activation Energy and Progress of Crosslinking

Isothermal storage and loss modulus curves are depicted in Figure 3 for EVA and POE cured at different temperatures ranging from 125 to 150 °C in 5 °C steps. For both encapsulants, the storage modulus was much more affected by the crosslinking than the loss modulus. Interestingly, EVA revealed a more pronounced change in the storage and loss modulus than POE. As long as the storage modulus is lower than the loss modulus (M′ < M″), the material is in the molten sol state. Upon onset of the crosslinking reaction, the storage modulus increases and crosses the loss modulus curve. This crossover point (M′ = M″) is called gelation and was used to evaluate the gelation time (t_{gel}). At the gelation point, a three-dimensional, weakly crosslinked gel is established. There are still linear and non-crosslinked polymer chains available in the encapsulant. The network formation increases as the crosslinking reaction proceeds. Above the gelation point, a solid crosslinked state is achieved, resulting in the leveling off of the storage modulus. This indicates the completion of the peroxide crosslinking reaction. While the lower and upper bound of the storage modulus were almost independent of the test temperature, the gelation time was significantly lower at the elevated temperatures.

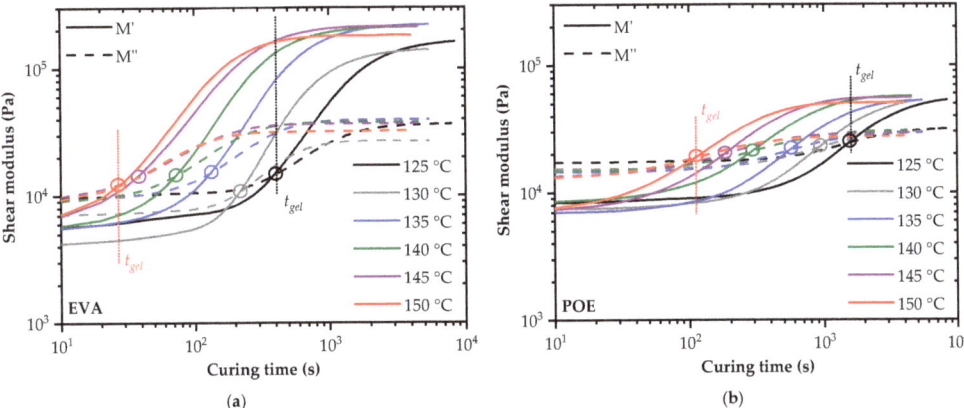

Figure 3. Storage and loss modulus curves of EVA (**a**) and POE (**b**) at different isothermal curing temperatures (t_{gel} ... gelation time).

Comparison of the storage and loss modulus data of the investigated UV-transparent, fast-cure EVA with the values reported in [20,24] for a UV-absorbing EVA grade revealed a good agreement in the thermoplastic, non-crosslinked state. In the cured state, a factor of about 1.5 lower values were obtained in this study. Schulze et al. (2010 and 2015) also used a plate-plate rheometer and performed experiments in controlled-stress mode. The diameter of the plates and the specimens was 20 mm, in contrast to 25 mm in this study.

In Figure 3, the gelation points are indicated with open circles. The deduced gelation times at various isothermal curing temperatures are summarized in Table 2. In agreement with the temperature-rise experiments, a significantly faster gelation was ascertained for the investigated UV-transparent EVA encapsulant. Slightly dependent on the testing temperature, the gelation time was a factor of four times higher for the examined UV-transparent POE grade. A potential reason for the differences in the crosslinking kinetics is the dependency of the reaction rate on the primary structure of EVA and POE. Furthermore, there might be differences in the formulation with the peroxides and the co-curing agents. While the amount of the comonomer content for EVA is well described in the literature [1] and ranges from 28 to 33 w%, no details are given for the POE encapsulants. However, as mentioned in [8], the comonomer content of POE is significantly lower than for EVA, which would result in a lower concentration of tertiary carbon atoms along the main chain, and hence less crosslinking reactivity. In contrast to the standard cure EVA grade modified with UV absorbers and investigated in [22,26], the gelation time at 140 °C was about one magnitude faster for the UV-transparent, fast-cure EVA encapsulant in this study. A comparison with the data provided in [20,24] for another fast-cure, UV-absorbing EVA encapsulant showed a deviation of about 20% in the gelation time at 140 °C (60 vs. 71 s), but a comparable value of about 220 s at 130 °C.

Table 2. Temperature-dependent gelation time, progress of crosslinking at gelation time and cure time to achieve a progress of crosslinking of 95% for UV-transparent EVA and POE.

Temperature, °C	Gelation Time t_{gel}, s		Progress of Crosslinking X at t_{gel}, %		Cure Time for X = 95%, s	
	EVA	POE	EVA	POE	EVA	POE
125	405	1559	6	36	5370	6320
130	224	945	6	35	2995	4120
135	136	582	5	34	2275	2975
140	71	297	5	29	1325	2040
145	39	177	5	26	900	1240
150	28	111	7	25	535	875

To model the dependency of the gelation time on temperature, an Arrhenius fit was used. According to Arrhenius, the temperature-dependent kinetics of chemical reactions for a reaction rate (k) can be written as:

$$k(T) = A\ exp\left(-\frac{E_a}{RT}\right) \quad (1)$$

where T is the temperature, A is a material constant, R is the gas constant and E_a is the activation energy. For many practical engineering implementations, the reaction rate k is insufficient to predict the activation energy. The gelation time or the time to achieve a specific threshold value is of higher interest. Hence, the Arrhenius equation can be re-written as:

$$lnt_{gel} = C - \left(\frac{E_a}{RT}\right) \quad (2)$$

where C is a specific material constant. Arrhenius variables alter when there is a change in the reaction mechanisms or a degradation of the polymeric network. It should be noted that the estimated reaction threshold time t_{gel} is only valid in a specific temperature range.

Using Equation (2), an Arrhenius plot was deduced (see Figure 4). The activation energy (E_a) was calculated from the slope of a linear fit of the Arrhenius plot. E_a values of 155 and 154 kJ/moles were obtained for the investigated UV-transparent EVA and POE grades, respectively. These values were found at the 0.99 goodness of the linear fitting coefficient (R^2). The comparable activation energy values are presumably related to the similar peroxide-curing and cyanurate co-curing agents added to the encapsulant formulations. The value obtained for the UV-transparent, fast-cure EVA grade of this study was significantly higher than the activation energy value of 124 kJ/moles reported in [20,24] for a UV-absorbing, fast-cure EVA encapsulant. While the gelation time for the investigated UV-transparent EVA grade was comparable at 130 °C, it was slower at 140 °C. Due to the unknown details of both of the commercially available EVA formulations, the ascertained differences could not be unambiguously attributed to the interactions of the curing and co-curing additives and aromatic UV absorbers or antioxidants. Nevertheless, as clearly evidenced in the literature, additives and stabilizers could have synergistic or antagonistic effects on peroxide-initiated crosslinking kinetics [17,18] and on long-term durability [30,31].

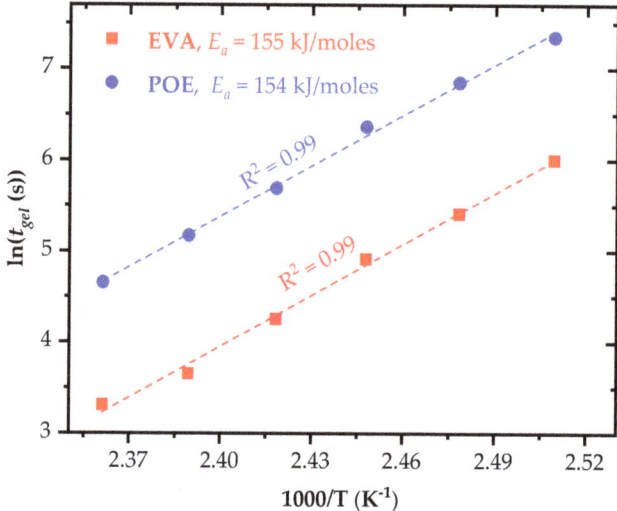

Figure 4. Arrhenius plots and activation energy for EVA and POE encapsulants.

To assess and describe the progress of the curing reaction as a function of time, the complex viscosity (η^*) was evaluated. In Figure 5, the complex viscosity of EVA and POE is plotted for different temperatures. While the increase in the complex viscosity is related to the curing reaction, the leveling off indicates the completion of the crosslinking. The shape of the complex viscosity curves was almost similar for the EVA and POE encapsulants. However, a factor of more than four higher complex viscosity values were obtained for EVA in the cured state. This is in agreement with the results of the temperature-rise DMA. Presumably, the higher viscosity in the cured state of EVA could be attributed to the higher amount of tertiary carbon atoms, the denser crosslinking structure and the higher gel content. The initial complex viscosity values were at a comparable level. The rate of decomposition of the peroxides and the curing reactions were much faster at higher temperatures. Hence, the shear modulus and complex viscosity curves were shifted to shorter curing times at higher temperatures. A comparison of the complex viscosity data of the investigated UV-transparent EVA grade with values reported in [20,24] again revealed a good agreement in the non-crosslinked state and slightly higher values for the cured elastomer. Presumably, the attainable gel content was lower for the UV-transparent EVA grade.

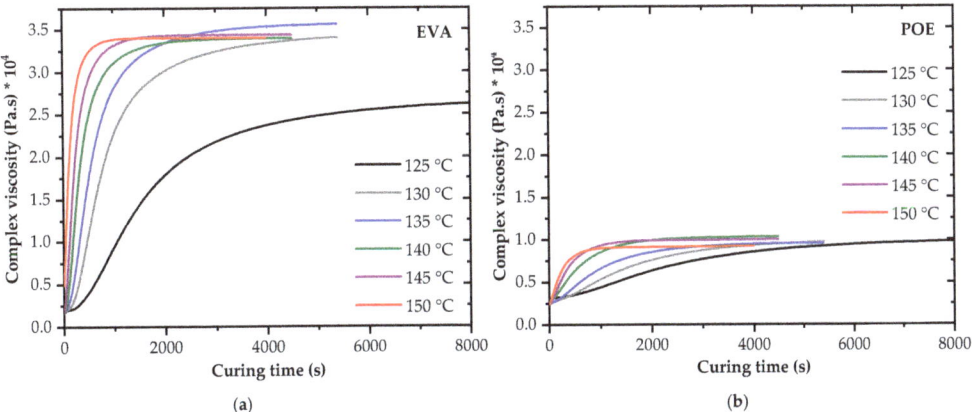

Figure 5. Complex viscosity of EVA (**a**) and POE (**b**) as a function of curing temperature and time.

The progress of the crosslinking reaction (X) (see Figure 6 and Table 2) was calculated from η^* by using Equation (3):

$$X = \frac{\eta^*(t) - \eta^*(t_o)}{\eta^*(t_f) - \eta^*(t_o)} \quad (3)$$

where $\eta^*(t)$ is the complex viscosity at time t, $\eta^*(t_o)$ is the complex viscosity at the beginning of the experiment time (t_o) and $\eta^*(t_f)$ is the final complex viscosity after reaching the saturation or levelled-off state.

The progress of the crosslinking reaction achieved at a specific gelation time was quite different for EVA and POE. For POE, a factor of more than six higher values of the progress of the crosslinking reaction values were deduced at the gelation time. Again, it was clearly confirmed that the crosslinking kinetics were significantly retarded for the investigated POE grade.

The required time to reach the recommended value of 95% of the progress of the crosslinking reaction [24] is summarized in Table 2. This control level of the crosslinking was achieved dependent on temperature by a factor of 1.2 to 1.6 times faster for the investigated EVA encapsulant. This retardation, which was more pronounced at higher temperatures, was not reflected unambiguously by the total time for the curing reaction obtained by the temperature-rise DMA (see Table 1).

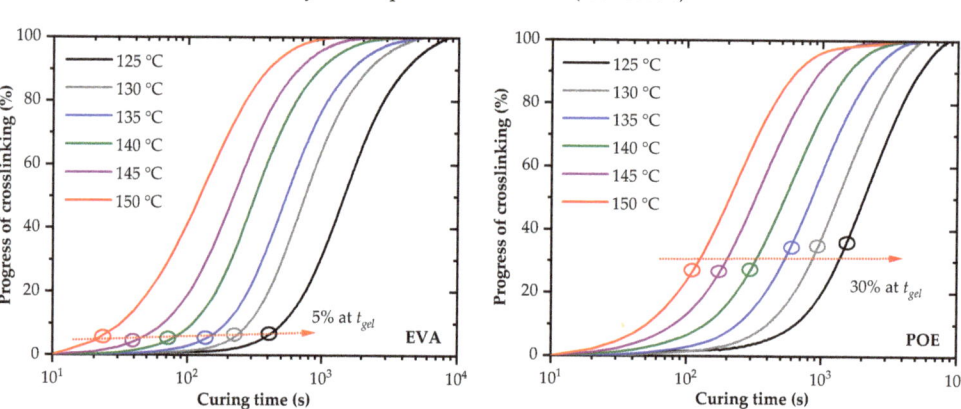

Figure 6. Progress of crosslinking of EVA (**a**) and POE (**b**) as a function of curing temperature and time (t_{gel} ... gelation time).

4. Conclusions

UV-transparent, fast-cure encapsulants based on EVA and POE were examined regarding their chemical structure, formulation and crosslinking kinetics. By using FTIR spectrophotometry in transmission mode, a significantly lower amount of a polar comonomer, most likely based on acrylates, was ascertained for the investigated POE encapsulant. Moreover, the absorption peaks of the carbonate-based peroxide-crosslinking agents and the unsaturated C=C bonds of the co-curing agents and vinyl silanes were clearly discernable. The investigated EVA and POE films were based on slightly differing curing agent formulations. To describe the crosslinking behavior, dynamic and isothermal DMA experiments were carried out in torsional mode using a plate-plate rheometer. The temperature-rise experiment revealed about a 10 °C lower onset and gelation temperature for EVA. Hence, EVA is gelating earlier during the heating-up process of the lamination cycle. The consequence of gelation is a pronounced increase in the storage modulus and viscosity associated with reduced flow capability. At the offset of the curing, which was slightly above 160 °C, a factor of more than four higher storage modulus values were obtained for EVA.

Isothermal DMA experiments were conducted at temperatures ranging from 125 to 150 °C in 5 °C steps. EVA exhibited a more significant change in the storage modulus or complex viscosity compared to POE. Moreover, the gelation time at a defined temperature was a factor of four times longer for POE, which was therefore characterized by a retarded crosslinking behavior. These differences are presumably related to the higher co-monomer content and more tertiary carbon atoms in EVA. Nevertheless, a quite similar activation energy value of about 155 kJ/moles was obtained for both of the materials. Hence, the consideration of just the activation energy to describe the crosslinking kinetics of the encapsulants is not meaningful. By evaluating the time and temperature-dependent complex viscosity, the progress of the crosslinking values was deduced. The time to achieve a progress of crosslinking of 95%, which is recommended in photovoltaic module lamination, was up to a factor of 1.6 longer for POE (at 150 °C: 8.9 min for EVA vs. 14.6 min for POE).

The crosslinking kinetics study revealed a significant difference between UV-transparent, fast-cure EVA and POE. The provided data are of high relevance for the definition of photovoltaic module lamination parameters. In future research, focus will be given to the establishment of the correlations between critical lamination process parameters and the crosslinking kinetics data deduced on an encapsulation film level. Such a fundamental understanding would allow for an efficient and reliable adjustment of the lamination parameters based on the materials data. Finally, it is emphasized that dynamic mechanical analysis is a highly efficient characterization method for the quality assurance and shelf-life testing of encapsulant films.

Author Contributions: Conceptualization, G.M.W. and R.P.; methodology, G.M.W., R.P. and B.A.; software, B.A.; validation, G.M.W. and R.P.; formal analysis, B.A.; investigation, B.A.; resources, F.R.C. and G.M.W.; data curation, B.A. and G.M.W.; writing—original draft preparation, B.A. and G.M.W.; writing—review and editing, F.R.C., S.M. and G.M.W.; visualization, B.A. and G.M.W.; supervision, F.R.C., S.M. and G.M.W.; project administration, G.M.W.; funding acquisition, G.M.W. All authors have read and agreed to the published version of the manuscript.

Funding: The financial support by the Austrian Federal Ministry for Digital and Economic Affairs, the National Foundation for Research, Technology and Development, and the Christian Doppler Research Association are gratefully acknowledged.

Institutional Review Board Statement: Not applicable.

Informed Consent Statement: Not applicable.

Data Availability Statement: Not applicable.

Conflicts of Interest: The authors declare no conflict of interest. The funders had no role in the design of the study; in the collection, analyses or interpretation of data; in the writing of the manuscript or in the decision to publish the results.

References

1. Czanderna, A.W.; Pern, F.J. Encapsulation of PV modules using an ethylene vinyl acetate copolymer as pottant: A critical review. *Sol. Energy Mater. Sol. Cells* **1996**, *43*, 101–181. [CrossRef]
2. Blieske, U.; Stollwerck, G. Glass and other encapsulation materials. *Semicond. Semimet.* **2013**, *28*, 199–258.
3. Peike, C.; Hädrich, I.; Weiß, K.-A.; Dürr, I. Overview of PV module encapsulation materials. *Photovolt. Int.* **2013**, *19*, 85–92.
4. De Oliveira, M.C.C.; Cardoso, A.S.A.D.; Viana, M.M.; Lins, V.F.C. The causes and effects of degradation of encapsulant ethylene vinyl acetate copolymer (EVA) in crystalline silicon photovoltaic modules: A review. *Renew. Sustain. Energy Rev.* **2018**, *81*, 2299–2317. [CrossRef]
5. Jin, J.; Chen, S.; Zhang, J. UV aging behaviour of ethylene-vinyl acetate copolymers (EVA) with different vinyl acetate contents. *Polym. Degrad. Stab.* **2010**, *95*, 725–732. [CrossRef]
6. López-Escalante, M.; Caballero, L.J.; Martín, F.; Gabás, M.; Cuevas, A.; Ramos-Barrado, J. Polyolefin as PID-resistant encapsulant material in 5PV6 modules. *Sol. Energy Mater. Sol. Cells* **2016**, *144*, 691–699. [CrossRef]
7. Lin, B.; Zheng, C.; Zhu, Q.; Xie, F. A polyolefin encapsulant material designed for photovoltaic modules: From perspectives of peel strength and transmittance. *J. Therm. Anal. Calorim.* **2019**, *140*, 2259–2265. [CrossRef]
8. Oreski, G.; Omazic, A.; Eder, G.C.; Voronko, Y.; Neumaier, L.; Mühleisen, W.; Hirschl, C.; Ujvari, G.; Ebner, R.; Edler, M. Properties and degradation behaviour of polyolefin encapsulants for photovoltaic modules. *Prog. Photovolt. Res. Appl.* **2020**, *28*, 1277–1288. [CrossRef]
9. Adothu, B.; Bhatt, P.; Zele, S.; Oderkerk, J.; Costa, F.R.; Mallick, S. Investigation of newly developed thermoplastic polyolefin encapsulant principle properties for the c-Si PV module application. *Mater. Chem. Phys.* **2020**, *243*, 122660. [CrossRef]
10. Barretta, C.; Oreski, G.; Feldbacher, S.; Resch-Fauster, K.; Pantani, R. Comparison of Degradation Behavior of Newly Developed Encapsulation Materials for Photovoltaic Applications under Different Artificial Ageing Tests. *Polymers* **2021**, *13*, 271. [CrossRef]
11. Kempe, M. Overview of Scientific Issues Involved in Selection of polymers for PV applications. In Proceedings of the 2011 37th IEEE Photovoltaic Specialists Conference, Seattle, WA, USA, 19–24 June 2011; pp. 85–90.
12. Chapuis, V.; Pélisset, S.; Raeis-Barnéoud, M.; Li, H.-Y.; Ballif, C.; Perret-Aebi, L.-E. Compressive-shear adhesion characterization of polyvinyl-butyral and ethylene-vinyl acetate at different curing times before and after exposure to damp-heat conditions. *Prog. Photovolt. Res. Appl.* **2014**, *22*, 405–414. [CrossRef]
13. Sung, Y.T.; Kum, C.K.; Lee, H.S.; Kim, J.S.; Yoon, H.G.; Kim, W.N. Effects of crystallinity and crosslinking on the thermal and rheological properties of ethylene vinyl acetate copolymer. *Polymer* **2005**, *46*, 11844–11848. [CrossRef]
14. Zhu, J.; Montiel-Chicharro, D.; Betts, T.R.; Gottschalg, R. Correlation of Degree of EVA Crosslinking with Formation and Discharge of Acetic Acid in PV Modules. In Proceedings of the 33rd European Photovoltaic Solar Energy Conference and Exhibition (PVSEC 2017), Amsterdam, The Netherlands, 25–29 September 2017; pp. 1795–1798.
15. Omazic, A.; Oreski, G.; Halwachs, M.; Eder, G.C.; Hirschl, C.; Neumaier, L.; Pinter, G.; Erceg, M. Relation between degradation of polymeric components in crystalline silicon PV module and climatic conditions: A literature review. *Sol. Energy Mater. Sol. Cells* **2019**, *192*, 123–133. [CrossRef]

16. Adothu, B.; Singh, A.K.; Kumar, S.; Zele, S.; Mallick, S. Effect of curing temperature on properties of ethylene vinyl-acetate used for crystalline silicon solar module encapsulation. In Proceedings of the 36th European Photovoltaic Solar Energy Conference and Exhibition (EU PVSEC 2019), Marseille, France, 9–13 September 2019; pp. 1204–1207.
17. Molloy, B.M.; Johnson, K.-A.; Ross, R.J.; Parent, J.S. Functional group tolerance of AOTEMPO-mediated peroxide cure chemistry. *Polymer* **2016**, *99*, 598–604. [CrossRef]
18. Twigg, C.; Ford, K.; Parent, J.S. Peroxide-initiated chemical modification of polyolefins: In search of a latent antioxidant. *Polymer* **2019**, *176*, 293–299. [CrossRef]
19. Wallner, G.M.; Grabmayer, K.; Oreski, G. Physical characterization of EVA embedding materials for solar cells. In Proceedings of the 25th European Photovoltaic Solar Energy Conference and Exhibition/5th World Conference on Photovoltaic Energy Conversion, Valencia, Spain, 6–10 September 2010; pp. 4033–4035.
20. Schulze, S.-H.; Ehrich, C.; Ebert, M.; Bagdahn, J. Mechanical Behavior and Lamination Issues of Solar Modules Containing Elastomeric and Amorphous Encapsulants. In Proceedings of the 25th European Photovoltaic Solar Energy Conference and Exhibition/5th World Conference on Photovoltaic Energy Conversion, Valencia, Spain, 6–10 September 2010; pp. 4064–4068.
21. Stark, W.; Jaunich, M. Investigation of ethylene/vinyl acetate copolymer (EVA) by thermal analysis DSC and DMA. *Polym. Test.* **2011**, *30*, 236–242. [CrossRef]
22. Stark, W.; Jaunich, M.; Bohmeyer, W.; Lange, K. Investigation of the crosslinking behaviour of ethylene vinyl acetate (EVA) for solar cell encapsulation by rheology and ultrasound. *Polym. Test.* **2012**, *31*, 904–908. [CrossRef]
23. Hirschl, C.; Biebl-Rydlo, M.; Debiasio, M.; Mühleisen, W.; Neumaier, L.; Scherf, W.; Oreski, G.; Eder, G.; Chernev, B.; Schwab, W.; et al. Determining the degree of crosslinking of ethylene vinyl acetate photovoltaic module encapsulants—A comparative study. *Sol. Energy Mater. Sol. Cells* **2013**, *116*, 203–218. [CrossRef]
24. Schulze, S.H.; Apel, A.; Daßler, D.; Ehrich, C. Cure state assessment of EVA-copolymers for PV-applications comparing dynamic-mechanical, dielectric and calorimetric properties. *Sol. Energy Mater. Sol. Cells* **2015**, *143*, 411–417. [CrossRef]
25. Hirschl, C.; Neumaier, L.; Puchberger, S.; Mühleisen, W.; Oreski, G.; Frank, M.K.R.; Tranitz, M.; Schoppa, M.; Wendt, M.; Bogdanski, N.; et al. Determination of the degree of ethylene vinyl acetate crosslinking via Soxhlet extraction: Gold standard or pitfall? *Sol. Energy Mater. Sol. Cells* **2015**, *143*, 494–502. [CrossRef]
26. Jaunich, M.; Bohning, M.; Braun, U.; Teteris, G.; Stark, W. Investigation of the curing state of ethylene/vinyl acetate copolymer (EVA) for photovoltaic applications by gel content determination, rheology, DSC and FTIR. *Polym. Test.* **2016**, *52*, 133–140. [CrossRef]
27. Schlothauer, J.C.; Grabmayer, K.; Wallner, G.M.; Röder, B. Correlation of spatially resolved photoluminescence and viscoelastic mechanical properties of encapsulating EVA in differently aged PV modules. *Prog. Photovolt. Res. Appl.* **2016**, *24*, 855–870. [CrossRef]
28. Oreski, G.; Rauschenbach, A.; Hirschl, C.; Kraft, M.; Eder, G.C.; Pinter, G. Crosslinking and post-crosslinking of ethylene vinyl acetate in photovoltaic modules. *J. Appl. Polym. Sci.* **2017**, *134*, 101–110. [CrossRef]
29. Hintersteiner, I.; Sternbauer, L.; Beißmann, S.; Buchberger, W.W.; Wallner, G.M. Determination of stabilisers in polymeric materials used as encapsulants in photovoltaic modules. *Polym. Test.* **2014**, *33*, 172–178. [CrossRef]
30. Beißmann, S.; Reisinger, M.; Grabmayer, K.; Wallner, G.M.; Nitsche, D.; Buchberger, W. Analytical evaluation of the performance of stabilization systems for polyolefinic materials. Part I: Interactions between hindered amine light stabilizers and phe-nolic antioxidants. *Polym. Degrad. Stab.* **2014**, *110*, 498–508. [CrossRef]
31. Beißmann, S.; Grabmayer, K.; Wallner, G.; Nitsche, D.; Buchberger, W. Analytical evaluation of the performance of stabilization systems for polyolefinic materials. Part II: Interactions between hindered amine light stabilizers and thiosynergists. *Polym. Degrad. Stab.* **2014**, *110*, 509–517. [CrossRef]
32. Wallner, G.M.; Oreski, G. Structure-infrared optical property-correlations of C,O,H-polymers for transparent insulation and greenhouse applications. *Monatsh. Chem.* **2006**, *137*, 899–910. [CrossRef]
33. Oreski, G.; Wallner, G.M. Structure–infrared optical property-correlations of polar ethylene copolymer films for solar applications. *Sol. Energy Mater. Sol. Cells* **2006**, *90*, 1208–1219. [CrossRef]
34. PubChem. Available online: https://pubchem.ncbi.nlm.nih.gov/ (accessed on 7 March 2022).
35. Jentsch, A.; Eichhorn, K.-J.; Voit, B. Influence of typical stabilizers on the aging behavior of EVA foils for photovoltaic applications during artificial UV-weathering. *Polym. Test.* **2015**, *44*, 242–247. [CrossRef]
36. Pacansky, J.; England, C.; Waltman, R.J. Complex Refractive Indexes for Polymers over the Infrared Spectral Region: Specular Reflection IR Spectra of Polymers. *J. Polym. Sci. Part B Polym. Phys.* **1987**, *25*, 901–933. [CrossRef]
37. Vacque, V.; Sombret, B.; Huvenne, J.P.; Legrand, P.; Suc, S. Characterisation of the O-O peroxide bond by vibrational spectroscopy. *Spectrochim. Acta Part A Mol. Spectrosc.* **1997**, *53*, 55–66. [CrossRef]
38. Fujii, A.; Iwasaki, A.; Yoshida, K.; Ebata, T.; Mikami, N. Infrared spectroscopy of (phenol)n+ (n = 2-4) and (phenol-benzene) + cluster ions. *J. Phys. Chem. A* **1997**, *101*, 1798–1803. [CrossRef]
39. Duh, Y.S.; Kao, C.S.; Hwang, H.H.; Lee, W.W.L. Thermal decomposition kinetics of cumene hydroperoxide. *Process. Saf. Environ. Prot.* **1998**, *76*, 271–276. [CrossRef]
40. Stelescu, M.D.; Manaila, E.; Craciun, G.; Zuga, N. Crosslinking and grafting ethylene vinyl acetate copolymer with accelerated electrons in the presence of polyfunctional monomers. *Polym. Bull.* **2012**, *68*, 263–285. [CrossRef]
41. NIST. Available online: https://webbook.nist.gov/cgi/cbook.cgi?Spec=C3006-82-4&Index=0&Type=IR&Large=on (accessed on 29 March 2022).

42. Sipaut, C.S.; Dayou, J. In situ FTIR analysis in determining possible chemical reactions for peroxide crosslinked LDPE in the presence of triallylcyanurate. *Funct. Compos. Struct.* **2019**, *1*, 025003. [CrossRef]
43. Smedberg, A.; Hjertberg, T.; Gustafsson, B. Crosslinking reactions in an unsaturated low density polyethylene. *Polymer* **1997**, *38*, 4127–4138. [CrossRef]

Article

Durability and Performance of Encapsulant Films for Bifacial Heterojunction Photovoltaic Modules

Marilena Baiamonte [1], Claudio Colletti [2], Antonino Ragonesi [2], Cosimo Gerardi [2] and Nadka Tz. Dintcheva [1,*]

- [1] Dipartimento di Ingegneria, Università di Palermo, Viale delle Scienze, Ed. 6, 90128 Palermo, Italy; marilena.baiamonte@unipa.it
- [2] 3SUN-Enel Green Power SpA Contrada Blocco Torrazze, Zona Industriale Catania, 95121 Catania, Italy; claudio.colletti@enel.com (C.C.); antonino.ragonesi@enel.com (A.R.); cosimo.geraradi@enel.com (C.G.)
- * Correspondence: nadka.dintcheva@unipa.it; Tel.: +39-091-23863704

Abstract: Energy recovery from renewable sources is a very attractive, and sometimes, challenging issue. To recover solar energy, the production of photovoltaic (PV) modules becomes a prosperous industrial certainty. An important material in PV modules production and correct functioning is the encapsulant material and it must have a good performance and durability. In this work, accurate characterizations of performance and durability, in terms of photo- and thermo-oxidation resistance, of encapsulants based on PolyEthylene Vinyl Acetate (EVA) and PolyOlefin Elastomer (POE), containing appropriate additives, before (pre-) and after (post-) lamination process have been carried out. To simulate industrial lamination processing conditions, both EVApre-lam and POEpre-lam sheets have been subjected to prolonged thermal treatment upon high pressure. To carry out an accurate characterization, differential scanning calorimetry, rheological and mechanical analysis, FTIR and UV-visible spectroscopy analyses have been performed on pre- and post-laminated EVA and POE. The durability, in terms of photo- and thermo-oxidation resistance, of pre-laminated and post-laminated EVA and POE sheets has been evaluated upon UVB exposure and prolonged thermal treatment, and the progress of degradation has been monitored by spectroscopy analysis. All obtained results agree that the lamination process has a beneficial effect on 3D-structuration of both EVA and POE sheets, and after lamination, the POE shows enhanced rigidity and appropriate ductility. Finally, although both EVA and POE can be considered good candidates as encapsulants for bifacial PV modules, it seems that the POE sheets show a better resistance to oxidation than the EVA sheets.

Keywords: encapsulants for PV modules; lamination process; EVA; POE

Citation: Baiamonte, M.; Colletti, C.; Ragonesi, A.; Gerardi, C.; Dintcheva, N.T. Durability and Performance of Encapsulant Films for Bifacial Heterojunction Photovoltaic Modules. *Polymers* **2022**, *14*, 1052. https://doi.org/10.3390/polym14051052

Academic Editor: Bożena Jarząbek

Received: 21 February 2022
Accepted: 2 March 2022
Published: 6 March 2022

Publisher's Note: MDPI stays neutral with regard to jurisdictional claims in published maps and institutional affiliations.

Copyright: © 2022 by the authors. Licensee MDPI, Basel, Switzerland. This article is an open access article distributed under the terms and conditions of the Creative Commons Attribution (CC BY) license (https://creativecommons.org/licenses/by/4.0/).

1. Introduction

Today, energy recovery from renewable sources and processes with less environmental and human health impacts, and the gradual release of traditional fossil fuel sources, due to their high carbon dioxide and pollutants production, are challenges and necessary issues. Therefore, the energy recovery considering sunlight, winds and tides is extremely attractive and specifically, the development of solar photovoltaic (PV) devices for efficient energy recovery is one of the most important research fields. As documented, the energy demand increases continuously, and is expected to reach around 778 exajoules (EJ) by 2035 [1].

Currently, 3SUN-ENEL Green Power (Catania, Italy) develops a new innovative device for efficient energy recovery, and particularly, in Figure 1a,b, a high reliable bifacial glass–glass heterojunction PV module is shown. This innovative heterojunction technology, combining amorphous and crystalline silicon, offers high performance and efficiency in energy recovery, even in extreme climatic conditions [2]. An important issue in PV modules construction and assembling is the use of appropriate encapsulant materials that can protect efficiently the active PV elements ensuring device high performance and durability [3,4]. The encapsulant polymer-based materials must protect PV modules efficiently against humidity, oxygen and other gas, must be transparent and flexible and must have a good adhesion with glass and solar cells [5–8]. Different encapsulant materials, such as polydimethylsiloxane (PDMS), poly-ethylene vinyl acetate (EVA), polyvinyl butyral (PVB), thermoplastic polyolefins (TPO), polyolefin elastomer (POE), have been considered suitable for industrial purpose [9–13].

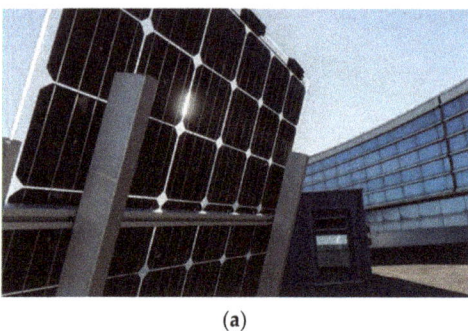
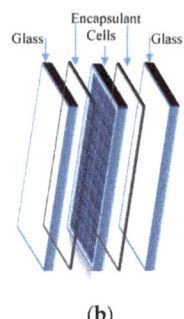

(a) (b)

Figure 1. (a) New novel high reliable bifacial heterojunction glass/glass PV module and (b) its schematic structure (by 3SUN-ENEL Green Power. Catania, Italy).

Considering the balance between costs and performance, the best polymer material as PV encapsulant is EVA, and furtherly to improve its environment resistance, the EVA is added with crosslinking agents and appropriate stabilizers [13]. Therefore, EVA degrades upon solar exposure, even if using crosslinking agents and stabilizers [14–16]. As documented, the EVA degradation proceeds with acetic acid development and the latter leads to encapsulants yellowing, compromising the PV module function [17–19].

However, 3SUN researchers and partners, related to compatibility assessment of commercial EVA, POE, TPO and Ionomer films as encapsulants for bifacial heterojunction PV modules, highlight that the polyolefin elastomers are more compatible to heterojunction technology than other considered commercial materials. Assembling full-size (72 cells) modules, no failures induced by the POE encapsulant are observed after 3000 h in damp heat conditions, 600 thermal cycles and a sequential test using 60 kWh/m^2 exposure [2].

Therefore, the published paper by Baiamonte et al. [20] proposes the formulation of encapsulants for bifacial heterojunction PV modules based on blends containing polyethylene vinyl acetate and polyolefin, i.e., EVA/PO, crosslinking agent and stabilizers, such as UV-adsorber, anti-oxidant and metal deactivator. All obtained results suggest that EVA/PO = 75/25 wt/wt%, containing crosslinking agent and stabilizers, show better mechanical behavior, optical properties and durability than that of neat EVA, suggesting a beneficial effect of the polyolefin presence at low amount. Besides, the photoxidation resistance of EVA/PO = 75/25 wt/wt% blend containing crosslinking agent and stabilizers is very similar to that experienced by neat EVA, highlighting that this blend is a good candidate as encapsulant material for bifacial PV modules.

In this work, the properties and performance of commercial EVA and POE sheets, before (pre-) and after (post-) lamination, as appropriate encapsulant materials for bifacial heterojunction PV modules are investigated. Accurate calorimetric, rheological and durability analysis, in terms of photo- and thermo- oxidation resistance, of both pre- and post-laminated EVA and POE are carried out, also considering the ability of these materials in heterojunction technology for PV modules assembling. However, commercial EVA and POE films are subjected to accelerated UVB exposure and prolonged thermal treatment, and their oxidation resistance is monitored by spectroscopic analysis in time.

Therefore, this proposed comparative study suggests and encourages further research studies regarding the formulation and discovery of PV encapsulants, with good performance, in terms of durability and oxidative resistance, and relatively low cost.

2. Materials and Methods

2.1. Materials

Commercial PolyEthylene Vinyl Acetate (EVA) and PolyOlefin Elastomer (POE) sheets, suitable for low UV cutoff, are purchased by Specialized Technology Resources Inc. Both EVA and POE contain appropriate crosslinking agent and stabilizing additives, such as antioxidants and hindered light amine stabilizers, as produced by manufacture. All additives have been added to EVA and POE during sheets formulation by producer. There are four different sheets considered: EVA pre-laminated (EVApre-lam), EVA post-laminated (EVApost-lam) and POE pre-laminated (POEpre-lam), POE post-laminated (POEpost-lam). The thicknesses of pre-laminated sheets are about 450 µm and to simulate industrially viable lamination process, the pre-laminated EVA and POE sheets have been subjected to pressure at 1 atm and temperature at 150 °C up to 20 min.

2.2. Characterizations

- Differential Scanning Calorimetry: The calorimetric data were evaluated by differential scanning calorimetry (DSC) using a DSC60-Shimadzu calorimeter. All experiments were performed under dry nitrogen on samples of about 10 mg in 40 µL sealed aluminum pans. For both EVA and POE, the calorimetric scans, heating: from −80 to 120 °C and cooling: from 120 to −80 °C, were performed for each sample at scanning heating/cooling rate of 10 °C/min. The values of heat flow have been normalized considering sample mass.
- Rheological analysis: Rheological tests were performed using a stress-controlled rheometer (Rheometric Scientific, SR5, mod. ARES G2 by TA Instrument, New Castle, DE, USA) in parallel plate geometry (plate diameter 25 mm). The complex viscosity (η^*), storage (G') and loss (G'') moduli were measured under frequency scans from $\omega = 10 - 1$ to 10^2 rad/s at T = 140 °C and T = 170 °C for EVA and POE, respectively. The strain amplitude was $\gamma = 5\%$, which preliminary strain sweep experiments proved to be low enough to be in the linear viscoelastic regime.
- FTIR Spectroscopy: A Fourier Transform Infrared Spectrometer (Spectrum One, Perkin Elmer) was used to record IR spectra using 16 scans at a resolution of 1 cm^{-1}. ATR-FTIR for some surface analysis has been also carried out, using 16 scans at a resolution of 1 cm^{-1}. The progress of both photo- and thermo-oxidation degradation for EVA and POE has been followed by running FTIR analysis with time and monitoring the variations in the hydroxyl range (3200–3600 cm^{-1}) and carbonyl range (1800–1500 cm^{-1}) in time, using Spectrum One software.
- UV-visible Spectrometer, (Specord®250 Plus, Analytikjena, Torre Boldone, Italy), was used to record UV-Vis spectra performing 8 scans between 200 and 1100 nm at a resolution of 1 nm. The values of linear attenuation coefficient (k) were calculated considering the measured absorption values (A) and sample thickness (D), using the formula: k = A/(2.3D).

2.3. Accelerated Weathering and Thermo-Oxidation

Photoxidation was carried out using a Q-UV/basic weatherometer (from Q-LAB, Westlake, OH, USA) equipped with UVB lamps (313 nm). The weathering conditions were a continuous light irradiation at T = 70 °C.

Thermo-oxidation was carried out in a ventilated oven at 70 °C a time up to ca. 3500 h for both EVA and POE post-laminated sheets.

The progress of both photo- and thermo-oxidative degradation was followed by FTIR spectroscopic technique.

3. Results

3.1. Differential Scanning Calorimetry (DSC) Characterization

The identification of the transition temperatures for both commercial EVA and POE sheets is performed through differential scanning calorimetry. In Figure 2a,b, the thermograms from −80 °C up to 120 °C of both pre-laminated and post-laminated EVA and POE materials are plotted, and in Table 1, the main identified transition temperatures are reported. In Figure 2a, the glass transition between −40 °C and −20 °C, i.e., Tg around −36 °C, for both EVApre-lam and EVApost-lam samples is detectable and this transition is well noticeable for post-laminated sample. It can be observed that EVApre-lam shows three endothermic peaks in the range from +30 °C up to +90 °C, see blue curve in Figure 2a. The first small peak at about +30 °C, probably, can be attributed to the presence of low molecular weight additives, with low temperature fusion transition. The other two fusion peaks, at about +55 °C and +85 °C, respectively, can be attributed to the fusion transition of two different crystalline structures of EVA sample. After the lamination, the thermogram of EVApost-lam appears slightly different, see red curve in Figure 2a, and there are two noticeable small exothermic peaks in the range +10 °C up to +35 °C, probably, due to the occurrence of crosslinking and additives dispersion during lamination upon prolonged thermal treatment at high pressure. Interestingly, the peak at about +30 °C is not well distinguished and a very broad shoulder in the range between 30 and 50 °C can be observed, highlighting a structural change in the organization of the low molecular weight additives and their interaction with EVA macromolecules. Besides, both fusion peaks at about +55 °C and +85 °C become well pronounced, pointing out the presence of two different polymer crystalline structures. Surprisingly, the fusion enthalpy for EVA increases ca. 1.6 times upon lamination process, suggesting the formation of better ordered 3D-structures, see last column in Table 1.

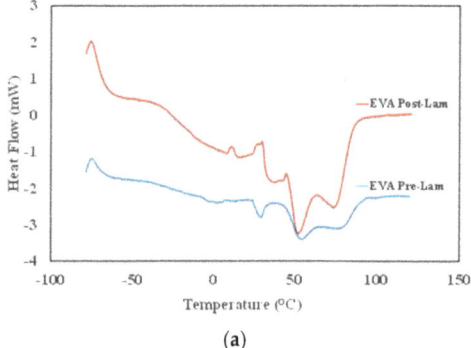
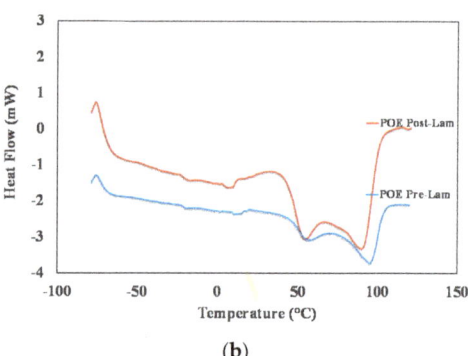

(a)   (b)

Figure 2. DSC thermograms of (**a**) EVApre-lam and EVApost-lam and (**b**) POEpre-lam and POEpost-lam commercial samples.

Table 1. Glass transition (Tg), exothermic and endothermic peaks (Tc and Tf) and fusion enthalpy (ΔH) of pre- and post-laminated EVA and POE samples.

	Glass Transition	Exothermic Phenomenon		Endothermic Phenomenon		
	Tg, °C	Tc, °C	Tf1, °C	Tf2, °C	Tf3, °C	ΔH, J/g
EVApre-lam	−66.9	5.27	29.92	54.78	78.77	12.32
EVApost-lam	−63.0	8.15/25.40 (*)	37.32	51.45	73.16	32.13
POEpre-lam	−67.4	12.02	29.26	58.36	78.11	12.54
POEpost-lam	−62.6	7.06	37.66	55.03	74.49	46.83

Note: (*) this exothermic peak appears as a complex peak and the temperature identification of is difficult.

In Figure 2b, the thermogram of POEpre-lam and POEpost-lam samples are plotted. Moreover, in this case, both POEpre-lam and POEpost-lam samples show a glass transition at around −25 °C and no significant different for glass transition of these samples is observed before and after lamination process. The POEpre-lam sample shows two well visible fusion peaks in the temperature range from +50 up to +100 °C, see blue curve in Figure 2b. It can be observed that after the lamination process both peaks at about +60 °C and +95 °C become well pronounced, pointing out again the presence of two different polymer crystalline structures. Interestingly, the increase of fusion enthalpy for POE upon lamination is ca. 2.7, suggesting the formation of better ordered 3D-structure also for POE, see last column in Table 1.

To sum up, it is worth noting that the glass transition and exothermic phenomena for both EVA and POE are almost uninfluenced by lamination process, while the fusion occurrence reveals that the lamination process could be considered responsible for the formation of a large amount of 3D-ordered crystalline structures. Specifically, the total peaks areas of EVApost-lam (from +25 °C up to 95 °C) and POEpost-lam (from +30 up to +110 °C) samples are higher ca. 1.6 times and 2.7 times than the peak areas of EVApre-lam and POEpre-lam samples, respectively. Based on these results, it can be supposed that the lamination process has a beneficial effect on the formation of 3D-ordered crystalline structures, and it seems that the final POE structure is better structured than the EVA one.

3.2. Rheological Characterization

In Figure 3, the trends of storage and loss moduli, G′ and G″, and complex viscosity, η*, as a function of the frequency of both pre-laminated and post-laminated EVA and POE materials are plotted. The rheological behavior of EVApre-lam and EVApost-lam are slightly different and it is worth noting that for both EVApre-lam and EVApost-lam, no Newtonian plateau is observed, and well pronounced shear thinning is visible, suggesting the presence of crosslinked 3D-structure. Unexpectedly, the values of both moduli G′ and G″ and complex viscosity are lower than the values of before the lamination process, and the latter could be understand considering that during prolonged lamination process, i.e., up to 20 min at high temperature and pressure, the EVA underwent thermal degradation, which leads to the formation of volatile acetic acid.

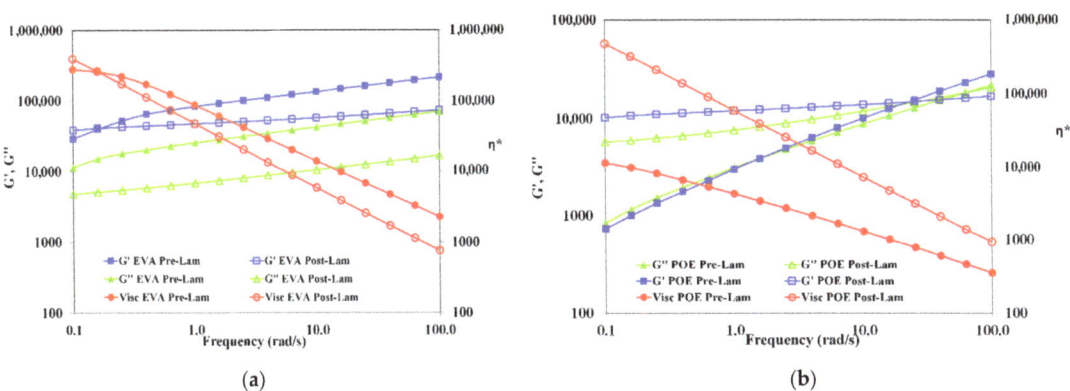

Figure 3. Storage and loss moduli (G′ and G″) and complex viscosity (η*) of (**a**) EVApre-lam (full symbols) and EVApost-lam (open symbols) and (**b**) POEpre-lam (full symbols) and POEpost-lam (open symbols) commercial samples.

Therefore, the elimination of acetic acid molecules during lamination causes decrease for both moduli and viscosity, although crosslinking also occurs. The latter is understandable considering that in the melt state, the macromolecules of EVApost-lam are able to move themself and the system rigidity is lower than EVApre-lam. Interestingly, the G′

and G″ trends remain almost parallel between them, i.e., no cross-over point is observed, for both EVApre-lam and EVApost-lam, suggesting the presence of crosslinked structure, which does not change significantly upon lamination.

Contrarily, the viscosity of POEpost-lam is significantly higher than the viscosity of POEpre-lam, i.e., the difference is more than one decade, and additionally, the slopes of trends are different, highlighting a beneficial effect of the lamination process on POE crosslinking. The change from solid-like to liquid-like behavior for pre-laminated and post-laminated POE occurs at different frequencies, i.e., the cross-over point changes from 1.58 rad/s for POEpre-lam to 25.11 rad/s for POEpost-lam. Therefore, the rheological behavior of pre-laminated POE sample reveals the existence of no well-3D-structured sample, and in this case also, no Newtonian plateau is noticed. The rheological behavior of POEpost-lam is significantly changed upon lamination process and there is a well-3D-structured crosslinked sample.

Based on the rheological behavior, it can be surmised that the lamination process has a well pronounced beneficial effect on 3D-structuration for POE, rather than for EVA. After the lamination, the EVA sample exhibits an affination of existing 3D-structure, without significant change in the melt state behavior. The POEpost-lam shows solid- to liquid-like transition at high frequency, a significant viscosity enhancement and well pronounced shear thinning in comparison to POEpre-lam, highlighting a very good 3D-structuration.

3.3. Mechanical Characterization

In Figure 4, typical stress-strain curves of pre-laminated and post-laminated EVA and POE samples are plotted, and in Table 2, obtained values of main mechanical properties, i.e., elastic modulus, E, tensile strength, TS, and elongation at break, EB, are reported. It is clearly noticeable that the lamination process has a positive effect on the rigidity of both EVA and POE, i.e., the values of elastic modulus after the lamination increase about 45% more than the values before lamination.

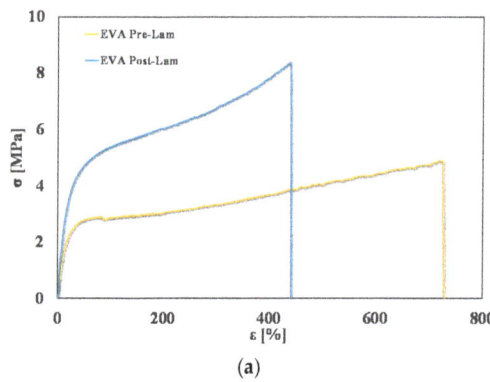
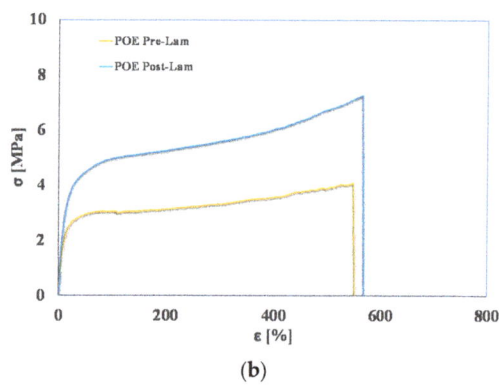

Figure 4. Stress-strain curves of (**a**) EVApre-lam and EVApost-lam and (**b**) POEpre-lam and POEpost-lam commercial samples.

Table 2. Main mechanical properties, i.e., elastic modulus (E), tensile strength (TS) and elongation at break (EB) of pre-laminated and post-laminated EVA and POE samples.

	E, MPa	TS, MPa	EB, %
EVApre-lam	11.6 ± 0.7	4.9 ± 0.3	725 ± 45
EVApost-lam	16.5 ± 1.2	8.3 ± 0.5	441 ± 27
POEpre-lam	21.4 ± 1.5	4.9 ± 0.3	550 ± 25
POEpost-lam	31.6 ± 2.5	7.3 ± 0.5	567 ± 25

As expected, for EVA sample, upon lamination, the tensile strength increases about 70%, while the elongation at break is reduced about 40%. Interestingly, for POE sample, upon lamination, the tensile strength increases about 48%, while the elongation at break remains almost unchanged. These results are understandable considering that during the lamination, the crosslinking process occurs, and this leads to an increase of rigidity, also according to the results by calorimetry and rheological analyses, above commented.

3.4. UV-Visible Characterization

In Figure 5a,b, the linear attenuation coefficient (K) of both pre- and post- laminated EVA and POE are plotted, respectively. The values of linear attenuation coefficient for all samples are calculated using the formula reported in the experimental part, i.e., considering the absorption values (A) and sample thicknesses (D). As known, the material is almost transparent when K value is close to zero. It is clearly noticeable that the EVApost-lam and POEpost-lam samples show K values lower than the EVApre-lam and POEpre-lam samples, especially in the visible range, although the thicknesses of both post-laminated samples are two times higher than the pre-laminated counterparts. This behavior is due to the lamination process having a beneficial effect on both occurrence of 3D-structuration for both EVA and POE and additives dispersion and distribution. Additionally, the small shoulders at about 290 nm in all K trends can be attributed to the presence of stabilizing molecules, and their presence is clearly noticeable before and after lamination.

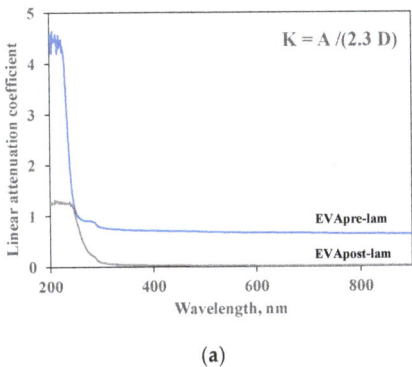
(a)

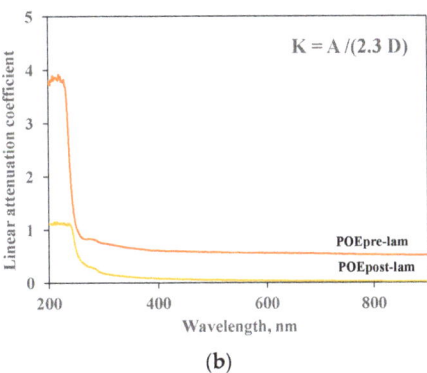
(b)

Figure 5. Linear attenuation coefficient (K) of (**a**) EVApre-lam and EVApost-lam and (**b**) POEpre-lam and POEpost-lam commercial samples.

3.5. FTIR Characterization

In Figure 6a,b, the FTIR spectra of both pre- and post- laminated EVA and POE are plotted, respectively. It is worth noting that the main absorption bands (ca. 2800–2900 cm^{-1}, due to CH stretching, ca. 1700 cm^{-1} due to carbonyl band stretching, and other bands in 1400–800 cm^{-1}, due to different chemical nature and structures) in FTIR spectra are saturated because there are thick original commercial samples. According to literature, the main representative FTIR ranges for polyolefins and polyolefin derivatives are both carbonyl (ca. 1600–1800 cm^{-1}) and hydroxyl (3200–3600 cm^{-1}) range, and additionally, the oxidation degradation of these polymers can be profitable following the monitoring of changes in these two main ranges. It is worth noting that in the spectra of EVApre-lam a small shoulder at ca. 1650 cm^{-1} is noticeable and this could be attributed to the presence of some unsaturation in this material. In the spectra of EVApost-lam, the shoulder at ca. 1650 cm^{-1} is not visible, also because the carbonyl bands appear larger due to higher sample thickness, while a small shoulder at ca. 1780 cm^{-1} appears, probably, due to the formation of some esters during prolonged lamination process. Besides, the hydroxyl bands in EVApost-lam spectra appear more pronounced than the bands in the spectra of EVApre-lam. Similar consideration can be made also for the spectra of POEpre-lam and POEpost-lam samples; upon the lamination process, in the spectra of POEpost-lam a small shoulder at ca. 1650 cm^{-1} appears and the hydroxyl bands are more pronounced in comparison to POEpre-lam.

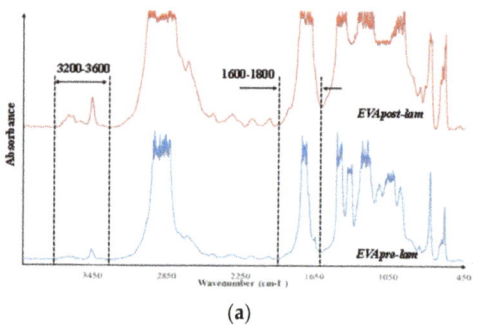
(a)

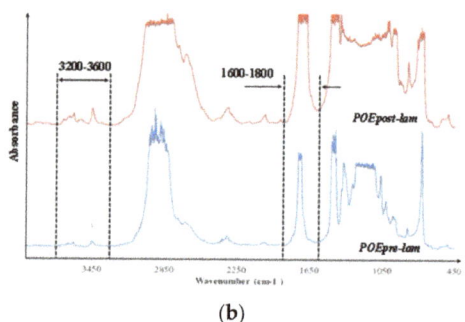
(b)

Figure 6. FTIR spectra of (**a**) EVApre-lam and EVApost-lam and (**b**) POEpre-lam and POEpost-lam commercial samples.

3.6. Photoxidation Resistance

To investigate the photoxidation resistance of EVA and POE, the original sheets have been subjected to UVB exposure and the degradation has been monitored by FTIR analysis in time. In Figure 7a–d, the FTIR of EVApre-lam, EVApost-lam, POEpre-lam and POEpost-lam commercial samples at different exposure times are plotted.

Therefore, according to the literature, EVA photodegradation proceeds with accumulation of oxidation products leading to the formation of new absorption bands in the carbonyl domain (shoulders at 1780 cm^{-1} and 1715 cm^{-1} in IR spectra), in the hydroxyl domain (3200–3600 cm^{-1}) and acetic acid formation, which leads to pH lowering and corrosion ability increasing. Moreover, EVA shows a very fast yellowing, due to the formation of oxidation products, and to avoid unwanted effects, the addition of stabilizers is imperative, especially for manufacturing in service upon sunlight [21,22]. However, as well known, the photodegradation of polyolefins and polyolefins-based polymers proceeds mainly with accumulation of groups in carbonyl domain (1600–1800 cm^{-1}) and hydroxyl domain (3200–3600 cm^{-1}), and subsequently, worsening of their macroscopic properties [20,23–25]. Considering all these issues and FTIR analysis of these commercial EVA and POE samples, reported above, the progress of photoxidation for both EVA and POE can be followed profitable accounting the changes in carbonyl and hydroxyl domains. Besides, as commend above, on the FTIR spectra, main absorption bands are saturated because there are thick original commercial samples. In Figure 7a,b, the changes in the hydroxyl domain for EVA sheets are significant and well appreciable, while, in the carbonyl domain, the presence of only a small shoulder at ca. 1780 cm^{-1} can be observed. Similarly, for POE sheets, the changes in the hydroxyl domain are noticeable, while, in the carbonyl domain a small shoulder at ca. 1640 cm^{-1}, due to presence of insaturations, is barely noticeable.

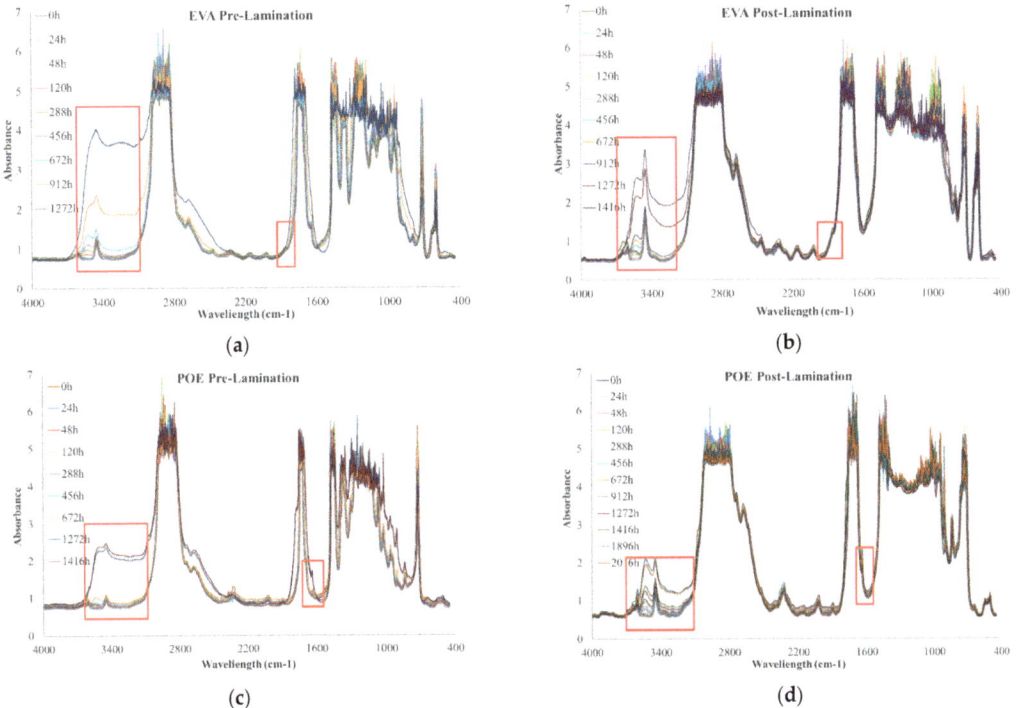

Figure 7. FTIR spectra of (**a**) EVApre-lam, (**b**) EVApost-lam, (**c**) POEpre-lam and (**d**) POEpost-lam commercial samples as a function of photo-oxidation time.

In Figure 8a,b, the variations of the total band areas in hydroxyl domains for EVA and POE are plotted, respectively. Worth noting that the EVApre-lam and POEpre-lam samples show larger hydroxyl accumulations than the EVApost-lam and POEpost-lam samples, especially at long exposure time. Moreover, in Figure 8c,d, the pre-laminated samples show more pronounced increases for the shoulders at 1780 cm^{-1} for EVA and at 1640 cm^{-1} for POE, rather than the post-laminated samples. All these results highlight that the lamination process has a beneficial effect also on photoxidation resistance and again, it seems that the POE show better photoxidation resistance than the EVA one.

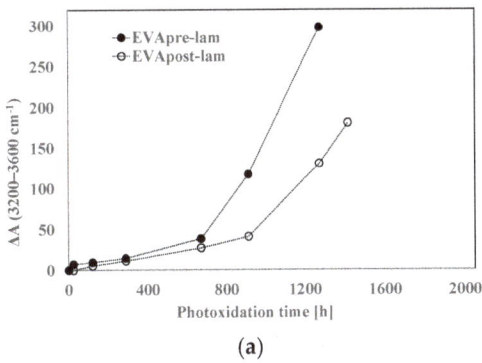

(**a**)

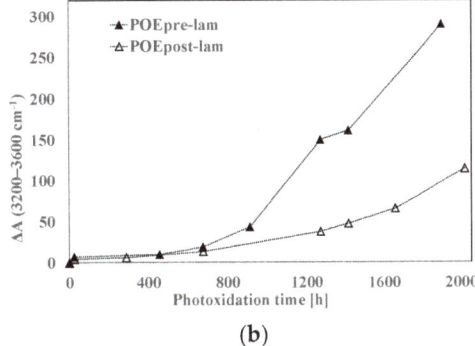

(**b**)

Figure 8. *Cont.*

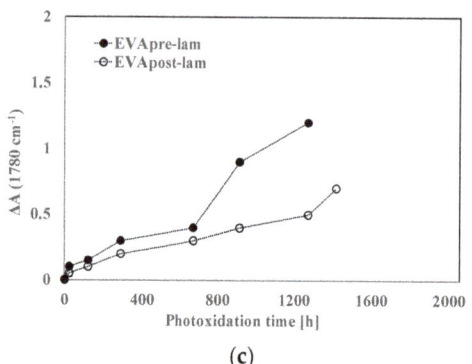
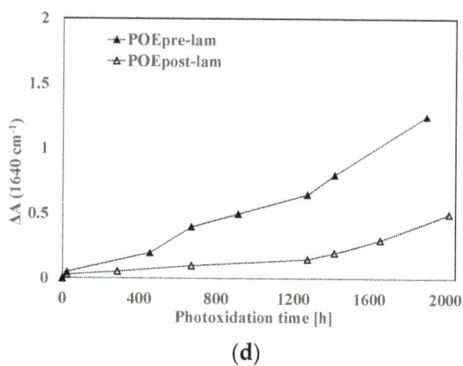

Figure 8. Variation of absorbance (**a**,**b**) at 3200–3600 cm^{-1} for EVA and POE samples, respectively, (**c**) at 1780 cm^{-1} for EVA and (**d**) at 1640 cm^{-1} for POE.

Further confirmation comes also by ATR-FTIR analysis of the investigated samples, in Figure 9a–d, the ATR-FTIR spectra of EVApre-lam, EVApost-lam, POEpre-lam and POEpost-lam commercial samples, before exposure and at maximum UVB exposure time are plotted. Therefore, to confirm the presence of some chemical species, the ATR-FTIR analysis can be considered suitable for qualitative surface analysis. The bands in both hydroxyl and carbonyl domains in the spectra of the four investigated samples at maximum exposure time appear larger than the same bands before exposure, and the latter is clearly exacerbated for the pre-laminated samples, confirming again the beneficial effect of the lamination process on the photoxidation resistance.

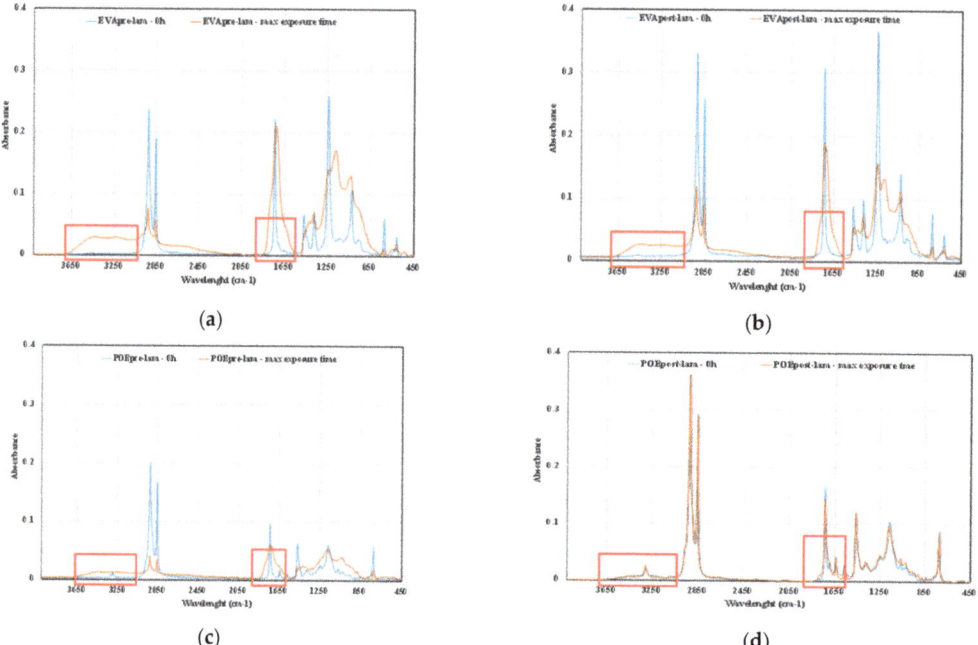

Figure 9. ATR-FTIR spectra of (**a**) EVApre-lam, (**b**) EVApost-lam, (**c**) POEpre-lam and (**d**) POEpost-lam commercial samples, before exposure and at maximum UVB exposure time.

3.7. Thermo-Oxidation Resistance

In Figure 10a,b, FTIR spectra of EVApost-lam and POEpost-lam as a function of thermo-oxidation time are plotted, respectively.

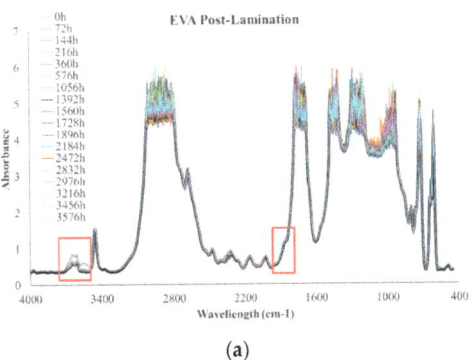
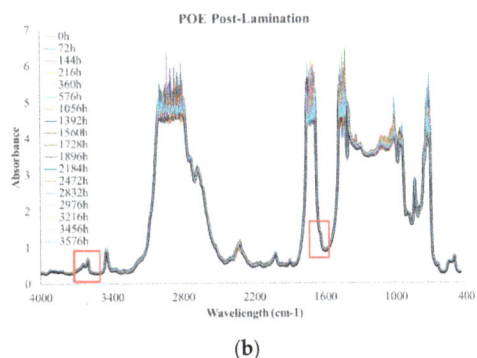

Figure 10. FTIR spectra of (**a**) EVApost-lam and (**b**) POEpost-lam as a function of thermo-oxidation time.

The monitoring of thermo-oxidation process was extended up to ca. 3500 h because no significant variations were noticeable prior. It is worth noting that EVApost-lam sample shows a slight increase of the absorption band in the hydroxyl range and there is the appearance of a small shoulder at 1780 cm^{-1}, suggesting the occurrence of oxidation phenomenon, see Figure 10a. Interestingly, the POEpost-lam sample is extremely stable up to ca. 3500 h thermo-oxidation, i.e., no significant variations in carbonyl and hydroxyl range are noticeable, highlighting no noteworthy occurrence of the oxidation phenomenon also at this prolonged thermal treatment, see Figure 10b. Considering this qualitative analysis, it can be summarized that the POEpost-lam sample is more stable and resistant to thermo-oxidation occurrence than EVApost-lam.

4. Conclusions

Accurate characterization of pre- and post-laminated EVA and POE industrial sheets was carried out by calorimetric, rheological and mechanical analysis. Obtained results suggest that the lamination process has a beneficial effect on the 3D-structuration on both polymers, though it seems that better results are obtained for POE sheets. Upon lamination process, in the melt state, the viscosity of POE increased, while the viscosity of EVA decreased. The latter could be understood considering that the EVA experiences degradation at high temperature with the formation of volatile acetic acids.

The durability, in terms of photo- and thermo-oxidation resistance, of EVA and POE sheets is evaluated monitoring the formation of new oxygen-containing species with absorption bands in the hydroxyl and carbonyl domain. The lamination process leads to the formation of more oxygen resistant sheets, and this is exacerbated for the POE sample.

Finally, to sum up, although both EVA and POE sheets can be considered suitable for encapsulant for bifacial heterojunction PV modules, the POEpost-lam sheet is better structured in the melt, it has good rigidity and ductility and is more stable, in terms of photo- and thermo-oxidation, than EVApost-lam.

Author Contributions: Data curation: M.B.; Investigation: M.B. and A.R.; Methodology: N.T.D., C.C. and C.G.; Validation: N.T.D., C.C., A.R. and C.G.; Supervision; N.T.D. and C.C.; Writing—original draft; Writing—review & editing: N.T.D. All authors have read and agreed to the published version of the manuscript.

Funding: This research received no external funding and APC was funded by N.T.D.

Institutional Review Board Statement: Not applicable.

Informed Consent Statement: Not applicable.

Data Availability Statement: Not applicable.

Acknowledgments: M.B. gives thanks to Italian Ministry of University and Research (MUR) for financial support in the field of PON 2017 (cod. DOT1320558).

Conflicts of Interest: The authors declare no conflict of interest.

References

1. Farrell, C.; Osman, A.I.; Zhang, X.; Murphy, A.; Doherty, R.; Morgan, K.; Rooney, D.W.; Harrison, J.; Coulter, R.; Shen, D. Assessment of the energy recovery potential of waste Photovoltaic (PV) modules. *Sci. Rep.* **2019**, *9*, 5267. [CrossRef] [PubMed]
2. Cattaneo, G.; Levrat, J.; Li, H.; Barth, V.; Sicot, L.; Richter, A.; Colletti, C.; Rametta, F.; Izzi, M.; Despeisse, M.; et al. Encapsulant Materials for High Reliable Bifacial Heterojunction Glass/Glass Photovoltaic Modules. In Proceedings of the 47th IEEE Photovoltaic Specialists Conference, PVSC 2020, Calgary, AB, Canada, 15 June–21 August 2020. [CrossRef]
3. Kempe, M. Evaluation of encapsulant materials for PV application. *Photovolt. Int. Pap.* **2010**, *9*. Available online: https://www.osti.gov/biblio/1049592 (accessed on 20 February 2022).
4. Pern, J. Module encapsulation materials, processing and testing. *APP Int. PV Reliab. Workshop* **2008**. Available online: https://www.nrel.gov/docs/fy09osti/44666.pdf (accessed on 20 February 2022).
5. Pern, F.J. EVA encapsulants for photovoltaic modules: Degradation and discoloration mechanisms and formulation modifi-cations for improved photostability. *Die Angew. Makromol. Chem.* **1997**, *252*, 195–216. [CrossRef]
6. Cuddihy, E.F.; Coulbert, C.; Gupla, A.; Liang, R. *Flat-Plate Solar Array Project Final Report*; Module Encapsulation; JPL Publication: Pasadena, CA, USA, 1986; Volume VII.
7. Cuddihy, E.F.; Coulbcrt, C.D.; Willis, P.; Baum, B.; Garcia, A.; Miming, C.; Ccbelein, C.G.; Williams, D.J.; Dcanin, R.D. *Polymers in Slur Energv Ufilizafion*; ACS: Washington, DC, USA, 1983; Chapter 22; pp. 353–366.
8. Gaume, J.; Taviot-Gueho, C.; Cros, S.; Rivaton, A.; Thérias, S.; Gardette, J.-L. Optimization of PVA clay nanocomposite for ultra-barrier multilayer encapsulation of organic solar cells. *Sol. Energy Mater. Sol. Cells* **2012**, *99*, 240–249. [CrossRef]
9. Kempe, M. Overview of Scientific Issues Involved in Selection of Polymers for PV Applications. In Proceedings of the 37th IEEE Photovoltaic Specialists Conference National Renewable Energy Laboratory, Seattle, WA, USA, 19–24 June 2011. [CrossRef]
10. Jorgensen, G.J.; McMahon, T.J. Accelerated and outdoor aging effects on photovoltaic module interfacial adhesion properties. *Prog. Photovolt. Res. Appl.* **2008**, *16*, 519–527. [CrossRef]
11. Green, M.A. Silicon photovoltaic modules: A brief history of the first 50 years. *Prog. Photovolt. Res. Appl.* **2005**, *13*, 447–455. [CrossRef]
12. Hirschl, C.; Biebl-Rydlo, M.; DeBiasio, M.; Mühleisen, W.; Neumaier, L.; Scherf, W.; Oreski, G.; Eder, G.; Chernev, B.; Schwab, W.; et al. Determining the degree of crosslinking of ethylene vinyl acetate photovoltaic module encapsulants—A comparative study. *Sol. Energy Mater. Sol. Cells* **2013**, *116*, 203–218. [CrossRef]
13. Chapuis, V.; Pelisset, S.; Raeis-Barneoud, M.; Li, H.-Y.; Ballif, C.; Perret-Aebi, L.-E. Compressive-shear adhesion characterization of PVB and EVA at different curing times before and after exposure to damp-heat conditions. *Prog. Photovolt. Res. Appl* **2014**, *22*, 405–414. Available online: https://onlinelibrary.wiley.com/doi/10.1002/pip.2270 (accessed on 20 February 2022). [CrossRef]
14. Pern, F. Factors that affect the EVA encapsulant discoloration rate upon accelerated exposure. *Sol. Energy Mater. Sol. Cells* **1996**, *41–42*, 587–615. [CrossRef]
15. Pern, F.J.; Glick, S.H. Thermal processing of EVA encapsulants and effects of formulation additives. In Proceedings of the Conference Record of the Twenty Fifth IEEE Photovoltaic Specialists Conference—1996, Washington, DC, USA, 13–17 May 1996. [CrossRef]
16. La Mantia, F.; Malatesta, V.; Ceraulo, M.; Mistretta, M.; Koci, P. Photooxidation and photostabilization of EVA and cross-linked EVA. *Polym. Test.* **2016**, *51*, 6–12. [CrossRef]
17. Kempe, M.D.; Jorgensen, G.J.; Terwilliger, K.M.; McMahon, T.J.; Kennedy, C.E.; Borek, T.T. Acetic Acid Production A and Glass Transition Concerns with Ethylene-Vinyl Acetate Used in Photovoltaic Devices. *Sol. Energy Mater. Sol. Cells* **2007**, *91*, 315–329. [CrossRef]
18. Wenger, H.; Schaefer, J.; Rosenthal, A.; Hammond, B.; Schlueter, L. Decline of the Carrisa Plains PV power plant: The impact of concentrating sunlight on flat plates. In Proceedings of the Conference Record of the Twenty-Second IEEE Photovoltaic Specialists Conference—1991, Las Vegas, NV, USA, 7–11 October 1991; pp. 586–592. [CrossRef]
19. Rosenthal, A.L.; Lane, C.G. Field test results for the 6 MW Carrizo solar photovoltaic power plant. *Sol. Cells* **1991**, *30*, 563–571. [CrossRef]
20. Baiamonte, M.; Therias, S.; Gardette, J.-L.; Colletti, C.; Dintcheva, N.T. Encapsulant polymer blend films for bifacial heterojunction photovoltaic modules: Formulation, characterization and durability. *Polym. Degrad. Stab.* **2021**, *193*, 109716. [CrossRef]
21. de Oliveira, M.C.C.; Cardoso, A.S.A.D.; Viana, M.M.; Lins, V.D.F.C. The causes and effects of degradation of encapsulant ethylene vinyl acetate copolymer (EVA) in crystalline silicon photovoltaic modules: A review. *Renew. Sustain. Energy Rev.* **2018**, *81*, 2299–2317. [CrossRef]

22. Jin, J.; Chen, S.; Zhang, J. UV aging behaviour of ethylene-vinyl acetate copolymers (EVA) with different vinyl acetate contents. *Polym. Degrad. Stab.* **2010**, *95*, 725–732. [CrossRef]
23. Dintcheva, N.T.; La Mantia, F.P.; Scaffaro, R.; Paci, M.; Acierno, D.; Camino, G. Reprocessing and restabilization of greenhouse films. *Polym. Degrad. Stab.* **2002**, *75*, 459–464. [CrossRef]
24. La Mantia, F.P.; Dintcheva, N.T. Re-Gradation of Photo-Oxidized Post-Consumer Greenhouse Films. *Macromol. Rapid Commun.* **2005**, *26*, 361–364. [CrossRef]
25. Allen, N.S. (Ed.) Photodegradation Processes in Polymeric Materials. In *Photochemistry and Photophysics of Polymer Materials*; John Wiley & Sons, Inc.: Hoboken, NJ, USA, 2010; pp. 569–601.

Article

Effects of Mechanical Deformation on the Opto-Electronic Responses, Reactivity, and Performance of Conjugated Polymers: A DFT Study

João P. Cachaneski-Lopes [1] and Augusto Batagin-Neto [1,2,*]

[1] POSMAT, School of Sciences, São Paulo State University (UNESP), Bauru 17033-360, SP, Brazil; cachaneski-lopes@unesp.br
[2] Institute of Science and Engineering, São Paulo State University (UNESP), Itapeva 18409-010, SP, Brazil
* Correspondence: a.batagin@unesp.br; Tel.: +55-(15)-3524-9100 (ext. 9159)

Abstract: The development of polymers for optoelectronic applications is an important research area; however, a deeper understanding of the effects induced by mechanical deformations on their intrinsic properties is needed to expand their applicability and improve their durability. Despite the number of recent studies on the mechanochemistry of organic materials, the basic knowledge and applicability of such concepts in these materials are far from those for their inorganic counterparts. To bring light to this, here we employ molecular modeling techniques to evaluate the effects of mechanical deformations on the structural, optoelectronic, and reactivity properties of traditional semiconducting polymers, such as polyaniline (PANI), polythiophene (PT), poly (*p*-phenylene vinylene) (PPV), and polypyrrole (PPy). For this purpose, density functional theory (DFT)-based calculations were conducted for the distinct systems at varied stretching levels in order to identify the influence of structural deformations on the electronic structure of the systems. In general, it is noticed that the elongation process leads to an increase in electronic gaps, hypsochromic effects in the optical absorption spectrum, and small changes in local reactivities. Such changes can influence the performance of polymer-based devices, allowing us to establish significant structure deformation response relationships.

Keywords: molecular modeling; stretching process; polymers; mechanical deformation; density functional theory

1. Introduction

Polymeric materials are very interesting compounds for several applications, playing significant roles in both applied and basic science. In particular, modern polymers have evolved into multifunctional systems wherein specific responses are expected from particular stimuli. In this context, once mechanical loads and deformations are practically inevitable, it is important to establish relationships between the mechanical properties and electronic responses of these compounds [1–4].

In fact, nowadays, mechano-responsive polymers are quite attractive for a number of applications. In these materials, mechanical forces can be used to transfer energy to chemical bonds and drive chemical reactions [5,6]. Although mechanochemistry has been known for years [5] and is already industrially employed [5,7–9], it has been marginally explored in organic materials compared to inorganic systems [10].

The use of organic materials for optoelectronic applications has achieved prominence; in this context, the improved mechanical properties of polymers have highlighted their possible application in organic-based flexible devices [11–16]. Composites based on elastomers and thermoplastics have provided high flexibility to these materials without considerable losses in conductivity [16–18], leading to devices with reasonable performance, low relative cost, and processing advantages [19,20].

As a matter of fact, among the different classes of organic compounds, conjugated polymers are promising materials for the development of flexible optoelectronic devices, such as organic solar cells (OSCs) and organic light-emitting diodes (OLEDs) [19,21,22]. However, additional studies are necessary to better understand the influence of mechanical deformations on their properties to identify relevant operating regimes/limits and obtain fully functional devices with broad applicability.

In fact, even after the insertion of flexible organic devices on the market, there is still no complete understanding of the influence of mechanical stresses on the intrinsic properties of the materials present in their active layers, which hinders the effective application of these devices. In general, the flexibility of these compounds is commonly associated with amorphous domains and interactions between chains, which rearrange themselves when subjected to external stresses [23–26]. On the other hand, the optoelectronic properties of these materials are commonly governed by planar sub-segments (and their interactions) [27–31], so that the influence of mechanical deformations on the response of a conjugated polymer involves a series of complex interactions [25].

A number of experimental data indicate the existence of a variety of effects on the performance of flexible devices under mechanical stresses. In general, successive deformation cycles lead to a reduction in the efficiency of OSCs [23,24,26], while high stability is reported for OLEDs [32,33]. However, the overall effects depend on the materials and processing methods [11,22,25,34].

Indeed, the influence of the stretching process on the optical properties of polymers is a well-known phenomenon [35] that has been revisited by several recent works focused on varied applications [36–38]. A number of reviews focused on mechanical-responsive functional devices and the causes of their deterioration have also been published [33,39]. For instance, Wang and collaborators [40] presented a survey of materials and techniques to obtain devices with equilibrated mechanical and electrical properties. A common point in these works is the search for an appropriate ratio between conductive and flexible materials, with minor discussions on the influence of the mechanical deformations on the intrinsic properties of the compounds.

In this context, molecular modeling techniques can be considered relevant tools to evaluate the variety of mechano-responsive processes, particularly to identify factors associated with the degradation of the optoelectronic properties, propose alternatives for their minimization, and guide the development of new materials with improved responses [41]. Such analyses are especially interesting for polymeric devices due to their high degree of structural flexibility and the strong relationship between conformational and optoelectronic properties [25,34,41,42].

Despite this, there is a scarcity of theoretical works on the effect of mechanical deformations on the optoelectronic properties of semiconducting polymers. Our previous studies at a moderate level of theory (semiempirical + density functional theory (DFT)) indicated that mechanical elongations of MEH-PPV and P3HT polymer chains lead to deleterious effects on the optoelectronic responses of the materials, which occurs at different regimes and levels [41]. However, the limitations imposed by the use of semiempirical approximations (for geometry optimization), and the observation of strong steric effects of the side groups, deserve further study.

To deepen such analyses, here we conduct a series of DFT-based calculations to investigate the influence of mechanical deformations on the intrinsic optoelectronic properties of four widely employed conjugated polymers. The results show that the main chain stretching leads to significant effects on the frontier energy levels of the systems, increasing the electronic gaps and leading to hypsochromic effects in the optical spectra. Such effects can influence the optoelectronic performance of polymer-based devices, mainly in systems with high molecular interactions and entanglements by polymer chains [16,22,34,43,44]. To exemplify such consequences of polymer main chain deformations, we analyze some important parameters of organic solar cells (OSCs); in particular, polymer chain deformations of around 7% and 15% can lead to non-functional PT- and PPV-based OSCs, respectively.

2. Materials and Methods

Figure 1 illustrates the basic structure of the compounds evaluated in this report.

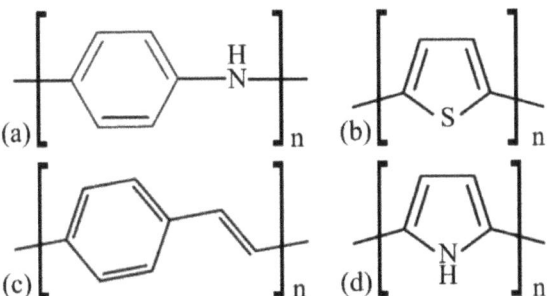

Figure 1. Basic structures of the polymeric materials. (**a**) polyaniline (PANI), (**b**) polythiophene (PT), (**c**) poly (*p*-phenylene vinylene) (PPV), and (**d**) polypyrrole (PPy).

The above presented polymers were chosen according to their high potential for applications in varied devices [45]. PANI is used in electro-rheological fluids, sensors, supercapacitors and rechargeable batteries [46–50]. PT is widely used in polymer-based OLEDs (PLEDs), OSCs, and chemical sensors due to its conductive, luminescence, and electrochromic properties, as well as its high versatility of synthesis [51–53]. PPV derivatives are also commonly used in PLEDs and OSCs [54–56]. PPy has diverse applications due to its low relative cost, electronic properties, ease of processing, and versatility of synthesis, being widely used in the active layer of gas sensors [57–60].

Planarized oligomeric systems with average sizes around the effective conjugation lengths of these materials have been considered as representative models of the polymers. Due to the strong influence of the effective conjugation length on the optoelectronic properties of these systems [28,61], representative oligomeric systems were considered in this study. In particular, studies conducted by our group and collaborators have shown that structures with 10–15 repeating units can reproduce the essential electronic and optical properties of these polymers [47,58,62–67]; for this reason, oligomers with 15 units were used.

Preliminary stretching studies conducted for amorphous structures show a convergence of folded to planarized structures after elongation (see Supplementary Materials—in particular, Figure S1 compared to elongated structures in Figure S2), with relatively small energetic costs (see Figure S3). Thus, folded polymers are supposed to converge to planarized conformations during stretching processes. This result, associated with the fact that the optoelectronic properties of conjugated polymers are commonly governed by planar subsegments [27–30], defined planar structures as the models of interest in the present work. Given the viscoelastic behavior of polymeric systems to mechanical deformations, Figure S1 can be also considered as an indicator of shrink effects.

The initial structures were designed with the aid of the Molden 5.0 computational package [68] (details regarding the amorphous structures and their planarization process can be seen in the Supplementary Materials). The geometry optimization and calculation of electronic properties were conducted in the framework of density functional theory (DFT), using the B3LYP hybrid XC functional [69–72] and 6-31G basis set on all the atoms. The choice of such an approach was made based on the well-known capability of B3LYP to describe the structural, electronic, and optical features of medium-size conjugated systems in comparison with other functionals [73], associated with preliminary calculations conducted for distinct approaches which indicated that the effects of mechanical deformations are not so sensitive to the XC functionals and basis sets employed (see Table S3 in the Supplementary Materials for additional calculations/considerations). In addition, given that we are not interested in the precise reproduction of absolute values, but rationalizing the relative changes induced by the deformations, we considered the B3LYP/6-31G approach to be a reasonable choice due to the relative cost–benefit ratio. All the calculations were conducted with the aid of the Gaussian 16 computational package [74].

The mechanical deformation processes were performed considering the fully optimized structures as starting geometries (without restrictions). The main chain stretches were then made by displacing the terminal carbons of the oligomeric structures (initially at distance d_0, see Figure 2), defining new relative positions between them (d_n) [75]. For instance, at Step 5, we obtained $d_5 = 1.05 d_0$, i.e., an imposed elongation of 5% of the chain in relation to the (pre-optimized) relaxed structure. The mechanical deformation process imposed on the oligomers is exemplified for PANI in Figure 2. All the structures were subjected to gradual elongation until their rupture.

$$d_n = d_0 \left(1 + \frac{n}{100}\right) \tag{1}$$

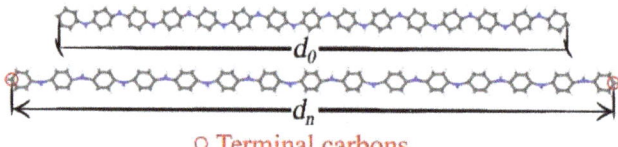

○ Terminal carbons

Figure 2. Illustration of the polymer chain deformation process (example for PANI).

The resulting structure was then allowed to relax (geometry optimizations), restricting the modified distances d_n (Equation (1)) (for n steps). The partial geometry relaxation of all the structures (restricting d_n for each step) was performed via a DFT/B3LYP/6-31G approach with the aid of the Gaussian 16 computational package.

The effects of structural changes on the opto-electronic properties of the systems were evaluated considering the: (i) electronic gaps, (ii) energetic and spatial distribution of the frontier molecular orbitals (FMOs); (iii) total density of states (DOS); (iv) optical absorption spectra; (v) exciton binding energies, vi) internal reorganization energies, and (vii) local reactivities.

The electronic gaps (E_{gap}), DOS, and FMO energies for the highest occupied (HOMO) and lowest unoccupied (LUMO) molecular orbitals were estimated from the Kohn–Sham orbitals. The theoretical optical absorption spectra were obtained via a time-dependent DFT (TD-DFT) approach using the same functional and basis set employed in the geometry optimizations. Ten transitions were evaluated, considering only single excitations. The exciton binding energies (E_X) (Equation (2)) were estimated from the difference between the fundamental gaps (from KS-FMOs) and the vertical transition energies (E_{vert}) resulting from TD-DFT calculations [41,76,77].

$$E_x = (E_{LUMO} - E_{HOMO}) - E_{vert} \tag{2}$$

The internal reorganization energies for electrons (λ_e) and holes (λ_h) were evaluated via Equations (3) and (4). These parameters indicate the energy penalties due to the structural relaxation of the molecules (internal contribution λ^{int}) and polarization effects on the surrounding medium (external contribution λ^{ext}) associated with the charge transfer processes between planar (conjugated) subsegments of the materials [78]. In fact, it is well known that, despite having some effect in charge transport, λ^{ext} is very small in relation to λ^{int}, being often neglected [65,79]. Thus, in this work the reorganization energy was approximated by $\lambda \approx \lambda^{int}$, i.e.,

$$\lambda_h = [E_T(\nu_{N-1}, N) - E_T(\nu_N, N)] - [E_T(\nu_N, N-1) - E_T(\nu_{N-1}, N-1)] \tag{3}$$

$$\lambda_e = [E_T(\nu_{N+1}, N) - E_T(\nu_N, N)] - [E_T(\nu_N, N+1) - E_T(\nu_{N+1}, N+1)] \tag{4}$$

For a system M with N electrons, $E_T(\nu_{N+k}, N+j)$ represents the total energy obtained from single-point calculations for the species M^{-j} (i.e., M with $N+j$ electrons) with structure (defined by ν_{N+k}) previously obtained from the optimization of the M^{-k} species (i.e., M with $N+k$ electrons) [41]. In general, low values of λ_e (λ_h) indicate that the transport of electrons (holes) is facilitated in the materials.

The local reactivities of the compounds were evaluated via condensed-to-atoms Fukui index (CAFI) values [80–83] and molecular electrostatic potential (MEP) maps. CAFI values were estimated from the finite difference of the atomic populations considering the Hirshfeld partition charge [84–86], and MEP maps were generated considering the CHelp partition charge scheme [87]. The graphical representations were created with the aid of the Jmol [88] and Gabedit [89] computational packages.

The mechanical parameters were estimated for all the levels of deformations considering the derivative of the change in the total energy ($\Delta Etotal = Etotal^{(n)} - Etotal^{(0)}$) in relation to the linear deformations ($x = d_n - d_0$). Thus, the amplitude of the forces ($|F|$) required for each level of deformation was estimated by $|F| = d\Delta E_{total}/dx$. The elastic constants (k) were estimated via linear fittings, by considering $|F| = k.x$ [90] (see the Supplementary Materials for details).

3. Results and Discussion

3.1. Structural Changes

Table 1 summarizes the structural data of the systems before and after the deformations. d_0 and d_n represent the distances between the terminal carbons of the chains before the deformations (equilibrium position) and at the point of imminent rupture. Δd_{max} represents the maximum percentage change before the oligomer break. PT and PPV presented slightly distorted (twisted or curved) initial structures (see Supplementary Materials).

Table 1. Structural data of the systems before the deformations and at the point of imminent rupture.

Compound	d_0 (Å)	d_u (Å)	Δd_{max} (%)
PANI	77.447	92.162	18.0
PT	57.192	69.774	21.0
PPV	94.046	110.974	17.0
PPy	52.588	62.054	17.0

Figure 3 illustrates how the chemical bonds of the systems (numbered in the insets) changed after successive stretches (the bond lengths at d_0 are compatible with those presented in the literature [91–94], see the Supplementary Materials). The relative variations are summarized in Figure 4.

The following order of bond lengths (BL) was noticed for non-modified PANI: $BL_1 > BL_4 \sim BL_5 > BL_2 \sim BL_3$. It is possible to note an exponential increase in BL_4 and BL_5 (with total changes of up to 0.22 Å), while the other connections show variations of up to 0.05 Å. We note that BL_1 tends to approach BL_2 and BL_3 during stretching, leading to uniform bonds in rings after 17%. For PT, minor increases are noted on BL_3 and BL_5. Although the oligomer disruption occurs at Bond 6, the highest stretches were noticed for BL_1 and BL_2 (up to 0.24 Å). Very small changes were observed for BL_4 (<0.05 Å). PPV presents significant changes only after the 4% stretch, after the planarization of the structures (see Figure S2c). Major changes were noticed for BL_4 and BL_6 (~0.21 Å). As noted in the case of PANI, for PPV, the bonds on the rings are uniform at high deformation levels. Finally, for PPy, we noticed an exponential increase in BL_6 (~0.31 Å), with a tendency of ring opening (increase in BL_1 and BL_2) and minor variations in BL_3, BL_4, and BL_5.

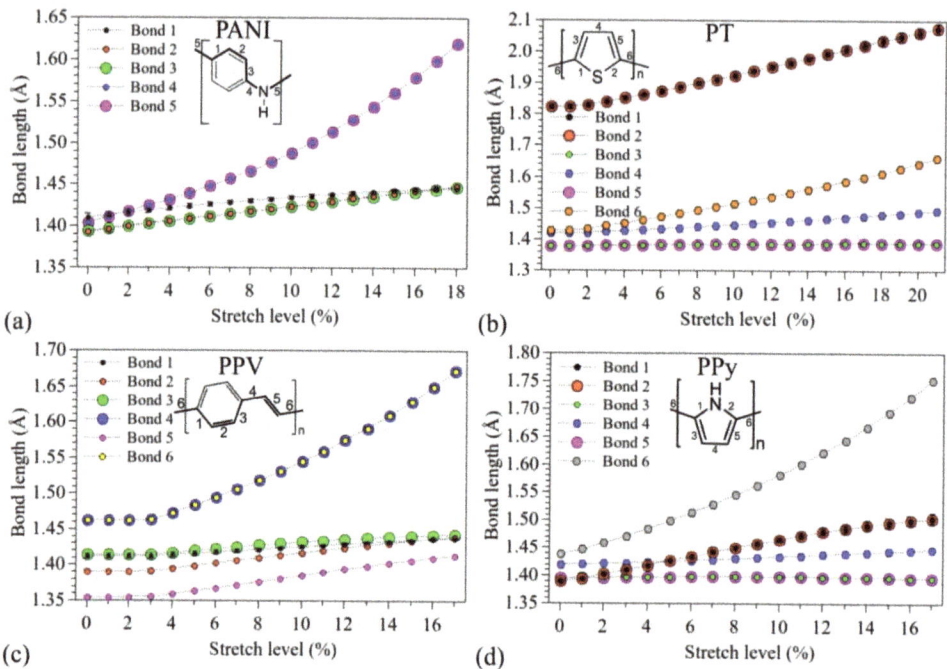

Figure 3. Changes in the oligomer bond lengths during the stretching processes: (**a**) PANI, (**b**) PT, (**c**) PPV, and (**d**) PPy.

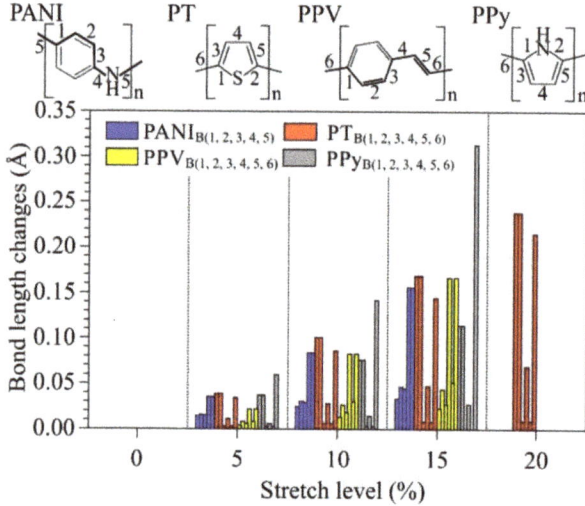

Figure 4. Summary of the changes in polymer bond lengths at distinct stretch levels.

Similar results were reported by Rodao et al. [41] for P3HT and MEH-PPV using semi-empirical methods for geometry optimization. In particular, for MEH-PPV (P3HT), an exponential increase was observed for single C–C bonds (C-S) with small variations in the benzene rings (C=C).

The above presented results allow us to underline the dominant degradation routes for each system. For instance, it is noted that the major stretches of PANI occur through nitrogen–carbon bonds (BL_4 and BL_5), while PPV presents significant stretches on the carbon bonds around the vinylene units; we note that the mechanical deformation of these systems does not significantly alter the C–C bonds inside the rings (i.e., $BL_i < 1.45$ Å). A tendency to ring opening was noticed for compounds containing five-membered heteroaromatic rings (PT and PPy); this is more evident for PT, where BL_1, BL_2, and BL_6 present very similar deformation rates. Although the PPy rings were not severely modified, significant changes are noted in relation to typical C–N bond lengths [95] (i.e., BL_1 and $BL_2 > 1.309$ Å).

Table 2 summarizes the relative forces required to deform the systems ($F = |F|$), as well as the elastic constants (k) estimated at each deformation level for distinct systems.

Table 2. Relative changes in the total energies (ΔE_{total}) and estimated values for the forces (F) and average elastic constants (k).

System	Δd_n (%)	$\|\Delta E_{total}\|$ (Joule × 10^{-19})	Force (nN)	Elastic Constant (nN/nm)
PANI	0%	—	—	~5.82
	6%	0.818	3.32	
	12%	2.995	6.01	
	17%	6.325	7.86	
PT	0%	—	—	k_1 = ~10.41 *
	7%	0.582	3.59	k_2 = ~5.66 *
	14%	2.700	4.53	k_{med} = ~8.04
	20%	5.801	8.37	
PPV	0%	—	—	~5.49
	5%	0.146	1.44	
	11%	1.937	4.79	
	16%	5.379	6.84	
PPy	0%	—	—	k_1 = ~13.87 *
	5%	0.462	3.64	k_2 = ~7.46 *
	11%	2.219	4.31	k_{med} = ~10.67
	16%	4.808	8.76	

* See Supplementary Materials.

We note that the forces required to break the polymer chain are compatible with those reported for covalent bonds [75,96–99]; in particular, the following order was noticed: PPy < PT < PANI < PPV, with F_{max} ranging from 6.84 to 8.76 nN (see Supplementary Materials for details).

3.2. Changes in Opto-Electronic Properties

Figure 5 illustrates how the energies of the FMO levels (E_{HOMO} and E_{LUMO}) vary as a function of the stretching levels. It also presents the changes induced in the electronic gaps ($E_{L-H} = E_{LUMO} - E_{HOMO}$) in relation to those for the unstretched system (ΔE_{L-H}). The numerical values for non-stretched structures are presented in the Supplementary Materials (Table S2).

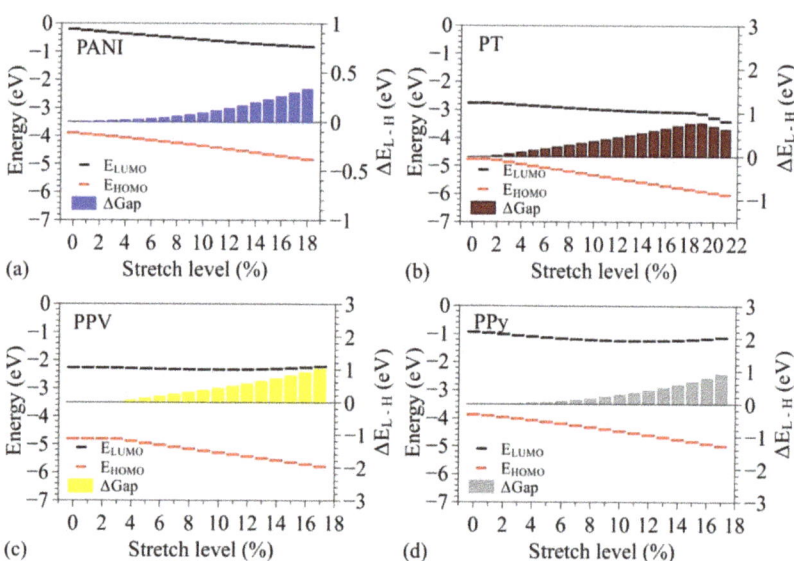

Figure 5. Evolution of the FMO energy levels with stretching of the oligomers: (**a**) PANI, (**b**) PT, (**c**) PPV, and (**d**) PPy.

In general, we noticed a reduction in the FMO energies with stretching for all the systems. The changes are more pronounced in the E_{HOMO} values, leading to an increase in the electronic gaps. The effects noticed for PT are partially compatible with the results reported for P3HT by Roldao et al. [41], mainly in relation to E_{LUMO}. We considered that the anomalous behavior of E_{HOMO} presented in ref [41] is associated with the steric effects of the side groups; these have a profound influence on the dihedral angles of the main chains, which is not present in our calculations. This interpretation is reinforced by the absence of transitions around 14% for PPV, which were reported for MEH-PPV in ref [41] and associated with the interaction between adjacent ramifications. For PT, a sharp drop was noted in E_{LUMO} after 20% stretching, which is associated with the saturation of the stretching process, whereafter the system behaves as a set of non-passivated thiophene units (see Figure S7).

For PANI, we noted that FMO levels vary linearly with the deformation level of the chains, presenting very similar rates for E_{HOMO} and E_{LUMO} changes, leading to small variation of the gaps (lower than 0.4 eV). Given the initial structural distortion of fully relaxed PT (see Figure S2b), their FMO levels remained unchanged until 2% stretching and then presented a decrease, leading to a gradual increase in the electronic gaps (of up to ~0.77 eV); a change in the behavior of the LUMO was noted after 19% stretch with a more pronounced decrease rate. Due to the initial distortion of fully relaxed structures of PPV, no variations could be observed for this compound until 4% stretching (see Figure S2c). After this, both the FMO levels were reduced with more expressive (and linear) effects on the HOMO, leading to increased electronic gaps (of up to 1 eV); after 11% stretching, we noted a change in the behavior of the LUMO, showing a tiny increase with stretching. Finally, for PPy, we initially noted a reduction of FMOs at similar rates; the behavior of the LUMO changed after 5% (lower decreasing rate) and 11% (when it started to increase).

Figure 6 shows the changes induced by the stretching on the density of states (from KS eigenvalues) around the FMOs.

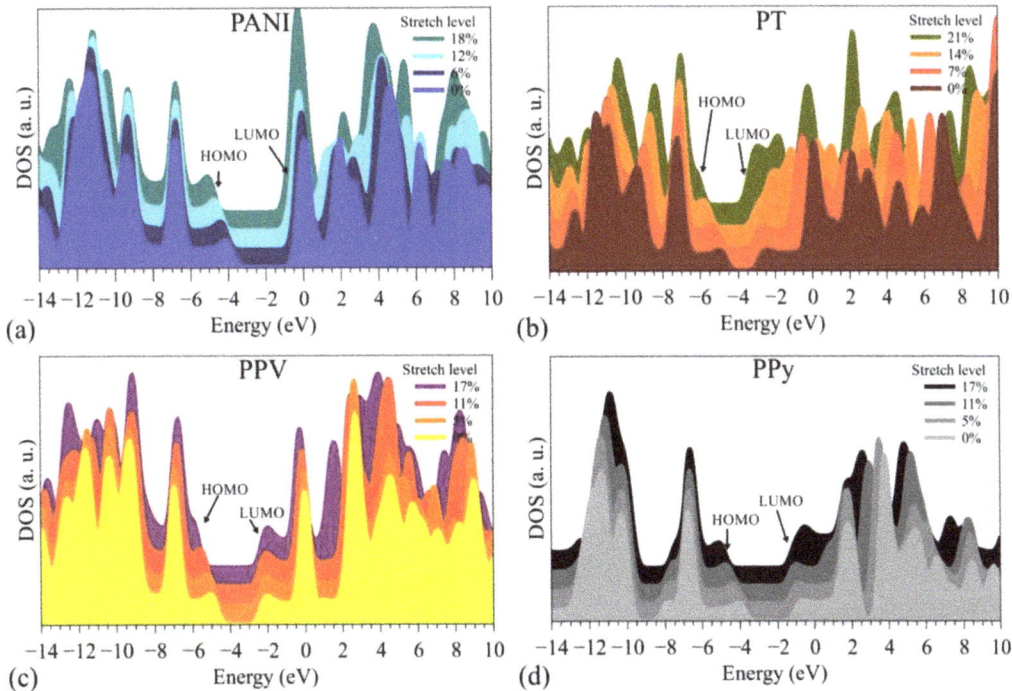

Figure 6. DOS for (**a**) PANI, (**b**) PT, (**c**) PPV, and (**d**) PPy at distinct stretch levels.

In general, significant variations were noted far away from the frontier levels, with subtle changes around the HOMO and LUMO (mainly in the LUMO) (see the Supplementary Materials for more details).

Figure 7 presents the spatial distribution of the HOMO and LUMO KS orbitals for some representative stretching levels.

For all the systems, we noted that HOMOs are transversely aligned to the polymer main axis, while the LUMO lobes are parallel to them. The stretching process does not have a strong influence on the spatial distribution of the FMOs, leading to significant changes (localization of the wavefunctions) only when rupture is imminent, mainly on the LUMO of PT, PPV, and PPy and on the HOMO of PANI, degrading their optoelectronic properties. An opposite effect was noticed for the HOMO (LUMO) of PT (PANI), for which higher delocalization was noticed in very distorted structures, indicating improved hole (electron) transport in the stretched systems. PPV is more sensitive to small deformations in relation to the other structures; however, it does not show intense degradation at high levels of deformation. A similar behavior of the LUMO was noticed in very distorted structures of PT and PPy (Figure 7b,d). In general, the results suggest that deterioration of the opto-electronic properties occurs for most of these systems after 11% stretching.

Figure 8 demonstrates the behavior of the optical absorption spectra of the compounds throughout the deformation process.

Hypsochromic effects were noticed for all the structures, and these are compatible with the relative increase in the electronic gaps presented in Figure 5. Stronger effects were noticed for PT ($\Delta\lambda_{max}$ ~233 nm) and PPV ($\Delta\lambda_{max}$ ~170 nm), with blueshifts that are approximately linear with regard to the deformation level; less intense (but non-linear) blueshifts were noticed for PPy ($\Delta\lambda_{max}$ ~124 nm) and PANI ($\Delta\lambda_{max}$ ~32nm) (see Figure S6). For all the cases, the main peaks are governed by HOMO–LUMO transitions along the deformation processes, except for PT at the point of imminent rupture (with a significant

HOMO–LUMO + 2 contribution). The blueshift of the main peaks is followed by slight increases in the relative amplitudes, which was continuous for PANI until rupture. The small effect on the oscillator forces is compatible with the small effects noticed on the spatial distribution of the FMOs (and then on their superpositions), as presented in Figure 7. In particular, it is possible to note the appearance of secondary absorption peaks at shorter wavelengths during the stretches (see Tables S4–S7 for details); these are associated with transitions involving a variety of levels around the FMOs (HOMO-n and LUMO+m) and are compatible with the intensification of the DOS around the frontier levels.

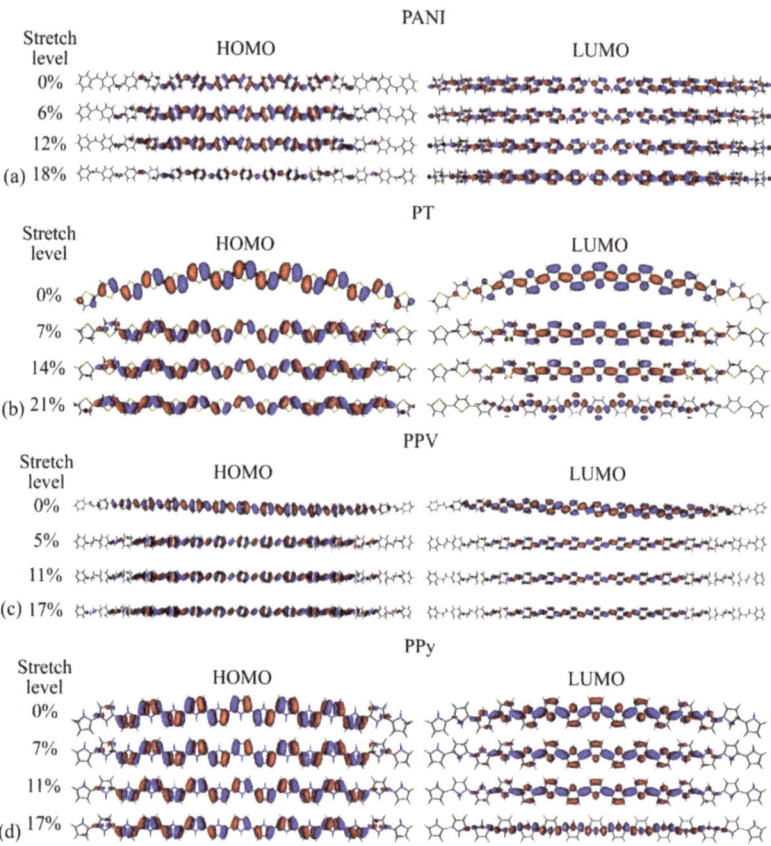

Figure 7. Effects of mechanical stretching on the spatial distribution of the KS-FMOs of (a) PANI, (b) PT, (c) PPV, and (d) PPy.

In fact, significant blueshift in the absorption spectra of polymer-based flexible devices has already been reported in the literature [35,95], being primarily associated with the disruption of aggregates in the active layer of the devices. Our results suggest that a similar effect can take place due to the reduced electronic coupling between adjacent units.

To better understand the effect of the main chain deformations on the optical properties of the compounds, Figure 9 presents the relationship between the chemical bond lengths (BL) that are broken during the rupture (highlighted in the insets) and E_{vert} values.

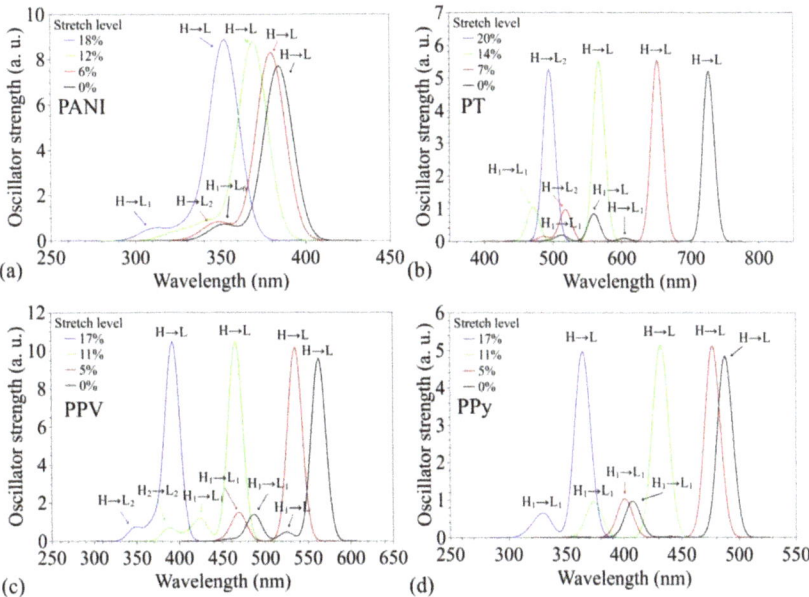

Figure 8. Summary of the changes on the optical absorption spectra of the polymers during the stretching process: (**a**) PANI, (**b**) PT, (**c**) PPV, and (**d**) PPy.

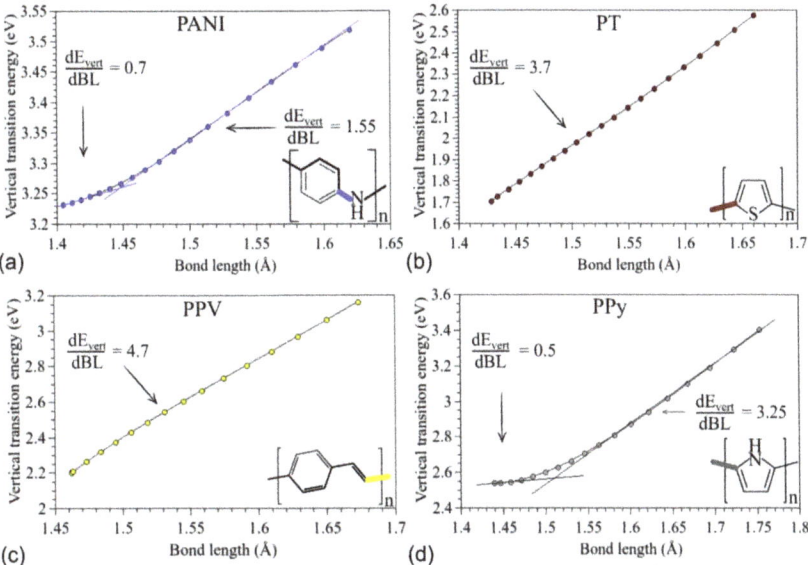

Figure 9. Relationships between the vertical transition energies and the most affected connections in the stretching process: (**a**) PANI, (**b**) PT, (**c**) PPV, and (**d**) PPy.

In general, there is a linear dependence between the vertical transition energies and the considered bond lengths. Such a dependence is more evident for PT and PPV. In the case of PANI and PPy, a transition is noted at ~1.45 Å and ~1.5 Å, respectively, from which point they start to present a linear behavior similar to that of the other systems. This effect, associated with those presented in Figure 8, suggests that the optical absorption of the systems can be modulated via mechanical deformations. In particular, deformations smaller than 10% can lead to changes of around −15 nm (PANI), −112 nm (PT), −97 nm (PPV), and −56 nm (PPy) in the position of the absorption mean peak, without significant changes in the optical performance.

Figure 10 presents the changes induced in the exciton binding energies due to stretching of the polymer main chains.

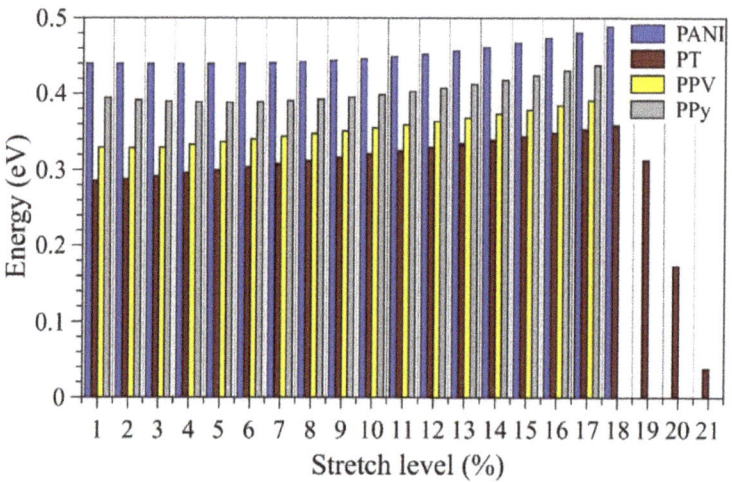

Figure 10. Evolution of the exciton binding energies during the stretching processes.

The E_X values of the non-modified systems are compatible with those expected for organic polymers (i.e., around 0.3 eV [100–102]); in particular, the order $E_X(PT) < E_X(PPV) < E_X(PPy) < E_X(PANI)$ was noticed, which evidences the applicability of PT and PPV in OSCs. For all the systems, we observed a gradual increase in the exciton binding energies, even with distinct rates, until 17%. In addition, for PT, we noticed a rapid decrease from 18% onwards, which is in line with the changes in the LUMO levels (Figure 5).

Figure 11 presents the changes induced in the internal reorganization energies during the stretching processes for the distinct polymers. For most systems, it shows the results obtained for up to 16% stretch, due to convergence problems in very distorted cationic structures.

The non-modified structures presented small reorganization energies (from 1 to 5 times the thermal energy at room temperature, kT_{300}). Very small energies were noticed for PANI, followed by PPV, PPy, and PT. For most systems, the λ_e values are slightly lower than λ_h, except for PPV. During the stretching process, we noticed a gradual reduction in λ_h values for all the systems, which was continuous for PPV and PPy until rupture. PANI and PT presented minimum values of λ_h at 11 and 17%, respectively, with a significant increase thereafter. On the other hand, for all the systems, we noticed a gradual increase in the λ_e values throughout the stretching process. These results indicate that mechanical deformations can improve (reduce) the hole (electron) transport in these materials.

CAFI and MEP analyses were conducted to evaluate the influence of mechanical stretching on the local chemical reactivity of the systems. CAFI indicates which molecular sites are prone to receive/donate electrons from/to the environment, defining which regions are susceptible to undergoing chemical reactions towards nucleophiles (f^+), electrophiles (f^-), and free radicals (f^0). MEPs, on the other hand, indicate the charge concentration on oligomer structures, which plays a relevant role in electrostatic interactions and drives effective molecular collisions for chemical reactions. The CAFI and MEP analyses during the stretching process allow us to evaluate how "soft–soft" (i.e., associated with deformations in the frontier orbitals) and "hard–hard" (guided by electrostatic effects) interactions [103] could change due to the continuous deformation of the systems. Figures 12 and 13 illustrate the CAFI and MEP colored maps for PANI, PT, PPV, and PPy. Red and blue regions define reactive (negatively charged) and non-reactive (positively charged) sites on the molecules via CAFI (MEP) analysis, respectively.

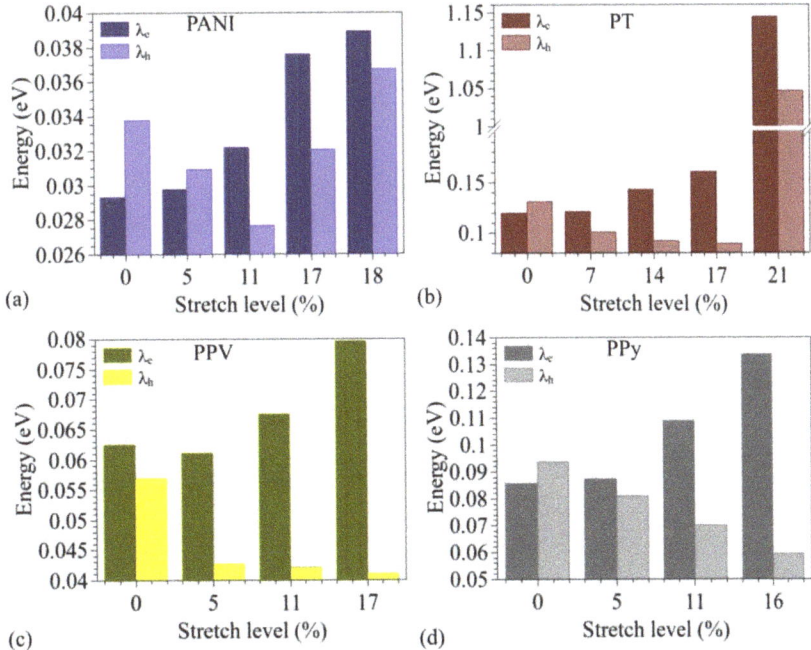

Figure 11. Evolution of the internal reorganization energies during the stretching process: (a) PANI, (b) PT, (c) PPV, and (d) PPy.

The influence of mechanical stretching on the stability of organic devices is generally discussed in terms of macroscopic/morphological features, such as delamination of the layers (active and transporting layers, as well as electrodes), the formation of punctures and cracks, strong lateral strains, limitations imposed by brittle electrodes, etc. [104]. However, a number of nanoscopic mechanisms for stress-induced polymer degradation have also been proposed [105]. In particular, the exponential behavior of the strongly stretched bonds (Figure 3) is compatible with Zhurkov-like evolution of the photochemical degradation rates of stretched polymers.

From the analysis of Figures 12 and 13, the reactivity of PANI (in relation to nucleophiles) is mainly located at terminal regions, which can be associated with the enhanced electropolymerization properties of this polymer [106]. High reactivity is observed on the nitrogen atoms, in relation to electrophiles, which is associated with the doping process of leucoemeraldine [107]. PT presents high reactivity on the sulphur atoms for both nucleophiles and electrophiles, which indicates the plausibility of p- and n-doping via this site; this is compatible with other works [108,109]. PPV presents high reactivity on the vinyl groups for all the reactions; this is not sensitive to stretching and can explain the well-known degradation routes of the polymer chains via varied mechanisms [62,110–112]. Finally, PPy shows high reactivity on positions 3–4 (towards electrophiles) and on position 2 (towards nucleophiles), which are compatible with the charge transfer mechanisms/doping processes proposed in the literature [108,113,114].

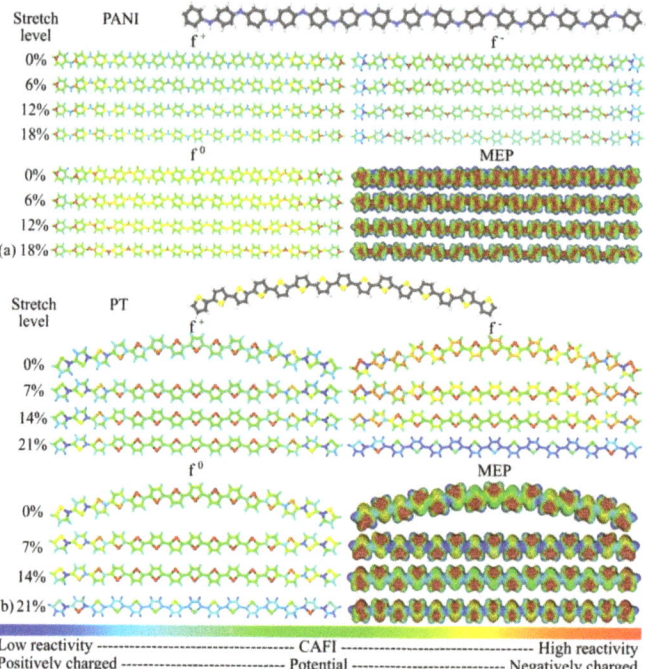

Figure 12. CAFI and MEP maps at distinct stretch levels: (**a**) PANI and (**b**) PT.

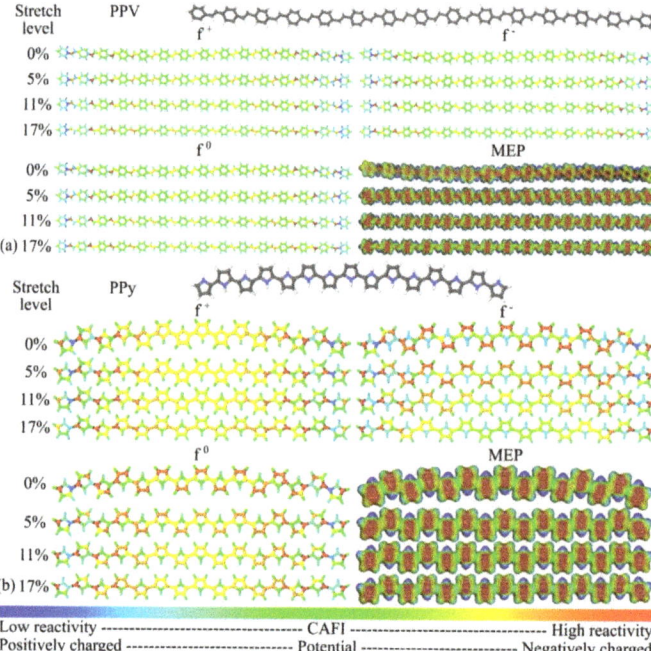

Figure 13. CAFI and MEP maps at distinct stretch levels: (**a**) PPV and (**b**) PPy.

In general, it was noticed that f^+ is very insensitive to mechanical stretching, indicating that chemical reactions towards nucleophilic agents are not changed due to the deformation of the polymer chains. On the other hand, a slight decrease in the reactivity for f^- and f^0 at central regions of the chains was noticeable during the stretching process, which influences the chemical reactivity towards electrophiles and free radicals. The most significant variation in the MEPs during the stretching process was the decrease in the electronic density between the monomeric units of the main chains.

3.3. Considerations Regarding the Effect of Stretching on the Performance of the Compounds in Devices and Identification of Operational Regimes

Several factors can influence the efficiency of organic devices, mainly in relation to charge transport, the formation of interfaces, solubility issues, oxidation stabilities, morphologies, processing, and material synthesis [41,115,116]. However, on a nanoscopic scale, some molecular parameters, particularly the FMO level alignment, are intrinsically associated with the performance of organic devices [116–118], governing a number of relevant mechanisms.

In recent years, a number of studies have been conducted to evaluate the effects of stretching on polymer-based OSCs [43,44,119,120]. Polymeric systems present a viscoelastic response to mechanical deformations which depends on the polymer structure, loading rates, and working temperatures. In particular, in these systems, there is competition between chain deformation and contraction due to the stress and entropic responses [34]. As a result, the dynamics of polymer-based thin films are generally governed by the relative slippage of adjacent chains. However, effective chain entanglements can also occur in the films (mainly those based on longer polymers), resulting in large plastic deformation prior to fractures, which are associated with polymer deformation, chain pullout, or even polymer scission [44,120]. In this context, to exemplify the effects of such mechanical deformations on the performance of OSCs, here we present a simple prediction of some electronic descriptors of interest.

Figures 14 and 15 show the influence of the structural deformation of PPV and PT (commonly used in BHJ-OSCs) on V_{OC}, ΔE_{HH}, and ΔE_{LL}, considering the frontier levels of typical electron acceptors, C_{60} and PCBM, computed at the same level of theory [121]. The dotted line indicates the limits for which the parameters ΔE_{HH} and ΔE_{LL} are lower than the typical exciton binding energies of the polymeric donors. Details on the changes induced in the relative alignments between the FMOs are presented in the Supplementary Materials (Figure S11).

As stated before, the mechanical deformation has a strong influence on the FMOs of the structures. In particular, considering that the electron donation process is mainly governed by the relative positions of the donor/acceptor LUMO levels (LUMO$_D$ and LUMO$_A$), the decrease induced in the LUMO$_D$ (see Figures 5 and S11) tends to reduce the ΔE_{LL} values, hindering the dissociation of the excitons in the systems [21,41]. In addition, the reduction of the donor HOMO (HOMO$_D$) tends to reduce the ΔE_{HH} values, facilitating the reassociation of electron–hole pairs. Such effects prevent the effective generation of free charge carriers in the devices, leading to loss of functionality for $\Delta E_{LL} < E_X$ and $\Delta E_{HH} < E_X$. In particular, it is noted that these limits are reached for deformations of around 13% and 7% for the PT/C_{60} and PT/PCBM systems, respectively. PPV/C_{60}-based systems are supposed to keep their functionality throughout the deformation process, while PPV/PCBM systems present operational conditions for up to 15% stretching.

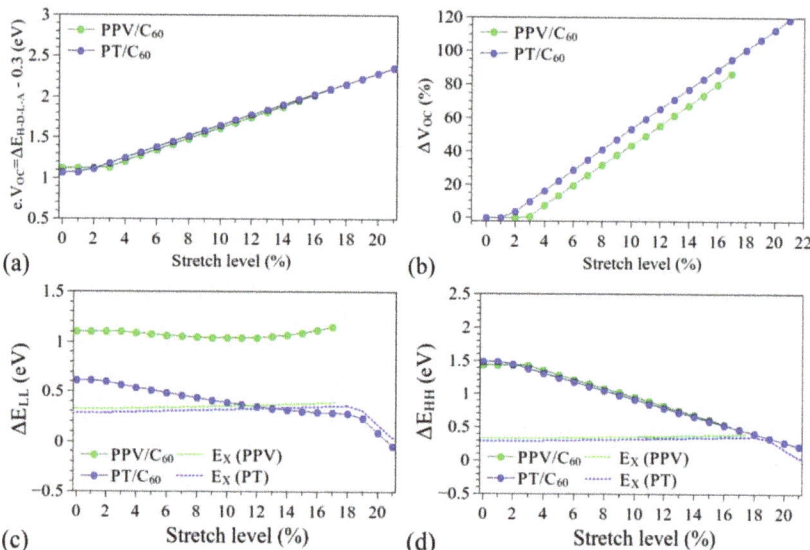

Figure 14. Evolution of the parameters (**a**) V_{OC}, (**b**) ΔV_{OC}, (**c**) ΔE_{LL}, and (**d**) ΔE_{HH} during the stretching processes for C_{60}-based devices.

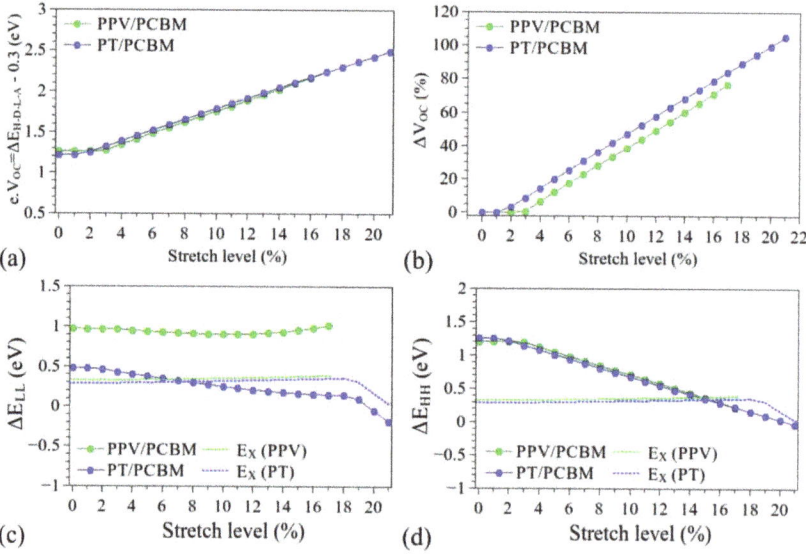

Figure 15. Evolution of the parameters (**a**) V_{OC}, (**b**) ΔV_{OC}, (**c**) ΔE_{LL}, and (**d**) ΔE_{HH} during the stretching processes for PCBM-based devices.

The structural changes can also lead to an increase in $\Delta E_{H\text{-}D\text{-}L\text{-}A}$ (i.e., LUMO$_A$-HOMO$_D$) which can lead to higher V_{OC} with stretching (of up to 100%). Such an interesting feature, however, is limited by the previously discussed difficulty of charge carrier generation ($\Delta E_{LL} < E_X$ and $\Delta E_{HH} < E_X$).

The increase in V_{OC} is compatible with the results reported for P3DDT/PCBM BHJ-OSCs. Savagatrup et al. [23] indicated improved performance (particularly with higher V_{OC}) of such PT derivative-based devices when they are subjected to stretching. Interestingly, an opposite effect was noticed for P3HT/PCBM, which is associated with the presence of morphological changes in the films (formation of visible cracks). In fact, such macroscopic defects are supposed to govern the performance of this system to the detriment of local deformation of the chains, degrading the electrical response of the devices. On the other hand, the presence of long side chains in P3DDT derivatives (dodecyl groups) improves the mechanical properties of the films, so that the effect of chain deformations (also associated with chain entanglements) can play a relevant role in P3DDT/PCBM devices. Savagatrup et al. [23] reported an improvement of 14% in the V_{OC} after stretching of 10% of such devices; according to our results, this is linked to an effective elongation of about 3-4% of the PT chains (see Figure S9b), for which we predict an operational BHJ-OSC (considering the ΔE_{LL}, ΔE_{HH}, and E_X relations).

In general, the effects of deformation on the LUMO of the polymers can lead to reduced charge injection barriers from electrodes. In particular, the pronounced effects for the HOMOs (0.95, 1.28, 0.98, and 1.11 eV for PANI, PT, PPV, and PPy, respectively) can lead to non-equilibrium electrode/organic layer alignments in very stretched systems. The significant changes noticed on the FMOs are also supposed to have a profound effect on the performance of these materials in chemical sensors, given the relative position of the electronic gaps of the polymers and the FMO levels of the analytes [122,123].

It is worth considering that despite our studies having been conducted until the rupture of the chains, much smaller deformations are supposed to occur in real devices, mainly due to the mechanical restrictions imposed by the electrodes [124] and relative flow of the polymer chains during the device deformation. The performance of flexible OSCs has been evaluated for varied stretch levels, from 7% [22] to 22% [24,125], which suggests that significant changes in the length of the polymer chains are somehow possible in real devices [41]. These changes can be even more pronounced with the use of elastomer-based composites [16].

4. Conclusions

The effects of mechanical deformations on the structural, optical, and electronic properties of polyaniline (PANI), polyethylene (PT), poly (*p*-phenylene vinylene) (PPV), and polypyrrole (PPy) were evaluated via DFT calculations.

The results suggest a strong influence of polymer chain stretching on the optoelectronic properties of the systems. In general, the stretching causes a decrease in the energy of the frontier molecular orbitals, with different ratios for the LUMO and HOMO, resulting in an increase in the electronic gaps. Significant hypsochromic effects are induced on the optical absorption spectra, which depend on the evolution of specific chemical bonds. Significant changes were also noticed on the exciton binding and reorganization energies, with small influences on the local reactivity of the systems.

These effects can lead to significant changes in the performance of the materials in devices where polymer elongation can take place, mainly those based on high-molecular-weight compounds and/or all polymer-based devices.

Supplementary Materials: The following supporting information can be downloaded at: https://www.mdpi.com/article/10.3390/polym14071354/s1, Table S1. Bond lengths observed in relaxed polymer structures (at d_0); Table S2. Summary of the optoelectronic properties of relaxed polymeric systems (at d_0); Table S3. Effect of XC functional and basis set on the description of optoelectronic properties of the polymers; Table S4. Optical absorption data associated with the main transitions of PANI at distinct levels of deformation (TD-DFT/B3LYP/6-31G approach); Table S5. Optical absorption data associated with the main transitions of PT at distinct levels of deformation (TD-DFT/B3LYP/6-31G approach); Table S6. Optical absorption data associated with the main transitions of PPV at distinct levels of deformation (TD-DFT/B3LYP/6-31G approach); Table S7. Optical absorption data associated with the main transitions of PPy at distinct levels of deformation (TD-DFT/B3LYP/6-31G approach); Figure S1. Deformation of amorphous structures: from amorphous to free optimization and from amorphous to stretched structures; Figure S2. (a) PANI, (b) PT, (c) PPV, and (d) PPy structures at the initial stages of deformation until stabilization; Figure S3. Total energy of the structures: amorphous, full optimization and disentanglement process. (a) PANI, (b) PT, (c) PPV, and (d) PPy; Figure S4. Force imposed to the systems during the disentanglement process: (a) PANI, (b) PT, (c) PPV, and (d) PPy; Figure S5. FMOs of the systems, (a) PANI, (b) PT, (c) PPV, and (d) PPy; Figure S6. Total energy variation (a) PANI, (d) PT; Strength and elastic constant (b) PANI and (e) PT; Figure S7. Total energy variation (a) PPV, (d) PPy, Strength and elastic constant (b) PPV and (e) PPy; Figure S8. DOS for (a) PANI, (b) PT, (c) PPV, and (d) PPy at all levels of stretching; Figure S9. Peak absorption wavelength at all stretch levels; Figure S10. Electronic gap estimated for PT at distinct stretching levels in comparison with Th and Th-2H units; Figure S11. Alignments between the FMO levels of PPV and PT in relation to C_{60} and PCBM. Reference [126–136] are cited in the supplementary materials.

Author Contributions: Conceptualization, methodology, formal analysis, investigation, writing—original draft preparation, J.P.C.-L. and A.B.-N.; writing—review and editing, supervision, project administration, and funding acquisition, A.B.-N. All authors have read and agreed to the published version of the manuscript.

Funding: Brazilian National Council of Scientific Research (CNPq): grants 448310/2014-7 and 420449/2018-3.

Institutional Review Board Statement: Not applicable.

Informed Consent Statement: Not applicable.

Data Availability Statement: Not applicable.

Acknowledgments: The authors thank the National Council of Scientific Research (CNPq) (grants 448310/2014-7 and 420449/2018-3) for their financial support. This research was also supported by resources supplied by the Center for Scientific Computing (NCC/GridUNESP) of São Paulo State University (UNESP).

Conflicts of Interest: The authors declare no conflict of interest.

References

1. Chen, J.; Peng, Q.; Peng, X.; Han, L.; Wang, X.; Wang, J.; Zeng, H. Recent Advances in Mechano-Responsive Hydrogels for Biomedical Applications. *ACS Appl. Polym. Mater.* **2020**, *2*, 1092–1107. [CrossRef]
2. Lee, S.; Kim, K.Y.; Jung, S.H.; Lee, J.H.; Yamada, M.; Sethy, R.; Kawai, T.; Jung, J.H. Finely Controlled Circularly Polarized Luminescence of a Mechano-Responsive Supramolecular Polymer. *Angew. Chem.* **2019**, *131*, 19054–19058. [CrossRef]
3. Uğur, G.; Chang, J.; Xiang, S.; Lin, L.; Lu, J. A Near-Infrared Mechano Responsive Polymer System. *Adv. Mater.* **2012**, *24*, 2685–2690. [CrossRef]
4. Schäfer, C.G.; Gallei, M.; Zahn, J.T.; Engelhardt, J.; Hellmann, G.P.; Rehahn, M. Reversible Light-, Thermo-, and Mechano-Responsive Elastomeric Polymer Opal Films. *Chem. Mater.* **2013**, *25*, 2309–2318. [CrossRef]
5. Willis-Fox, N.; Rognin, E.; Aljohani, T.; Daly, R. Polymer Mechanochemistry: Manufacturing Is Now a Force to Be Reckoned with. *Chem* **2018**, *4*, 2499–2537. [CrossRef]
6. Takacs, L. The historical development of mechanochemistry. *Chem. Soc. Rev.* **2013**, *42*, 7649–7659. [CrossRef]
7. Crawford, D.E.; Casaban, J. Recent Developments in Mechanochemical Materials Synthesis by Extrusion. *Adv. Mater.* **2016**, *28*, 5747–5754. [CrossRef] [PubMed]
8. de Oliveira, P.F.M.; Torresi, R.M.; Emmerling, F.; Camargo, P.H.C. Challenges and opportunities in the bottom-up mechanochemical synthesis of noble metal nanoparticles. *J. Mater. Chem. A* **2020**, *8*, 16114–16141. [CrossRef]
9. Szczęśniak, B.; Borysiuk, S.; Choma, J.; Jaroniec, M. Mechanochemical synthesis of highly porous materials. *Mater. Horiz.* **2020**, *7*, 1457–1473. [CrossRef]
10. Boldyreva, E. Mechanochemistry of inorganic and organic systems: What is similar, what is different? *Chem. Soc. Rev.* **2013**, *42*, 7719–7738. [CrossRef]
11. Ma, R.; Feng, J.; Yin, D.; Sun, H.-B. Highly efficient and mechanically robust stretchable polymer solar cells with random buckling. *Org. Electron.* **2017**, *43*, 77–81. [CrossRef]
12. Ostroverkhova, O. Organic Optoelectronic Materials: Mechanisms and Applications. *Chem. Rev.* **2016**, *116*, 13279–13412. [CrossRef]
13. Lee, S.-M.; Kwon, J.H.; Kwon, S.; Choi, K.C. A Review of Flexible OLEDs Toward Highly Durable Unusual Displays. *IEEE Trans. Electron Devices* **2017**, *64*, 1922–1931. [CrossRef]
14. Lee, H.B.; Jin, W.-Y.; Ovhal, M.M.; Kumar, N.; Kang, J.-W. Flexible transparent conducting electrodes based on metal meshes for organic optoelectronic device applications: A review. *J. Mater. Chem. C* **2018**, *7*, 1087–1110. [CrossRef]
15. Verboven, I.; Deferme, W. Printing of flexible light emitting devices: A review on different technologies and devices, printing technologies and state-of-the-art applications and future prospects. *Prog. Mater. Sci.* **2020**, *118*, 100760. [CrossRef]

16. Boratto, M.H.; Nozella, N.L.; Ramos, R.A., Jr.; Da Silva, R.A.; Graeff, C.F.O. Flexible conductive blend of natural rubber latex with PEDOT: PSS. *APL Mater.* **2020**, *8*, 121107. [CrossRef]
17. Shen, J.; Fujita, K.; Matsumoto, T.; Hongo, C.; Misaki, M.; Ishida, K.; Mori, A.; Nishino, T. Mechanical, Thermal, and Electrical Properties of Flexible Polythiophene with Disiloxane Side Chains. *Macromol. Chem. Phys.* **2017**, *218*, 1700197. [CrossRef]
18. Chen, Q.; Wang, X.; Chen, F.; Zhang, N.; Ma, M. Extremely strong and tough polythiophene composite for flexible electronics. *Chem. Eng. J.* **2019**, *368*, 933–940. [CrossRef]
19. Etxebarria, I.; Ajuria, J.; Pacios, R. Solution-processable polymeric solar cells: A review on materials, strategies and cell architectures to overcome 10%. *Org. Electron.* **2015**, *19*, 34–60. [CrossRef]
20. Grimsdale, A.C.; Chan, K.L.; Martin, R.E.; Jokisz, P.G.; Holmes, A.B. Synthesis of Light-Emitting Conjugated Polymers for Applications in Electroluminescent Devices. *Chem. Rev.* **2009**, *109*, 897–1091. [CrossRef]
21. Zhang, Z.; Liao, M.; Lou, H.; Hu, Y.; Sun, X.; Peng, H. Conjugated Polymers for Flexible Energy Harvesting and Storage. *Adv. Mater.* **2018**, *30*, e1704261. [CrossRef] [PubMed]
22. Kim, T.; Kim, J.-H.; Kang, T.E.; Lee, C.; Kang, H.; Shin, M.; Wang, C.; Ma, B.; Jeong, U.; Kim, T.-S.; et al. Flexible, Highly Efficient All-Polymer Solar Cells. *Nat. Commun.* **2015**, *6*, 8547. [CrossRef]
23. Savagatrup, S.; Makaram, A.S.; Burke, D.J.; Lipomi, D.J. Mechanical Properties of Conjugated Polymers and Polymer-Fullerene Composites as a Function of Molecular Structure. *Adv. Funct. Mater.* **2013**, *24*, 1169–1181. [CrossRef]
24. Kaltenbrunner, M.; White, M.; Głowacki, E.D.; Sekitani, T.; Someya, T.; Sariftci, N.S.; Bauer, S. Ultrathin and lightweight organic solar cells with high flexibility. *Nat. Commun.* **2012**, *3*, 770. [CrossRef]
25. Lee, Y.; Mongare, A.; Plant, A.; Ryu, D. Strain–Microstructure–Optoelectronic Inter-Relationship toward Engineering Mechano-Optoelectronic Conjugated Polymer Thin Films. *Polymers* **2021**, *13*, 935. [CrossRef] [PubMed]
26. Fukuda, K.; Yu, K.; Someya, T. The Future of Flexible Organic Solar Cells. *Adv. Energy Mater.* **2020**, *10*, 2000765. [CrossRef]
27. Schwartz, B.J. Conjugated Polymers as Molecular Materials: How Chain Conformation and Film Morphology Influence Energy Transfer and Interchain Interactions. *Annu. Rev. Phys. Chem.* **2003**, *54*, 141–172. [CrossRef]
28. Meier, H.; Stalmach, U.; Kolshorn, H. Effective conjugation length and UV/vis spectra of oligomers. *Acta Polym.* **1997**, *48*, 379–384. [CrossRef]
29. Rissler, J. Effective conjugation length of π-conjugated systems. *Chem. Phys. Lett.* **2004**, *395*, 92–96. [CrossRef]
30. He, J.; Crase, J.L.; Wadumethrige, S.H.; Thakur, K.; Dai, L.; Zou, S.; Rathore, R.; Hartley, C.S. Ortho-Phenylenes: Unusual Conjugated Oligomers with a Surprisingly Long Effective Conjugation Length. *J. Am. Chem. Soc.* **2010**, *132*, 13848–13857. [CrossRef]
31. Wiebeler, C.; Gopalakrishna Rao, A.; Gärtner, W.; Schapiro, I. The Effective Conjugation Length is Responsible for the Red/Green Spectral Tuning in the Cyanobacteriochrome Slr1393g3. *Angew. Chem. Int. Ed.* **2019**, *58*, 1934–1938. [CrossRef]
32. Kwak, K.; Cho, K.; Kim, S. Stable Bending Performance of Flexible Organic Light-Emitting Diodes Using IZO Anodes. *Sci. Rep.* **2013**, *3*, 2787. [CrossRef]
33. Ma, R.; Chou, S.-Y.; Xie, Y.; Pei, Q. Morphological/nanostructural control toward intrinsically stretchable organic electronics. *Chem. Soc. Rev.* **2019**, *48*, 1741–1786. [CrossRef]
34. Caruso, M.M.; Davis, D.A.; Shen, Q.; Odom, S.A.; Sottos, N.R.; White, S.R.; Moore, J.S. Mechanically-Induced Chemical Changes in Polymeric Materials. *Chem. Rev.* **2009**, *109*, 5755–5798. [CrossRef]
35. Rubner, M.F. Novel optical properties of polyurethane-diacetylene segmented copolymers. *Macromolecules* **1986**, *19*, 2129–2138. [CrossRef]
36. Chen, J.-Y.; Hsieh, H.-C.; Chiu, Y.-C.; Lee, W.-Y.; Hung, C.-C.; Chueh, C.-C.; Chen, W.-C. Electrospinning-induced elastomeric properties of conjugated polymers for extremely stretchable nanofibers and rubbery optoelectronics. *J. Mater. Chem. C* **2019**, *8*, 873–882. [CrossRef]
37. Chow, P.C.Y.; Someya, T. Organic Photodetectors for Next-Generation Wearable Electronics. *Adv. Mater.* **2019**, *32*, e1902045. [CrossRef]
38. Sagara, Y.; Traeger, H.; Li, J.; Okado, Y.; Schrettl, S.; Tamaoki, N.; Weder, C. Mechanically Responsive Luminescent Polymers Based on Supramolecular Cyclophane Mechanophores. *J. Am. Chem. Soc.* **2021**, *143*, 5519–5525. [CrossRef]
39. Onorato, J.; Pakhnyuk, V.; Luscombe, C.K. Structure and design of polymers for durable, stretchable organic electronics. *Polym. J.* **2016**, *49*, 41–60. [CrossRef]
40. Wang, M.; Baek, P.; Akbarinejad, A.; Barker, D.; Travas-Sejdic, J. Conjugated polymers and composites for stretchable organic electronics. *J. Mater. Chem. C* **2019**, *7*, 5534–5552. [CrossRef]
41. Roldao, J.C.; Batagin-Neto, A.; Lavarda, F.; Sato, F. Effects of Mechanical Stretching on the Properties of Conjugated Polymers: Case Study for MEH-PPV and P3HT Oligomers. *J. Polym. Sci. Part B Polym. Phys.* **2018**, *56*, 1413–1426. [CrossRef]
42. Brédas, J.L.; Street, G.B.; Thémans, B.; André, J.M. Organic polymers based on aromatic rings (polyparaphenylene, polypyrrole, polythiophene): Evolution of the electronic properties as a function of the torsion angle between adjacent rings. *J. Chem. Phys.* **1985**, *83*, 1323–1329. [CrossRef]
43. Lee, I.; Rolston, N.; Brunner, P.-L.; Dauskardt, R.H. Hole-Transport Layer Molecular Weight and Doping Effects on Perovskite Solar Cell Efficiency and Mechanical Behavior. *ACS Appl. Mater. Interfaces* **2019**, *11*, 23757–23764. [CrossRef] [PubMed]
44. Kim, W.; Ma, B.S.; Kim, Y.H.; Kim, T.-S. Mechanical properties of organic semiconductors for flexible electronics. In *Organic Flexible Electronics*; Elsevier: Amsterdam, The Netherlands, 2021; pp. 199–223. [CrossRef]

45. Rasmussen, S.C. Conjugated and Conducting Organic Polymers: The First 150 Years. *ChemPlusChem* **2020**, *85*, 1412–1429. [CrossRef]
46. Bhadra, S.; Singha, N.K.; Khastgir, D. Electrochemical synthesis of polyaniline and its comparison with chemically synthesized polyaniline. *J. Appl. Polym. Sci.* **2007**, *104*, 1900–1904. [CrossRef]
47. Mandú, L.O.; Batagin-Neto, A. Chemical sensors based on N-substituted polyaniline derivatives: Reactivity and adsorption studies via electronic structure calculations. *J. Mol. Model.* **2018**, *24*, 157. [CrossRef]
48. Xing, J.; Liao, M.; Zhang, C.; Yin, M.; Li, D.; Song, Y. The effect of anions on the electrochemical properties of polyaniline for supercapacitors. *Phys. Chem. Chem. Phys.* **2017**, *19*, 14030–14041. [CrossRef]
49. Hu, Z.; Zu, L.; Jiang, Y.; Lian, H.; Liu, Y.; Li, Z.; Chen, F.; Wang, X.; Cui, X. High Specific Capacitance of Polyaniline/Mesoporous Manganese Dioxide Composite Using KI-H2SO4 Electrolyte. *Polymers* **2015**, *7*, 1939–1953. [CrossRef]
50. Wang, H.; Lin, J.; Shen, Z.X. Polyaniline (PANi) based electrode materials for energy storage and conversion. *J. Sci. Adv. Mater. Devices* **2016**, *1*, 225–255. [CrossRef]
51. Kaloni, T.P.; Giesbrecht, P.K.; Schreckenbach, G.; Freund, M.S. Polythiophene: From Fundamental Perspectives to Applications. *Chem. Mater.* **2017**, *29*, 10248–10283. [CrossRef]
52. Kanal, I.Y.; Owens, S.G.; Bechtel, J.S.; Hutchison, G. Efficient Computational Screening of Organic Polymer Photovoltaics. *J. Phys. Chem. Lett.* **2013**, *4*, 1613–1623. [CrossRef]
53. Huong, V.T.T.; Nguyen, H.T.; Tai, T.B.; Nguyen, M.T. π-Conjugated Molecules Containing Naphtho[2,3-b]thiophene and Their Derivatives: Theoretical Design for Organic Semiconductors. *J. Phys. Chem. C* **2013**, *117*, 10175–10184. [CrossRef]
54. Nimith, K.; Satyanarayan, M.; Umesh, G. Enhancement in fluorescence quantum yield of MEH-PPV:BT blends for polymer light emitting diode applications. *Opt. Mater.* **2018**, *80*, 143–148. [CrossRef]
55. Farjamtalab, I.; Sabbaghi-Nadooshan, R. Current density of anodes, recombination rate and luminance in MEH-PPV, MDMO-PPV, and P3HT polymers in polymer light-emitting diodes. *Polym. Sci. Ser. A* **2016**, *58*, 726–731. [CrossRef]
56. Cernini, R.; Li, X.-C.; Spencer, G.; Holmes, A.; Moratti, S.; Friend, R. Electrochemical and optical studies of PPV derivatives and poly(aromatic oxadiazoles). *Synth. Met.* **1997**, *84*, 359–360. [CrossRef]
57. Yussuf, A.; Al-Saleh, M.; Al-Enezi, S.; Abraham, G. Synthesis and Characterization of Conductive Polypyrrole: The Influence of the Oxidants and Monomer on the Electrical, Thermal, and Morphological Properties. *Int. J. Polym. Sci.* **2018**, *2018*, 1–8. [CrossRef]
58. Coleone, A.P.; Lascane, L.G.; Batagin-Neto, A. Polypyrrole derivatives for optoelectronic applications: A DFT study on the influence of side groups. *Phys. Chem. Chem. Phys.* **2019**, *21*, 17729–17739. [CrossRef]
59. Bibi, S.; Ullah, H.; Ahmad, S.M.; Ali Shah, A.-U.; Bilal, S.; Tahir, A.A.; Ayub, K. Molecular and Electronic Structure Elucidation of Polypyrrole Gas Sensors. *J. Phys. Chem. C* **2015**, *119*, 15994–16003. [CrossRef]
60. Ghoorchian, A.; Alizadeh, N. Chemiresistor gas sensor based on sulfonated dye-doped modified conducting polypyrrole film for high sensitive detection of 2,4,6-trinitrotoluene in air. *Sens. Actuators B Chem.* **2018**, *255*, 826–835. [CrossRef]
61. Gierschner, J.; Cornil, J.; Egelhaaf, H.-J. Optical Bandgaps of π-Conjugated Organic Materials at the Polymer Limit: Experiment and Theory. *Adv. Mater.* **2007**, *19*, 173–191. [CrossRef]
62. Bronze-Uhle, E.S.; Batagin-Neto, A.; Lavarda, F.C.; Graeff, C. Ionizing radiation induced degradation of poly (2-methoxy-5-(2'-ethyl-hexyloxy) -1,4-phenylene vinylene) in solution. *J. Appl. Phys.* **2011**, *110*, 073510. [CrossRef]
63. Batagin-Neto, A.; Oliveira, E.F.; Graeff, C.; Lavarda, F.C. Modelling polymers with side chains: MEH-PPV and P3HT. *Mol. Simul.* **2013**, *39*, 309–321. [CrossRef]
64. de Oliveira, E.F.; Camilo, A., Jr.; da Silva-Filho, L.C.; Lavarda, F.C. Effect of chemical modifications on the electronic structure of poly(3-hexylthiophene). *J. Polym. Sci. Part B Polym. Phys.* **2013**, *51*, 842–846. [CrossRef]
65. Oliveira, E.F.; Lavarda, F.C. Reorganization energy for hole and electron transfer of poly(3-hexylthiophene) derivatives. *Polymer* **2016**, *99*, 105–111. [CrossRef]
66. Oliveira, G.P.; Barboza, B.H.; Batagin-Neto, A. Polyaniline-based gas sensors: DFT study on the effect of side groups. *Comput. Theor. Chem.* **2021**, *1207*, 113526. [CrossRef]
67. Coleone, A.P.; Barboza, B.H.; Batagin-Neto, A. Polypyrrole derivatives for detection of toxic gases: A theoretical study. *Polym. Adv. Technol.* **2021**, *32*, 4464–4478. [CrossRef]
68. Schaftenaar, G.; Noordik, J.H. Molden: A pre- and post-processing program for molecular and electronic structures. *J. Comput. Aided Mol. Des.* **2000**, *14*, 123–134. [CrossRef]
69. Becke, A.D. Density-functional thermochemistry. III. The role of exact exchange. *J. Chem. Phys.* **1993**, *98*, 5648. [CrossRef]
70. Lee, C.; Yang, W.; Parr, R.G. Development of the Colle-Salvetti correlation-energy formula into a functional of the electron density. *Phys. Rev. B* **1988**, *37*, 785–789. [CrossRef]
71. Vosko, S.H.; Wilk, L.; Nusair, M. Accurate spin-dependent electron liquid correlation energies for local spin density calculations: A critical analysis. *Can. J. Phys.* **1980**, *58*, 1200–1211. [CrossRef]
72. Stephens, P.J.; Devlin, F.J.; Chabalowski, C.F.; Frisch, M.J. Ab Initio Calculation of Vibrational Absorption and Circular Dichroism Spectra Using Density Functional Force Fields. *J. Phys. Chem.* **1994**, *98*, 11623–11627. [CrossRef]
73. McCormick, T.M.; Bridges, C.R.; Carrera, E.I.; DiCarmine, P.M.; Gibson, G.L.; Hollinger, J.; Kozycz, L.M.; Seferos, D.S. Conjugated Polymers: Evaluating DFT Methods for More Accurate Orbital Energy Modeling. *Macromolecules* **2013**, *46*, 3879–3886. [CrossRef]

74. Frisch, M.J.; Trucks, G.W.; Schlegel, H.B.; Scuseria, G.E.; Robb, M.A.; Cheeseman, J.R.; Scalmani, G.; Barone, V.; Petersson, G.A.; Nakatsuji, H.; et al. *Gaussian 16 Revision A.03*; Gaussian Inc.: Wallingford, CT, USA, 2016.
75. Beyer, M.K. The mechanical strength of a covalent bond calculated by density functional theory. *J. Chem. Phys.* **2000**, *112*, 7307–7312. [CrossRef]
76. Alves, G.G.; Oliveira, E.F.; Batagin-Neto, A.; Lavarda, F.C. Molecular modeling of low bandgap diblock co-oligomers with π-bridges for applications in photovoltaics. *Comput. Mater. Sci.* **2018**, *152*, 12–19. [CrossRef]
77. Zhu, L.; Yi, Y.; Chen, L.; Shuai, Z. Exciton binding energy of electronic polymers: A first principles study. *J. Theor. Comput. Chem.* **2008**, *7*, 517–530. [CrossRef]
78. Organic Electronics. *Advances in Polymer Science*; Grasser, T., Meller, G., Li, L., Eds.; Springer: Berlin/Heidelberg, Germany, 2010; Volume 223, ISBN 978-3-642-04537-0.
79. Hutchison, G.R.; Ratner, M.A.; Marks, T.J. Hopping Transport in Conductive Heterocyclic Oligomers: Reorganization Energies and Substituent Effects. *J. Am. Chem. Soc.* **2005**, *127*, 2339–2350. [CrossRef]
80. Yang, W.; Mortier, W.J. The use of global and local molecular parameters for the analysis of the gas-phase basicity of amines. *J. Am. Chem. Soc.* **1986**, *108*, 5708–5711. [CrossRef]
81. Geerlings, P.; De Proft, A.F.; Langenaeker, W. Conceptual Density Functional Theory. *Chem. Rev.* **2003**, *103*, 1793–1874. [CrossRef]
82. Maia, R.A.; Ventorim, G.; Batagin-Neto, A. Reactivity of lignin subunits: The influence of dehydrogenation and formation of dimeric structures. *J. Mol. Model.* **2019**, *25*, 1–11. [CrossRef]
83. Alves, G.G.; Lavarda, F.C.; Graeff, C.F.; Batagin-Neto, A. Reactivity of eumelanin building blocks: A DFT study of monomers and dimers. *J. Mol. Graph. Model.* **2020**, *98*, 107609. [CrossRef]
84. Hirshfeld, F.L. Bonded-atom fragments for describing molecular charge densities. *Theor. Chim. Acta* **1977**, *44*, 129–138. [CrossRef]
85. De Proft, F.; Van Alsenoy, C.; Peeters, A.; Langenaeker, W.; Geerlings, P. Atomic charges, dipole moments, and Fukui functions using the Hirshfeld partitioning of the electron density. *J. Comput. Chem.* **2002**, *23*, 1198–1209. [CrossRef] [PubMed]
86. Roy, R.K.; Pal, S.; Hirao, K. On non-negativity of Fukui function indices. *J. Chem. Phys.* **1999**, *110*, 8236–8245. [CrossRef]
87. Chirlian, L.E.; Francl, M. Atomic charges derived from electrostatic potentials: A detailed study. *J. Comput. Chem.* **1987**, *8*, 894–905. [CrossRef]
88. Herráez, A. *How to Use Jmol to Study and Present Molecular Structures*; Lulu.com: Morrisville, NC, USA, 2007; ISBN 978-1-84799-259-8.
89. Allouche, A.-R. Gabedit-A graphical user interface for computational chemistry softwares. *J. Comput. Chem.* **2010**, *32*, 174–182. [CrossRef]
90. Wang, Z.; Su, K.; Fan, H.; Hu, L.; Wang, X.; Li, Y.; Wen, Z. Mechanical and electronic properties of C60 under structure distortion studied with density functional theory. *Comput. Mater. Sci.* **2007**, *40*, 537–547. [CrossRef]
91. Salaneck, W.R.; Clark, D.T.; Samuelsen, E.J. *Science and Applications of Conducting Polymers: Papers from the Sixth European Industrial Workshop*; CRC Press: Trondheim, Norway, 2019; ISBN 978-1-00-011224-5.
92. Chen, X.; Liang, Q.H.; Jiang, J.; Wong, C.K.Y.; Leung, S.Y.; Ye, H.; Yang, D.G.; Ren, T.L. Functionalization-induced changes in the structural and physical properties of amorphous polyaniline: A first-principles and molecular dynamics study. *Sci. Rep.* **2016**, *6*, 20621. [CrossRef]
93. Salzner, U.; Lagowski, J.; Pickup, P.; Poirier, R. Comparison of geometries and electronic structures of polyacetylene, polyborole, polycyclopentadiene, polypyrrole, polyfuran, polysilole, polyphosphole, polythiophene, polyselenophene and polytellurophene. *Synth. Met.* **1998**, *96*, 177–189. [CrossRef]
94. Fu, Y.; Shen, W.; Li, M. Theoretical analysis on the electronic structures and properties of PPV fused with electron-withdrawing unit: Monomer, oligomer and polymer. *Polymer* **2008**, *49*, 2614–2620. [CrossRef]
95. Cumper, C.W.N. The structures of some heterocyclic molecules. *Trans. Faraday Soc.* **1958**, *54*, 1266–1270. [CrossRef]
96. Roöhrig, U.F.; Frank, I. First-principles molecular dynamics study of a polymer under tensile stress. *J. Chem. Phys.* **2001**, *115*, 8670–8674. [CrossRef]
97. Garnier, L.; Gauthier-Manuel, B.; Van Der Vegte, E.W.; Snijders, J.; Hadziioannou, G. Covalent bond force profile and cleavage in a single polymer chain. *J. Chem. Phys.* **2000**, *113*, 2497–2503. [CrossRef]
98. Grandbois, M.; Beyer, M.; Rief, M.; Clausen-Schaumann, H.; Gaub, H.E. How Strong Is a Covalent Bond? *Science* **1999**, *283*, 1727–1730. [CrossRef]
99. Smalø, H.S.; Rybkin, V.V.; Klopper, W.; Helgaker, T.; Uggerud, E. Mechanochemistry: The Effect of Dynamics. *J. Phys. Chem. A* **2014**, *118*, 7683–7694. [CrossRef]
100. Ruini, A.; Caldas, M.J.; Bussi, G.; Molinari, E. Solid State Effects on Exciton States and Optical Properties of PPV. *Phys. Rev. Lett.* **2002**, *88*, 206403. [CrossRef]
101. Brédas, J.-L.; Beljonne, D.; Coropceanu, V.; Cornil, J. Charge-Transfer and Energy-Transfer Processes in π-Conjugated Oligomers and Polymers: A Molecular Picture. *Chem. Rev.* **2004**, *104*, 4971–5004. [CrossRef]
102. Nayak, P.K. Exciton binding energy in small organic conjugated molecule. *Synth. Met.* **2013**, *174*, 42–45. [CrossRef]
103. Melin, J.; Aparicio, F.; Subramanian, V.; Galván, M.; Chattaraj, P.K. Is the Fukui Function a Right Descriptor of Hard–Hard Interactions? *J. Phys. Chem. A* **2004**, *108*, 2487–2491. [CrossRef]
104. Duan, L.; Uddin, A. Progress in Stability of Organic Solar Cells. *Adv. Sci.* **2020**, *7*, 1903259. [CrossRef]

105. Tyler, D.R. Mechanistic Aspects of the Effects of Stress on the Rates of Photochemical Degradation Reactions in Polymers. *J. Macromol. Sci. Part C Polym. Rev.* **2004**, *44*, 351–388. [CrossRef]
106. Li, Z.; Ye, B.; Hu, X.; Ma, X.; Zhang, X.; Deng, Y. Facile electropolymerized-PANI as counter electrode for low cost dye-sensitized solar cell. *Electrochem. Commun.* **2009**, *11*, 1768–1771. [CrossRef]
107. Malhotra, B.; Dhand, C.; Lakshminarayanan, R.; Dwivedi, N.; Mishra, S.; Solanki, P.; Venkatesh, M.; Beuerman, R.W.; Ramakrishna, S.; Mayandi, V. Polyaniline-based biosensors. *Nanobiosensors Dis. Diagn.* **2015**, *4*, 25–46. [CrossRef]
108. Jayasundara, W.S.R.; Schreckenbach, G. Theoretical Study of p- and n-Doping of Polythiophene- and Polypyrrole-Based Conjugated Polymers. *J. Phys. Chem. C* **2020**, *124*, 17528–17537. [CrossRef]
109. Kaloni, T.P.; Schreckenbach, G.; Freund, M.S. Structural and Electronic Properties of Pristine and Doped Polythiophene: Periodic versus Molecular Calculations. *J. Phys. Chem. C* **2015**, *119*, 3979–3989. [CrossRef]
110. Sezen, M.; Plank, H.; Fisslthaler, E.; Chernev, B.; Zankel, A.; Tchernychova, E.; Blümel, A.; List, E.J.W.; Grogger, W.; Pölt, P. An investigation on focused electron/ion beam induced degradation mechanisms of conjugated polymers. *Phys. Chem. Chem. Phys.* **2011**, *13*, 20235–20240. [CrossRef]
111. Chambon, S.; Rivaton, A.; Gardette, J.-L.; Firon, M. Photo- and thermal degradation of MDMO-PPV:PCBM blends. *Sol. Energy Mater. Sol. Cells* **2007**, *91*, 394–398. [CrossRef]
112. Chambon, S.; Rivaton, A.; Gardette, J.-L.; Firon, M. Durability of MDMO-PPV and MDMO-PPV: PCBM blends under illumination in the absence of oxygen. *Sol. Energy Mater. Sol. Cells* **2008**, *92*, 785–792. [CrossRef]
113. Le, T.-H.; Kim, Y.; Yoon, H. Electrical and Electrochemical Properties of Conducting Polymers. *Polymers* **2017**, *9*, 150. [CrossRef]
114. Ohtsuka, T. Corrosion Protection of Steels by Conducting Polymer Coating. *Int. J. Corros.* **2012**, *2012*, 1–7. [CrossRef]
115. Li, Y. Molecular Design of Photovoltaic Materials for Polymer Solar Cells: Toward Suitable Electronic Energy Levels and Broad Absorption. *Acc. Chem. Res.* **2012**, *45*, 723–733. [CrossRef]
116. Brédas, J.-L.; Norton, J.E.; Cornil, J.; Coropceanu, V. Molecular Understanding of Organic Solar Cells: The Challenges. *Acc. Chem. Res.* **2009**, *42*, 1691–1699. [CrossRef]
117. Dang, M.T.; Hirsch, L.; Wantz, G.; Wuest, J.D. Controlling the Morphology and Performance of Bulk Heterojunctions in Solar Cells. Lessons Learned from the Benchmark Poly(3-hexylthiophene):[6,6]-Phenyl-C61-butyric Acid Methyl Ester System. *Chem. Rev.* **2013**, *113*, 3734–3765. [CrossRef]
118. Proctor, C.; Kuik, M.; Nguyen, T.-Q. Charge carrier recombination in organic solar cells. *Prog. Polym. Sci.* **2013**, *38*, 1941–1960. [CrossRef]
119. Kim, J.-H.; Noh, J.; Choi, H.; Lee, J.-Y.; Kim, T.-S. Mechanical Properties of Polymer–Fullerene Bulk Heterojunction Films: Role of Nanomorphology of Composite Films. *Chem. Mater.* **2017**, *29*, 3954–3961. [CrossRef]
120. Kim, W.; Choi, J.; Kim, J.-H.; Kim, T.; Lee, C.; Lee, S.; Kim, M.; Kim, B.J.; Kim, T.-S. Comparative Study of the Mechanical Properties of All-Polymer and Fullerene–Polymer Solar Cells: The Importance of Polymer Acceptors for High Fracture Resistance. *Chem. Mater.* **2018**, *30*, 2102–2111. [CrossRef]
121. Cachaneski-Lopes, J.P.; Batagin-Neto, A. Effects of Mechanical Deformation on the Optoelectronic Properties of Fullerenes: A DFT Study. *J. Nanostructure Chem.* **2021**, 1–17. [CrossRef]
122. Barboza, B.H.; Gomes, O.P.; Batagin-Neto, A. Polythiophene derivatives as chemical sensors: A DFT study on the influence of side groups. *J. Mol. Model.* **2021**, *27*, 1–13. [CrossRef]
123. Lascane, L.G.; Oliveira, E.F.; Galvão, D.S.; Batagin-Neto, A. Polyfuran-based chemical sensors: Identification of promising derivatives via DFT calculations and fully atomistic reactive molecular dynamics. *Eur. Polym. J.* **2020**, *141*, 110085. [CrossRef]
124. Savagatrup, S.; Printz, A.D.; O'Connor, T.F.; Zaretski, A.V.; Rodriquez, D.; Sawyer, E.J.; Rajan, K.M.; Acosta, R.I.; Root, S.E.; Lipomi, D.J. Mechanical degradation and stability of organic solar cells: Molecular and microstructural determinants. *Energy Environ. Sci.* **2014**, *8*, 55–80. [CrossRef]
125. Lipomi, D.J.; Tee, B.C.-K.; Vosgueritchian, M.; Bao, Z. Stretchable Organic Solar Cells. *Adv. Mater.* **2011**, *23*, 1771–1775. [CrossRef]
126. Geniès, E.M.; Boyle, A.; Lapkowski, M.; Tsintavis, C. Polyaniline: A Historical Survey. *Synth. Met.* **1990**, *36*, 139–182. [CrossRef]
127. Ullah, H. Inter-Molecular Interaction in Polypyrrole/TiO2: A DFT Study. *J. Alloys Compd.* **2017**, *692*, 140–148. [CrossRef]
128. Greenham, N.C.; Friend, R.H. Semiconductor Device Physics of Conjugated Polymers. In *Solid State Physics*; Elsevier: Amsterdam, The Netherlands, 1996; Volume 49, pp. 1–149, ISBN 978-0-12-607749-0.
129. McCall, R.P.; Ginder, J.M.; Leng, J.M.; Ye, H.J.; Manohar, S.K.; Masters, J.G.; Asturias, G.E.; MacDiarmid, A.G.; Epstein, A.J. Spectroscopy and Defect States in Polyaniline. *Phys. Rev. B* **1990**, *41*, 5202–5213. [CrossRef]
130. Arjmandi, M.; Arjmandi, A.; Peyravi, M.; Pirzaman, A.K. First-Principles Study of Adsorption of XCN (X = F, Cl, and Br) on Surfaces of Polyaniline. *Russ. J. Phys. Chem.* **2020**, *94*, 2148–2154. [CrossRef]
131. Hosseini, H.; Mousavi, S.M. Density Functional Theory Simulation for Cr(VI) Removal from Wastewater Using Bacterial Cellulose/Polyaniline. *Int. J. Biol. Macromol.* **2020**, *165*, 883–901. [CrossRef] [PubMed]
132. Ullah, H.; Shah, A.-H.A.; Bilal, S.; Ayub, K. DFT Study of Polyaniline NH_3, CO_2, and CO Gas Sensors: Comparison with Recent Experimental Data. *J. Phys. Chem. C* **2013**, *117*, 23701–23711. [CrossRef]
133. Furukawa, Y. Electronic Absorption and Vibrational Spectroscopies of Conjugated Conducting Polymers. *J. Phys. Chem.* **1996**, *100*, 15644–15653. [CrossRef]
134. Zhang, L.; Colella, N.S.; Cherniawski, B.P.; Mannsfeld, S.C.B.; Briseno, A.L. Oligothiophene Semiconductors: Synthesis, Characterization, and Applications for Organic Devices. *ACS Appl. Mater. Interfaces* **2014**, *6*, 5327–5343. [CrossRef]

135. Kaneto, K.; Yoshino, K.; Inuishi, Y. Electrical and Optical Properties of Polythiophene Prepared by Electrochemical Polymerization. *Solid State Commun.* **1983**, *46*, 389–391. [CrossRef]
136. Bredas, J.L.; Silbey, R.; Boudreaux, D.S.; Chance, R.R. Chain-Length Dependence of Electronic and Electrochemical Properties of Conjugated Systems: Polyacetylene, Polyphenylene, Polythiophene, and Polypyrrole. *J. Am. Chem. Soc.* **1983**, *105*, 6555–6559. [CrossRef]

Article

A Refined Prediction Parameter for Molecular Alignability in Stretched Polymers and a New Light-Harvesting Material for AlGaAs Photovoltaics

Manuel Hohgardt, Franka Elisabeth Gädeke, Lucas Wegener and Peter Jomo Walla *

Department for Biophysical Chemistry, Institute for Physical and Theoretical Chemistry, Technische Universität Braunschweig, 38106 Braunschweig, Germany; m.hohgardt@tu-braunschweig.de (M.H.); f.gaedeke@tu-braunschweig.de (F.E.G.); lucas.wegener@tu-braunschweig.de (L.W.)
* Correspondence: p.walla@tu-braunschweig.de

Citation: Hohgardt, M.; Gädeke, F.E.; Wegener, L.; Walla, P.J. A Refined Prediction Parameter for Molecular Alignability in Stretched Polymers and a New Light-Harvesting Material for AlGaAs Photovoltaics. *Polymers* **2022**, *14*, 532. https://doi.org/10.3390/polym14030532

Academic Editor: Bożena Jarząbek

Received: 22 December 2021
Accepted: 25 January 2022
Published: 28 January 2022

Publisher's Note: MDPI stays neutral with regard to jurisdictional claims in published maps and institutional affiliations.

Copyright: © 2022 by the authors. Licensee MDPI, Basel, Switzerland. This article is an open access article distributed under the terms and conditions of the Creative Commons Attribution (CC BY) license (https://creativecommons.org/licenses/by/4.0/).

Abstract: Light-harvesting concentrators have a high potential to make highly efficient but precious energy converters, such as multijunction photovoltaics, more affordable for everyday applications. They collect sunlight, including diffusively scattered light, on large areas and redirect it to much smaller areas of the highly efficiency solar cells. Among the best current concepts are pools of randomly oriented light-collecting donor molecules that transfer all excitons to few aligned acceptors reemitting the light in the direction of the photovoltaics. So far, this system has only been realized for the 350–550 nm wavelength range, suitable for AlGaInP photovoltaics. This was achieved by using acceptor molecules that aligned during mechanical stretching of polymers together with donors, that stay random in that very same material and procedure. However, until recently, very little was known about the factors that are responsible for the alignability of molecules in stretched polymers and therefore it was difficult to find suitable donors and acceptors, as well as for other spectral ranges. Recently, a structural parameter was introduced with a high predictivity for the alignability of molecules that contain rigid band-like structures or linear aromatic π-systems. However, for light concentrators in more red spectral ranges, molecular systems often contain larger and extended, planar-like π-systems for which the previously reported parameter is not directly applicable. Here, we present a refined prediction parameter also suitable for larger plane-like structures. The new parameter depends on the number of in-plane atoms divided by out-of-plane atoms as determined by computational geometry optimization and additionally the planar aspect ratio for molecules that contain only in-plane atoms. With the help of this parameter, we found a new system that can efficiently collect and redirect light for the second 500–700 nm AlGaAs layer of current world-record multijunction photovoltaics. Similarly, as the previously reported system for the blue-green layer, it has also overall absorption and re-directing quantum efficiencies close to 80–100%. Both layers, together, already cover about 75% of the energy in the solar spectrum.

Keywords: artificially light-harvesting; luminescent solar concentrators; molecular alignability prediction; redirecting diffuse light

1. Introduction

Solar energy is one of the renewable energy sources with the greatest potential. However, the efficiency of conventional silicon solar cells is limited by the Shockley–Queisser limit, so that a theoretical efficiency of ~30% cannot be exceeded here [1,2]. Therefore, in recent years, work has been carried out on more effective photovoltaic cells and multijunction solar cells with much higher levels of efficiency have been developed. However, the materials used are very expensive [3–5]. One solution to make such a type of photovoltaics usable for more everyday applications would be to collect the light on a large area with more affordable material that redirects it towards much smaller areas of precious high efficiency solar cells. When only direct irradiation of sunlight would reach the earth's surface, lenses would very easily fulfil this task; however, very often the sunlight is diffusively scattered by clouds or other surfaces and there is no way to refocus the light scattered by standard ray optics. As an alternative, diffusive light can be concentrated by luminescent solar concentrators (LSC) but, unfortunately, no such system was capable to really redirect nearly 100% of the photons onto much smaller photovoltaic areas in the past [6].

If a LSC is used that consists of only one type of pigments, there are high intrinsic losses termed escape-cone or reabsorption losses. The former is due to the fact that light absorbed by molecules is usually reemitted in a similar direction as it came from, instead of being redirected to the photovoltaics. The latter occurs because light is reabsorbed within the waveguide due to the high concentration necessary for absorption of all sunlight. Both lead to losses of energy that are intrinsically higher in single-pigment LSCs than the performance advantage of high efficiency photovoltaics [7–11]. A great deal of research has been carried out in recent years to find LSCs with the necessary better efficiencies [12–22], but so far the necessary level of near to 100% light re-direction quantum yield was not reached.

One intrinsic loss mechanism of conventional LSCs, the high reabsorption losses, can be overcome by donor–acceptor fluorophore systems. This requires, however, that the donors are present in excess in order to collect all sunlight on short optical path lengths that are necessary for a high concentration factor (=input surface/output surface, Figure 1) and transfer it also with near to 100% efficiency to the acceptors. The acceptors, in turn, must be far less concentrated to avoid reabsorption on the necessary long pathway to the photovoltaics, which is also necessary for a high concentration factor. However, even then, the acceptors would emit light in any direction, resulting in losses of those rays, that are not directed into the direction of the PV cell (escape-cone losses).

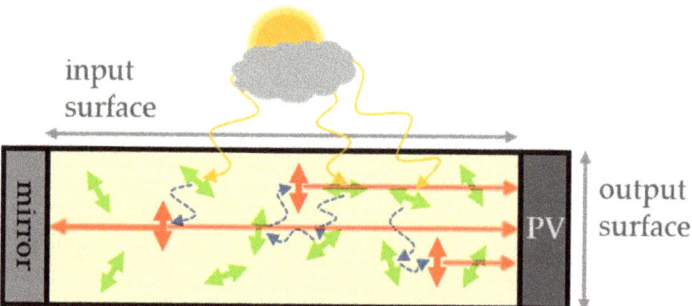

Figure 1. In the FunDiLight LSC concept randomly oriented donor molecule pools (green) absorb diffuse sunlight from all directions (yellow arrows) and transfer the energy (blue arrows) to few aligned acceptor molecules (red). The oriented acceptor molecules then emit specifically in the direction of the photovoltaic (PV, gray). The potential concentration factor is the ratio of input and output surface. The FunDiLight LSC concept avoids intrinsic re-absorption and escape-cone loss mechanisms of previous LSC concepts as it allows for high light absorption with the highly concentrated donor molecules on short optical path lengths (and, consequently, small output surfaces) while avoiding reabsorption due to the low concentrations of the emitting acceptors and escape-cone losses as the oriented acceptors emit almost all light in directions towards the PV. This concept enabled for the first time overall light-re-directioning quantum efficiencies on the order of 90% while still allowing for reasonable high concentration factors (ratio of input surface/output surface): the output surface can be small as all light is absorbed by the highly concentrated donors on short optical path lengths and the input surface can be large as only little light is lost due to reabsorption by the low concentrated, oriented acceptors on the longer path to the output surface.

Therefore, the best system so far is a two-component system that has a larger pool of randomly oriented light-donor molecules and a much lower concentration of acceptors that are all aligned parallel to the photovoltaics (Figure 1). The highly concentrated and randomly oriented donors collect the sunlight from all directions and pass it on to the aligned acceptors via radiationless energy transfer. The aligned receptors, in turn, then radiate specifically in the direction of the PV cell. This minimizes both the escape-cone losses and the losses due to reabsorption in on system, which is also called FunDiLight LSC (funnelling diffuse light re-directioning LSC) [23,24].

Realizing the necessary alignment of the acceptors and the random orientation of the donors in the very same material was possible because, under mechanical stretching of certain polymers, some types of molecules were aligned and others were not in that very same material and procedure. In the first case of a FunDiLight LSC poly(vinyl alcohol) (PVA) was used as a waveguide. However, until recently it was very unclear as to why some molecules are aligned in the polymer during such a procedure whereas others stay random. This knowledge, however, is crucial to developing high-efficiency FunDiLight LSCs for other spectral ranges. Research into the reasons for this found a parameter with a high predictive factor for a molecules alignability in the polymer PVA that depend on the number of atoms that lie within rigid band-like structures of a molecules structure versus the number of atoms that lie outside this band [24]. However, this parameter is only well defined for molecules that contain a single rigid band-like structure. For larger π-systems that are often necessary for light-harvesting in red spectral ranges, and that contain structures that cannot be easily assigned to a single band, this parameter was not well defined. Therefore, it was also more difficult to find appropriate donor and acceptor dyes for highly efficient FunDiLight LSCs in the red spectral range [23,24].

Here, we present a refined and very well-defined parameter that is also capable to predict the molecular alignability of larger π-systems and planar structures in stretched polymers. The new parameter depends on the number of in-plane atoms divided by out-of-plane atoms as determined by computational geometry optimization and additionally the planar aspect ratio for molecules that contain only in-plane atoms. Using this parameter, we found a new FunDiLight LSC that can efficiently collect and redirect light also in the more reddish AlGaAs band gap of current world-record multijunction photovoltaics [3]. Similarly, as the previously reported system for the blue-green layer, it has also an overall absorption and re-directioning quantum efficiency on the order of 90%. Both layers together already cover about 75% of the energy of the solar spectrum.

2. Materials and Methods

2.1. Sample Preparation for Screening the Alignability of Dyes Containing Larger π-Systems

For screening of the alignability of all dyes investigated in this study, first, 2.2 g poly(vinyl alcohol) (PVA) and the corresponding dye are weighed and dissolved in 20 mL dimethyl sulfoxide, so that a 5×10^{-4} M dye-solution is created. This solution is heated for 3 h at 70 °C under a nitrogen atmosphere while stirring. About 4 g are weighed into a Petri plate and dried for 2 days at 60 °C under 200 mbar. The resulting film is stretched 500% so that one can see which molecules are aligned. For the three-dimensional single-molecule polarization measurements a foil with a concentration of 10^{-10} M was prepared.

2.2. Sample Preparation of the Red FunDiLight LSC

As described before, 2.2 g PVA and 38.56 mg Lumogen F Red 300 and 1.73 mg Oxazine 170 perchlorate are weighed out and dissolved together in 20 mL dimethyl sulfoxide (DMSO). This solution is also under a nitrogen atmosphere while stirring heated for 3 h to 70 °C. About 4 g are weighed into a Petri plate and dried for 2 days at 60 °C under 200 mbar. The resulting film is stretched by 500%, with the Oxazine 170 aligning and the Lumogen F Red 305 molecules remaining randomly oriented.

2.3. Fluorescence Spectroscopy to Dertermine the Alignability

A similar setup has already been described [24]. Briefly, the polymers are placed in the fluorescence spectrometer. In addition, polarization filters are built in so that different combinations of polarizations can be measured. From this, the intensities of the parallel and perpendicular components can be determined. For more details, see [24].

2.4. Three-Dimensional Single-Molecule Orientation Microscopy

A similar setup has already been described [24]. This time only, the Coherent Chameleon laser was used, together with an optical parametric oscillator, to set the wavelengths used. The polarization of the laser is rotated by a motorized half-wave plate during the entire measurement. Since light is only absorbed parallel to the transition dipole moment, a modulation of individual dyes can be seen. The dyes are measured from three directions. This is achieved in that two wedge prisms arranged contrary to one another move the laser beam laterally to the optical axis. When the laser hits the objective, it leaves it at an angle of approximately 35° to the optical axis. The fluorescence of the wide field microscope setup is detected with an EMCCD camera. For more details, see [24].

2.5. Pump-Probe Spectroscopy

A similar setup has already been described [23,24]. A high repetitive laser system (Coherent OPA/Coherent RegA operated at 120 kHz, pumped by Coherent Verdi) was used for the pump-probe experiments. The pump wavelength was set to 580 nm for all measurements and a BP 580/10 (Thorlabs) was placed before the sample. A linear gradient filter and a miniature spectrometer USB2000+ (Ocean Optics) were used to obtain certain wavelengths from the OPA white light for the probe laser. To improve the signal to noise ratio an ultrafast photodiode (provided by Prof. D. Schwarzer) with a set of 10 bandpass filters for every used wavelength BP600/10-690/10 (Thorlabs) was used directly behind the sample. One half-wave plate (achromatic half-wave plate, 400–800 nm, THORLABS) in each beam-path was used for the polarization dependent measurements. A special sample holder was used to minimize photobleaching by rotating the foil while maintaining its orientation, similar to Pieper et al. [23]. The analysis of the data was completed similarly to in Willich, Wegener et al. [24].

3. Results

3.1. An Improved Prediction Parameter and Scheme for the Alignability of Molecules in Stretched Poylmers

To characterize the experimental alignability of the molecules investigated here, first fluorescence excitation spectra were recorded with different polarization filters and the ratio in the fluorescence intensities detected with polarization filters parallel or perpendicular to the polymer stretching direction, $\Delta I_{\parallel}/I_{\perp}$, was determined, as has already been described in Willich, Wegener et al. [24]. In case of no molecular alignment, $\Delta I_{\parallel}/I_{\perp} = 1$ because the fluorescence emission is the same parallel and perpendicular to mechanical stretching direction. When more molecules are oriented along the stretching direction, the detected intensity in the parallel direction increases, $\Delta I_{\parallel}/I_{\perp} > 1$. The larger $\Delta I_{\parallel}/I_{\perp}$, the higher the alignment. Values below 1 do not usually occur or mean an alignment of the transition dipole moments orthogonal to the direction of stretching.

As already mentioned in the introduction, a parameter with a good predictive power for the alignability of organic dyes in mechanically stretched PVA matrix has already been introduced in Willich, Wegener et al. [24]. This parameter, η, was determined as follows. First, the longest rigid and planar band in the molecule was identified. This band can consist of a linear chain of aromatic rings, such as in Acridine Yellow G, or can be also generated by other structural factors, such as in Coumarin 6, in which molecular rotation of a single bond is inhibited by interactions between certain intramolecular groups. In cases that were not immediately clear from the structure itself, a simple computational geometry optimization (Chem3D, minimize energy (as MM2 Calculation)) was used to determine structural parts that are planar, rigid bands. For example, Coumarin 6 is planar throughout the entire band-like structural part denoted by red color in Figure 2. From the structure itself, it was not entirely clear whether steric effects tilt both groups connected by the single bond but the computational geometry optimization confirmed that the red part is planar, likely due to stabilizing interactions between the sulfur and the neighboring nitrogen atom. All the atoms within the plane of this rigid band were then counted as intraband atoms, N_{Band}, whereas all atoms outside this rigid band, i.e., flexible side chains or rigid groups that point in different directions than the rigid core band, are counted as out-of-band atoms, $N_{OutOfBand}$ (See, for example, red bands in Figure 2). Hydrogen atoms were simply counted with the assigned atoms they were attached to. The ratio of these atoms,

$$\eta = N_{Band} / N_{OutOfBand} \qquad (1)$$

had a very high predictive power for the alignability, $\Delta I_\parallel / I_\perp$, of the molecules in stretched polymers. This empiric observation can likely be explained by molecular forces aligning longer rigid bands in polymers during stretching with all atoms pointing out of this band in any three-dimensional directions, $N_{OutOfBand}$, rather hindering this alignment (For more examples see Figure 4 in [24]). Generally, it was observed that an η-parameter of >1 typically predicts a good alignability of $\Delta I_\parallel / I_\perp > 1.5$.

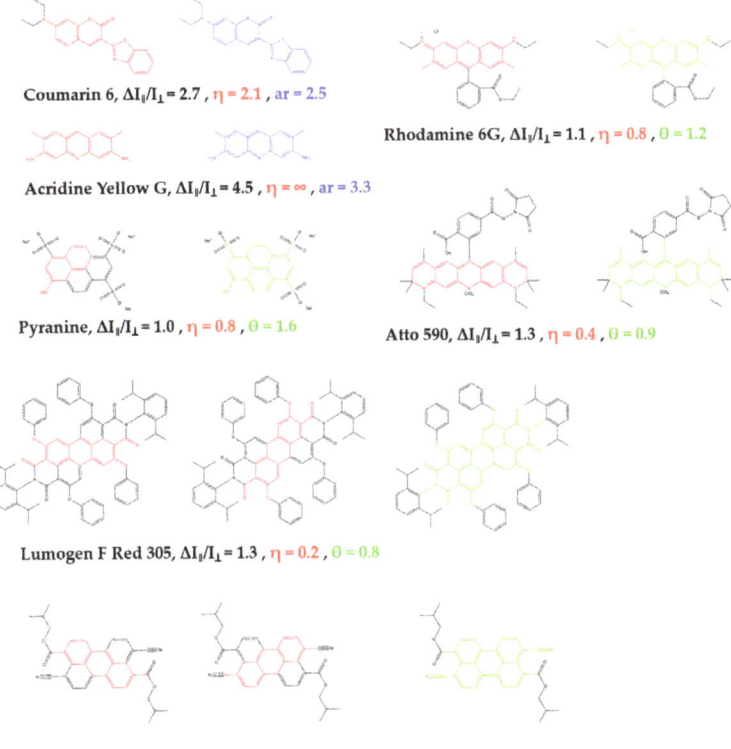

Figure 2. Exemplary structures of molecules denoted with atoms counted for different predictors for the experimentally observed alignability, $\Delta I_\parallel / I_\perp$, in stretched polymers. Green: Atoms counted as in-plane atoms, $N_{InPlane}$, for the predictor θ presented in this work. Red: Atoms countable as intra-band atoms, $N_{InBande}$, for the predictor η presented in [24]. Blue: For molecules in which all atoms lie in one plane, the aspect ratio (ar) serves as robust predictor.

However, this simple prediction parameter, η, is not well defined for larger molecules with multiple potential bands. For example, in molecules such as Lumogen F Red 305, it is not as straightforward to identify the dominating band, even though η still predicts the low alignability reasonability well, regardless in which structural the dominating band is identified (e.g., Figure 2 shows two possibilities of red band assignments in Lumogen F Red 305).

In order to provide a more precise definition for a prediction parameter also for larger molecules consisting, for example, of more extended π-systems we refined the prediction parameter in the following way. First, we consistently did a geometry optimization for all molecules investigated. Then, all atoms were counted in this optimized geometry that were within the molecular plane of these molecules, $N_{InPlane}$, rather than just those in one band. All other atoms were counted as $N_{OutOfPlane}$. Figure 2 (green) illustrates this count for a couple of molecules. Hydrogen atoms were not included, as they play only a minor steric role but do often occur more dominantly in out of plane groups. This refined parameter

$$\theta = N_{InPlane}/N_{OutOfPlane} \qquad (2)$$

provides similarly well estimates for the alignability as our previous parameter η, (compare green θ values in Figure 3 with experimentally observed alignabilities $\Delta I_\parallel / I_\perp$ in black) but is unambiguous for more planar molecular structures.

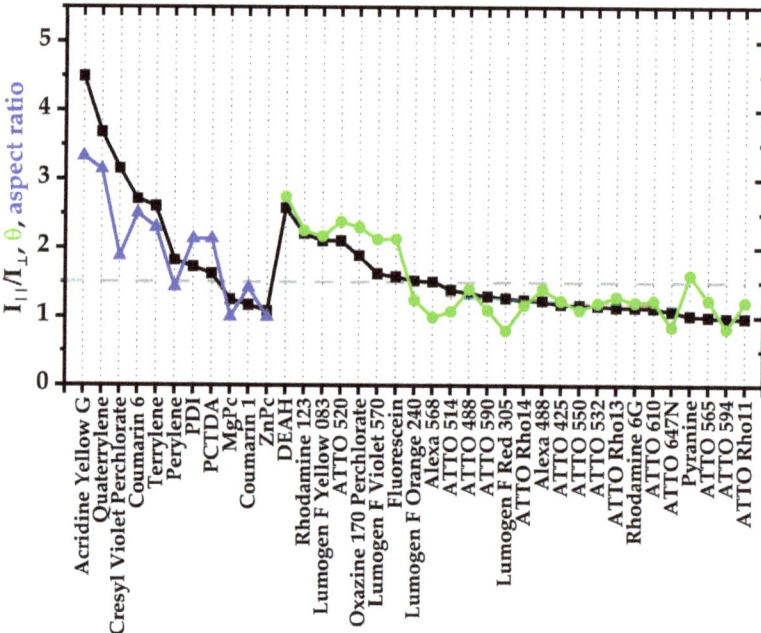

Figure 3. Correlation between the experimentally alignability (black curve, $\Delta I_\parallel / I_\perp$) and the alignability prediction parameter θ (green curve, Equation (2)) and the aspect ratio (blue curve, Equation (3)) as predictor for entirely planar molecules according to molecular geometry optimization calculations.

In Lumogen F Red 305, for example, the total number of atoms in the molecule is 82 without hydrogens. There are 36 atoms inside a plane and 46 outside, which results in θ = 0.8 (Figure 2), and corresponds well with the experimental observation of little alignability. Another important example is Lumogen F Yellow 083. This is also a molecule with a larger π-system that does not easily allow to identify the largest rigid band. Indeed, the rigid-band bases parameter η = 0.5 does not predict any alignment. However, the new parameter presented here is precisely defined also for such molecules and predicts with a value of θ = 2.1 very well the experimentally observed alignability of $\Delta I_\parallel / I_\perp \sim 2.2$.

In cases, however, in which *all* atoms are in plane according to geometry optimization, $N_{OutOfPlane}$ becomes zero and Equation (2) is not defined anymore. Important examples are rylenes, such as perylene, terrylene, and quaterrylene (structures are shown in Figure 4). Still, better alignment was observed experimentally for structures that resemble rigid band like structures, similar as our previous empirical observation that led to the band-based prediction parameter η, (Equation (1)). To account for this observation, we considered the aspect ratio for such completely planar molecules that do not contain any out-of-plane atoms, $N_{OutOfPlane}$ = 0. To do so, the rigid planar parts of the molecules were first simplified as simple geometric shapes (e.g., benzene ≡ hexagon) with all bonds assumed to be of approximately equal length. With this simplification, the length of the long and short axis of the rigid, planar part of the molecules was then computed in units of the length of one bond, a. Figure 5 illustrates this exemplarily for perylene. Due to geometric consideration the aspect ratio, ar, is in this case

$$ar = l_{longaxis}/l_{shortaxis} = 5a/(4a \times \cos(30°)) = 1.44 \quad (3)$$

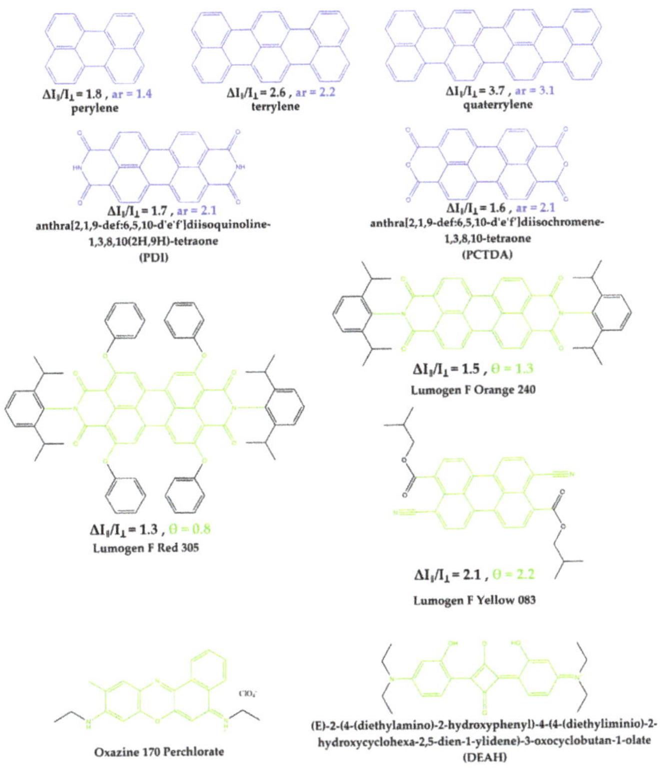

Figure 4. Structures of rylenes and rylene derivates exemplary for planar molecular structures that either do or do not contain any out-of-plane atoms. When there was no out-of-plane atom (blue structures) the aspect ratio (Equation (3)) is a good predictor for the experimentally observed alignabilities, $\Delta I_{\parallel}/I_{\perp}$. Otherwise, the θ–parameter (Equation (2)), that is determined from the ratio of in-plane atom (green) number over the number of out-of-plane atoms (black), is a good predictor.

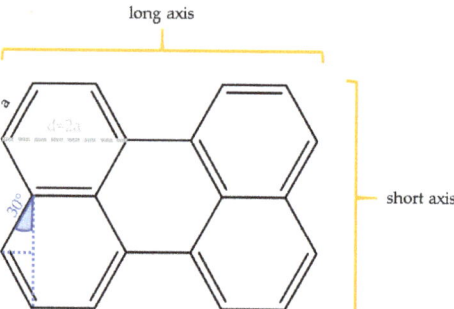

Figure 5. In planar molecules that do not contain any out-of-plane atoms the aspect ratio is a good predictor for the alignability in stretched polymers. The figures show exemplarily how the aspect ratio can be determined from the molecular structure of perylene. For details see text.

A comparison of the aspect ratio-based parameter computed for these molecules (blue in Figure 3) demonstrates also a high predictive power for the alignability of such molecules in polymers (black curve in Figure 3).

The general prediction scheme for the alignability of molecules in polymers is illustrated in Figure 6. First a simple geometry optimization is performed to identify all atoms, expect hydrogens, that are either within the molecules plane, $N_{InPlane}$, or outside the molecular plan, $N_{OutOfPlane}$. If $N_{OutOfPlane} \neq 0$, the alignability can be predicted by the ratio defined in Equation (2) (green data in Figure 3). If $N_{OutOfPlane} = 0$, the alignability can be predicted by the aspect ratio of the plane defined in Equation (3) (Figure 2 and blue data in Figure 3). These predictions correspond very well with the experimentally observed alignabilities, $\Delta I_\parallel / I_\perp$ (black data in Figure 3).

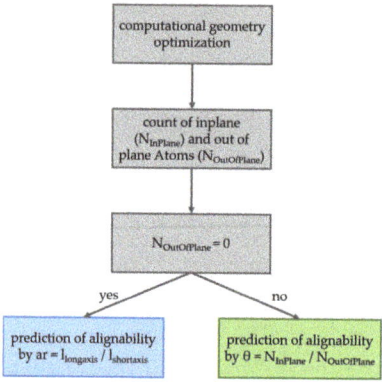

Figure 6. The diagram shows how to estimate the alignability of a molecules transition dipole moment in stretched polymers (PVA) with either the predictor parameter θ (Equation (2)) or the aspect ratio for case of entirely planar molecules (Equation (3)).

An experimental observation that has so far not been considered in the proposed prediction parameters is the influence of chemical polarization on the alignability. An interesting comparison is that of the molecules PCTDA, PDI and terrylene that all have similar spatial dimensions but are decreasingly polar (structures are shown in Figure 4). All molecules show an alignment but it is noticeable that the $\Delta I_\parallel / I_\perp$ value decreases with greater polarity. The value of $\Delta I_\parallel / I_\perp$ for PCTDA is only half as high as that for terrylene. We suspect that the polar molecules are wedged between polymer chains due to hydrogen bonds, while the non-polar molecules are well placed between the chains using the weaker

Van der Waals forces. Therefore, when using the aspect ratio to predict alignability, it must be considered that the polarity has an additional influence.

3.2. A New Light-Harvesting Solar Concentrator for the AlGaAs Layer of High-Efficiency Photovoltaics

So far, high efficiency light-harvesting materials based on the scheme shown in Figure 1 have only been demonstrated for the blue AlGaInP layer of current high efficiency photovoltaics. High efficiency light-harvesting materials for the next, AlGaAs layer, of such high efficiency photovoltaics have not been reported so far, partly due to the above-described difficulties in the ability of predicting the alignability of larger light-harvesting donors and light-redirecting acceptors, that shall either stay randomly oriented or align in that same material during stretching. In order to build a FunDiLight-LSC for the second band gap of the currently best solar cell by Geisz et al. now a suitable donor or acceptor can be taken from Figure 3 [3]. Lumogen F Red 305 is ideal as a donor. No alignability ($\theta = 0.8$) is predicted which is also confirmed experimentally ($\Delta I_\parallel / I_\perp \sim 1.3$). In addition, the spectral requirements fit (Figure 7) and a high fluorescence quantum yield of almost 100% suggests a very good energy transfer [25].

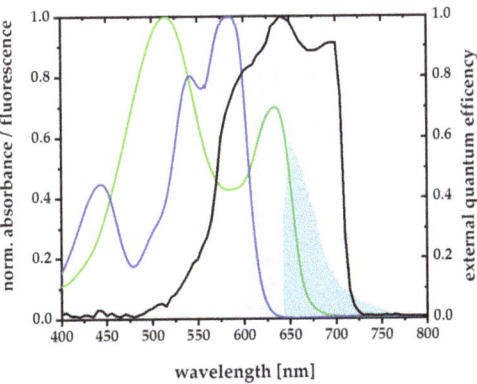

Figure 7. Absorption and emission spectra of Oxazine 170 (green) and Lumogen F Red 305 (blue), respectively. The fluorescence spectrum of Oxazine 170 (green, filled) fits perfectly the EQE spectrum and thus band gap of AlGaAs cells (black) (Data from [3]). The fluorescence Lumogen F Red 305 (filled in blue) overlaps perfectly with the absorption spectrum of Oxazine 170 (green).

In order to find a suitable acceptor, the steric alignment requirements for the molecule must be met in addition to spectral requirements for effective energy transfer and photovoltaics bandgap match, as well as highest possible fluorescence quantum yields. These requirements are met, among others, by the squaraine dye DEAH and the dye Oxazine 170. These two are predicted to be alignable by θ-parameters of $\theta = 2.75$ (DEAH) and $\theta = 2.3$ (Oxazine 170), and is confirmed by the experimentally measurement alignability. Even though squaraine dye has a better fluorescence quantum yield of 86% compared to the Oxazine with 63% in solution, it photodegraded very quickly. Therefore, Oxazine 170 was selected as the acceptor. In addition, it is also observed very often that the fluorescence quantum yields in solid environments, such as polymers, are significantly larger than in aqueous solution. To generate a high-efficiency light harvesting system with Lumogen F Red 305® and Oxazine 170, we first calculated optimized donor/acceptor ratios as well as concentrations using our previously published computational ray tracing tool [26] and improved the overall quantum efficiencies experimentally, thereafter. The following characterization of the systems performance was similarly completed as reported previously [27,28].

First, the angle distribution of aligned Oxazine 170 was examined in a single molecule 3D orientation microscope. In such a microscope, the stretched or unstretched polymer containing Oxazine 170 is illuminated from three different directions (Figure 8a) while the polarization of the light is rotated. Since light is best absorbed with a polarization vector parallel to the transition dipole moment of the fluorophore, differently oriented dyes are excited at different times. Therefore, the fluorescence traces of the single dyes also show modulation. The 3D orientation of each individual Oxazine 170 molecule can then be determined from these modulations observed from the three different directions. Figure 8b,c shows a typical microscope image from individual Oxazine 170 molecules in PVA. The angular distribution of the flat azimuth angles is shown in polar plot histograms in Figure 8d,e.

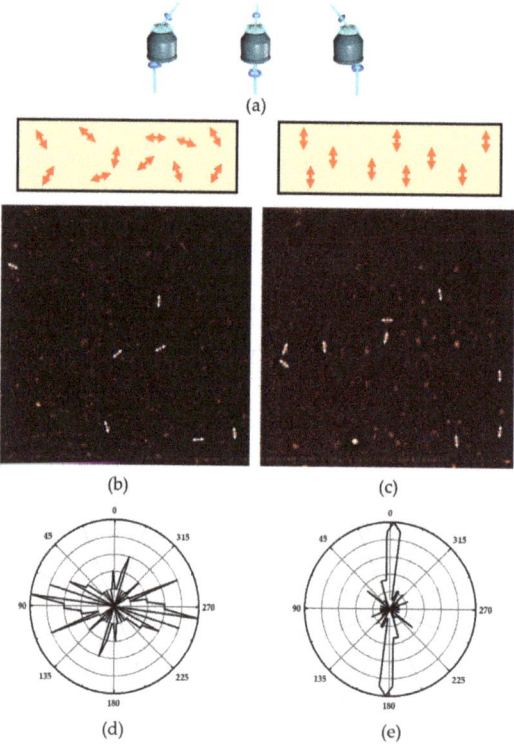

Figure 8. In 3D single molecule orientation microscopy, the sample is illuminated from three different directions with rotating polarization (**a**). The analysis of the observed fluorescence intensity modulations allows to determine the 3D orientation of single molecules in the unstretched (**b**) and stretched polymer (**c**). Polar plots of histograms of the azimuth angles show a clear alignment along the polymer stretching direction (**e**) that is not present in the unstretched control (**d**).

Although a very random distribution can be seen in unstretched polymers in Figure 8d, in Figure 8e there is a clear majority of molecules aligned in the stretched polymer.

Figure 9 shows a more detailed representation of all single molecules investigated along with linear presentations of the azimuth and polar angle histograms. After stretching, more than half of all molecules lie flat in the plane within 10°. In addition, over half of the molecules are approximately 20° around the direction of stretching. Gaussian fits show a half-width of the azimuth angle distribution of 14.5° and a half-width of 10.8° for the polar angles (Figure 9d).

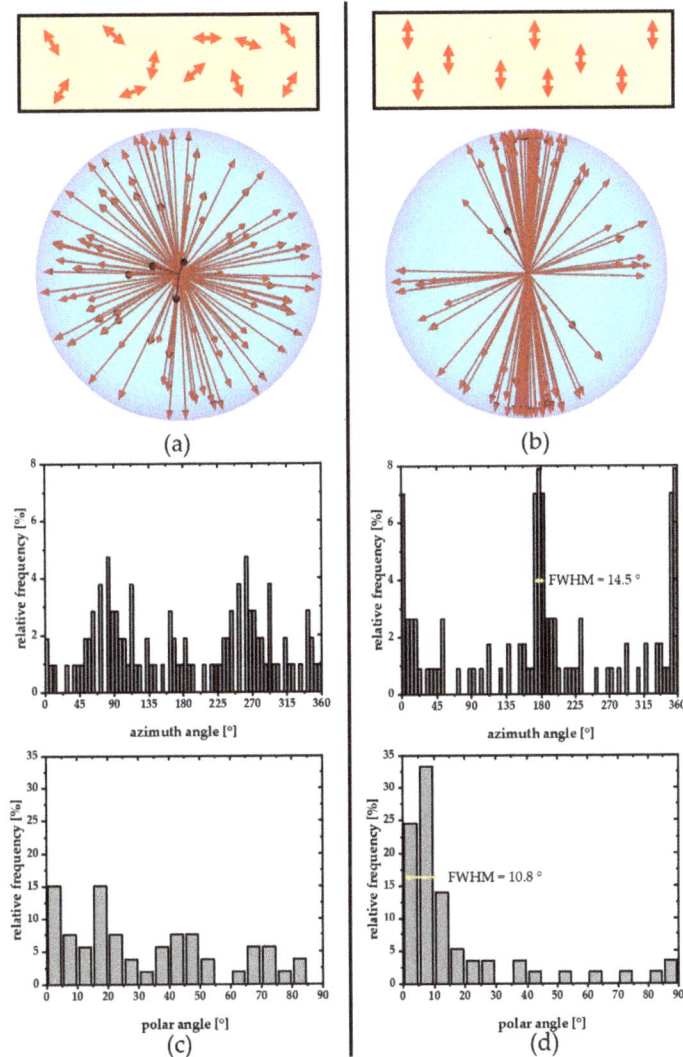

Figure 9. Detailed representation of the 3D angular distribution of the single molecules in an unstretched (**a**) and stretched polymers (**b**) as spherical plot, along with the corresponding azimuth (**b**,**c**) and polar angle distributions (**c**,**d**). Note that the measured polar angle distribution has always a tendency to overrepresent flat angles, even in random orientation distributions (**c**) as upright molecules emit most of their light perpendicular to the microscope objective and are, therefore, hard to detect.

Based on the microscopic data, the orientation of the acceptor could be reliably verified. This is essential for a precise FunDiLight LSC. However, efficient and fast energy transfer is at least as important for overall high light-re-directing quantum efficiencies.

Therefore, pump-probe measurements were completed for the donor–acceptor system to investigate the dynamics of energy migration and dipole reorientation in more detail.

First, a spectrum with different probe wavelengths was recorded after pumping the donor at its absorption maximum, λ_{exc} = 590 nm. The highest signal was obtained at a wavelength of λ_{det} = 650 nm (Figure 10a,b), corresponding well with the acceptor absorption and thus arising very likely from acceptor ground state bleaching after receiving energy from the donors. During the temporal evolution, a decrease in the signal can be seen at bluer wavelengths, which is likely indicative of intramolecular vibrational relaxation processes in the acceptor after receiving the energy from the donors in higher vibrational acceptor states. At red wavelengths, an increase can be seen after this first quick step. (This suggests that donors continue to deliver energy to acceptors).

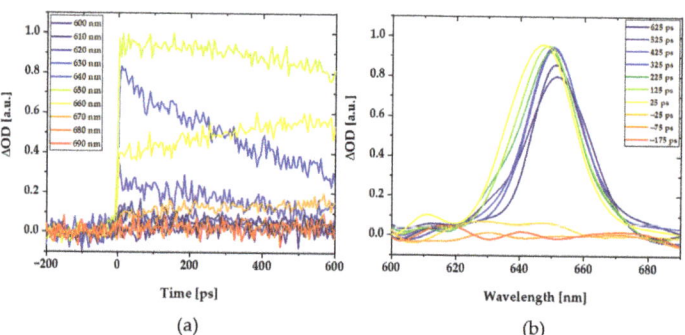

Figure 10. Pump-probe data of the unstretched donor–acceptor polymer samples observed after pumping the donors at λ_{exc} = 590 nm and detecting the acceptor kinetics at various probe wavelengths (**a**) as well as the transient pump-probe-spectra converted therefrom (**b**).

Kinetic signals where either fitted by mono- or biexponential rise terms, depending on which better described the observed signals. Since the donor molecules are at different distances from each other and from the acceptor, different time scales, and kinetics are expected. Before the donors close to the acceptor molecules transfer their energy, excitation energy migration and transition dipole reorientation occurs in the larger donor pools. To dissect these processes, polarization-dependent pump-probe measurements were also carried out (Figure 11a). Four different polarizations were measured, with combinations of pump and probe polarizations parallel or perpendicular to the direction of stretching of the polymers. The rise time of these curves reflect direct energy transfer from the (closest) donors to the acceptors and is with about 6 ps is comparable to that what one expects from Förster Theory for a single donor to acceptor transfer at the closest distance of about 2.6 nm between the pigments. With donor pool pump polarization perpendicular to the acceptor probe polarization (and stretching direction), additional kinetic rising components were observed (green in Figure 11a) that are not visible when directly pumping and probing parallel to the stretching direction (violet in Figure 11a). This is due to the additional time necessary to rotate the initial perpendicular transition dipole orientations into transition dipole orientations parallel to the acceptors during energy migration from the donors to the acceptors. A difference spectrum can be formed from these two measurements and a biexponential function can be fitted to this difference spectrum (Figure 11b). The biexponential rise term gives a time constant for the intra donor-pool energy migration and dipole moment reorientation on the order of ~27 ps, as well as a decay time constant of approximately 400 ps, after which the energy transfer from the donor-pool to the acceptors is completed. Overall, the times are all well below the ns lifetime of the donor (7.9 ns [29]), which is why an almost perfect energy transfer efficiency c with a quantum efficiency close to unity can be assumed. In addition, the significantly lower amplitude with perpendicular polarization of the pump and probe beam to the stretching direction compared to corresponding parallel pump and probe beam polarization data once more confirm the acceptor alignment parallel to the polymer stretching directions (Figure 11a red and violet curves).

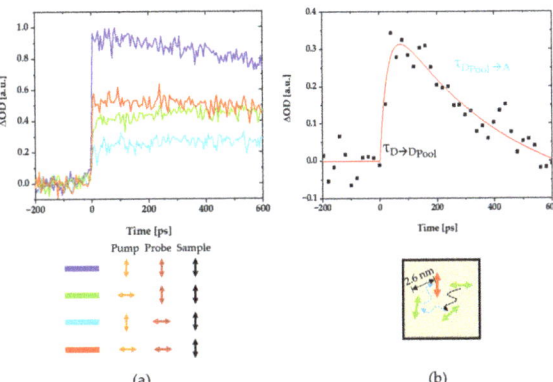

Figure 11. Polarization-dependent pump-probe measurements of the stretched sample with various pump and probe polarizations parallel or perpendicular to the stretching direction (**a**). The time constants for the energy migration and dipole-reorientation dynamics within the donor pool and the subsequent transfer to the acceptors can be determined through the biexponential fit of the difference spectrum of the violet and green curve (**a**) from the stretched sample (**b**). For details of this analysis see text and [24].

4. Discussion

With the present study, structural factors that lead to an alignability of molecules in polymers—at least in PVA—become clearer. Key factors are obviously the size of rigid planar parts in a molecules structure, as well as the aspect ratio of this plane and the size and number of structural groups that point outside this plane and/or that are flexible. Based on these observations, we provide a refined prediction parameter, θ, and scheme (Figure 6) that is based on the ratio in the numbers of in-plane and out-of-plane atoms observed in a simple geometry optimization calculation (Equation (2), green in Figures 2 and 4)) and the aspect ratio for planar molecules (Equation (3), blue in Figures 2 and 4), that do not contain any out-of-plan atoms at all. These parameters and the scheme predict the alignability (e.g., for light re-directing acceptor molecules, red in Figure 1) or non-alignability (e.g., for randomly oriented light harvesting donor molecules in the same material, green in Figure 1) at least as good as our previously reported parameter, but allows to better predict the alignability of larger molecules (Figure 3). In addition, we found an indication that the alignability decreases with greater polarity. We suspect that this is due to increasing distorting interactions with polar groups of PVA during the stretching. However, this observation is not included in our alignability estimation parameters yet, as it needs more experimental verification.

Obviously, the alignability is a largely steric phenomenon. We find that small hydrogen atoms can be neglected for a good alignment prediction. We found experimental alignment when the θ-parameter was greater than 1.5 or in other words when 1.5 times more atoms are in the plane than outside of it. We suspect that the polymer chains start to orient in one preferred direction when they are mechanically stretched and that the molecules between the ordered chains are aligned by shear forces (Figure 12a). Molecules with larger numbers of rigid or non-rigid groups and atoms pointing outside this plane are more likely to wedge between the chains, hindering alignment in that same direction. We also suspect that this more likely, when more heteroatoms are present that make the molecules more polar. Since the polymer PVA is already polar itself, the wedging of the molecules could be increased during the polymer stretching.

In the case of completely flat molecules, the aspect ratio is also important. We suspect that while all flat molecules align themselves, a rotation around the transverse axis of the molecule allows also for transition emission dipole moments perpendicular to the stretching direction, and, therefore, no alignment of the light emission is observed. With elongated molecules of higher aspect ratios, such rotation becomes less likely or in other words the molecules dipole moment does align in just one direction parallel to the stretching direction. We found that emission dipole moment alignment can be typically found when the aspect ratio was approximately higher than 1.5.

With these insights we were able to seek suitable molecules that act as randomly oriented light-harvesting donor pools (green in Figure 1) and light redirecting acceptor molecules (red in Figure 1) also for other spectral ranges than our previously published system for the blue-green AlGaInP spectral range of high-efficiency photovoltaic [23,24]. Before the present study, this was difficult as the necessary alignment or non-alignment of larger molecules in the very same polymer and stretching procedure could not be predicted as easily with our previous prediction parameter, η, that rather relied on the size of rigid band size structure in smaller pigment molecules.

For the more reddish spectral range necessary for the AlGaAs layer of high efficiency photovoltaics, the relatively flat and long dye Oxazine 170 was found as a suitable example of a fluorophore that aligns very well in polymers when mechanically stretched. It has a prediction parameter of θ = 2.3 and indeed showed an experimental alignability of $\Delta I_{\parallel}/I_{\perp}$ ~ 1.9. Together with the bulky Lumogen F Red 305, that stays with θ = 0.8 and $\Delta I_{\parallel}/I_{\perp}$ ~ 1.3 randomly oriented in the very same polymer and stretching procedure, it forms a FunDiLight system as illustrated in Figure 1 for the spectral range of the AlGaAs layer. The Donor Lumogen F Red 305 collects exactly the spectral range of the light not covered by the previously published blue layer, and can pass it on to the acceptor Oxazine 170 with high yield, due to perfect spectral overlap. The aligned acceptor can purposefully redirect the energy and the emission of Oxazine 170 fits perfectly with the AlGaAs band gap of high efficiency photovoltaics. Microscopic 3D single molecule orientation measurements confirmed that the light re-directing acceptors in this system are very well aligned, with 85% of the dyes within 25° of the stretching direction.

The highly efficient light-harvesting donor to light-redirecting acceptor energy transfer is confirmed by efficient, ultrafast energy transfer unveiled by polarized pump-probe spectroscopy (Figures 10 and 11). These experiments also provided valuable insights into the donor pool energy migration and emission dipole moment reorientation on timescales on the order of approximately 6–400 ps. The excitation energy is transferred from the primarily excited light harvesting donors to the donor-pool on a timescale of about 27 ps. From there, the energy is gradually passed on to the acceptors. The final ultrafast one-step donor to acceptor energy transfer step from the nearest donor in the light-harvesting pool to the light-redirecting acceptors time occurs in about 6 ps. Even if the transition dipole moments of the excited donors are very different for the light-redirecting acceptors, they still transferred efficiently all excitons after about 400 ps.

5. Conclusions and Perspective

In summary, these results confirm that about 99.9% of light in the 420–660 nm spectral range of the light-harvesting donor is collected in a single foil of 50 μm thickness, at least 98% of these excitations is transferred to the light-redirecting acceptors, and that the acceptors emit about 80% of the light in directions suitable for effective total internal reflection waveguiding to, for example, high-efficiency photovoltaics. The percentage of absorbed light was inferred from the absorption spectrum of a single foil in the spectral range of 420–660 nm (Figure 7), the efficiency of the energy transfer from a direct comparison of the donor fluorescence intensity in the presence and the absence of acceptors [30], as well as the observed ultrafast energy transfer on timescales from 6 to 400 ps (Figures 10 and 11) in comparison to the donor lifetime (7.9 ns [29]) and the emission angle range from the 3D orientation single molecule experiments (Figures 8 and 9) in a similar manner as described in Pieper et al. and Willich, Wegener et al. [23,24]. Therefore, the overall efficiency of the new funneling light-harvesting system for the reddish AlGaAs layer of high-efficiency photovoltaics is as similarly high as our previously reported systems for the blue-green AlGaInP layer [23,24]. Together with the previously proposed system for the blue-green spectral range, these two layers already cover about 75% of the total energy of the solar light (denoted by grey color in Figure 12c).

For the future development we envision two layers of our high-efficiency light-harvesting systems together with the corresponding two layers of high efficiency photovoltaics. This principle is shown in Figure 12b. One of our future aims is to realize such a real word system including all components that are necessary for a high efficiency light harvesting together with high efficiency photovoltaics and a concentration factor (input surface/output surface, see Figure 1) as large as possible and the very little loss mechanisms provided by our FunDiLight approach.

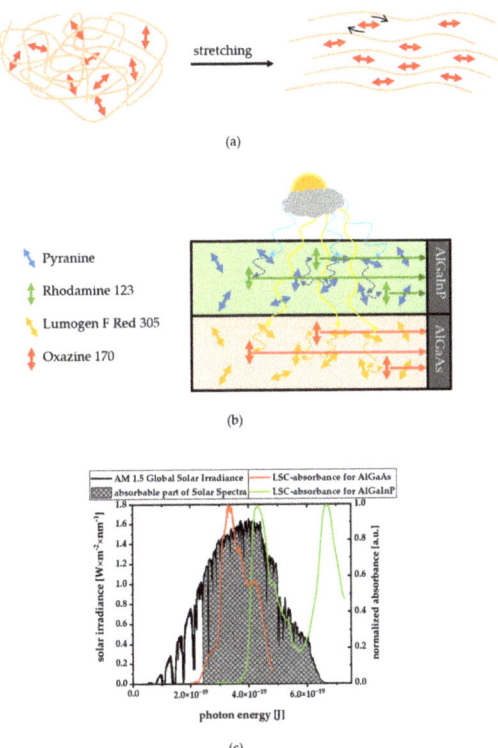

Figure 12. Polymer chains (beige) are likely ordered by the stretching procedure. We suspect that shear forces (small black arrows) can align the dyes (red arrows) between the polymer chains when they have the corresponding structural requirements (see Figures 2–6) (a). In the envisioned two-layer light-harvesting and energy conversion system light in the wavelength range between 275 and 500 nm is first collected in the upper layer by the randomly oriented Pyranine donor molecules (blue) and transferred to the Rhodamine 123 acceptors (green) aligned parallel to the AlGaInP photovoltaics (grey). Wavelengths longer than 500 nm pass through the first layer and are then absorbed by the Lumogen F Red 305 donors (yellow) and transferred on to the aligned Oxazine 170 acceptors (red). These emit correspondingly in the direction on the AlGaAs PV cell material. (b) The part of the solar spectrum ([31,32], Data from [32]) that are entirely harvested by the Pyranine and Rhodamine 123 molecules (green absorption spectrum) and Lumogen F Red 305 and Oxazine 170 molecules (red absorption spectrum) of such a two-layer system is marked by grey color and corresponds already to about 75% of the total solar irradiation energy (c).

6. Patents

The University of Braunschweig (P.J.W.) filed a patent for parts of this work.

Author Contributions: Conceptualization, M.H. and P.J.W.; methodology, M.H. and L.W.; software, M.H. and F.E.G.; validation, M.H., F.E.G. and L.W.; formal analysis, M.H., F.E.G. and L.W.; investigation, M.H., F.E.G. and L.W.; resources, P.J.W.; data curation, M.H.; writing—original draft preparation, M.H.; writing—review and editing, M.H., P.J.W.; visualization, M.H.; supervision, P.J.W.; project administration, P.J.W.; funding acquisition, P.J.W. All authors have read and agreed to the published version of the manuscript.

Funding: This research was funded by the German science foundation (Grants INST 188/334-1 FUGG and GRK2223/1), by the Open Access Publication Funds of Technische Universität Braunschweig and the Cluster of Excellence PhoenixD (DFG–EXS 2122).

Institutional Review Board Statement: Not applicable.

Informed Consent Statement: Not applicable.

Data Availability Statement: The data presented in this study are available on request from the corresponding author. All study data are included in the article.

Conflicts of Interest: The authors declare no conflict of interest.

References

1. Shockley, W.; Queisser, H.J. Detailed balance limit of efficiency of p-n junction solar cells. *J. Appl. Phys.* **1961**, *32*, 510–519. [CrossRef]
2. Zdanowicz, T.; Rodziewicz, T.; Zabkowska-Waclawek, M. Theoretical analysis of the optimum energy band gap of semiconductors for fabrication of solar cells for applications in higher latitudes locations. *Sol. Energy Mater. Sol. Cells* **2005**, *87*, 757–769. [CrossRef]
3. Geisz, J.F.; France, R.M.; Schulte, K.L.; Steiner, M.A.; Norman, A.G.; Guthrey, H.L.; Young, M.R.; Song, T.; Moriarty, T. Six-junction III–V solar cells with 47.1% conversion efficiency under 143 Suns concentration. *Nat. Energy* **2020**, *5*, 326–335. [CrossRef]
4. Dimroth, F.; Grave, M.; Beutel, P.; Fiedeler, U.; Karcher, C.; Tibbits, T.N.D.; Oliva, E.; Siefer, G.; Schachtner, M.; Wekkeli, A.; et al. Wafer bonded four-junction GaInP/GaAs//GaInAsP/GaInAs concentrator solar cells with 44.7% efficiency. *Prog. Photovolt. Res. Appl.* **2014**, *22*, 277–282. [CrossRef]
5. Dimroth, F.; Tibbits, T.N.D.; Niemeyer, M.; Predan, F.; Beutel, P.; Karcher, C.; Oliva, E.; Siefer, G.; Lackner, D.; Fus-Kailuweit, P.; et al. Four-junction wafer-bonded concentrator solar cells. *IEEE J. Photovolt.* **2016**, *6*, 343–349. [CrossRef]
6. Batchelder, J.S. The Luminescent Solar Concentrator. Ph.D. Thesis, California Institute of Technology, Pasadena, CA, USA, 4 August 1981.
7. Fang, L.; Parel, T.S.; Danos, L.; Markvart, T. Photon reabsorption in fluorescent solar collectors. *J. Appl. Phys.* **2012**, *111*, 076104. [CrossRef]
8. Olson, R.W.; Loring, R.F.; Fayer, M.D. Luminescent solar concentrators and the reabsorption problem. *Appl. Opt.* **1981**, *20*, 2934–2940. [CrossRef]
9. Ten Kate, O.M.; Hooning, K.M.; van der Kolk, E. Quantifying self-absorption losses in luminescent solar concentrators. *Appl. Opt.* **2014**, *53*, 5238–5245. [CrossRef] [PubMed]
10. Batchelder, J.S.; Zewail, A.H.; Cole, T. Luminescent solar concentrators. 1: Theory of operation and techniques for performance evaluation. *Appl. Opt.* **1979**, *18*, 3090–3110. [CrossRef]
11. McDowall, S.; Butler, T.; Bain, E.; Scharnhorst, K.; Patrick, D. Comprehensive analysis of escape-cone losses from luminescent waveguides. *Appl. Opt.* **2013**, *52*, 1230–1239. [CrossRef]
12. Zhang, B.; Gao, C.; Soleimaninejad, H.; White, J.M.; Smith, T.A.; Jones, D.J.; Ghiggino, K.P.; Wong, W.W.H. Highly efficient luminescent solar concentrators by selective alignment of donor–emitter fluorophores. *Chem. Mater.* **2019**, *31*, 3001–3008. [CrossRef]
13. Li, M.; Chen, J.-S.; Cotlet, M. Efficient light harvesting biotic–abiotic nanohybrid system incorporating atomically thin van der Waals transition metal dichalcogenides. *ACS Photonics* **2019**, *6*, 1451–1457. [CrossRef]
14. Li, M.; Chen, J.-S.; Cotlet, M. Light-induced interfacial phenomena in atomically thin 2D van der Waals material hybrids and heterojunctions. *ACS Energy Lett.* **2019**, *4*, 2323–2335. [CrossRef]
15. Mazzaro, R.; Vomiero, A. The renaissance of luminescent solar concentrators: The role of inorganic nanomaterials. *Adv. Energy Mater.* **2018**, *8*, 1801903. [CrossRef]
16. Im, S.W.; Ha, H.; Yang, W.; Jang, J.H.; Kang, B.; Seo, D.H.; Seo, J.; Nam, K.T. Light polarization dependency existing in the biological photosystem and possible implications for artificial antenna systems. *Photosynth. Res.* **2020**, *143*, 205–220. [CrossRef] [PubMed]

17. Lee, H.; Sriramdas, R.; Kumar, P.; Sanghadasa, M.; Kang, M.G.; Priya, S. Maximizing power generation from ambient stray magnetic fields around smart infrastructures enabling self-powered wireless devices. *Energy Environ. Sci.* **2020**, *13*, 1462–1472. [CrossRef]
18. Kumari, B.; Singh, A.; Jana, P.; Radhakrishna, M.; Kanvah, S. White light emission in water through admixtures of donor–π–acceptor siblings: Experiment and simulation. *New J. Chem.* **2019**, *43*, 11701–11709. [CrossRef]
19. Yuan, Z.; Wang, Z.; Guan, P.; Wu, X.; Chen, Y.-C. Lasing-encoded microsensor driven by interfacial cavity resonance energy transfer. *Adv. Opt. Mater.* **2020**, *8*, 1901596. [CrossRef]
20. Scholes, G.D.; Fleming, G.R.; Olaya-Castro, A.; van Grondelle, R. Lessons from nature about solar light harvesting. *Nat. Chem.* **2011**, *3*, 763–774. [CrossRef]
21. Verbunt, P.P.C.; Kaiser, A.; Hermans, K.; Bastiaansen, C.W.M.; Broer, D.J.; Debije, M.G. Controlling Light Emission in Luminescent Solar Concentrators Through Use of Dye Molecules Aligned in a Planar Manner by Liquid Crystals. *Adv. Funct. Mater.* **2009**, *19*, 2714–2719. [CrossRef]
22. Minkowski, C.; Calzaferri, G. Förster-Type Energy Transfer along a Specified Axis. *Angew. Chem.* **2005**, *117*, 5459–5463. [CrossRef]
23. Pieper, A.; Hohgardt, M.; Willich, M.; Gacek, D.A.; Hafi, N.; Pfennig, D.; Albrecht, A.; Walla, P.J. Biomimetic light-harvesting funnels for re-directioning of diffuse light. *Nat. Commun.* **2018**, *9*, 666. [CrossRef] [PubMed]
24. Willich, M.M.; Wegener, L.; Vornweg, J.; Hohgardt, M.; Nowak, J.; Wolter, M.; Jacob, C.R.; Walla, P.J. A new ultrafast energy funneling material harvests three times more diffusive solar energy for GaInP photovoltaics. *Proc. Natl. Acad. Sci. USA* **2020**, *117*, 32929–32938. [CrossRef]
25. Wilson, L.; Richards, B. Measurement method for photoluminescent quantum yields of fluorescent organic dyes in polymethyl methacrylate for luminescent solar concentrators. *Appl. Opt.* **2009**, *48*, 212–220. [CrossRef] [PubMed]
26. Albrecht, A.; Pfennig, D.; Nowak, J.; Grunwald, M.; Walla, P.J. On the efficiency limits of artificial and ultrafast light-funnels. *Nano Sel.* **2020**, *1*, 525–538. [CrossRef]
27. Law, K.-Y. Squaraine Chemistry. Effects of Structural Changes on the Absorption and Multiple Fluorescence Emission of Bis[4-(dimethylamino)phenyl]squaraine and Its Derivates. *J. Phys. Chem.* **1987**, *91*, 5184–5193. [CrossRef]
28. Sens, R.; Drexhage, K.H. Fluorescence quantum yield of oxazine and carbazine laser dyes. *J. Lumin.* **1918**, *24–25*, 709–712. [CrossRef]
29. Green, A.P.; Buckley, A.R. Solid state concentration quenching of organic fluorophores in PMMA. *Phys. Chem. Chem. Phys.* **2015**, *17*, 1435–1440. [CrossRef]
30. Walla, P.J. *Modern Biophysical Chemistry Detection and Analysis of Biomolecules*, 2nd ed.; Wiley-VCH Verlag GmbH & Co. KGaA: Weinheim, Germany, 2014.
31. Thuillier, G.; Hersé, M.; Foujols, T.; Peetermans, W.; Gillotay, D.; Simon, P.C.; Mandel, H. The solar spectral irradiance from 200 to 2400 nm as measured by the solspec spectrometer from the atlas and eureka missions. *Solar Phys.* **2003**, *214*, 1–22. [CrossRef]
32. Honsberg, C.B.; Bowden, S.G. Photovoltaics Education Website. 2019. Available online: www.pveducation.org (accessed on 21 December 2021).

Article

Thin-Film Luminescent Solar Concentrator Based on Intramolecular Charge Transfer Fluorophore and Effect of Polymer Matrix on Device Efficiency

Fahad Mateen [1], Namcheol Lee [1], Sae Youn Lee [2,*], Syed Taj Ud Din [3], Woochul Yang [3], Asif Shahzad [2], Ashok Kumar Kaliamurthy [4], Jae-Joon Lee [4] and Sung-Kyu Hong [1,*]

[1] Department of Chemical and Biochemical Engineering, Dongguk University, Seoul 04620, Korea; fahadmateen@dongguk.edu (F.M.); lnc9428@naver.com (N.L.)
[2] Department of Energy and Material Engineering, Dongguk University, Seoul 04620, Korea; asifshzd8@dongguk.edu
[3] Department of Physics, Dongguk University, Seoul 04620, Korea; tajuddins.phy@gmail.com (S.T.U.D.); wyang@dongguk.edu (W.Y.)
[4] Research Center for Photoenergy Harvesting & Conversion Technology (phct), Department of Energy and Material Engineering, Dongguk University, Seoul 04620, Korea; ashoksjc88@gmail.com (A.K.K.); jjlee@dongguk.edu (J.-J.L.)
* Correspondence: saeyounlee@dongguk.edu (S.Y.L.); hsk5457@dongguk.edu (S.-K.H.)

Citation: Mateen, F.; Lee, N.; Lee, S.Y.; Taj Ud Din, S.; Yang, W.; Shahzad, A.; Kaliamurthy, A.K.; Lee, J.-J.; Hong, S.-K. Thin-Film Luminescent Solar Concentrator Based on Intramolecular Charge Transfer Fluorophore and Effect of Polymer Matrix on Device Efficiency. *Polymers* **2021**, *13*, 3770. https://doi.org/10.3390/polym13213770

Academic Editors: Bożena Jarząbek and Muhammad Salahuddin Khan

Received: 23 September 2021
Accepted: 27 October 2021
Published: 31 October 2021

Publisher's Note: MDPI stays neutral with regard to jurisdictional claims in published maps and institutional affiliations.

Copyright: © 2021 by the authors. Licensee MDPI, Basel, Switzerland. This article is an open access article distributed under the terms and conditions of the Creative Commons Attribution (CC BY) license (https://creativecommons.org/licenses/by/4.0/).

Abstract: Luminescent solar concentrators (LSCs) provide a transformative approach to integrating photovoltaics into a built environment. In this paper, we report thin-film LSCs composed of intramolecular charge transfer fluorophore (DACT-II) and discuss the effect of two polymers, polymethyl methacrylate (PMMA), and poly (benzyl methacrylate) (PBzMA) on the performance of large-area LSCs. As observed experimentally, DACT-II with the charge-donating diphenylaminocarbazole and charge-accepting triphenyltriazine moieties shows a large Stokes shift and limited reabsorption losses in both polymers. Our results show that thin-film LSC (10 × 10 × 0.3 cm^3) with optimized concentration (0.9 wt%) of DACT-II in PBzMA gives better performance than that in the PMMA matrix. In particular, optical conversion efficiency (η_{opt}) and power-conversion efficiency (η_{PCE}) of DACT-II/PBzMA LSC are 2.32% and 0.33%, respectively, almost 1.2 times higher than for DACT-II/PMMA LSC.

Keywords: luminescent solar concentrator; polymer matrix; organic fluorophore; intramolecular charge transfer; light harvesting

1. Introduction

Due to rapid urbanization, a considerable increase in global energy consumption has been observed over the past several decades. Currently, buildings utilize around 30% of energy worldwide, due to cooling, heating, and artificial-lighting loads [1,2]. To meet this huge energy demand, substantial attention has been paid to clean and renewable energy technologies, especially grid-free building-integrated photovoltaics (BIPVs) [3]. Among many BIPVs, luminescent solar concentrators (LSCs) offer a cost-effective solution to harness solar energy, while warranting their compatibility with the existing and new infrastructures [4]. Typically, LSCs are fabricated in two simple architectures, namely bulk and thin-film LSC. In the case of bulk LSC, light-emissive fluorophores are embedded in the optically transparent slab of polymer, while a thin-film LSC consists of fluorophores mixed with the polymer matrix to form a thin film on the haze-free glass. In both cases, fluorophores absorb incident sunlight and re-emit it at longer wavelengths. The re-emitted photons are trapped within the polymer slab or glass substrate due to the total internal reflection (TIR) process and are directed to its edges, where they are transformed into electricity by attached PV cells (Figure 1) [5,6]. Recently, it has been suggested that LSCs' application is not just limited to BIPVs, but they can be applied to various other platforms, such as greenhouses [7], noise barriers [8], indoor decorative elements [9,10], medical devices [7,11], indoor light-harvesting glass [12], and sunroofs of vehicles [13].

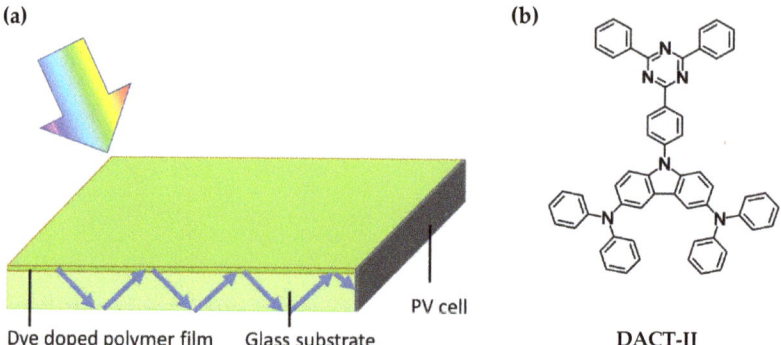

Figure 1. (**a**) Schematic representation of thin-film LSC. (**b**) Chemical structure of DACT-II employed in our study.

Dating back to the 1970s, LSCs were first introduced as an inexpensive alternative to traditional photovoltaics [14]. However, the recent LSCs still offer stability issues [15,16] and reduced efficiency mainly due to re-absorption losses that occur due to the small Stokes shift of fluorophores [17,18], limited fluorophore–polymer compatibility [19], and low photoluminescent quantum yield (PLQY) of fluorophores [20]. Major research efforts devoted to LSCs include (1) the use of suitable fluorophores, such as organic dyes, e.g., coumarins [21], perylenes [22], aggregation induce emissive molecules [23–25], π-conjugated polymers [26,27], rare earth complexes [28], and semiconducting quantum dots QDs (core/shell, carbon, and silicon) [29–32]; (2) different designs, e.g., plasmonic LSCs [33,34], fibers structures [35,36], and multi-layer LSCs [36–38]; and (3) identifying the appropriate host polymer. The most reported polymer for LSCs is polymethyl methacrylate (PMMA), while other examples include crosslinked fluoro-polymers [39], polycyclic hexyl methacrylate [40], polysiloxanes [41], L-poly(lactic acid) [42], fluorescent proteins [43], cellulose crystals [44], and unsaturated polyesters [45].

In recent work by our group [46,47], bulk PMMA LSCs were fabricated by utilizing thermally activated delayed fluorescence (TADF) dye, 1,2,3,5-tetrakis(carbazol-9-yl)-4,6-dicyanobenzene (4CzIPN) [48]. Moreover, 4CzIPN shows the intramolecular charge transfer (ICT) features between the carbazole and dicyanobenzene moieties that lead to a drastic increase in the Stokes shift. Reduced re-absorption losses, high photostability due to strong steric hindrance, and high PLQY of 4CzIPN make it an excellent candidate for the large-area LSCs.

In this study, we investigated the effect of the host polymer matrix on the performance of LSC incorporating intramolecular charge transfer fluorophore. We fabricated large-area thin-film LSCs ($10 \times 10 \times 0.3$ cm^3) based on another TADF dye, 9-[4-(4,6-diphenyl-1,3,5-triazin-2-yl)phenyl]-N,N,N′,N′-tetraphenyl-9H-carbazole-3,6-diamine, denoted as DACT-II (Figure 1b) [49]. DACT-II consists of electron donor diphenylaminocarbazole and electron-acceptor triphenyltriazine moieties and exhibits ICT characteristics. In particular, PMMA and poly (benzyl methacrylate) (PBzMA) were investigated as host polymer matrices for DACT-II-based thin-film LSCs. Besides synthesis of DACT-II, we report the optical properties and photovoltaic performance of DACT-II-based thin-film LSCs employing PMMA and PBzMA matrices. Our results suggest that the DACT-II-based thin-film LSC with the PBzMA matrix shows an optical efficiency of 2.32%, which is 1.2 times higher than that with the PMMA matrix.

2. Materials and Methods

2.1. Materials

For the synthesis of DACT-II, all reagents were acquired from Tokyo Chemical Industry (TCI) and Sigma-Aldrich. For the fabrication of LSCs, PMMA and PBzMA were purchased from Sigma-Aldrich.

2.2. Synthesis

2.2.1. Synthesis of 3,6-dibromo-9-(4-(4,6-diphenyl-1,3,5-triazin-2-yl)phenyl)-9H-carbazole (1)

First, 2-(4-bromophenyl)-4,6-diphenyl-1,3,5-triazine (0.4 g, 1.0 mmol), 3,6-dibromo-9H-carbazole (0.33 g, 1.0 mmol), bis(tri-tert-butylphosphine)palladium(0) (0.026 g, 0.05 mmol) and sodium tert-butoxide (0.25 g, 2.6 mmol) were dissolved in anhydrous toluene (13 mL) under a nitrogen atmosphere. The mixture was refluxed for 4 h. After cooling down to room temperature, the solution was poured into chloroform and distilled water for extraction. The chloroform layer was washed with distilled water several times and dried over magnesium sulfate. The crude product was filtered by using Celite 545 and purified via column chromatography on silica gel (eluent:dichloromethane/hexane, 1:4, v/v). The product was dried in a vacuum oven to give a white powder (yield = 0.10 g, 10%). ^{1}H NMR (500 MHz, CDCl$_3$): δ 9.04 (d, J = 8.5 Hz, 2 H), 8.83 (d, J = 6.5 Hz, 4 H), 8.24 (s, J = 2.0 Hz, 2 H), 7.76 (d, J = 8.5 Hz, 2 H), 7.67–7.55 (m, 6 H), 7.57 (dd, J = 9.0 Hz, 2.0 Hz, 2 H), 7.41 (d, J = 9.0 Hz, 2 H).

2.2.2. Synthesis of DACT-II

First, (0.10 g, 0.16 mmol), diphenylamine (0.06 g, 0.35 mmol), tris(dibenzylideneacetone)-dipalladium(0)-chloroform adduct (0.004 g, 0.004 mmol), 2-dicyclohexylphosphino-2′,4′,6′-triisopropyl-biphenyl (0.01 g, 0.016 mmol) and sodium *tert*-butoxide (0.037 g, 0.384 mmol) were dissolved in anhydrous toluene (5 mL) under a nitrogen atmosphere. The mixture was refluxed for 12 h. After cooling down to room temperature, the solution was poured into chloroform and distilled water for extraction. The chloroform layer was washed with distilled water several times and dried over magnesium sulfate. The crude product was filtered by using Celite 545 and purified via column chromatography on silica gel (eluent:dichloromethane/hexane, 1:2.5, v/v). The product was dried in a vacuum oven to give a yellow powder (yield = 0.09 g, 69%). ^{1}H NMR (500 MHz, DMSO-d$_6$): δ 9.03 (d, J = 8.5 Hz, 2 H), 8.80 (d, J = 7.0 Hz, 4 H), 8.05 (s, J = 2.5 Hz, 2 H), 8.00 (d, J = 8.5 Hz, 2 H), 7.76 (t, J = 7.0 Hz, 2 H), 7.71 (t, J = 7.5 Hz, 4 H), 7.60 (d, J = 9.0 Hz, 2 H), 7.27 (t, J = 8.5 Hz, 10 H), 7.00 (d, J = 7.5 Hz, 8 H), 6.96 (t, J = 7.5 Hz, 4 H).

2.3. Fabrication of Thin-Film LSCs

To fabricate DACT-II-based thin-film LSCs with PMMA matrix, 10 wt% solutions of PMMA in chloroform were prepared. The solution was then blended with various concentrations of as-synthesized DACT-II (0.1–1.3 wt%). After proper mixing of final solutions, the doctor-blade coating technique was used to make thin film (~60 µm film thickness) on transparent glass substrates of different sizes. Chloroform was slowly evaporated by keeping the sample under room conditions. The same procedure was applied to fabricate DACT-II-based thin-film LSCs with PBzMA matrix.

2.4. Measurements

UV–visible spectrophotometer (Perkin Elmer Lambda 35) and fluorescence spectrophotometer (JASCO, FP-8600) were used to obtain absorbance and emission of the samples. To obtain the spectra of edge emitted photons, an integrating sphere connected to a spectrometer (Avantes, ULS2048) was employed. For PV measurements, crystalline silicon (c-Si) PV cells were purchased locally. Highly transparent adhesive (United Adhesives, OE 1582) was used to attach the PV cells with all edges of the fabricated LSCs (10 × 10 × 0.3 cm^3). Current–voltage measurements were obtained by illuminating the surface of LSCs with a solar simulator (Mc-Science) having a Xenon arc lamp of 160 W equipped with filters to approximate AM 1.5 G spectrum. The irradiance of the illumination source was calibrated before and found to be 100 mW cm^{-2}.

3. Results and Discussion

The synthesis of DACT-II was performed by using a two-step approach (Scheme 1). In detail, Compound 1 was synthesized by 3,6-dibromo-9H-carbazole, bis(tri-tert-butylphosphine) palladium(0) and sodium tert-butoxide under inert environment. Then, diphenylamine, tris(dibenzylideneacetone)dipalladium (0)-chloroform adduct, 2-dicyclohexylphosphino-2′,4′,6′-triisopropyl-biphenyl, and sodium tert-butoxide were reacted in anhydrous toluene under a nitrogen atmosphere, delivering DACT-II in 69% yield after crystallization. The obtained DACT-II was consisted of chemically bonded diphenylaminocarbazole (charge donor) and triphenyltriazine (charge acceptor) moieties.

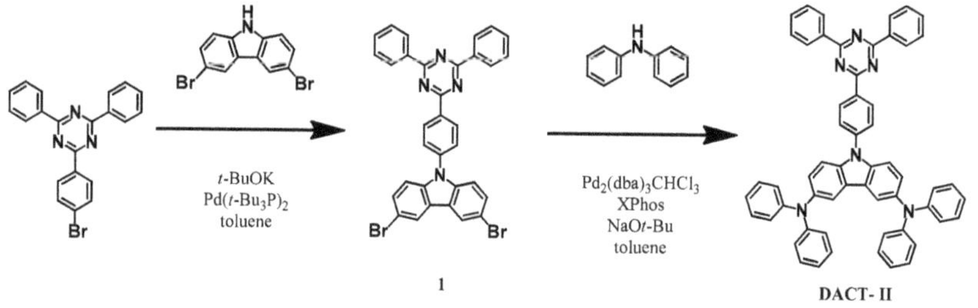

Scheme 1. Synthesis of DACT-II.

The optical properties of synthesized DACT-II were investigated in PMMA and PBzMA matrices. PBzMA is highly transparent and amorphous that makes it an excellent alternative to the commonly employed PMMA matrix in LSCs. Normalized absorbance and emission spectra of DACT-II in PMMA and PBzMA are displayed in Figure 2. The absorbance range covered the entire ultraviolet (UV) and near UV region, i.e., from 300 to 450 nm. As observed, absorbance is low in the 350–450 nm range, however, this issue can be solved by using a higher concentration of DACT-II in thin-film LSCs. As evident from Figure 2, the absorbance of DACT-II was nearly the same in both polymer matrices. DACT-II exhibited a broad emission with the peak values at 490 and 507 nm in PBzMA and PMMA films, respectively. The expected blue shift in the case of PBzMA is due to the modest polarity of the lateral benzyl group compared to the methyl ester substitution in PMMA. Stokes shift is an important factor in designing an efficient LSC device. Figure 2 also confirms that DACT-II exhibited a large Stokes shift, i.e., less overlap between absorbance and emission spectra in both polymers. Such a large Stokes shift limits the re-absorption losses, even at higher concentrations of DACT-II; thus, it helps improve the LSC efficiency.

To obtain the optimum concentration of DACT-II in PMMA and PBzMA, thin-film LSCs (5 × 2.5 × 0.1 cm^3) employing different concentrations, ranging from 0.1 to 1.3 wt%, were fabricated as explained in the experimental section. In Figure 3a, the effect of DACT-II concentration in PMMA film on the emission intensity is reported. For 0.1–0.9 wt% DACT-II loading, a gradual increase in the emission intensity was detected at the excitation wavelength of 350 nm. However, a further rise in the concentration caused a decrease in emission which could be associated with the formation of the aggregate. These aggregates offer sites for non-radiative relaxations of excited state electrons, leading to emission reduction [50]. The limited solubility of DCAT-II in the PMMA matrix is another factor that leads to lowered emission at higher concentrations of DACT-II. Moreover, these spectra were also characterized by a negligible red shift, suggesting modest re-absorption losses. Identical to what was observed for DACT-II/PMMA films, emission intensity (for DACT-II/PBzMA films) improved linearly with DACT-II concentrations up to 0.9 wt% (Figure 3b). Beyond 0.9 wt%, a decrease in the DACT-II emission, along with the progressive red shift, was observed. The comparison of DACT-II in both polymer matrices shows that emission intensity remained higher in PBzMA than PMMA for all the concentrations, making PBzMA a superior alternative to the commonly used PMMA matrix. Our experimental investigations can be justified by the fact that the reduced polarity of PBzMA helps create not only a better dispersion of the DACT-II but also improves the radiative decay channels. To deeply understand the emission mechanism of the DACT-II in PMMA and PBzMA matrices, time-resolved photoluminescent measurements were performed. A single-exponential decay model was employed to fit the photoluminescence decay curve and calculate the excited-state lifetimes. The emission of DACT-II in PMMA at 507 nm decayed with an average excited-state lifetime of 8.6 ns. On the other hand, the excited state lifetime of DACT-II in PBzMA at 490 nm was 9.6 ns (Figure 3c). The decreased excited-state lifetime in the case of the PMMA matrix indicates the possible formation of alternative non-radiative decay channels.

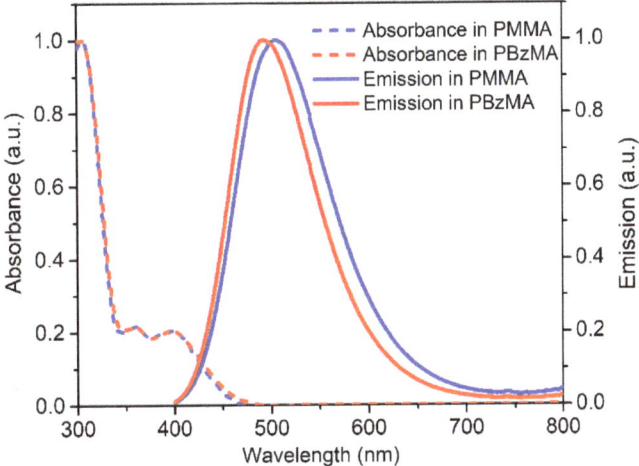

Figure 2. Normalized absorbance and emission spectra of DACT-II in PMMA and PBzMA.

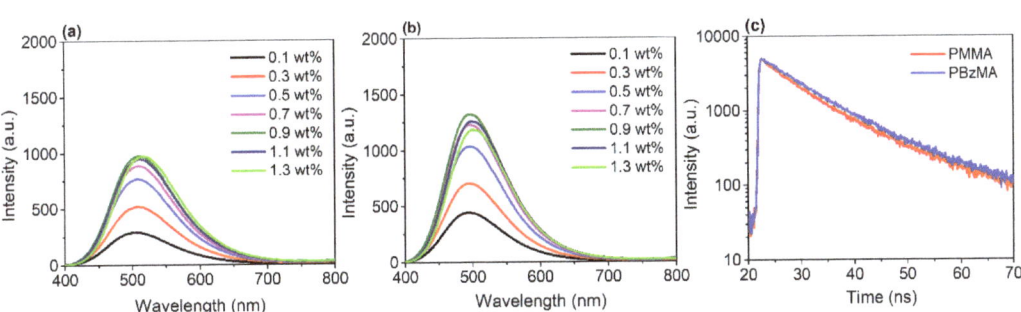

Figure 3. Emission spectra of DACT-II in (a) PMMA and (b) PBzMA films at the excitation wavelength of 350 nm. (c) Time-resolved photoluminescence spectra of DACT-II in PMMA and PBzMA films.

To investigate the effect of LSC size on the edge emission, we prepared the square-dimensioned thin-film LSCs (DCAT-II concentration 0.9 wt%) with different lengths, and edge emitted photons were obtained by using the integrating sphere method. Figure 4a shows the edge emitted photons spectra of different sized DACT-II/PMMA-film LSCs. The number of edges emitted photons increased linearly with the lengths, which is obvious because, when the size of LSC increases, the total number of incident photons will increase. The same trend was observed in the case of DACT-II based LSC with PBzMA matrix

(Figure 4b). Notably, for all the lengths, the total number of photons emitted by the edges of DACT-II-based LSC with PBzMA matrix remained higher than that of the device with PMMA matrix. This trend is consistent with the front-facing emission measurements (Figure 3a,b). To our surprise, a red shift was observed when the size of the LSCs was increased from 2.5 to 15 cm. Peak wavelengths of the edge emission spectra are also presented in Figure 4c. For 2.5 cm length, the peak emission wavelength was 509 and 498 nm for LSCs with PMMA and PBzMA matrices, respectively. Meanwhile, the values changed to 517 and 507 nm for 15 cm–long respective devices. Generally, increment in the size of LSC is accompanied by the escape cone losses, reabsorption losses, and red-shifted edge emissions. The same phenomena have been also noted for LSCs with various designs and using other fluorophores [51,52].

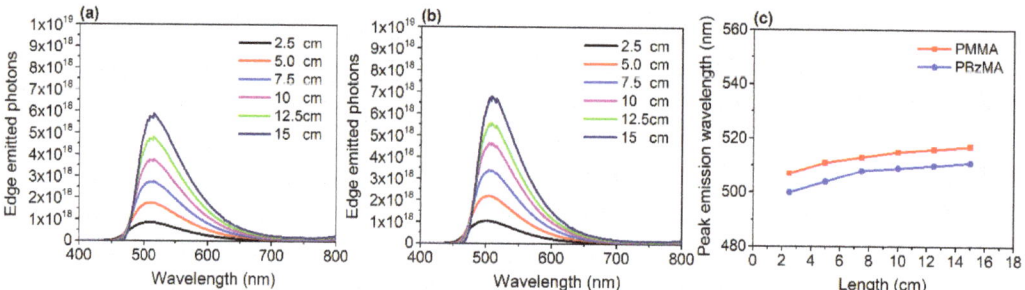

Figure 4. (a) Edge emitted photons spectra of DACT-II/PMMA-film LSCs at various lengths. (b) Edge emitted photons spectra of DACT-II/PBzMA-film LSCs at various lengths. (c) Peak emission wavelengths of LSCs with different lengths. All the LSCs were square-shaped so that length = width.

The potential of DACT-II-based thin-film LSCs as power-producing windows was determined by obtaining optical-conversion efficiency (η_{opt}) and power-conversion efficiency (η_{PCE}) of the large-area LSCs (dimension: $10 \times 10 \times 0.3$ cm^3) having 0.9 wt% of DACT-II in PMMA and PBzMA matrices. Moreover, η_{opt} is described as the ratio of LSC edge emitted photons to the total incident photons, while η_{PCE} is the ratio of the electrical output to the solar power input. The formula of η_{opt} and η_{PCE} is given in Equations (1) and (2), respectively.

$$\eta_{opt} = \frac{I_{LSC} \times A_{Edges}}{I_{PV\,cell} \times A_{LSC}} \qquad (1)$$

$$\eta_{PCE} = \frac{I_{LSC} \times V_{OC} \times FF}{A_{LSC} \times F_{IN}} \qquad (2)$$

where I_{LSC} (mA) and $I_{PV\,cell}$ (mA) are short-circuit current obtained by LSC connected PV cell and short-circuit current of bare PV cell(without LSCs attached). A_{Edges} (cm^2) and A_{LSC} (cm^2) are the area of LSC edges where PV cells are attached and surface area of LSC. While in Equation (2), V_{OC} (V), FF, F_{IN} (mWcm^{-2}) are the open-circuit voltage, fill factor, and the incident solar power density, respectively. It is important to note that PV cell was connected to only one edge while other three edges were masked, and overall LSC was then corrected by multi-plying the current density by 4. Current density–voltage (J–V) curves taken by the DACT-II-based LSC with PMMA and PBzMA matrices are depicted in Figure 5, while the values of other PV parameters and values of η_{opt} and η_{PCE} are listed in Table 1. It is evident from the results that DACT-II-based LSC with PBzMA matrix outperformed and gave the η_{opt} and η_{PCE} of 2.32 and 0.33%, respectively. These values are 1.2 times higher than the LSC with the PMMA matrix.

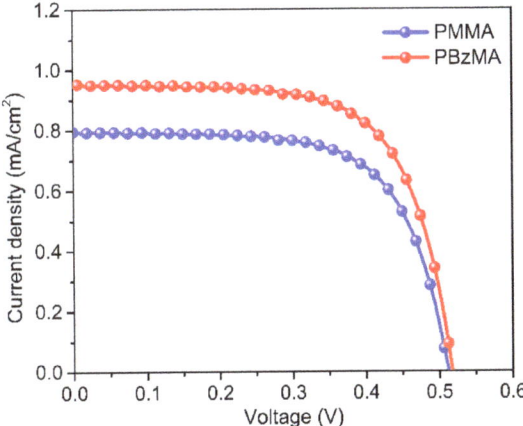

Figure 5. J–V curves of DACT-II based LSC with different polymer matrices.

Table 1. Photovoltaic parameters of DACT-II-based LSC (10 × 10 × 0.3 cm^3) with different polymer matrices.

Samples	Voc (V)	Isc (mA)	FF (%)	η_{opt} (%)	η_{PCE} (%)
DCAT-II/PMMA-based LSC	0.51	79.44	66.57	1.92	0.28
DCAT-II/PBzMA-based LSC	0.51	97.10	66.70	2.32	0.33

Additionally, an analytical model (Equation (3)) [53] was used to estimate the optical efficiency (η_{opt}) of large-area LSCs (up to 10,000 cm^2, for Length = 100 cm) utilizing DACT-II in PMMA and PBzMA matrices.

$$\eta_{opt} = (1-R)\, \eta_{abs} \cdot \eta_{int} \qquad (3)$$

where R denotes the reflection losses, which are approximately 4% in the case of polymers with a refractive index of 1.5. Note that PMMA and PBzMA show same refractive index. Moreover, η_{abs} and η_{int} are the absorption efficiency (Equation (4)) and internal quantum efficiency (Equation (5)), respectively.

$$\eta_{abs} = \frac{\int_{280}^{1100} P_{in}(\lambda)\left[1 - e^{-\alpha(\lambda)t}\right] d\lambda}{\int_{280}^{1100} P_{in}(\lambda) d\lambda} \qquad (4)$$

$$\eta_{int} = \frac{\int_0^\infty \frac{\eta_{QY}\, \eta_{trap}}{1+\beta\alpha(\lambda)\frac{L}{D}(1-\eta_{QY}\, \eta_{trap})}\, I_{PL}(\lambda) d\lambda}{\int_0^\infty I_{PL}(\lambda) d\lambda} \qquad (5)$$

In Equation (4), α is the absorption coefficient of DACT-II in polymeric films, t is the thickness of the film and P_{in} is the incident photon flux. In Equation (5), η_{QY} is the PLQY of DACT-II (49 and 56% in PMMA and PBzMA, respectively); η_{trap} is a light-trapping efficiency, which is around 75% for a given system; β is a numerical factor and is equal to 1.4 [37]; I_{PL} is an emission intensity; and D and L represent the thickness and length of a whole LSC device. Moreover, LSCs are assumed to be square such that the length of the LSC is equal to the width. As shown in Figure 6, η_{opt} drops with the increasing length of LSCs with PMMA and PBzMA matrices. In the case of DACT-II-based LSC with PMMA matrix, the calculated η_{opt} was 2.25 and 1.85% for the length 2.5 and 100 cm, respectively. When the PBzMA matrix was employed, η_{opt} soared to 2.64% for 2.5 cm and 2.05% for the 100 cm long LSCs. The calculated η_{opt} was based on the emissions from all edges of the LSC device as shown in the insert of Figure 6. A drop in η_{opt} is expected, since the re-emitted photons are more susceptible to optical losses at higher lengths of LSCs, as can be seen in other studies [25,29]. Although the overall η_{opt} of our fabricated LSCs is low, which is due to the low absorption range (300–450 nm), our results confirm that PBzMA can be applied as a potential alternative to commonly employed PMMA matrix for most of the organic fluorophores based LSCs.

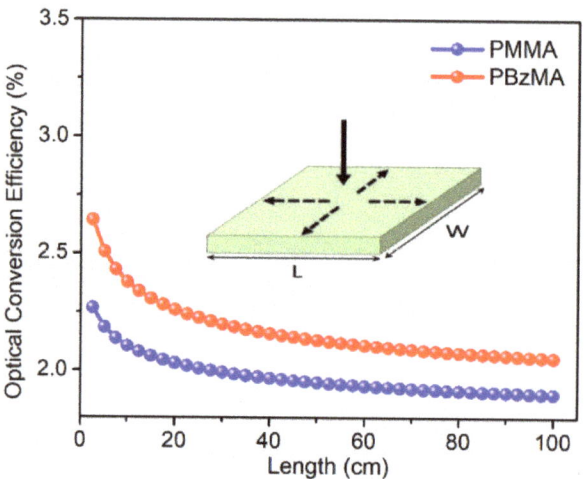

Figure 6. Calculated optical conversion efficiency of DACT-II-based LSCs with PMMA and PBzMA matrices. For all η_{opt} values, the LSCs were assumed to be in square shape (length = width).

4. Conclusions

In summary, we demonstrated the effect of polymer matrices on the DACT-II-based LSCs. First, we synthesized the DACT-II, a TADF dye with intramolecular charge transfer characteristics, and blended it with PMMA and PBzMA to make large-area thin-film LSCs. At the optimized concentration (0.9 wt%), DACT-II-based LSC with PBzMA matrix showed 2.32 and 0.33% of η_{opt} and η_{PCE}, respectively. Conversely, the η_{opt} and η_{PCE} of the device with PMMA matrix were 1.92 and 0.28%, respectively. Better efficiencies in the case of PBzMA are attributed to more efficient dispersion of the DACT-II in PBzMA, which makes PBzMA a better choice for the fabrication of thin-film LSCs.

Author Contributions: F.M., conceptualization, investigation, writing—original draft, validation and methodology; N.L., data curation; S.Y.L., conceptualization, validation and resources; S.T.U.D., methodology and data curation; W.Y., resources; A.S., data curation; A.K.K., data curation; J.-J.L., resources; S.-K.H., project administration, funding acquisition and supervision. All authors have read and agreed to the published version of the manuscript.

Funding: This research was supported by the National Research Foundation of Korea (NRF) grant funded by the Korea government (MSIT) (No. 2019R1A2C1005805) and (2020R1F1A1065891), the Korea institute of energy technology evaluation and planning (KETEP) and the Ministry of Trade, Industry and Energy (MOTIE) of the Korea (20194030202320), and Brain Pool Program through NRF funded by the MSIT, Korea (2019H1D3A1A01071183).

Institutional Review Board Statement: Not applicable.

Informed Consent Statement: Not applicable.

Data Availability Statement: Not applicable.

Conflicts of Interest: The authors declare no conflict of interest.

References

1. Ghosh, A. Potential of building integrated and attached/applied photovoltaic (BIPV/BAPV) for adaptive less energy-hungry building's skin: A comprehensive Review. *J. Clean. Prod.* **2020**, *276*, 123343. [CrossRef]
2. Saeed, M.A.; Kim, S.H.; Baek, K.; Hyun, J.K.; Lee, S.Y.; Shim, J.W. PEDOT: PSS: CuNW-based transparent composite electrodes for high-performance and flexible organic photovoltaics under indoor lighting. *Appl. Surf. Sci.* **2021**, *567*, 150852. [CrossRef]
3. Lee, J.-H.; You, Y.-J.; Saeed, M.A.; Kim, S.H.; Choi, S.-H.; Kim, S.; Lee, S.Y.; Park, J.-S.; Shim, J.W. Undoped tin dioxide transparent electrodes for efficient and cost-effective indoor organic photovoltaics (SnO_2 electrode for indoor organic photovoltaics). *NPG Asia Mater.* **2021**, *13*, 1–10. [CrossRef]
4. Bergren, M.R.; Makarov, N.S.; Ramasamy, K.; Jackson, A.; Guglielmetti, R.; McDaniel, H. High-performance $CuInS_2$ quantum dot laminated glass luminescent solar concentrators for windows. *ACS Energy Lett.* **2018**, *3*, 520–525. [CrossRef]
5. Mateen, F.; Saeed, M.A.; Shim, J.W.; Hong, S.-K. Indoor/outdoor light-harvesting by coupling low-cost organic solar cell with a luminescent solar concentrator. *Sol. Energy* **2020**, *207*, 379–387. [CrossRef]
6. Roncali, J. Luminescent Solar Collectors: Quo Vadis? *Adv. Energy Mater.* **2020**, *10*, 2001907. [CrossRef]
7. Makarov, N.S.; Ramasamy, K.; Jackson, A.; Velarde, A.; Castaneda, C.; Archuleta, N.; Hebert, D.; Bergren, M.R.; McDaniel, H. Fiber-coupled luminescent concentrators for medical diagnostics, agriculture, and telecommunications. *ACS Nano* **2019**, *13*, 9112–9121. [CrossRef]
8. Kanellis, M.; de Jong, M.M.; Slooff, L.; Debije, M.G. The solar noise barrier project: 1. Effect of incident light orientation on the performance of a large-scale luminescent solar concentrator noise barrier. *Renew. Energy* **2017**, *103*, 647–652. [CrossRef]
9. Reinders, A.; Kishore, R.; Slooff, L.; Eggink, W. Luminescent solar concentrator photovoltaic designs. *Jpn. J. Appl. Phys.* **2018**, *57*, 08RD10. [CrossRef]
10. ter Schiphorst, J.; Cheng, M.L.; van der Heijden, M.; Hageman, R.L.; Bugg, E.L.; Wagenaar, T.J.; Debije, M.G. Printed luminescent solar concentrators: Artistic renewable energy. *Energy Build.* **2020**, *207*, 109625. [CrossRef]
11. Papakonstantinou, I.; Portnoi, M.; Debije, M.G. The hidden potential of luminescent solar concentrators. *Adv. Energy Mater.* **2021**, *11*, 2002883. [CrossRef]
12. Li, Y.; Sun, Y.; Zhang, Y. Luminescent solar concentrators performing under different light conditions. *Sol. Energy* **2019**, *188*, 1248–1255. [CrossRef]
13. Yang, C.; Liu, D.; Renny, A.; Kuttipillai, P.S.; Lunt, R.R. Integration of near-infrared harvesting transparent luminescent solar concentrators onto arbitrary surfaces. *J. Lumin.* **2019**, *210*, 239–246. [CrossRef]
14. Weber, W.; Lambe, J. Luminescent greenhouse collector for solar radiation. *Appl. Opt.* **1976**, *15*, 2299–2300. [CrossRef] [PubMed]
15. Corsini, F.; Tatsi, E.; Colombo, A.; Dragonetti, C.; Botta, C.; Turri, S.; Griffini, G. Highly emissive fluorescent silica-based core/shell nanoparticles for efficient and stable luminescent solar concentrators. *Nano Energy* **2021**, *80*, 105551. [CrossRef]
16. Griffini, G.; Brambilla, L.; Levi, M.; Del Zoppo, M.; Turri, S. Photo-degradation of a perylene-based organic luminescent solar concentrator: Molecular aspects and device implications. *Sol. Energy Mater. Sol. Cells* **2013**, *111*, 41–48. [CrossRef]
17. Tummeltshammer, C.; Taylor, A.; Kenyon, A.J.; Papakonstantinou, I. Losses in luminescent solar concentrators unveiled. *Sol. Energy Mater. Sol. Cells* **2015**, *144*, 40–47. [CrossRef]
18. Li, S.; Liu, H.; Chen, W.; Zhou, Z.; Wu, D.; Lu, R.; Zhao, B.; Hao, J.; Yang, L.; Yang, H. Low reabsorption and stability enhanced luminescent solar concentrators based on silica encapsulated quantum rods. *Sol. Energy Mater. Sol. Cells* **2020**, *206*, 110321. [CrossRef]
19. Li, Y.; Zhang, X.; Zhang, Y.; Dong, R.; Luscombe, C.K. Review on the Role of Polymers in Luminescent Solar Concentrators. *J. Polym. Sci. Part A-1 Polym. Chem.* **2018**, *57*, 201–215. [CrossRef]
20. Zhao, Y.; Meek, G.A.; Levine, B.G.; Lunt, R.R. Near-infrared harvesting transparent luminescent solar concentrators. *Adv. Opt. Mater.* **2014**, *2*, 606–611. [CrossRef]
21. Mateen, F.; Oh, H.; Jung, W.; Lee, S.Y.; Kikuchi, H.; Hong, S.-K. Polymer dispersed liquid crystal device with integrated luminescent solar concentrator. *Liq. Cryst.* **2018**, *45*, 498–506. [CrossRef]
22. Li, Y.; Olsen, J.; Nunez-Ortega, K.; Dong, W.-J. A structurally modified perylene dye for efficient luminescent solar concentrators. *Sol. Energy* **2016**, *136*, 668–674. [CrossRef]
23. Corsini, F.; Nitti, A.; Tatsi, E.; Mattioli, G.; Botta, C.; Pasini, D.; Griffini, G. Large-Area Semi-Transparent Luminescent Solar Concentrators Based on Large Stokes Shift Aggregation-Induced Fluorinated Emitters Obtained Through a Sustainable Synthetic Approach. *Adv. Opt. Mater.* **2021**, *9*, 2100182. [CrossRef]
24. Zhang, B.; Banal, J.L.; Jones, D.J.; Tang, B.Z.; Ghiggino, K.P.; Wong, W.W. Aggregation-induced emission-mediated spectral downconversion in luminescent solar concentrators. *Mater. Chem. Front.* **2018**, *2*, 615–619. [CrossRef]
25. Mateen, F.; Hwang, T.G.; Boesel, L.F.; Choi, W.J.; Kim, J.P.; Gong, X.; Park, J.M.; Hong, S.K. Luminescent solar concentrator utilizing energy transfer paired aggregation-induced emissive fluorophores. *Int. J. Energy Res.* **2021**, *45*, 17971–17981. [CrossRef]
26. Lyu, G.; Kendall, J.; Meazzini, I.; Preis, E.; Bayseç, S.; Scherf, U.; Clément, S.B.; Evans, R.C. Luminescent Solar Concentrators Based on Energy Transfer from an Aggregation-Induced Emitter Conjugated Polymer. *ACS Appl. Polym. Mater.* **2019**, *1*, 3039–3047. [CrossRef]
27. Li, Y.; Sun, Y.; Zhang, Y.; Li, Y.; Verduzco, R. High-performance hybrid luminescent-scattering solar concentrators based on a luminescent conjugated polymer. *Polym. Int.* **2021**, *70*, 475–482. [CrossRef]

28. Wang, T.; Zhang, J.; Ma, W.; Luo, Y.; Wang, L.; Hu, Z.; Wu, W.; Wang, X.; Zou, G.; Zhang, Q. Luminescent solar concentrator employing rare earth complex with zero self-absorption loss. *Sol. Energy* **2011**, *85*, 2571–2579. [CrossRef]
29. Zhao, H.; Benetti, D.; Tong, X.; Zhang, H.; Zhou, Y.; Liu, G.; Ma, D.; Sun, S.; Wang, Z.M.; Wang, Y. Efficient and stable tandem luminescent solar concentrators based on carbon dots and perovskite quantum dots. *Nano Energy* **2018**, *50*, 756–765. [CrossRef]
30. Mateen, F.; Ali, M.; Oh, H.; Hong, S.-K. Nitrogen-doped carbon quantum dot based luminescent solar concentrator coupled with polymer dispersed liquid crystal device for smart management of solar spectrum. *Sol. Energy* **2019**, *178*, 48–55. [CrossRef]
31. Li, H.; Wu, K.; Lim, J.; Song, H.-J.; Klimov, V.I. Doctor-blade deposition of quantum dots onto standard window glass for low-loss large-area luminescent solar concentrators. *Nat. Energy* **2016**, *1*, 16157. [CrossRef]
32. Liu, G.; Zhao, H.; Diao, F.; Ling, Z.; Wang, Y. Stable tandem luminescent solar concentrators based on CdSe/CdS quantum dots and carbon dots. *J. Mater. Chem. C* **2018**, *6*, 10059–10066. [CrossRef]
33. Mateen, F.; Oh, H.; Jung, W.; Binns, M.; Hong, S.-K. Metal nanoparticles based stack structured plasmonic luminescent solar concentrator. *Sol. Energy* **2017**, *155*, 934–941. [CrossRef]
34. Liu, X.; Benetti, D.; Rosei, F. Semi-transparent Luminescent Solar Concentrator based on Plasmonic-enhanced carbon dots. *J. Mater. Chem. A* **2021**, *9*, 23345–23352.
35. Banaei, E.-H.; Abouraddy, A.F. Fiber luminescent solar concentrator with 5.7% conversion efficiency. In Proceedings of the High and Low Concentrator Systems for Solar Electric Applications VIII, San Diego, CA, USA, 27–28 August 2013; p. 882102.
36. Mateen, F.; Ali, M.; Lee, S.Y.; Jeong, S.H.; Ko, M.J.; Hong, S.-K. Tandem structured luminescent solar concentrator based on inorganic carbon quantum dots and organic dyes. *Sol. Energy* **2019**, *190*, 488–494. [CrossRef]
37. Wu, K.; Li, H.; Klimov, V.I. Tandem luminescent solar concentrators based on engineered quantum dots. *Nat. Photonics* **2018**, *12*, 105–110. [CrossRef]
38. Wang, J.; Wang, J.; Xu, Y.; Jin, J.; Xiao, W.; Tan, D.; Li, J.; Mei, T.; Xue, L.; Wang, X. Controlled Synthesis of Long-Wavelength Multicolor-Emitting Carbon Dots for Highly Efficient Tandem Luminescent Solar Concentrators. *ACS Appl. Energy Mater.* **2020**, *3*, 12230–12237. [CrossRef]
39. Griffini, G.; Levi, M.; Turri, S. Novel crosslinked host matrices based on fluorinated polymers for long-term durability in thin-film luminescent solar concentrators. *Sol. Energy Mater. Sol. Cells* **2013**, *118*, 36–42. [CrossRef]
40. Ostos, F.J.; Iasilli, G.; Carlotti, M.; Pucci, A. High-Performance Luminescent Solar Concentrators Based on Poly (Cyclohexylmethacrylate)(PCHMA) Films. *Polymers* **2020**, *12*, 2898. [CrossRef]
41. Buffa, M.; Carturan, S.; Debije, M.; Quaranta, A.; Maggioni, G. Dye-doped polysiloxane rubbers for luminescent solar concentrator systems. *Sol. Energy Mater. Sol. Cells* **2012**, *103*, 114–118. [CrossRef]
42. Fattori, V.; Melucci, M.; Ferrante, L.; Zambianchi, M.; Manet, I.; Oberhauser, W.; Giambastiani, G.; Frediani, M.; Giachi, G.; Camaioni, N. Poly (lactic acid) as a transparent matrix for luminescent solar concentrators: A renewable material for a renewable energy technology. *Energy Environ. Sci.* **2011**, *4*, 2849–2853. [CrossRef]
43. Sadeghi, S.; Melikov, R.; Bahmani Jalali, H.; Karatum, O.; Srivastava, S.B.; Conkar, D.; Firat-Karalar, E.N.; Nizamoglu, S. Ecofriendly and efficient luminescent solar concentrators based on fluorescent proteins. *ACS Appl. Mater. Interfaces* **2019**, *11*, 8710–8716. [CrossRef] [PubMed]
44. Chowdhury, F.I.; Dick, C.; Meng, L.; Mahpeykar, S.M.; Ahvazi, B.; Wang, X. Cellulose nanocrystals as host matrix and waveguide materials for recyclable luminescent solar concentrators. *RSC Adv.* **2017**, *7*, 32436–32441. [CrossRef]
45. Geervliet, T.A.; Gavrila, I.; Iasilli, G.; Picchioni, F.; Pucci, A. Luminescent solar concentrators based on renewable polyester matrices. *Chem. Asian J.* **2019**, *14*, 877–883. [CrossRef] [PubMed]
46. Mateen, F.; Lee, S.Y.; Hong, S.-K. Luminescent solar concentrators based on thermally activated delayed fluorescence dyes. *J. Mater. Chem. A* **2020**, *8*, 3708–3716. [CrossRef]
47. Mateen, F.; Li, Y.; Saeed, M.A.; Sun, Y.; Zhang, Y.; Lee, S.Y.; Hong, S.-K. Large-area luminescent solar concentrator utilizing donor-acceptor luminophore with nearly zero reabsorption: Indoor/outdoor performance evaluation. *J. Lumin.* **2021**, *231*, 117837. [CrossRef]
48. Uoyama, H.; Goushi, K.; Shizu, K.; Nomura, H.; Adachi, C. Highly efficient organic light-emitting diodes from delayed fluorescence. *Nature* **2012**, *492*, 234–238. [CrossRef] [PubMed]
49. Kaji, H.; Suzuki, H.; Fukushima, T.; Shizu, K.; Suzuki, K.; Kubo, S.; Komino, T.; Oiwa, H.; Suzuki, F.; Wakamiya, A. Purely organic electroluminescent material realizing 100% conversion from electricity to light. *Nat. Commun.* **2015**, *6*, 1–8. [CrossRef]
50. Lucarelli, J.; Lessi, M.; Manzini, C.; Minei, P.; Bellina, F.; Pucci, A. N-alkyl diketopyrrolopyrrole-based fluorophores for luminescent solar concentrators: Effect of the alkyl chain on dye efficiency. *Dye. Pigm.* **2016**, *135*, 154–162. [CrossRef]
51. Inman, R.; Shcherbatyuk, G.; Medvedko, D.; Gopinathan, A.; Ghosh, S. Cylindrical luminescent solar concentrators with near-infrared quantum dots. *Opt. Express* **2011**, *19*, 24308–24313. [CrossRef] [PubMed]
52. Jakubowski, K.; Huang, C.-S.; Gooneie, A.; Boesel, L.F.; Heuberger, M.; Hufenus, R. Luminescent solar concentrators based on melt-spun polymer optical fibers. *Mater. Des.* **2020**, *189*, 108518. [CrossRef]
53. Klimov, V.I.; Baker, T.A.; Lim, J.; Velizhanin, K.A.; McDaniel, H. Quality factor of luminescent solar concentrators and practical concentration limits attainable with semiconductor quantum dots. *ACS Photonics* **2016**, *3*, 1138–1148. [CrossRef]

Article

Impact-Resistant and Tough 3D Helicoidally Architected Polymer Composites Enabling Next-Generation Lightweight Silicon Photovoltaics Module Design and Technology

Arief Suriadi Budiman [1,2,*], Rahul Sahay [2,*], Komal Agarwal [2], Gregoria Illya [3], Ryo Geoffrey Widjaja [1], Avinash Baji [4] and Nagarajan Raghavan [5]

1. Industrial Engineering Department, BINUS Graduate Program-Master of Industrial Engineering, Bina Nusantara University, Jakarta 11480, Indonesia; ryo.widjaja@binus.ac.id
2. Xtreme Materials Lab, Engineering Product Development, Singapore University of Technology and Design (SUTD), Singapore 487372, Singapore; komal_agarwal@alumni.sutd.edu.sg
3. Physics Department, Matana University, Tangerang 15810, Indonesia; gregoria.illya@matanauniversity.ac.id
4. Mechanical Engineering, La Trobe University, Melbourne, VIC 3086, Australia; A.Baji@latrobe.edu.au
5. Nano-Macro Reliability Lab, Engineering Product Development Pillar, Singapore University of Technology and Design (SUTD), Singapore 487372, Singapore; nagarajan@sutd.edu.sg
* Correspondence: suriadi@alumni.stanford.edu (A.S.B.); rahul@sutd.edu.sg (R.S.)

Abstract: Lightweight photovoltaics (PV) modules are important for certain segments of the renewable energy markets—such as exhibition halls, factories, supermarkets, farms, etc. However, lightweight silicon-based PV modules have their own set of technical challenges or concerns. One of them, which is the subject of this paper, is the lack of impact resistance, especially against hailstorms in deep winter in countries with four seasons. Even if the front sheet can be made sufficiently strong and impact-resistant, the silicon cells inside remain fragile and very prone to impact loading. This leads to cracks that significantly degrade performance (output power) over time. A 3D helicoidally architected fiber-based polymer composite has recently been found to exhibit excellent impact resistance, inspired by the multi-hierarchical internal structures of the mantis shrimp's dactyl clubs. In previous work, our group demonstrated that via electrospinning-based additive manufacturing methodologies, weak polymer material constituents could be made to exhibit significantly improved toughness and impact properties. In this study, we demonstrate the use of 3D architected fiber-based polymer composites to protect the silicon solar cells by absorbing impact energy. The absorbed energy is equivalent to the energy that would impact the solar cells during hailstorms. We have shown that silicon cells placed under such 3D architected polymer layers break at substantially higher impact load/energy (compared to those placed under standard PV encapsulation polymer material). This could lead to the development of novel PV encapsulant materials for the next generation of lightweight PV modules and technology with excellent impact resistance.

Keywords: 3D helicoidal architecture; fiber-based polymer composite; impact resistance; lightweight photovoltaics (PV); integrated PV rooftop

1. Introduction

Among other renewable energy sources, PV can be considered as the most versatile—it can be used in highly urbanized areas, as well as in the most remote areas, and can even float in water (oceans, lakes, etc.). Therefore, it is critical to develop PV modules and technologies that are appropriate for the particular uses and their unique circumstances/conditions. A single PV module design and technology may not be appropriate for PV modules used in various applications. Lightweight photovoltaics (PV) modules, for instance, are important for certain segments of renewable energy markets. Many large-scale buildings—such as exhibition halls, factories, supermarkets, farms, etc.—have huge footprints, with a limited number of supporting pillars, hence the roof structure has low-load bearing capacity. Such roofs require lightweight PV modules, otherwise the expensive reinforcement of such building structures required before the installation of the heavy conventional glass-based PV modules would render the whole renewable energy project (building plus the PV power source) uneconomical and make it unattractive to potential business interests [1–3]. Lightweight PV as part of building-integrated PV and mostly for urban building applications has been discussed quite extensively and comprehensively elsewhere [3–7].

However, in the present manuscript, we propose another important role of lightweight PV that could contribute to the climate and sustainability challenge in the world through use in unique geography of countries like Indonesia. Lightweight PV modules would also be critical for accelerating the adoption of renewable energy in certain geographies, such as archipelagic Indonesia—where many underdeveloped areas (which need renewable energy the most) are located in very remote locations and on thousands of separate islands. Heavy conventional PV modules would again render the prospects of building green renewable energy parks in such places uneconomical and unattractive due to very high transportation and installation costs of such PV systems.

When it comes to accelerating the adoption of renewable energy to meet the climate sustainability challenge facing the world, Indonesia is an interesting case in point. The country is blessed with diverse and abundant energy sources—both renewable (wind, hydro, photovoltaic, geothermal) and fossil. However, Indonesia's geographic conditions are less than ideal for efficient energy distribution. Indonesia is an archipelagic country with large, sprawling geographic regions, typically lacking electrical infrastructure in very remote, underdeveloped, and outermost areas, which are often separated by seas [8–11]. Centralized energy sources are not the ideal option for such a geography; independent, decentralized power generation based on small wind turbines or micro-hydropower plants combined with photovoltaic (PV) plants are. Small villages in remote areas are currently either cut off from a centralized power supply or run diesel generators. Such independent renewable energy systems, particularly in these remote areas, have a strategic importance for Indonesia as a country, and perhaps more importantly as an integrated part of the global economic and environmental ecosystem for sustainability. This challenge represents an opportunity to collectively transform the energy sector in Indonesia into a sustainable and environmentally friendly energy economy.

Lightweight PV can play a crucial role in setting up such independent energy supply systems that are needed in remote, rural areas, such as Indonesia, by saving transportation and installation costs of solar PV systems. The PV power supply systems must be lightweight to facilitate transportation to very remote areas that often lack road and mobility infrastructures. Therefore, lightweight PV modules are needed—whether for applications in urban buildings in most advanced countries (especially in Europe) [3–6], or for easy installation in the most remote and underdeveloped areas in unique geographies such as archipelagic Indonesia— for accelerating the adoption of renewable energy for the world.

Since silicon is likely to be the mainstream PV technology for quite some time [3,4], we need to enable lightweight silicon-based PV modules. One of the major technical problems in designing lightweight PV modules is impact resistance and structural strength, especially against hailstorms and strong winds in countries with four seasons, such as Germany (or Europe in general) and in North America [12,13]. Although, the concept of lightweight flexible PV is very appealing, nevertheless, it is still not a viable solution due to issues with low module stiffness, and structural reliability [3–5,12,13]. Many commercial lightweight PV solutions (including ones following IEC/UL standards) have a limited operational lifetime [4,5]. Nevertheless, recent studies with significant material development and clever design have enabled wonderful enhancements in impact resistance of many polymer substrates used as the frontsheet (instead of glass) in existing lightweight

PV modules [13–16], although the silicon cells inside remain fragile and highly susceptible to particular impact loads.

Upon impact loading (such as from hailstorms against the frontsheet of the PV module), the (non-glass) frontsheet itself maybe strong enough and does not break, but the energy is passed directly to the underlying materials, i.e., first encapsulant (ethylene vinyl acetate/EVA is the most typical in PVs), then eventually to the fragile silicon cells. Receiving the impact energy from the frontsheet, the EVA would just comply (it has high compliance) and thus the energy was transmitted further down to the silicon cells. The fragile silicon material is especially prone to such point impact loads, and thus either cracks occur (nucleates) or extend further than their propagation points [5,13,14]. Consequently, electrical performance (i.e., power output) will either reduce gradually (degrade over time) or drastically—possibly leading to hotspots and potential safety issues (such as fires, etc.). The standard PV test for this hail resistance of a PV module is known as the IEC 61215/61646 clause 10.17.

Natural structural materials found in mantis shrimp, nacre, and shells were recently reported to exhibit superior mechanical and especially impact characteristics [17–19]. For instance, the 3D architecture with helicoidal geometry found in the dactyl club of the mantis shrimp can dissipate energy through quasi-plastic compressive reactions, forming a fracture toughening obstruction to the propagation of microcracks during repeated impacts [20–23]. Our group's recent publications reported higher impact performance/resistance of such materials [24,25]. The layered geometry consisting of 3D helicoidally aligned fibers of such materials would efficiently absorb the impact energy and transfer very little energy to the fragile silicon solar cells. This would enable a novel lightweight PV module designs (based on polymer front and backsheets) with enhanced impact resistance and structural integrity/reliability (especially against cracks in the silicon cell). Our aim in the present study is to present the evidence for the basic feasibility of the proposed concept, i.e., using 3D-architected layered polymer structures consisting of helicoidally aligned fibers to provide protection of the silicon solar cells (in lightweight PV module designs) especially against the initiation/propagation of cracks due to impact loads (from hailstorms, for instance). Building on our previous research investigations on these novel materials [24–27], we extend our methodologies to enable this feasibility study for the application of lightweight PV technologies. In addition, the design of lightweight PV modules would allow PV integration with curved or surfaces with contours (such as for automobiles or boats) thus enabling more aesthetic design for the integration of PV into urban structures or buildings in cities.

2. Materials and Methods
2.1. Materials

Polyvinyl alcohol (PVA) having M_W = 98,000, polyvinylidene fluoride-co-hexafluoropropylene (PVDF-HFP) having M_W = 400,000, acetone and dimethylacetamide (DMAc) were purchased from Merck, Singapore. These chemicals were then used without further purification. PVDF-HFP was then dissolved in 1:3 solvent ratio (wt/wt) of acetone and dimethylacetamide, which was then stirred overnight (temperature = 45 °C) to prepare 35 wt% PVDF-HFP (DMAc/acetone) solution be used later for electrospinning. The rheological properties of the PVDF-HFP (DMAc/acetone) solution allowed morphology change from round to broad and flatter fibers.

2.1.1. Choice of Fiber Material

PVDF-HFP polymer possesses properties, such as higher solubility [28], greater free volume [29], better mechanical properties [30], ease of processing, and flexibility [31].

2.1.2. Choice of Matrix Material

Several polymers were considered for embedding the PVDF-HFP fibers in the matrix. However, polymers like polyurethane, polyurethane acrylate, epoxy, etc. are viscous polymers and form thicker films in comparison to the fibers. In some of the polymers, the polymer is dissolved in a solvent to form a liquid matrix, which in turn destroys the structural design of the fibers. While the thermoset matrix materials deal with temperature curing or UV curing which just increases the number of optimization processes and also end up dominating the mechanical properties of the composites. For the amount of fiber samples that the equipment allows to be made in the laboratory setting, it is very important to choose the matrix material very wisely such that its viscosity can be altered to form thin-films without affecting the structure of the fibers.

Therefore, after several trials with varieties of matrix materials, PVA was chosen because of its solubility in water (H_2O), and its ability to form uniform films. Further, PVA also provides the capability to control its rheological properties to control the weight ratio between the fibers and matrix material in the fabricated composite. PVA is also more compatible with several other polymer materials due to its hydrophilic nature and is also a transparent polymer. Thin-films can be prepared by simple water evaporation with no requirements of any external factors [24]. The choices of these materials are simply as model materials with the aim to demonstrate the basic feasibility of the concept of enhanced impact resistance through 3D architected encapsulant enabled by electrospinning-based additive manufacturing methodologies.

2.2. Electrospinning-Based Additive Manufacturing

Near-field electrospinning (NFES, designed and assembled at SUTD, Singapore [24]) has been used as an additive manufacturing technique to fabricate helicoidally aligned fibrous layers (HA-FLs). In NFES, a high voltage is applied between a hemispherical polymer drop impregnated by a needle and the metallic collector plate [32,33]. NFES deposits one-dimensional fibers in a controlled manner at precise locations that allow obtaining of 3D structures by stacking the fibers layer by layer with angular offsets, as shown in Figure 1. A 5 mL syringe (make: Terumo Corporation, Tokyo, Japan) with a 25 G needle was filled with PVDF-HFP (DMAc/acetone) solution. The solution was then dispensed using a syringe pump (model no: EQ -500SP-H, make: Premier Solution Pte Ltd., Singapore). The flow rate of the PVDF-HFP (DMAc/acetone) solution was maintained as 1 mL/h. During NFES, voltage is applied to the needle whereas the aluminum collector plate was grounded. The collector plate is used for collecting fibers. The motion of the collector plate was controlled by placing it on programmable XY stage (PI High Precision XY stage) (model no: C-891, make: Physik Instrumente (PI) GmbH & Co. KG, Karlsruhe, Germany). The motion of the collection plate is then controlled to obtain layers of electrospun fibers. The layers of electrospun fibers were then stacked at different angular orientations to obtain HA-FLs. The details of the fabrication of HA-FLs with different angular orientations can be found in Komal et al. [24]. The speed of the motorized XY stage (as well as the collector) was fixed at 200 mm/s. The initial distance between the collector plate, and the needle was 7 mm, which was later decreased with a decrement of 0.5 mm with the collection of each fiber layer. The initial applied voltage during electrospinning was 2.4 kV for the deposition of the first four layers, which was then increased to 2.6 kV for the deposition of the next four layers.

The helicoidal structure was then achieved by stacking layers of electrospun fibers at different angular orientations on top of each other (see Figure 1). In the case of 45° HA-PVDFs with 45° angular offsets, the fiber layers were deposited at 0°, 45°, 90°, 135°, 180°, 225°, 270°, and 315°. In this case, the final HA-FLs consist of eight fused layers. The fabricated HA-FLs were then dried in a controlled environment inside an oven for 1 h at 50 °C to remove residual solvents. Three such 8-layered HA-FLs were then stacked onto each other to obtain a 24-layered helicoidally arranged HA-FLs.

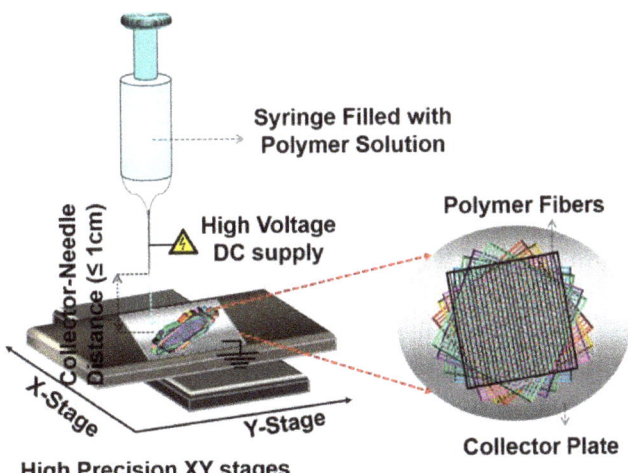

Figure 1. Schematic of NFES for depositing electrospun fibrous layers on a collector plate to fabricate 3D electrospun helicoidally aligned fibrous layers (HA-FLs).

2.3. Fabrication of Helicoidally-Aligned Synthetic Structural Composites (HA-SSCs)

The fabricated 24-layered HA-FLs were then embedded into PVA matrix solution to fabricate HA-SSCs. A 5 wt% PVA solution (doped with a surfactant) was sprayed onto the HA-FLs. The spraying of PVA solution was performed at a fixed flow rate along the length of the HA-FLs to obtain HA-SSCs. The surfactant (Triton X-100, Merck KGaA, Darmstadt, Germany) is added to the PVA solution to enhance the interfacial adhesion between HA-FLs and the PVA matrix. The samples were then dried at room temperature (with 75% humidity) for 72 h to fabricate opaque HA-SSCs, which, of course, is not yet appropriate for full integration into PV module design applications. An optically transparent material would be needed for real PV application with comparable transmission of sunlight (in terms of intensity and range of suitable wavelengths). However, as explained in the Introduction section as well as in the beginning of the Materials and Methods section, the focus of the present study is to demonstrate the basic feasibility of the concept of enhanced impact resistance through 3D architected encapsulant, not the full technological integration in PV module design.

The thickness of the HA-SSCs samples ranged from 230 to 250 μm. For a more detailed description of the synthesis of the composites, see our earlier report [24]. Of the many types of HA-SSC samples, we reported there [24], we used only HA-SSCs15 and HA-SSCs45—with fiber alignment in rotational angle every 15° and 45°, respectively, from layer to layer—in the present study.

2.4. Impact Testing

The impact tests were performed using the ball drop method, which was also employed by Chen et al. for their electrospun nylon (fiber pattern)-epoxy (matrix)-based composite [34] as well as own previous work [24,25]. The ball drop method typically finds the height at which the PV solar cell protected by the sample would break upon impact.

Customized setup used for determining the impact strength of the specimens is shown in Figure 2. A standard steel ball (diameter = 7.14 mm, weight = 1.4 g) was dropped under gravity onto the sample to measure the height required to facture solar cell. The height was increased in 5 cm increments to determine the final height required to fracture the solar cell placed beneath the samples. A digital oscilloscope (model no: DL1620 −200 MS/s 200 MHz make: Yokogawa Electric Corporation, Tokyo, Japan) with a resolution of 0.5 V/division and a charge meter (make: Kistler, model: 5015) with a sensor sensitivity of −4.060 pC/N were used to measure the impact force required to fracture solar cell. The samples were glued to solar cells at 2 diagonal edge points to restrict their motion during impact measurements. An unprotected bare solar cell was used as a control sample during impact measurements. The solar cells used are typical commercially available monocrystalline type without any interconnect (a new cell was used for each ball dropping experiment/test).

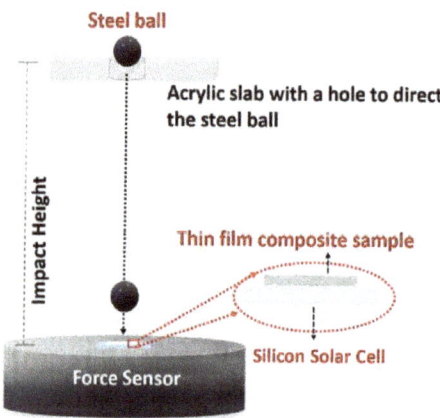

Figure 2. Schematic showing the impact test setup.

This employed impact test methodology is used due to the absence of a quantitative impact fracture mechanics methodology for small samples such as polymeric composites. Typically, standard impact fracture mechanics tests such as Charpy, ballistic, and Izod tests are suitable only for relatively large samples [35–37]. The current impact test methodology is used to determine the fracture height, and subsequently, the specific potential energy required to fracture solar cells. [24,25]. All the values of height and calculated specific potential energy at which the silicon solar cell breaks are compared to gain deeper insights about the impact properties of bare silicon solar cells (without any protection), silicon solar cells under EVA (ethylene vinyl acetate—typical encapsulant materials in PV technology), and silicon solar cells under HA-SSCs (15° and 45°).

3. Results and Discussion

3.1. Physical Characteristics of the HA-SSCs

The impact properties of the electrospun HA-SSCs can be modulated by varying the offset angle of the stacked fibrous layers. Here, two samples (HA-SSC15 and HA-SSC45) were tested for their impact properties. In the cases of HA-SSC45 and HA-SSC15, after the deposition of the first fibrous layer, the next layer was deposited with its longitudinal axis rotated by angular offset of 45° and 15°, respectively. HA-FLs were then obtained by stacking eight layers of fibers at a fixed offset angle. The offset angle was 45° for HA-SSC45 resulting in fibrous layers located at 0°, 45°, 90°, 135°, 180°, 225°, 270°, and 315°, Similarly, an offset angle 15° for HA-SSC15 resulted in fibrous layers located at 0°, 15°, 30°, 45°, 60°, 75°, 900, and 105°. Three such 8-layered HA-FLs were then stacked onto each other to obtain a 24-layered HA-FLs, which were later embedded in a PVA solution to obtain HA-SSCs.

SEM and optical microscope images (see Figure 3) show the microstructure of HA-SSCs produced using NFES. The SEM images show that the formation of broad ribbons, as compared to circular fibers, were found to have better contact area than circular fibers. The samples shown in Figure 3 belong to the same batch, which led to the improvement in the fibers' production, fiber adhesion as described in detail in our previous report [24]. Figure 3(a1,b1) (optical images) clearly show the helicoidal arrangement of the ribbons in HA-SSCs. Figure 3(a2,b2) (SEM images) also show 24-layered HA-SSC. SEM images clearly show that ribbons were deposited at certain offset angles within the HA-SSC. The red dashed lines in Figure 3 show the angular sequence between each layer of the ribbons. For example, Figure 3(a1,a2) show that layers of the ribbons were oriented with 45° offset angles to obtain HA-SSC45. Similarly, Figure 3(b1,b2) show that layers of the ribbons were oriented with 15° offset angles to obtain HA-SSC15.

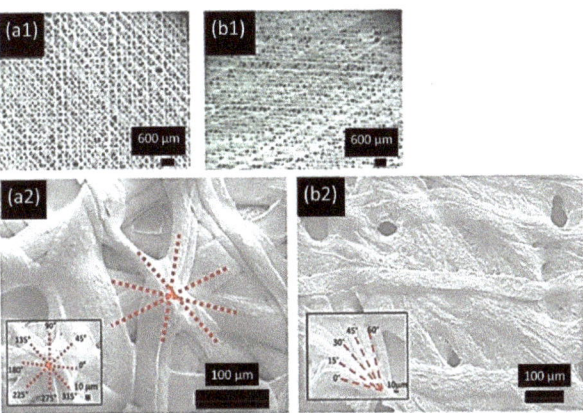

Figure 3. Optical micrographs at 35× of HA-SSC fabricated with offset angles: (**a1**) 45° and (**b1**) 15°. SEM images of HA-SSC fabricated with offset angle: (**a2**) 45° (at 270×) and (**b2**) 15° (at 190×). Reproduced with permission from American Chemical Society (ACS) [24].

3.2. Impact Test of Photovoltaic (PV) Cells

Silicon solar cells are fragile and are highly susceptible to impact load. Therefore, a customized impact testing setup (see Figure 2) was used to determine the impact resistance of such solar cells when protected by the samples [35]. A bare solar cell was chosen as a control group to compare with the solar cells protected with the fabricated samples. A standard steel ball (diameter = 7.14 mm, weight = 1.4 g) was dropped under gravity onto the specimens during impact test. At least six impact tests were performed for each type of sample (cross-sectional area = ~1.1 × 1.1 cm^2). The height from which the ball was dropped was increased in increments of 5 cm to determine the fracture height of the solar cell placed under the sample. When the tentative fracture height for the samples was identified, the tests were then performed with 1 cm increments for careful estimation of the final fracture height. Table 1 summarizes the results of the impact tests of bare silicon solar cells (without any protection), silicon solar cells under EVA (ethylene vinyl acetate—typical encapsulant materials in PV technology), and silicon solar cells under HA-SSCs (15° and 45°) when impacted with the steel ball. The error bars in Table 1 show the range of the mean impacted height accumulated by performing experiments on at least six samples for each test condition.

Table 1. Impact resistance properties of the samples protecting silicon solar cell.

Samples	Height at Which the Silicon Cell Breaks (cm)	Specific Gravitational Potential Energy (10^{-2} Jcm^3/g)
Silicon solar cell	25 ± 5	-
EVA on Si cell	50 ± 4	7.3 ± 0.5
HA-SSC15 on Si cell	69 ± 2	9.4 ± 0.2
HA-SSC45 on Si cell	82 ± 4	11.2 ± 0.2

The fracture height means obtained experimentally are shown in Table 1. The nominal height of 25 cm in the bare Si solar cell group, for instance, was the actual height at which all silicon cells broke in the minimum six times we repeated the tests (often we conducted the ball drop tests on more than 6 samples at this height—up to 11 samples). When we did the test at a height of 20 cm (i.e., =25 − 5 cm), at least 50% of the silicon cells broke (out of the minimum 6 tests). At height of 30 cm (i.e., =25 + 5 cm), all silicon solar cells broke (out of the minimum 6 tests). Thus, the fracture data above merely shows the heights at which fractures began to be observed. Obviously, larger heights than the reported data above would break the silicon cells.

The statistical summary of the fracture height data of all the ball dropping experiments when fracture was observed in the underlying silicon solar cells are listed in Table 2. Each of the ball dropping tests was done with at least six samples and each sample consisted of a new silicon solar cell and protection layer each time. The means here were calculated considering the frequency at which fracture of the silicon cells happened at each height the experiment was conducted. The data here (in Table 2) shows an excellent agreement with the more discrete data shown in Table 1 (following the discrete heights as observed in the experiments).

Table 2. Summary of the statistical analysis of the fracture height data.

	Bare Si Cell	Si Cell + EVA	Si + HA-SSC15	Si + HA-SSC45
Mean (cm)	25.39	48.97	70.05	82.06
Standard Deviation (cm)	3.19	2.71	2.19	2.29

Further, we conducted a statistical F-test with ANOVA (analysis of variance) method using level of significance = 0.05 to split observed aggregate variability found in the fracture height data into two parts: systematic factors and random factors (i.e., errors). The ANOVA test suggests that the treatments (i.e., the different protection layer placed on top of the silicon solar cell) are responsible for 98% of the variability found in the experiment, which we consider significant. This strongly suggests that the means of each group are significantly different from each other, and most of the variations found in the experiments were due to the different treatments in each group. In addition to this global ANOVA test, we also conducted individual tests comparing each group with each other group—one on one. The results suggest consistent findings with the global test. Hence, within the level of significance taken (i.e., 0.05) in this statistical analysis, each group was found to be significantly different from the others—even between HA-SSC15 vs. HA-SSC45 (which appeared to be the closest to each other amongst all other groups). This is certainly to be expected given the data as shown in Tables 1 and 2.

3.3. Fracture of the Silicon Photovoltaic (PV) Cells

The optical microscope images after the impact testing of the representative silicon solar cells (front and back surfaces) are shown in Figures 4–6. Each time the ball dropping experiment was performed, a new silicon solar cell and a new protection layer were used. All silicon monocrystalline solar cells used in the present study came from the same batch of 200 cells. Within the expected manufacturing variability specifications, all the silicon solar cells used in the present impact test/experiment may reasonably be assumed to have uniform mechanical strength, structural integrity, and fracture toughness. Thus, any differences we see after the impact experiments (as shown in Table 1, as well as in Figures 4–6) may be associated with the protection/encapsulation samples (either EVA or HA-SSCs or nothing) placed on top of the silicon solar cells.

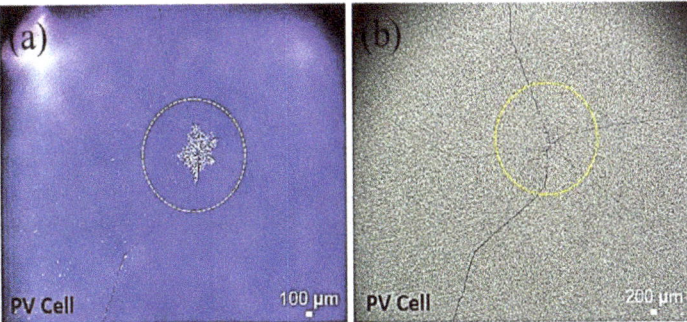

Figure 4. Microscope images taken after impact test: (**a**) front and (**b**) back of silicon monocrystalline solar cell after impact with a steel ball. These images were taken from a representative silicon solar cell broken at the nominal height as shown in Table 1 (in this case, this particular cell was broken when the steel ball dropped from a height of 25 cm). The circles on the silicon solar cell surfaces indicate the approximate impact contact area with the steel ball during impact test.

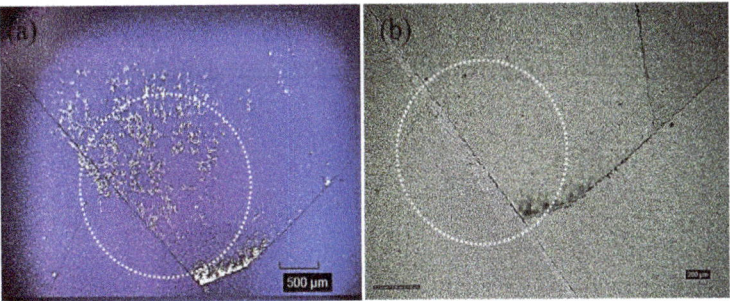

Figure 5. Microscope images taken after impact test: (**a**) front and (**b**) back of silicon monocrystalline solar cell after impact with a steel ball. The silicon cell was covered with EVA on top for protection against impacting ball. These images were taken from a representative silicon solar cell broken at the nominal height as shown in Table 1 (in this case, this particular cell was broken when the steel ball dropped from a height of 50 cm). The circles on the silicon solar cell surfaces indicate the approximate impact contact area with the steel ball during impact test.

Both the results shown in Table 1 (the fracture height/specific potential energy data and the ensuing statistical analysis as summarized above) as well as the images shown in Figures 4–6 clearly suggest that HA-SSC composite materials could provide better protection or encapsulation of the bare silicon solar cell against impact loads, compared to the standard PV EVA encapsulation layer. Both the HA-SSC composite materials (HA-SSC15 and HA-SSC45) allow substantially higher fracture heights and associated specific potential energies (significantly beyond the experimental error bars/uncertainties) before the silicon solar cells under them start breaking or initiating/propagating catastrophic fracture events upon the ball-dropping impact testing experiments.

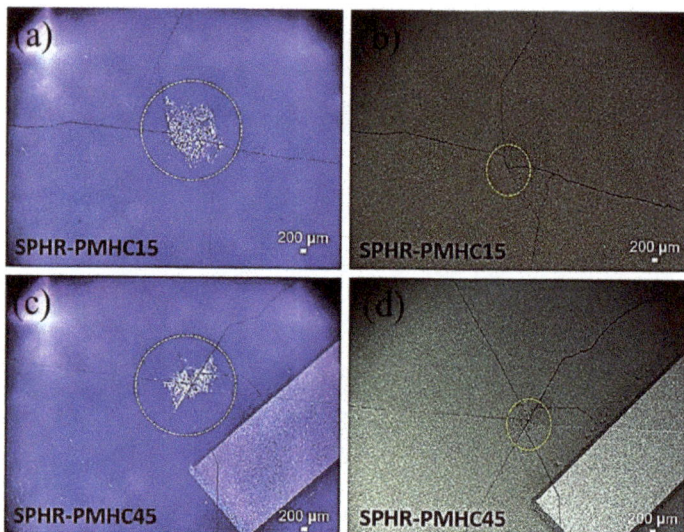

Figure 6. Microscope images taken after impact test: (**a,c**) front and (**b,d**) back of silicon cells after impact with the steel ball. The silicon cell was covered with HA-SSC on top for protection against impacting ball. For (**a,b**) the HA-SSC used was HA-SSC15 and (**c,d**) the HA-SSC used was HA-SSC45. These images were taken from the representative silicon solar cell broken at the nominal height as shown in Table 1. In this case, the particular cell in (**a,b**) was broken when the steel ball was dropped from a height of 69 cm, while the particular cell in (**c,d**) was broken when the steel ball was dropped from a height of 82 cm. The circles on the silicon solar cell surfaces indicate the approximate impact contact area with the steel ball during the impact test.

When the EVA receives the impact energy from the ball drop test, it would simply comply due to its very low elastic modulus and thus the energy (and damage) is passed down to the silicon cells. The fragile silicon material is especially prone to such point impact loads, and thus cracks occur and extend through the thickness of the silicon solar cell almost instantaneously at relatively low height of 50 ± 4 cm. This is also evident from the images shown in Figure 5, where most of the impact energy was absorbed by creating only one main line (through the thickness of the cell) of a very long crack (almost across the whole area of the cell, which is often considered to be the more serious concern in terms of its effect on electrical power output degradation).

In contrast, upon receiving the impact energy, both the HA-SSC membranes owing to their helicoidally aligned fiber-reinforced layered structures would dissipate the impact energy very efficiently and deflect the crack laterally by following the helical path of the fiber directions instead of breaking instantaneously in a straight fracture line across the thickness of the membranes. In fact, the crack transitions between the matrix material and the ribbons at different angles in each layer—layer by layer. Every time the crack tip extends into the next layer, it finds the fiber aligned at different angle of orientation and thus gets deflected laterally in every layer, and it gradually loses its primary driving force/energy to extend through the thickness of the membrane. This avoids the catastrophic failure of the HA-SSC membrane, but also absorbs more energy to deflect the crack along different directions and planes in every layer, and due to regular encounters with laminate modulus changes, much less impact energy/damage is transmitted to the underlying silicon solar cell. These results are consistent with the load dissipation mechanism proposed in our previous reports [24,25] of this novel 3D helicoidally fiber-aligned composite materials, which is particularly effective and suggestive of more effective dissipating of energy such that only when the steel ball is dropped from the height of 69 ± 2 cm and 82 ± 4 cm for the HA-SSC15 and HA-SSC45, respectively, it would break the silicon solar cell underneath it.

The efficient mechanism of load dissipation and effective absorption of damage/energy are also evident from the images shown in Figure 6. Upon finally breaking the silicon solar cells underneath them at much higher heights, both the HA-SSC membrane materials exhibit several crack lines on the silicon cells indicating highly dissipated impact energy and transfer of load to the sideways (evident from the crack lines at different angles on the surfaces of the silicon cells) as received by the silicon cells. These cracks appear to follow the preferred crystallographic directions of <110> associated with the weakest crystallographic planes of {111} in typical silicon monocrystalline wafers/solar cells, as has been widely reported in the literature for both PV and other silicon-based energy devices [38–45].

These crack deflection mechanisms, which are evident from the impact tests in the present study as well as complete mechanical and microstructural characterization studies conducted in our previous reports [24,25], suggest a higher toughness and higher material's resistance against crack propagation. Upon impact, the ribbons of the HA-SSC, which are built at different angles from one layer to the next, orient themselves along the direction of the applied force. During the deformation caused by the impact, the helicoidally arranged ribbons show an efficient transfer of the load to the adjacent ribbons by sliding and continuously pulling the ribbons until they break. This explains the higher energy absorbed by the HA-SSC protecting solar cells. The collective effect of the structural design and the variations in the moduli leads to a higher specific toughness of the HA-SSCs [24,25,34].

The HA-SSC45 samples seem to be the most effective in continuously deflecting crack or impact damage along different angular directions. It is evident from the fracture height data that fracture of the silicon solar cells occurs at 82 ± 4 cm—the highest amongst the materials tested in the present study. The crack repeatedly encounters differences in the modulus of the ribbons and the matrix material. In HA-SSC45 the crack continues along one ribbon, encounters a deviation in modulus due to the existence of the matrix material, continues to penetrate the matrix, then comes across another ribbon that is in a different direction and plane, and then deviates from its initial path to follow a different path. The crack propagates in different planes and directions, turning and twisting inside and outside the ribbon and matrix phases, further stretching the ribbon short of the final catastrophic events of specimen failure. It follows that due to the presence of helicoidal network of ribbons in the composites, this continuous deflection of the cracks requires a higher energy to break, thus transmitting a minimum of impact energy/damage to the underlying silicon solar cells.

However, our impact experiments suggest a slightly different picture for the HA-SSC15. Fracture of the silicon solar cells underneath these samples occurs at lower heights (69 ± 2 cm) and thus lower specific potential energy (indicating lower impact energy/damage absorption rate). In HA-SSC15, the ribbons are closer together than in HA-SSC45, which is due to lower offset angle. When a crack begins to form, its path is hindered when it encounters an alteration in direction more frequently compared to the composites with larger offset angle (in the case of HA-SSC45) due to the variation in the modulus through the thickness of the sample. When the angles between the layers are larger, the crack can propagate in a straight path in the matrix until the next ribbon prevents it. Because of the smaller offset angle, the crack encounters variations in phase modulus much more frequently, which should delay the fracture events even more and absorb even more energy. These results were shown in our previous report [24] when we placed a piece of glass under the materials. The results there [24] showed that the 15° helicoidally aligned samples could absorb much higher impact energy compared to the 45° samples. Since we followed the same full procedure of the ball-dropping experiment with [24,25], including the same size of the steel ball, we believe that this deviation (in the impact damage absorption capacity between the 15 vs. 45 HA-SSC samples) maybe associated with the silicon monocrystalline solar cells placed under the HA-SSC samples. Compared to glass slip, the silicon monocrystalline samples have crystallographic dependence of mechanical properties, including fracture preferred occurrences. However, the full correlation between the helicoidal orientation and the fracture of the monocrystalline silicon solar cells needs further investigation and is beyond the scope of the present study.

3.4. Enabling Next-Gen Lightweight Photovoltaic (PV) Module Technology

It is clear from the results presented so far, that the HA-SSC composites/membranes are very effective in absorbing and dissipating impact energy/damage and thus could protect the fragile silicon solar cells under them from point impact loads that silicon PV is very prone to. The present manuscript aims to provide evidence of the feasibility of using HA-SSC polymer films for PV encapsulation materials to protect the fragile silicon solar cells, especially in the design of lightweight PV modules that are particularly vulnerable to impact damage, such as hailstorms, as illustrated by the IEC 61215/61646 Clause 10.17. While many recent studies have shown that many light polymer-based materials [12–16] may be used to increase the overall structural integrity and mechanical strength (including fracture and impact resistance) of lightweight PV technology, the encapsulants used in those studies are invariably EVA—maybe with different thicknesses, or a slight variation, such as low cure EVA [46].

Obviously, EVA is the encapsulant of choice in the previous studies due to its cost implication and manufacturing readiness of the overall lightweight PV module. Our findings as reported in the present manuscript offer preliminary evidence, from the basic technological feasibility point of view, to use other kinds of novel polymer films with unique 3D architecture to enable stronger and higher resistant (including against fracture and impact loading) silicon-based lightweight PV module design. The manufacturing and economic implications are beyond the scope of the present manuscript.

However, other technological issues need to be addressed to fully enable this novel concept. First, the HA-SSCs fabricated in the present study were not transparent. It is certainly a must to have transparent protective layers on top of the silicon cells in PV modules. It should be noted that these experiments were conducted to verify the basic feasibility of incorporating HA-SSC thin-films into PV modules. Now that our findings have confirmed the basic feasibility, using the same electrospinning-based additive manufacturing (AM) methodology, we could find other polymers which we can be fabricated into transparent forms, such as nylon [47] and poly(methyl methacrylate) (PMMA) (or PMMA-based composites) [48,49]. Moreover, the interfacial adhesion must be good not only with the frontsheet, but also with the silicon solar cell itself [46]. Lastly, the novel materials may need to be further developed to maintain their 3D architecture upon lamination process [46,50]. This could be the path forward for future studies to further develop these novel 3D-architected polymer composites/materials for enhancing the use of the silicon-based lightweight PV modules and technology.

4. Conclusions

Three-dimensional helicoidally architected fiber-based polymer composites synthesized in the present study using an electrospinning-based additive manufacturing (AM) methodology were shown to have excellent impact energy/damage absorption and dissipation rates. Such helicoidally aligned synthetic structural composites (HA-SSCs), when placed over monocrystalline silicon solar cells, may provide superior protection to silicon solar cells against impact loads, which is done via a ball-drop experiment in the present study. During the ball-drop experiment, both HA-SSC composite materials (HA-SSC15 and HA-SSC45) allow substantially higher fracture heights of 69 ± 2 cm and 82 ± 4 cm, respectively, in compassion to 25 ± 5 cm, and 50 ± 4 cm for unprotected Si cells and EVA protected Si cells, respectively. The present results provide preliminary evidence of a basic technological feasibility of using the novel materials for PV encapsulation to enable the design and technology of lightweight silicon-based PV modules. Full economic considerations of the implementations of this concept remain to be conducted in future investigations. Further, these results provide a promising framework for the development of strong, impact resistant, and resilient tunable polymeric composites suitable for technological applications ranging from aerospace to flexible electronic devices.

Author Contributions: Conceptualization, A.S.B. and R.S.; methodology, R.S., K.A. and A.B.; software, K.A. and R.G.W.; validation, A.S.B., R.S. and K.A.; formal analysis, A.S.B.; investigation, R.S. and K.A.; resources, N.R.; data curation, R.S., R.G.W. and K.A.; writing—original draft preparation, R.S. and A.S.B.; writing—review and editing, A.S.B. and G.I.; visualization, R.S., K.A. and G.I.; supervision, A.S.B., N.R. and A.B.; project administration, N.R.; funding acquisition, N.R. and A.B. All authors have read and agreed to the published version of the manuscript.

Funding: The authors would like to acknowledge the funding from the Ministry of Education (MOE) Academic Research Funds MOE2017-T2-2-175 titled "Materials with Tunable Impact Resistance via Integrated Additive Manufacturing as well as MOE2019-T2-1-197 titled "Monte Carlo Design and Optimization of Multicomponent Polymer Nano-composites". This work is supported by a La Trobe University Leadership RFA grant, a La Trobe University Start-up grant and the Collaboration and Research Engagement (CaRE) grant offered by the School of Engineering and Mathematical Sciences (SEMS), La Trobe University. The authors also acknowledge receipt of funding support from Temasek Labs@SUTD Singapore, through its SEED grant program for the project IGDSS1501011 and SMART (Singapore-MIT Alliance for Research and Technology) through its Ignition grant program for the project SMART ING-000067 ENG IGN. Nagarajan Raghavan acknowledges the support of grant No. IGIPAMD1801 for payment of article processing charges (APC).

Institutional Review Board Statement: Not applicable.

Informed Consent Statement: Not applicable.

Data Availability Statement: The data presented in this study are available upon request from the corresponding author.

Acknowledgments: The authors gratefully acknowledge the critical support and infrastructure provided by the Singapore University of Technology and Design (SUTD). A.S.B. gratefully acknowledges PT Impack Pratama Indonesia for providing materials, exploratory supports and technical discussions related to possible implementations of the materials development for the applications in building-integrated PV (photovoltaics) technology/design.

Conflicts of Interest: The authors declare no conflict of interest.

References

1. Berger, K.; Cueli, A.B.; Boddaert, S.; Del Buono, M.; Delisle, V.; Fedorova, A.; Frontini, F.; Hendrick, P.; Inoue, S.; Ishii, H.; et al. International definitions of BIPV. In *IEA 597 Photovoltaic Power Systems Programme*; Report IEA-PVPS T15-04; IEA: Paris, France, 2018. Available online: https://iea-pvps.org/key-topics/international598definitions-of-bipv/ (accessed on 11 May 2021).
2. Zhang, F.; Deng, H.; Margolis, R.; Su, J. Analysis of distributedgeneration photovoltaic deployment, installation time and cost, market barriers, and policies in China. *Energy Policy* **2015**, *81*, 43–55. [CrossRef]
3. Martins, A.C.; Chapuis, V.; Virtuani, A.; Li, H.Y.; Perret-Aebi, L.E.; Ballif, C. Thermomechanical stability of lightweight glass-free photovoltaic modules based on a composite substrate. *Sol. Energy Mater Sol. Cells* **2018**, *187*, 82–90. [CrossRef]
4. Martins, A.C.; Chapuis, V.; Virtuani, A.; Ballif, C. Light and durable: Composite structures for building-integrated photovoltaic modules. *Prog. Photovolt. Res. Appl.* **2018**, *26*, 718–729. [CrossRef]
5. Martins, A.C.; Chapuis, V.; Virtuani, A.; Ballif, C. Robust Glass-Free Lightweight Photovoltaic Modules with Improved Resistance to Mechanical Loads and Impact. *IEEE J. Photovolt.* **2019**, *9*, 245–251. [CrossRef]
6. Ballif, C.; Perret-Aebi, L.E.; Lufkin, S.; Rey, E. Integrated thinking for photovoltaics in buildings. *Nat. Energy* **2018**, *3*, 438–442. [CrossRef]
7. Kajisa, T.; Miyauchi, H.; Mizuhara, K.; Hayashi, K.; Tokimitsu, T.; Inoue, M.; Hara, K.; Masuda, A. Novel lighter weight crystalline silicon photovoltaic module using acrylic-film as a cover sheet. *Jpn. J. Appl. Phys.* **2014**, *53*, 092302. [CrossRef]
8. Handara, V.; Illya, G.; Tippabhotla, S.K.; Shivakumar, R.; Budiman, A.S. Novel and Innovative Solar Photovoltaics Systems Design for Tropical and Near-Ocean Regions—An Overview and Research Directions. *Proc. Eng.* **2016**, *139*, 22. [CrossRef]
9. Illya, G.; Handara, V.; Yujing, L.; Shivakumar, R.; Budiman, A.S. Backsheet Degradation under Salt Damp Heat Environments—Enabling Novel and Innovative Solar Photovoltaics Systems Design for Tropical Regions and Sea-Close Areas. *Proc. Eng.* **2016**, *139*, 7. [CrossRef]
10. Illya, G.; Handara, V.; Siahandan, M.; Nathania, A.; Budiman, A.S. Mechanical Studies of Solar Photovoltaics (PV) Backsheets under Salt Damp Heat Environments. *Proc. Eng.* **2017**, *215*, 238–245. [CrossRef]

11. Handara, V.; Radchenko, I.; Tippabhotla, S.K.; Narayanan, K.; Llya, G.; Kunz, M.; Tamura, N.; Budiman, A.S. Probing Stress and Fracture Mechanism in Encapsulated Thin Silicon Solar Cells by Synchrotron X-ray Microdiffraction. *Sol. Energy Mater. Solar Cells* **2017**, *162*, 30–40. [CrossRef]
12. Dhere, N.G. Flexible packaging for PV modules. In *Reliability of Photovoltaic Cells, Modules, Components, and Systems*; International Society for Optics and Photonics: Bellingham, WA, USA, 2008; Volume 7048, p. 70480.
13. Wright, A.; Lee, E.J. Impact Resistant Lightweight Photovoltaic Modules. U.S. Patent WO/2016/077402, 19 May 2016.
14. Martins, A.C.; Chapuis, V.; Virtuani, A.; Perret-Aebi, L.E.; Ballif, C. Hail resistance of BIPV composite-based lightweight modules. In Proceedings of the 33rd European Photovoltaic Solar Energy Conference Exhibition, Amsterdam, The Netherlands, 25–29 September 2017; pp. 2604–2608.
15. Gaume, J.; Quesnel, F.; Guillerez, S.; LeQuang, N.; Williatte, S.; Goaer, G. Solight: A new lightweight PV module complying IEC standards. In Proceedings of the 33rd European Photovoltaic Solar Energy Conference Exhibition, Amsterdam, The Netherlands, 25–29 September 2017.
16. Boulanger, A.; Gaume, J.; Quesnel, F.; Ruols, P.; Rouby, F. Operasol: A light photovoltaic panel with integrated connectors. In Proceedings of the 33rd European Photovoltaic Solar Energy Conference Exhibition, Amsterdam, The Netherlands, 25–29 September 2017.
17. Yang, W.; Chen, I.H.; Gludovatz, B.; Zimmermann, E.A.; Ritchie, R.O.; Meyers, M.A. Natural Flexible Dermal Armor. *Adv. Mater.* **2013**, *25*, 31–48. [CrossRef] [PubMed]
18. Naleway, S.E.; Porter, M.M.; McKittrick, J.; Meyer, M.A. Structural Design Elements in Biological Materials: Application to Bioinspiration. *Adv. Mater.* **2015**, *27*, 5455–5476. [CrossRef] [PubMed]
19. Wegst, U.G.K.; Bai, H.; Saiz, E.; Tomsia, A.P.; Ritchie, R.O. Bioinspired Structural Materials. *Nat. Mater.* **2015**, *14*, 23–36. [CrossRef] [PubMed]
20. Patek, S.N.; Caldwel, R.L. Extreme Impact and Cavitation Forces of a Biological Hammer: Strike Forces of the Peacock Mantis Shrimp *Odontodactylus Scyllarus*. *J. Exp. Biol.* **2005**, *208*, 3655–3664. [CrossRef]
21. Cronin, T.W.; Marshall, N.J.; Quinn, C.A.; King, C.A. Ultraviolet Photoreception in Mantis Shrimp. *Vis. Res.* **1994**, *34*, 1443–1452. [CrossRef]
22. Amini, S.; Tadayon, M.; Idapalapati, S.; Miserez, A. The Role of Quasi-Plasticity in the Extreme Contact Damage Tolerance of the Stomatopod Dactyl Club. *Nat. Mater.* **2015**, *14*, 943–950. [CrossRef] [PubMed]
23. Fratzl, P.; Weinkamer, R. Nature's Hierarchical Materials. *Prog. Mater. Sci.* **2007**, *52*, 1263–1334. [CrossRef]
24. Agarwal, K.; Sahay, R.; Baji, A.; Budiman, A.S. Impact-Resistant and Tough Helicoidally Aligned Ribbon Reinforced Composites with Tunable Mechanical Properties via Integrated Additive Manufacturing Methodologies. *ACS Appl. Polym. Mater.* **2020**, *2*, 3491–3504. [CrossRef]
25. Sahay, R.; Agarwal, K.; Subramani, A.; Raghavan, N.; Budiman, A.S.; Baji, A. Helicoidally arranged polyacrylonitrile fiber-reinforced strong and impact-resistant thin polyvinyl alcohol film enabled by electrospinningbased additive manufacturing. *Polymers* **2020**, *12*, 2376. [CrossRef]
26. Agarwal, K.; Zhou, Y.; Ali, H.P.A.; Radchenko, I.; Baji, A.; Budiman, A.S. Additive Manufacturing Enabled by Electrospinning for Tougher Bio-Inspired Materials. *Adv. Mater. Sci. Eng.* **2018**, *2018*, 8460751. [CrossRef]
27. Agarwal, K.; Sahay, R.; Baji, A.; Budiman, A.S. Biomimetic Tough Helicoidally Structured Material Through Novel Electrospinning Based Additive Manufacturing. *MRS Adv.* **2019**, *4*, 2345–2354. [CrossRef]
28. García-Payo, M.C.; Essalhi, M.; Khayet, M. Effects of PVDF-HFP concentration on membrane distillation performance and structural morphology of hollow fiber membranes. *J. Membr. Sci.* **2010**, *347*, 209–219. [CrossRef]
29. Lalia, B.S.; Guillen-Burrieza, E.; Arafat, H.A.; Hashaikeh, R. Fabrication and characterization of polyvinylidenefluoride-co-hexafluoropropylene (PVDF-HFP) electrospun membranes for direct contact membrane distillation. *J. Membr. Sci.* **2013**, *428*, 104–115. [CrossRef]
30. Stephan, A.M.; Nahm, K.S.; Kulandainathan, M.A. Poly (vinylidene fluoride-hexafluoropropylene)(PVdF-HFP) based composite electrolytes for lithium batteries. *Eur. Polym. J.* **2006**, *42*, 1728–1734. [CrossRef]
31. Yang, L.; Zhao, Q.; Hou, Y. Flexible polyvinylidene fluoride based nanocomposites with high and stable piezoelectric performance over a wide temperature range utilizing the strong multi-interface effect. *Compos. Sci. Technol.* **2019**, *174*, 33–41. [CrossRef]
32. Sahay, R.; Teo, C.J.; Chew, Y.T. New correlation formulae for the straight section of the electrospun jet from a polymer drop. *J. Fluid Mech.* **2013**, *735*, 150–175. [CrossRef]
33. Zheng, J.Y.; Zhuang, M.F.; Yu, Z.J. The effect of surfactants on the diameter and morphology of electrospun ultrafine nanofiber. *J. Nanomater.* **2014**, *2014*, 8. [CrossRef]
34. Chen, R.; Liu, J.; Yang, C. Transparent Impact-Resistant Composite Films with Bioinspired Hierarchical Structure. *ACS Appl. Mater. Interfaces* **2019**, *11*, 23616–23622. [CrossRef] [PubMed]
35. Artabiei, S.N.A.; Sultan, M.T.H.; Jawaid, M.; Jayakrishna, K. Impact behaviour of hybrid composites for structural applications: A review. *Compos. Part B Eng.* **2018**, *133*, 112–121.
36. Armstrong, R.W.; Walley, S.M. High strain rate properties of metals and alloys. *Int. Mater. Rev.* **2008**, *53*, 105–128. [CrossRef]
37. Tabiei, A.; Nilakantan, G. Ballistic impact of dry woven fabric composites: A review. *Appl. Mech. Rev.* **2008**, *61*, 10801. [CrossRef]

38. Song, W.J.R.; Tippabhotla, S.K.; Tay, A.A.O.; Budiman, A.S. Numerical Simulation of the Evolution of Stress in Solar Cells during the Entire Manufacturing Cycle of a Conventional Silicon Wafer-based Photovoltaic Laminate. *IEEE J. Photovolt.* **2018**, *8*, 210–217. [CrossRef]
39. Budiman, A.S.; Illya, G.; Handara, V.; Caldwell, W.A.; Bonelli, C.; Kunz, M.; Tamura, N.; Verstraeten, D. Enabling Thin Silicon Technologies for Next Generation c-Si Solar PV Renewable Energy Systems using Synchrotron X-ray Microdiffraction as Stress and Crack Mechanism Probe. *Sol. Energy Mater. Sol. Cells* **2014**, *130*, 303–308. [CrossRef]
40. Tippabhotla, S.K.; Radchenko, I.; Song, W.J.R.; Illya, G.; Handara, V.; Kunz, M.; Tamura, N.; Tay, A.A.O.; Budiman, A.S. From Cells to Laminate: Probing and Modeling Residual Stress Evolution in Thin Silicon Photovoltaic Modules using Synchrotron X-ray Micro-Diffraction Experiments and Finite Element Simulations. *Prog. Photovolt.* **2017**, *25*, 791–809. [CrossRef]
41. Tippabhotla, S.K.; Song, W.J.R.; Tay, A.A.O.; Budiman, A.S. Effect of Encapsulants on the Thermomechanical Residual Stress in the Back-Contact Silicon Solar Cells of Photovoltaic Modules—A Constrained Local Curvature Model. *Sol. Energy* **2019**, *182*, 134–147. [CrossRef]
42. Tian, T.; Morusupalli, R.; Shin, H.; Son, H.; Byun, K.; Joo, Y.; Caramto, R.; Smith, L.; Shen, Y.; Kunz, M.; et al. On the Mechanical Stresses of Cu Through-Silicon Via (TSV) Samples Fabricated by SK Hynix vs. SEMATECH—Enabling Robust and Reliable 3-D Interconnect/Integrated Circuit (IC) Technology. *Proc. Eng.* **2016**, *139*, 101–111. [CrossRef]
43. Ali, I.; Tippabhotla, S.K.; Radchenko, I.; Al-Obeidi, A.; Stan, C.V.; Tamura, N.; Budiman, A.S. Probing Stress States in Silicon Nanowires during Electrochemical Lithiation using In Situ Synchrotron X-ray Microdiffraction. *Front. Energy Res.* **2018**, *6*, 19. [CrossRef]
44. Tippabhotla, S.K.; Radchenko, I.; Stan, C.V.; Tamura, N.; Budiman, A.S. Stress Evolution in Silicon Nanowires during Electrochemical Lithiation Using In Situ Synchrotron X-ray Microdiffraction. *J. Mater. Res.* **2019**, *34*, 1622–1631. [CrossRef]
45. Budiman, A.S.; Shin, H.-A.-S.; Kim, B.-J.; Hwang, S.-H.; Son, H.-Y.; Suh, M.-S.; Chung, Q.-H.; Byun, K.-Y.; Tamura, N.; Kunz, M.; et al. Measurements of Stresses in Cu and Si around Through-Silicon Via by Synchrotron X-Ray Microdiffraction for 3-Dimensional Integrated Circuits. *Microelectron. Rel.* **2012**, *52*, 530–533. [CrossRef]
46. Budiman, A.S.; Illya, G.; Anbazhagan, S.; Tippabhotla, S.K.; Song, W.J.R.; Sahay, R.; Tay, A.A.O. Enabling lightweight PC-PC Photovoltaics Module Technology—Enhancing Integration of Silicon Solar Cells into Aesthetic Design for Greener Building and Urban Structures. *Sol. Energy* **2021**, *227*, 38–45.
47. Kim, I.; Kim, T.; Lee, S.; Matei, B. Extremely Foldable and Highly Transparent Nanofiber-Based Electrodes for Liquid Crystal Smart Devices. *Sci. Rep.* **2018**, *8*, 11517. [CrossRef] [PubMed]
48. Matei, E.; Busuioc, C.; Evanghelidis, A.; Zgura, I.; Enculescu, M.; Beregoi, M.; Enculescu, I. Hierarchical Functionalization of Electrospun Fibers by Electrodeposition of Zinc Oxide Nanostructures. *Appl. Surf. Sci.* **2018**, *458*, 555–563. [CrossRef]
49. Wu, M.; Wu, Y.; Liu, Z.; Liu, H. Optically Transparent Poly(methyl methacrylate) Composite Films Reinforced with Electrospun Polyacrylonitrile Nanofibers. *J. Compos. Mater.* **2012**, *46*, 2731–2738. [CrossRef]
50. Ali, H.P.A.; Radchenko, I.; Li, N.; Budiman, A.S. The Roles of Interfaces and Other Microstructural Features in Cu/Nb Nanolayers as Revealed by In Situ Beam Bending Experiments Inside a Scanning Electron Microscope (SEM). *Mater. Sci. Eng. A* **2018**, *738*, 253–263.

Article

Enhancement of Light Amplification of CsPbBr$_3$ Perovskite Quantum Dot Films via Surface Encapsulation by PMMA Polymer

Saif M. H. Qaid [1,2,*], Hamid M. Ghaithan [1], Khulod K. AlHarbi [1], Bandar Ali Al-Asbahi [1,3] and Abdullah S. Aldwayyan [1,4,5]

[1] Physics and Astronomy Department, College of Science, King Saud University, Riyadh 11451, Saudi Arabia; hghaithan@ksu.edu.sa (H.M.G.); 438204190@student.ksu.edu.sa (K.K.A.); balasbahi@ksu.edu.sa (B.A.A.-A.); dwayyan@ksu.edu.sa (A.S.A.)
[2] Department of Physics, Faculty of Science, Ibb University, Ibb 70270, Yemen
[3] Department of Physics, Faculty of Science, Sana'a University, Sana'a 12544, Yemen
[4] King Abdullah Institute for Nanotechnology, King Saud University, Riyadh 11451, Saudi Arabia
[5] K.A. CARE Energy Research and Innovation Center at Riyadh, Riyadh 11451, Saudi Arabia
* Correspondence: sqaid@ksu.edu.sa

Citation: Qaid, S.M.H.; Ghaithan, H.M.; AlHarbi, K.K.; Al-Asbahi, B.A.; Aldwayyan, A.S. Enhancement of Light Amplification of CsPbBr$_3$ Perovskite Quantum Dot Films via Surface Encapsulation by PMMA Polymer. *Polymers* **2021**, *13*, 2574. https://doi.org/10.3390/polym13152574

Academic Editor: Bożena Jarząbek

Received: 9 July 2021
Accepted: 28 July 2021
Published: 2 August 2021

Publisher's Note: MDPI stays neutral with regard to jurisdictional claims in published maps and institutional affiliations.

Copyright: © 2021 by the authors. Licensee MDPI, Basel, Switzerland. This article is an open access article distributed under the terms and conditions of the Creative Commons Attribution (CC BY) license (https://creativecommons.org/licenses/by/4.0/).

Abstract: Photonic devices based on perovskite materials are considered promising alternatives for a wide range of these devices in the future because of their broad bandgaps and ability to contribute to light amplification. The current study investigates the possibility of improving the light amplification characteristics of CsPbBr$_3$ perovskite quantum dot (PQD) films using the surface encapsulation technique. To further amplify emission within a perovskite layer, CsPbBr$_3$ PQD films were sandwiched between two transparent layers of poly(methyl methacrylate) (PMMA) to create a highly flexible PMMA/PQD/PMMA waveguide film configuration. The prepared perovskite film, primed with a polymer layer coating, shows a marked improvement in both emission efficiency and amplified spontaneous emission (ASE)/laser threshold compared with bare perovskite films on glass substrates. Additionally, significantly improved photoluminescence (PL) and long decay lifetime were observed. Consequently, under pulse pumping in a picosecond duration, ASE with a reduction in ASE threshold of ~1.2 and 1.4 times the optical pumping threshold was observed for PQDs of films whose upper face was encapsulated and embedded within a cavity comprising two PMMA reflectors, respectively. Moreover, the exposure stability under laser pumping was greatly improved after adding the polymer coating to the top face of the perovskite film. Finally, this process improved the emission and PL in addition to enhancements in exposure stability. These results were ascribed in part to the passivation of defects in the perovskite top surface, accounting for the higher PL intensity, the slower PL relaxation, and for about 14 % of the ASE threshold decrease.

Keywords: CsPbBr$_3$ perovskite QDs; amplified spontaneous emission (ASE); light amplification; surface passivation; photostability

1. Introduction

Powerful and intense photoluminescence (PL), low non-radiative recombination rates, and long carrier lifetimes in pure and mixed perovskites have expanded their application in optoelectronic devices, such as light-emitting diodes (LEDs), lasers, and photodetectors [1–9]. Despite these features, controlling the morphology and thickness of perovskite films remain crucial to achieving high efficiency in a LED or in electricity production as defects and traps are essential to the movement of carriers in the material. Although inorganic cations (e.g., CsPbX$_3$) show relatively improved stability compared with those of organic–inorganic hybrid counterparts (e.g., MAPbX$_3$ and FAPbX$_3$), CsPbX$_3$ PQDs in practical operation are still very sensitive to polar solvents and moisture, anion exchange reactions, and heating. All of these are due to the low formation energy of the crystal lattice and the high decentralization activity of surface ions [2–7]. CsPbX$_3$ nanocrystals (NCs) have received attention for their remarkable optoelectronic properties [8,9]. QDs, as well as other nanomaterials, have a large specific surface area, which greatly affects their intrinsic properties. Their inherent instability impedes further development and future application of CsPbX$_3$ PQDs in optoelectronics and in other fields. Hence, it is crucial to explore an effective pathway to enhance the stability of CsPbX$_3$ perovskite QDs.

To enhance the stability of PQDs, different protective strategies have been proposed by either modifying the surface ligand molecules of the PQDs [3,10–16] or by encapsulating PQDs into inorganic dielectric materials. Some examples of these strategies include ligand engineering, shell design, overcoating, and compositing PQDs with other materials. Composite materials, such as polymers, oxides, metallic ions, and other inorganics and organic options, passivate the PQD surface and form a protective layer [16–18]. All of these strategies and methods are practical and promising. For example, a surface modification strategy, by modifying ligand molecules, can passivate surface defects or dangling bonds to improve material stability. These encapsulation strategies can protect PQDs from exposure to corrosive exogenous species and enhance the stability of PQDs and other properties expected to improve performance in photonics, electronics, sensors, and other fields. The encapsulating can involve encapsulation by inorganic dielectric materials, for example, SiO$_2$ [15,16,19,20], TiO$_2$ [18,21,22], and organic poly(methyl methacrylate) (PMMA) polymer matrixes [23–26]. There are other methods of encapsulating using diverse structures, such as shelling the QDs at the single-particle level, encapsulating QDs in a broad matrix, loading QDs on a surface, ionic doping in the lattice of halide perovskite quantum dots, or forming halide perovskite QDs/QD nanocomposites. In addition to using these strategies to protect perovskites, the strategies contribute to the light amplification process. The light amplification and lasing properties of CsPbX$_3$ PQDs can be enhanced through both high-quality surface passivation and high perovskite NC filling factors. There is feasibility of NC-based amplifiers and lasers tunable in the visible range because of their large energy gap (>1.75 eV), but these NCs cannot be used for light amplification in the infrared spectral range. Chemically synthesized NCs exhibit a wide range of size-controlled tunability of emission color and high PL quantum yields. These properties make NCs attractive materials for light-emitting applications ranging from bio-labeling and solid-state lighting to optical amplification and lasing [27].

The ASE in the CsPbX$_3$ QDs were stimulated at ~7–15 µJ cm^{-2} and ~450 µJ cm^{-2} thresholds for excitation by femtosecond and nanosecond laser pulses, respectively [28]. Additionally, the ASE thresholds of CsPbBr$_3$ thin films under femtosecond laser system were reported and two-photons using a femtosecond laser system as pumping lasers were 192 µJ cm^{-2} and 12 mJ cm^{-2}, respectively [28,29]. Another group studies the role of the ligand in improvements of stability problem of CsPbBr$_3$ QDs in air [14]. The ASE thresholds of CsPbBr$_3$ thin films were demonstrated under picosecond laser excitation around 22.5 µJ cm^{-2}, which confirmed the role of laser pulse duration in determined the ASE threshold [30]. In one of the previously published works, the ASE threshold in CsPbBr$_3$ quantum dot films was reducing through controlling the TiO$_2$ compact layer under perovskite film by the reduced roughness of the obtained films to less than 5 nm with 50 nm TiO$_2$ substrate [31].

In view of these reports, which listed different ways to improve the properties of ASE, the accelerated research activity has allowed many groups interested in studying perovskite materials and their applications in light-emitting, to propose approaches to improve the ASE and lasing properties that address a number of aspects, such as reducing the threshold of ASE/laser, laser cavity geometry, wavelength range engineering, and stability improvement. However, the inherent instability of perovskite materials hampers further development and future application of these materials in optoelectronics and in other fields.

Herein, a simple strategy will be demonstrated for improving optical properties by modifying the upper and lower surfaces of a CsPbBr$_3$ film by coating it with PMMA polymer, which leads to improving the perovskites materials stability and light amplification at the same time. Then, the basic physics of light amplification in the CsPbBr$_3$ PQD with emission energies in the visible range was analyzed. For this reason, high-quality films were fabricated directly from CsPbBr$_3$ PQD in powder form. The top face of the perovskite layer was coated by PMMA, and the CsPbBr$_3$ PQD films were placed between two transparent PMMA layers. These simple methods result in the formation of a highly flexible ultra-light PMMA/PQD/PMMA waveguide film configuration used to examine the optical response of perovskite films after PMMA surface passivation and waveguide fabrication.

2. Materials and Methods

2.1. Materials

Cesium lead bromide quantum dot powders was purchased from Quantum Solutions Company (Thuwal, Saudi Arabia, www.qdot.inc (accessed on January 2021)). N-Hexane analytical reagent solution was purchased from (Avonchem, Cheshire, UK). Poly(methyl methacrylate) (PMMA) with an average molecular weight of ~120,000 g/mol was purchased from Sigma-Aldrich (Saint Louis, MO, USA). All chemicals were used as received, without further purification.

2.1.1. Fabrication of CsPbBr$_3$ QD Solution and Thin Films

The powdered CsPbBr$_3$ PQDs were directly dispersed into hexane (25 mg/mL) for suspension. Then, the suspension was left overnight to ensure complete dispersion before thin film fabrication. To fabricate thin films, CsPbBr$_3$ PQDs were coated onto pre-cleaned microscope glass (1 × 2 cm^2) substrates. The PQD mixture (50 µL/cm^2) was dropped onto the substrate and spin-coated at 4000 rpm for 30 s. Then, the films were dried under vacuum for 1 h. The CsPbBr$_3$ PQDs film thickness could be adjusting to 300 nm (estimated from a Dektak 150 stylus profiler (Bruker Corp, Tucson, AZ, USA)) in all configurations to compare the difference on ASE performance.

2.1.2. Preparation of PMMA Solution and Modification of the Perovskite Surface for Encapsulation

First, the PMMA stock solution was prepared in toluene (25 mg/mL). For the encapsulation of PQDs, PMMA thin film was prepared by depositing 25 µL/cm^2 of the PMMA solution onto PQD films using a spin-coating procedure (6500 rpm for 30 s) under ambient conditions. Next, the PMMA/CsPbBr$_3$ PQD films were dried in ambient air also. Finally, for PMMA thickness measurements, pure PMMA films were condensed from the stock solution onto a clean glass substrate by the spin-coating procedure. The film thickness could be adjusting to the required PMMA layer thickness (100 nm). After that section, the thickness will be referred to the PQD structure only. Figure 1 shows a schematic for all configurations CsPbBr$_3$/glass, PMMA/CsPbBr$_3$/glass, and PMMA/CsPbBr$_3$/PMMA/glass, respectively.

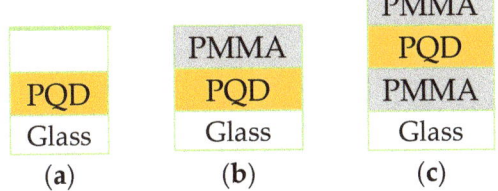

Figure 1. Configuration schematic of the (**a**) PQD; (**b**) PMMA/PQD; (**c**) PMMA/PQD/PMMA; All configurations fabricated on glass substrates.

2.2. Characterization

Structural characterization: Quantum dot structure and morphology of the $CsPbBr_3$ perovskite were analyzed via transmission electron microscopy (TEM; JEOL JEM-1011, JEOL, Tokyo, Japan). The samples were prepared by adding a few drops of dilute PQD solution onto TEM grids. The crystallization structures and the crystal phase of $CsPbBr_3$ PQDs were characterized using X-ray diffraction (XRD) analysis (Miniflex 600 XRD, Rigaku, Japan) with a copper Kα radiation source (λ = 1.5418 Å). The scanning range was 2θ = 10°–80° for a scan rate of 3° min^{-1} with a step size of 0.02°.

2.2.1. Optical Characterization

Absorption and photoluminescence (PL) measurements of the PQD thin films were recorded in the 350–700 nm spectral range using a V-670 UV-vis spectrophotometer (JASCO Corp., Tokyo, Japan) and a fluorescence spectrophotometer (Lumina, Thermo Fisher Scientific, Madison, WI, USA), respectively. In both measurements, a portion of the $CsPbBr_3$ PQD suspension was dispersed onto a microscopic slide with a thickness of approximately 300 nm. The resulting films were checked via observation by naked eye under a UV lamp (model XX15NF, Spectroline, ME, USA) at 365 nm. Furthermore, steady-state measurements and time-resolved PL (TRPL) were performed using a Shamrock SR-500i spectrometer (Andor Technology Co, Belfast, UK) equipped with an MS257 ICCD detector (Lot Oriel Instruments, Stratford, CT, USA). For sample excitation, a pulsed laser was used via the third harmonic generation of a Q-switched Nd: YAG nanosecond laser (Solara, LPS 1500, 3rd harmonic, wavelength: 355 nm, pulse width: 11 ns, repetition rate: 100 Hz, energy density: 1.5 µJ cm^{-2}). To collect the laser excitation pulse from the detector and select the wavelength emitted from the sample, special filters will be used for this purpose. Moreover, it will be use a lens to collect and focus light emitted from the samples. The resulting emission is spectrally resolved using a spectrograph and detected by a gated intensified and a sufficiently sensitive ICCD camera. By a sequential shift of the gate window (to change time delay over a range from 1 ns up to 1 ms) with respect to the excitation, it is possible to measure the spectrally resolved decay of the photoluminescence, providing information about the excited state. A schematic diagram of the ICCD setup experimental used to measure the TRPL is shown in Figure 2a.

2.2.2. Laser Experiments and ASE Measurements

To investigate the ASE characteristics, energy-dependent ASE intensity spectra were collected at the sample edges, in particular near the ends of the excitation strips. A LOTUS II Q-switched Nd:YAG picosecond laser (LOTIS, Belarus) with a pulse duration of 70–80 ps at a repetition rate of 15 Hz was used for excitation while using an LT-2215-OPG optical parametric generator (OPG) with a tunable range of 425–2300 nm. Then, a cylindrical lens was used as a focusing tool to create a narrow excitation stripe with a 100 µm width of variable length on the sample surface to be sure that the collection efficiency and intensity profile were effectively constant across the lengths of the stripe used. The light emitted by the samples was detected from the edge of the waveguides by using an optical fiber connected to A QE65 Pro spectrograph (Ocean Optics, Inc., Dunedin, FL, USA). To enable the study of the threshold dependence on energy density, the laser energy density was attenuated using a variable neutral density filter wheel and the energy was read by an LM-P-209 coherent thermal sensor head. Finally, to analyze the data collected from ASE experiments and to obtain gaussian fits of dual PL and ASE emission peaks, a custom python-based program, developed by our research group, was used. A scheme diagram of the laser setup experimental used to investigate the presence of stimulated emission is shown in Figure 2b.

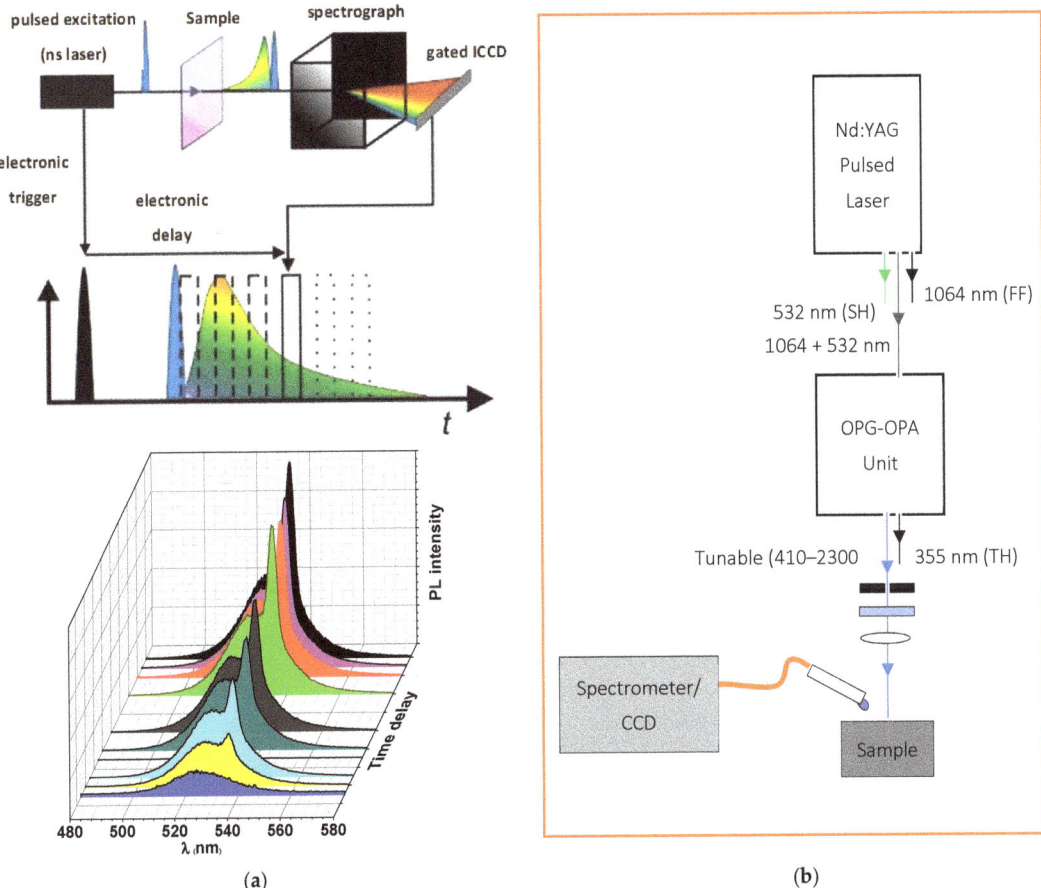

Figure 2. (**a**) The upper part shows the schematic of the ICCD setup with the beam path and the shifted integration window compared to the excitation window. The lower graph shows the building up of the measured signal over several single measurements; (**b**) Schematic showing the High Power Picosecond Pulsed Time Integrated PL Setup for characterizing the presence of stimulated emission in thin-film configuration.

3. Results

3.1. Structural Characteristics

The TEM image in Figure 3a shows the structure of the CsPbBr$_3$ PQDs material and reveals that the CsPbBr$_3$ PQDs have a uniform shape and homogeneous size distribution. The particle sizes range from ~4 to 11 nm with an average particle size of ~7.5 nm. Figure 3b shows the XRD patterns of perovskite films with and without PMMA polymer coating. The XRD patterns have characteristic peaks at ($2\theta = 15.54°$, $21.90°$, $31.09°$, and $51.56°$), which correspond to diffraction from (100), (110), (200), and (311) crystal planes, respectively. All peaks were indexed to cubic phase in the Pm-3m space group (221) and XRD pattern samples could be indexed to the pure cubic phase of CsPbBr$_3$ (JCPDS card no. 01-075-0412), with slight peak shifts. Slight shifts in peak positions (~0.6°) were consistent with alloy formation and the results are well in line with previous reports [31]. The peak shifts decrease when the top surface of the PQD was modified by the PMMA polymer; they revert to appear as in the bare surface of PQD when the PQD top and bottom surface are modified. Although, the XRD peak at 51.56° was apparent from pristine-CsPbBr$_3$ PQDs and was not apparent from polymer due to the polymers did not readily absorb in the

X-ray region [26,32,33], as evidenced by the XRD patterns of the experimental samples. The strong peak after adding the polymer may be attributed to the film quality improvement. Diffraction from the (200) plane was apparent, along with the secondary diffraction peak of the (100) plane, indicating the existence of a very pure and crystalline cubic phase, without any defects. The appearance of the peak at 28.56° in pristine-CsPbBr$_3$ PQDs exhibited a mixture of predominant cubic phase and a minor portion of the orthorhombic phase, which maybe be attributed to the stored age of PQDs under ambient conditions in this experiment is 6 months from production [34].

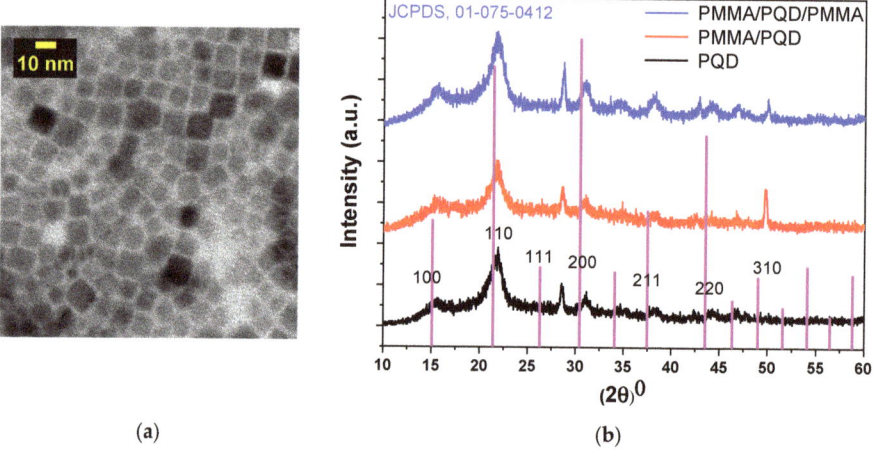

Figure 3. (a) TEM image of CsPbBr$_3$ PQDs; the CsPbBr$_3$ QDs sample for TEM investigation was prepared by the dilution of CsPbBr$_3$ QDs solution to (125 µg/1 mL), followed by placing several drops on a carbon-coated copper grid (b) X-ray diffraction patterns of CsPbBr$_3$ PQDs films for all configurations fabricated on glass substrates.

The Scherrer formula ($D = 0.9 \frac{\lambda}{\beta \cos \theta}$) was used to estimate the crystallite size (D). Additionally, the dislocation density (δ) and lattice strain (ε) are given by $\delta = \frac{1}{D^2}$ and $\varepsilon = \beta \cos \theta / 4$, respectively [30,35]. β represents peak broadening (FWHM), λ is the wavelength of the incident X-ray (0.154 nm), and k is a constant (~0.9). Table 1 lists the values of XRD parameters.

Table 1. X-ray diffraction (XRD) parameters for various samples.

Sample	FWHM (Degrees)	D (nm)	Lattice Strain $\varepsilon \times 10^{-3}$	Dislocation Density $\delta \times 10^{-3}$ (nm)$^{-2}$
(PMMA/PQDs/PMMA)	1.35	6.0	5.80	27.98
(PMMA/PQDs)	1.22	6.6	5.22	22.69
Pure PQDs	1.40	5.8	5.99	29.87

The crystallite sizes were estimated from the Scherrer formula to be 5.8, 6.6, and 6 nm for pure PQDs, PMMA/PQDs, and PMMA/PQDs/PMMA, respectively. Compared with the TEM image, the XRD results were consistent and broadly in agreement. The gradual formation of the perovskite layer reduced the strain. The narrow linewidth of the diffraction peaks in the XRD patterns (FWHMs) indicated the lowering of residual stress in the crystals and a low dislocation density; generally, a high-quality perovskite film with a low density of defect states [35]. Since PMMA is a highly transparent amorphous polymer, it does not exhibit any sharp diffraction peaks in the XRD spectra due to not readily absorb in the X-ray region [36]. The agglomeration of PQDs on the polymer segments is the reason for the observed increase in grain size. The sandwiched PQD by PMMA shows a lower grain size than PMMA/PQD. The reason behind this may be due to the confined of the QDs between the PMMA layer. Additionally, grain growth rates increase in the film because of stresses in the film, which are created within the film and substrate and usually with no dislocations at the interface.

3.2. UV-Vis Absorption and Steady-State Photoluminescence Properties

Figures 4 and 5a show the UV-vis absorption and steady-state photoluminescence (PL) spectra of the bare perovskite film and that modified by PMMA polymer in one and two faces. These results correspond well to perovskite film results reported for $CsPbBr_3$ films, with only a change in intensity [30,31]. Additionally, the high crystallinity of the perovskite film was confirmed by the observation of narrow-band emission (FWHM) located at 516 nm, which indicates a low density of defect states as shown in XRD results.

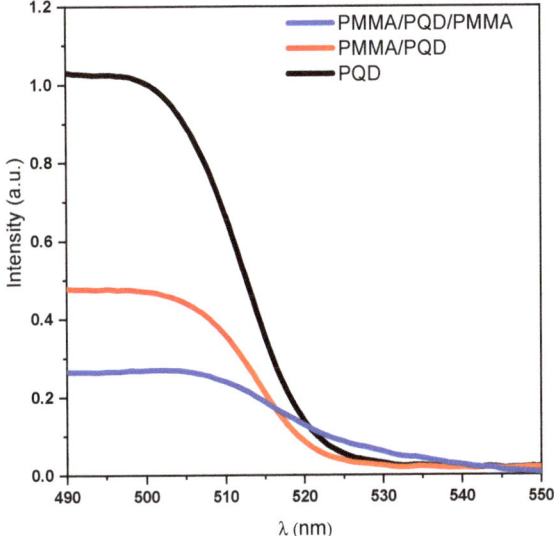

Figure 4. Optical absorption spectra of $CsPbBr_3$ PQDs films at room temperature.

Thus, the high-quality of the perovskite film came from the smoothing of the surface of the film after covering it with a polymer layer, which is evident from the increase in the emission intensity. After covering the top of the perovskite film with a polymer layer, the PL intensity increased at the same pump fluence compared with that of the bare film (Figure 5a). This is expected due to the fact that the polymer layer improves the interface and thus the smooth surface reduces the loss of pumping light incident at the air interface with the PQD thin film [31,33,37]. Although this increase is reduced when the perovskite film is sandwich between two polymer layers. The role of the layers that sandwiched the perovskite films plays is to redirect the emission to propagate along the path inside the perovskite to go out from the edge. This behavior is attributed to the steady-state photoluminescence (PL) measurements taken at a 45° angle in the PL steady-state. Therefore, the losses of the output light from the PMMA/PQD/PMMA sample will be much higher than the other samples. Whereas the emission is guided by the two PMMA polymer layers; this high emission taken from the edge will be discussed later in the ASE studies. The inset of Figure 5a shows a photograph of the film taken under UV light (λ_{ex} = 365 nm).

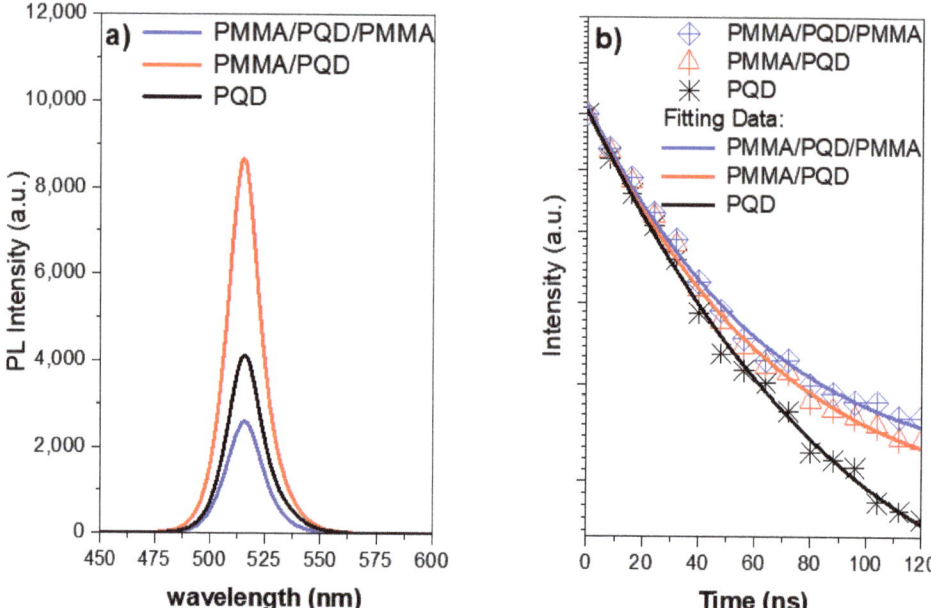

Figure 5. (a) Steady-state PL spectra of CsPbBr$_3$ PQD films measured at room temperature under 410 nm excitation. The inset shows a real-color image CsPbBr$_3$ PQD excited under 365 nm UV lamp; (b) TRPL decay plots of CsPbBr$_3$ PQD thin films at low pump energy densities.

Additionally, to further understand the effect of the surface passivation layer in the perovskite top surface, time-resolved PL (TRPL) studies were conducted when created photo-generated carriers after the laser excitation pulses as can be illustrated in the characterization section and shown in Figure 5b. The PL lifetimes and decay component can be measured from studies of the emission intensity with decay time, which can be deduced from the PL decay curve. The PL decay profile was fitted at the peak position using single-exponential decay with $I(t) = A \exp(-t/\tau)$, where τ is the average lifetime. The PL decay curve with a component, shown in Figure 5b, revealed an emission with a PMMA/PQD/PMMA and PMMA/PQD decay time of ~17.0 ns and 16.4, respectively, which is longer than that of the PQD (~14.0 ns), which arose because of surface passivation which the passivation of defects in the top of perovskite surface [37,38]. So, the average lifetime increased with surface passivation and the passivation surface have a lower surface recombination velocity than that obtained from the bare surface. Moreover, the intensity of the coated film was considerably higher than that caused by the bare film due to slow PL relaxation dynamics (long decay time). These results were ascribed in part to the passivation of defects in the perovskite top surface.

3.3. Light Amplification and ASE Properties

Evaluation of photoluminescence with optical pumping: As demonstrated in previously published work [23], the threshold properties and gain characteristics of PMMA/perovskite can be controlled by changing the polymer thickness. Here, Figure 6 shows the dependence of the excitation energy density of PL spectra in a wide energy range (low and high excitation energy). This figure shows the pump-dependent emission from three configurations, CsPbBr$_3$/glass, PMMA/CsPbBr$_3$/glass, and PMMA/CsPbBr$_3$/PMMA/glass, through a range of excitation energy densities at room temperature (T = 300 K).

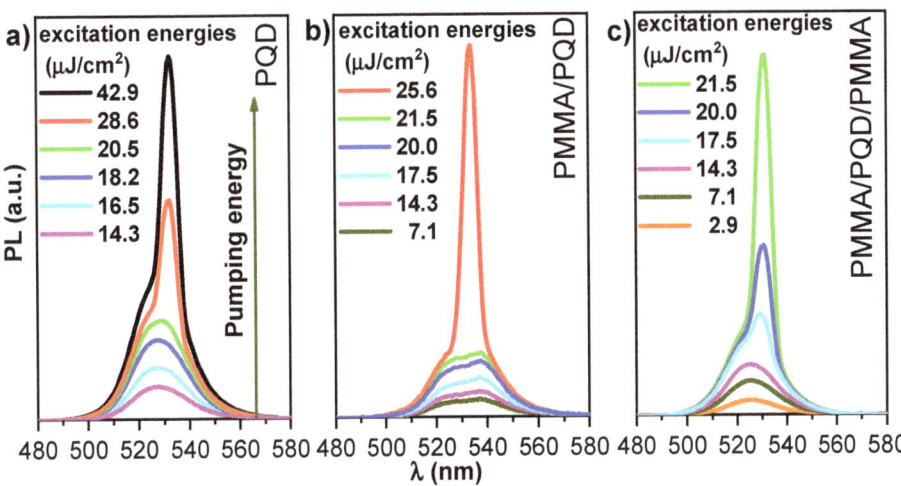

Figure 6. PL spectra measured at room temperature versus excitation energy density for (**a**) bare CsPbBr$_3$ PQDs; (**b**) PMMA/PQDs, and (**c**) PMMA/PQDs/PMMA with pulsed excitation (410 nm, 70 ps pulses, 15 Hz repetition rate).

In the CsPbBr$_3$/glass configuration, the transition from a broad PL spectrum to a narrower ASE feature, with the appearance of a narrow band peaked at 535 nm, occurred at 22.2 µJ/cm^2. By contrast, the configurations PMMA/CsPbBr$_3$ and PMMA/CsPbBr$_3$/PMMA show broad PL, even at the lowest excitation energy density, at 19.3 and 16.4 µJ/cm^2, respectively. Thus, the presented results confirmed that the ASE feature was observed for all configurations, but the use of a PMMA passivation layer consistently yielded stronger ASE density, lower ASE thresholds, and high emission control. At low excitation energy density, the PL peak at 525 nm has an FWHM of ~16.5 nm. Line–shape variation can be observed as the excitation density increases. Increasing the excitation energy density results in a transition from a broad PL spectrum to a narrower ASE feature with a narrow band peaked at 535 nm. At low pumping energy, the PL appeared at a broad peak, but when pumping energy increase until the pump energy reach threshold, the sharp peak appeared near the long wavelength. The broad peak dispersed at the ASE became dominant in this state. However, when the pump energy increases above the ASE threshold, a redshifted peak has multiple causes, such as defect transitions, thermal effects [39], and reabsorption effect arising from the overlap of the absorption band edge with the PL emission (spontaneous emission spectrum) (Figure 4), the self-absorption effect should contribute to the ASE state [40]. Moreover, as suggested by band gap renormalization in the highly excited perovskite crystal [41], the band gap is redshifted by hole–electron interactions under high population conditions. The peak transition from a broad PL spectrum to a narrower ASE feature also reflects the bandgap behavior. The wavelength of the PL peak shows a progressive red shift (~10 nm) up to the maximum investigated excitation energy density. Then, the emission changes from broad PL (FWHM of ~17 nm) at low fluence to ASE (FWHM of ~6 nm) at high fluence. The transition from a broad PL spectrum to a narrower ASE feature occurs at a threshold fluence of ~16–23 µJ/cm^2. The visible excitation energy density, estimated by determining the mean value of the minimum pump energy density that allows observation of the ASE regime, was measured in three different positions on the sample and was approximately 22.2, 19.3, and 16.4 µJ cm^{-2} for the CsPbBr$_3$/glass, PMMA/CsPbBr$_3$/glass, and PMMA/CsPbBr$_3$/PMMA configurations, respectively. The ASE threshold was estimated from PL excitation density dependence analysis by fitting the low excitation density data point and the high excitation density portion of the data with a constant and an increasing straight line, respectively, while considering the threshold crossing point between the two fitting lines (Figure 7). Additionally, the PL intensity of the coating film was stronger than the intensity of the bare film, as can be seen from Figure 7. This mechanism is often referred to as a smoother surface after covering the PMMA polymer layer, the passivation of defects in the perovskite top surface, and as index guiding since the refractive index discontinuity between the active (CsPbBr$_3$ PQD) and cladding (PMMA) layers is responsible for mode confinement through total internal reflection occurring at the interface. Here, the refractive index of cladding is almost identical to that of the glass substrate (n = 1.5) and smaller than that of active layer (n ≈ 2); at λ= 410 nm [42–44]. The inset of Figure 7 shows that the PMMA/CsPbBr$_3$/PMMA configuration has a higher quantum efficiency (as can be deduced from its larger slope in the linear region). Indeed, the strong contrast between the three configurations clearly illustrates the importance of the enhancement of the film surface and light guiding in the optical and gain characteristics of the perovskite films. The PMMA-coated film showed slower PL relaxation dynamics (long decay time). These results were ascribed in part to the passivation of defects in the top of perovskite surface, accounting for the higher PL intensity, the slower PL relaxation and for about 14 % of the ASE threshold decrease. The remaining ASE threshold decrease was instead ascribed to improved waveguiding, which was facilitated by the realization that an almost symmetric glass-perovskite-PMMA waveguide results in higher mode confinement in the perovskite layer concerning the asymmetric glass–perovskite layer. This suggests that the PMMA layer redirects emission that propagates along the out-of-plane and oblique paths back into the bulk. Consequently, the optical path length in the medium is increased, resulting in a lower ASE threshold compared with the bare PQDs. Here, the PMMA layer is effectively improving the waveguiding capability of the perovskite film, these findings compatible with the previous studies in the open literature [23,31,37,38,45,46]. Additionally, in all of these, the ASE threshold could further reduce with one or both reflectivity/encapsulation substrates.

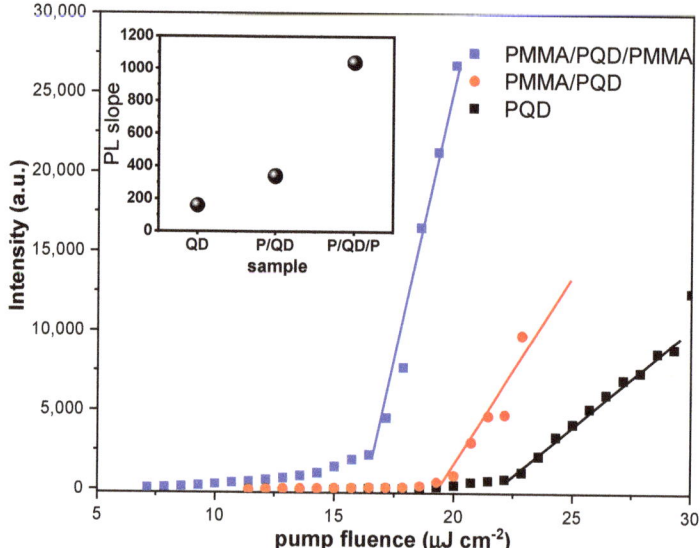

Figure 7. Output light-input light curve of the emission spectra versus excitation pump energy. The inset shows the quantum efficiency behavior.

Figure 8 depicts the ASE intensity dependence as a function of carrier density (n). The carrier densities are calculated from the absorbance spectra (Figure 4) and film thickness (estimated from a Dektak 150 stylus profiler). The calculation method has been explained in detail in previously published work [31,47]. The transition from spontaneous emission (SE) to ASE occurs, with a clear onset, at ~1.8–6.7 × 10^{18} cm^{-3} as shown from spectra that were magnified at the onset of ASE as shown as in the inset of Figure 8. The carrier density threshold was estimated from PL excitation density dependence analysis by fitting the low excitation density data point and the high excitation density portion of the data with a constant and an increasing straight line, respectively, while considering the threshold crossing point between the two fitting lines. From Figure 8, CsPbBr$_3$/glass shows a sharp SE to ASE transition at a pump fluence of 22.2 µJ/cm^2. This fluence value corresponds to an ASE threshold carrier density of ~6.7 × 10^{18} cm^{-3}. This sample showed slower ASE growth with increasing pump fluence. A sharp SE to ASE transition for PMMA/CsPbBr$_3$/glass and PMMA/CsPbBr$_3$/PMMA/glass appeared at a pump fluence of 19.3 and 16.4 µJ cm^{-2}, respectively. These threshold values correspond to a threshold carrier density of ~2.8 × 10^{18} cm^{-3} and 1.8 × 10^{18} cm^{-3}, respectively. Thus, the PMMA/CsPbBr$_3$/PMMA/glass configuration is the best for gain and light amplification applications. This is because, for very high carrier densities, the Coulomb interaction between electrons and holes can induce additional effects. Auger recombination is one effect that can severely reduce the gain. The bandgap energy and exciton binding energy are both pushed to transfer to the lower values at carrier density pushed to higher values. This process is called renormalization, whereas when all excitons in the electron hole plasma are ionized at the threshold point of carrier density, a "Mott transition" occurs.

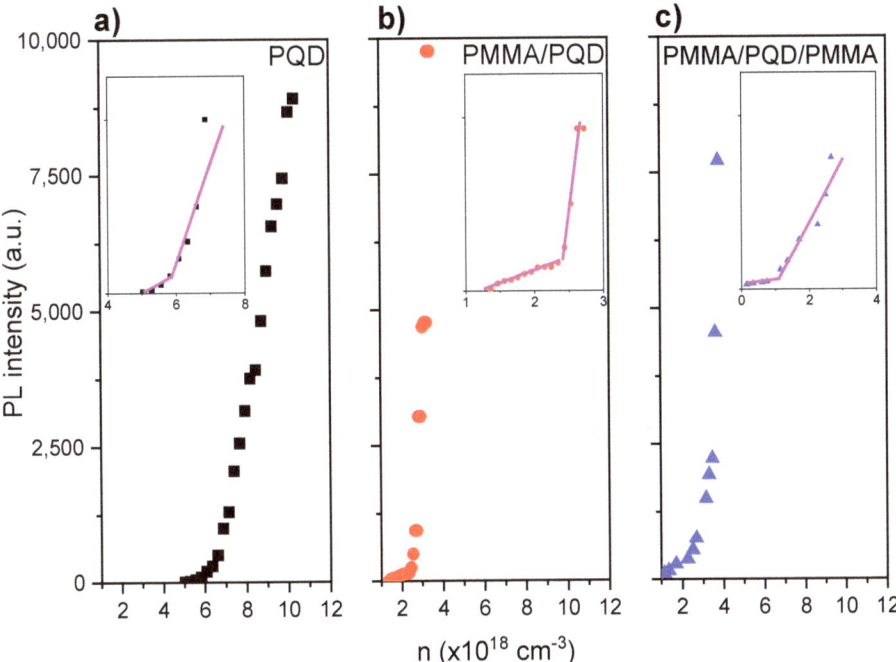

Figure 8. Integrated PL versus carrier density, n. (**a**) bare CsPbBr$_3$ PQDs, (**b**) PMMA/PQDs, and (**c**) PMMA/PQDs/PMMA films deposited on the microscopic glass.

Photostability studies: The time-dependent ASE intensity of synthesized films, before and after PMMA coating, with over 120 min laser excitation lasts without interruption (for 108,000 excitation shots). Figure 9 shows the ASE stability under long excitation by picosecond laser pulses (pulse width is ≈70 ps and repetition rate is 15 Hz). For an excitation wavelength of 410 nm, the excitation energy is adjusted above the threshold (~twice the energy threshold) with a run at room temperature and under atmospheric conditions. It is noted that, for the film coated by PMMA, the intensity output does not deviate from the original output until more than 110,000 excitation shots. By contrast, the ASE, even after 80,000 shots, from the bare sample without coating, can remain at approximately 95% of its original emission intensity. Thus, the ASE photostability of synthesized films is significantly improved after PMMA coating. The improved stability can be ascribed to ligands engineering (lengths of ligands—the short branched chains) of the $CsPbBr_3$ PQD [48,49] and also by incorporation into hydrophobic polymer matrices [45]. The short branched chains will be increase the binding energy between the ligands and QDs and the lengths of ligands, which is related to the strength of the van der Waals (VDW) interactions among the ligands, and the strength is dominant to determine the crystalline structure and follow the optical properties of PQDs [49–52].

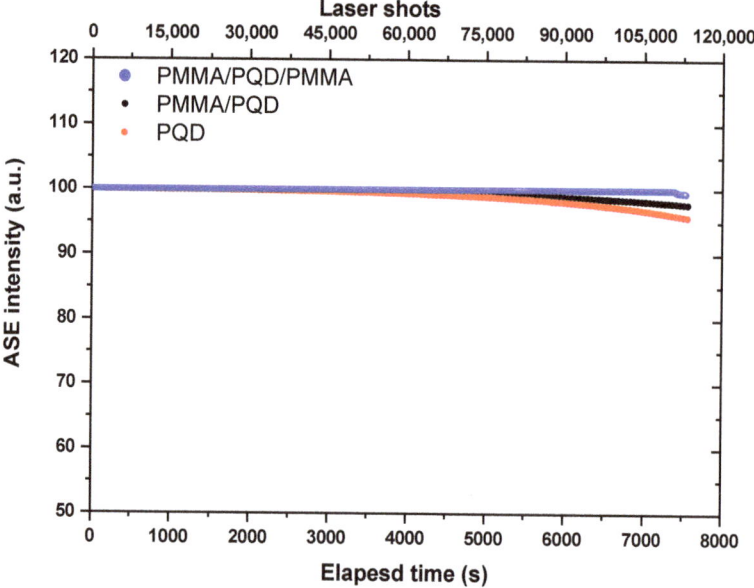

Figure 9. ASE intensity for $CsPbBr_3$ PQDs stability studies under a constant pulsed excitation density of 40 µJ/cm^2 for over 120 min in an ambient atmosphere.

4. Discussion

At the beginning, the TEM image and XRD patterns for three configurations, $CsPbBr_3$/glass, PMMA/$CsPbBr_3$/glass, and PMMA/$CsPbBr_3$/PMMA/glass (Figure 3a,b) confirm that the grown of $CsPbBr_3$ PQD films are of high phase purity and the PQD have a uniform shape and homogeneous size distribution. All XRD peaks were indexed to cubic phase in the Pm-3m space group (221) and XRD pattern samples could be indexed to the predominant cubic phase with slight peak shifts, which are consistent with alloy formation [30,31]. The peak shifts decrease when the top surface of the PQD was modified by the PMMA polymer; they revert to appear as in the bare surface of PQD when the PQD top and bottom surface are modified. Although the PMMA polymer is highly transparent amorphous and do not readily absorb in the X-ray region, the strong peak after adding the polymer may be attributed to the film quality improvement. The UV-vis absorption results (Figure 4) correspond well to perovskite film without any effect for polymer the change the $CsPbBr_3$ PQD band gap, with only a change in intensity [30,31].

The steady-state photoluminescence (PL) spectra of the bare perovskite film and that modified by PMMA polymer in one and two faces also was investigated. From the PL investigation, the PL intensity increased at the same pump fluence after covering the top of the perovskite film with a polymer layer compared with that of the bare film. This is expected due to the fact that the polymer layer improves the interface and thus the smooth surface reduces the loss of pumping light incident at the air interface with the PQD thin film [31,33,37]. This increase is reduced when the perovskite film is sandwich between two polymer layers (Figure 5a).

Distinctly, the PL decay profile was fitted at the peak position using single-exponential decay. As shown in Figure 5b. A PMMA/PQD and PMMA/PQD/PMMA were revealed decay times of ~17.0 ns and 16.4, respectively, which is longer than that of the PQD (~14.0 ns), which arose because of surface passivation which the passivation of defects in the perovskite top surface [37]. Figure 6 shows the pump-dependent emission from three configurations, through a range of excitation energy densities. The transition from a broad PL spectrum to a narrower ASE feature has occurred at 22.2, 19.3, and, 16.4 µJ/cm^2 for the configurations PQD, PMMA/CsPbBr$_3$, and PMMA/CsPbBr$_3$/PMMA, respectively (Figure 7). This improved PL and the subsequent slower emission are the essences of the observed reduction in the ASE threshold. The remaining ASE threshold decrease was instead ascribed to improved waveguiding, which was facilitated by the realization that an almost symmetric PMMA/perovskite/glass and PMMA/perovskite/PMMA/glass waveguide results in higher mode confinement in the perovskite layer concerning the asymmetric perovskite/glass layer.

Figure 8 depicts the ASE intensity dependence as a function of carrier density (n). The carrier densities are calculated from the absorbance spectra (Figure 4) and film thickness (estimated from a Dektak 150 stylus profiler). The calculation method has been explained in detail in previously published work [31,47]. The transition from spontaneous emission (SE) to ASE occurs, with a clear onset, at ASE threshold carrier density of ~6.7 × 10^{18} cm^{-3}, 2.8 × 10^{18} cm^{-3} and 1.8 × 10^{18} cm^{-3}, CsPbBr$_3$/glass, PMMA/ CsPbBr$_3$/glass, and PMMA/CsPbBr$_3$/PMMA/glass, respectively. Thus, the PMMA/CsPbBr$_3$/PMMA/glass configuration is the best for gain and light amplification applications. This is because, for very high carrier densities, the Coulomb interaction between electrons and holes can induce additional effects. Figure 9 shows ASE stability of the PQD films, before and after PMMA coating under long excitation with over 120 min laser excitation lasts without interruption (for 108,000 excitation shots) by picosecond laser pulses. It is noted that, for the film coated by PMMA, the intensity output does not deviate from the original output until more than 110,000 excitation shots. The improved stability can be ascribed to the incorporation into hydrophobic polymer matrices [45]. By contrast, the ASE, even after 80,000 shots, from the bare sample without coating, can remain at approximately 95% of its original emission intensity. Thus, the ASE photostability of synthesized films is significantly improved after PMMA coating.

For the film with coating, the much stronger and nearly invariant output ASE intensity suggests good optical stability. Such improvement is a result of reduced ASE threshold and increased ASE intensity at a certain pump density, which implies that less heat is generated during operation. As discussed above, it does not only reduce the ASE threshold but also improves the photostability of the films through a simple polymer coating process. This simple technique provides a pathway to improve the photostability of perovskite materials for a sustainable laser. Additionally, CsPbBr$_3$ PQDs are verified to be sufficiently stable as optical materials to achieve laser devices. Finally, the stable and continuous laser operation observed here is promising for future applications as it indicates both high photostability and thermal stability (thermal produce by laser pulse) under environmental conditions. Through the coating of a ~100 nm PMMA layer, perovskite films show remarkably enhanced PL and a prolonged decay lifetime. Most importantly, the ASE threshold of the perovskite films is significantly reduced, from 22.2 to 16.4 µJ/cm^2, with greatly improved light exposure stability. Then, the PMMA polymer layer plays a role that coincides with both surface passivation and symmetric waveguiding have been confirmed. A lower ASE threshold in perovskite films is conducive to stable and sustained output of laser light.

5. Conclusions

In conclusion, we have demonstrated that our simple strategy can improve the optical properties of CsPbBr$_3$ PQDs by modifying the upper surfaces of the CsPbBr$_3$ PQD film or both its upper and lower surfaces. High-quality films were directly synthesized from CsPbBr$_3$ PQD powder. The perovskite layer was coated to examine the optical response of the perovskite film after passivation of the surface by PMMA. PMMA acts as a reflector to enhance line narrowing and ASE from perovskite films toward lasing in the PQD structures. These advantages also suggest the great potential of inorganic perovskite films to support stimulated emission. The coating of two perovskite film faces resulted in an ultra-flexible film. The stability of CsPbBr$_3$ PQDs films was greatly improved by PMMA coating because of strict isolation from air and moisture in the atmosphere. The ASE thresholds was found of ~16.4 µJ/cm^2 on flexible PMMA reflectors, which are lower than values reported elsewhere under picosecond laser excitation for bare CsPbBr$_3$ PQDs films under picosecond excitation. This work suggests a promising pathway to flexible substrates that may additionally act as the reflector. This work has also confirmed that the PMMA layer plays the roles of both the surface passivation layer and symmetric waveguide. The lowering of the ASE threshold in perovskite films will result in stable, continuous production of perovskite laser light.

Author Contributions: Conceptualization, S.M.H.Q. and A.S.A.; methodology, S.M.H.Q.; software, S.M.H.Q.; validation, S.M.H.Q. and A.S.A.; formal analysis, S.M.H.Q.; investigation, S.M.H.Q., H.M.G., K.K.A., B.A.A.-A. and A.S.A.; resources, S.M.H.Q. and A.S.A.; data curation, S.M.H.Q.; writing—original draft preparation, S.M.H.Q.; writing—review and editing, H.M.G., A.S.A., S.M.H.Q., K.K.A. and B.A.A.-A.; visualization, S.M.H.Q.; supervision, S.M.H.Q., A.S.A.; project administration, S.M.H.Q. and A.S.A.; funding acquisition, S.M.H.Q. All authors have read and agreed to the published version of the manuscript.

Funding: The authors thank the Deanship of Scientific Research at King Saud University for funding this work through Research Group No. RG-1440-038.

Institutional Review Board Statement: Not applicable.

Informed Consent Statement: Not applicable.

Data Availability Statement: The data presented in this study are available on request from the corresponding author.

Conflicts of Interest: The authors declare no conflict of interest.

References

1. Chen, J.; Zhou, S.; Jin, S.; Li, H.; Zhai, T.; Gaál, R.; Magrez, A.; Forró, L.; Horváth, E.; Tang, J.; et al. Crystal organometal halide perovskites with promising optoelectronic applications. *J. Mater. Chem. C* **2016**, *4*, 11–27. [CrossRef]
2. Wei, Y.; Cheng, Z.; Lin, J. An overview on enhancing the stability of lead halide perovskite quantum dots and their applications in phosphor-converted LEDs. *Chem. Soc. Rev.* **2019**, *48*, 310–350. [CrossRef] [PubMed]
3. Levchuk, I.; Osvet, A.; Tang, X.; Brandl, M.; Perea, J.D.; Hoegl, F.; Matt, G.J.; Hock, R.; Batentschuk, M.; Brabec, C.J. Brightly Luminescent and Color-Tunable Formamidinium Lead Halide Perovskite FAPbX$_3$ (X = Cl, Br, I) Colloidal Nanocrystals. *Nano Lett.* **2017**, *17*, 2765–2770. [CrossRef]
4. Yang, W.; Gao, F.; Qiu, Y.; Liu, W.; Xu, H.; Yang, L.; Liu, Y. CsPbBr3-Quantum-Dots/Polystyrene@Silica Hybrid Microsphere Structures with Significantly Improved Stability for White LEDs. *Adv. Opt. Mater.* **2019**, *7*, 1900546. [CrossRef]
5. Wang, Y.; Ding, G.; Mao, J.Y.; Zhou, Y.; Han, S.T. Recent advances in synthesis and application of perovskite quantum dot based composites for photonics, electronics and sensors. *Sci. Technol. Adv. Mater.* **2020**, *21*, 278–302. [CrossRef]
6. Yang, D.; Li, X.; Wu, Y.; Wei, C.; Qin, Z.; Zhang, C.; Sun, Z.; Li, Y.; Wang, Y.; Zeng, H. Surface Halogen Compensation for Robust Performance Enhancements of CsPbX$_3$ Perovskite Quantum Dots. *Adv. Opt. Mater.* **2019**, *7*, 1900276. [CrossRef]
7. Nedelcu, G.; Protesescu, L.; Yakunin, S.; Bodnarchuk, M.I.; Grotevent, M.J.; Kovalenko, M.V. Fast Anion-Exchange in Highly Luminescent Nanocrystals of Cesium Lead Halide Perovskites (CsPbX$_3$, X = Cl, Br, I). *Nano Lett.* **2015**, *15*, 5635–5640. [CrossRef]
8. Kumawat, N.K.; Gupta, D.; Kabra, D. Recent Advances in Metal Halide-Based Perovskite Light-Emitting Diodes. *Energy Technol.* **2017**, *5*, 1734–1749. [CrossRef]
9. Akkerman, Q.A.; Motti, S.G.; Kandada, A.R.S.; Mosconi, E.; D'Innocenzo, V.; Bertoni, G.; Marras, S.; Kamino, B.A.; Miranda, L.; De Angelis, F.; et al. Solution Synthesis Approach to Colloidal Cesium Lead Halide Perovskite Nanoplatelets with Monolayer-Level Thickness Control. *J. Am. Chem. Soc.* **2016**, *138*, 1010–1016. [CrossRef]

10. Liu, Z.; Zhang, Y.; Fan, Y.; Chen, Z.; Tang, Z.; Zhao, J.; Lv, Y.; Lin, J.; Guo, X.; Zhang, J.; et al. Toward Highly Luminescent and Stabilized Silica-Coated Perovskite Quantum Dots through Simply Mixing and Stirring under Room Temperature in Air. *ACS Appl. Mater. Interfaces* **2018**, *10*, 13053–13061. [CrossRef]
11. Sun, C.; Zhang, Y.; Ruan, C.; Yin, C.; Wang, X.; Wang, Y.; Yu, W.W. Efficient and Stable White LEDs with Silica-Coated Inorganic Perovskite Quantum Dots. *Adv. Mater.* **2016**, *28*, 10088–10094. [CrossRef]
12. Yassitepe, E.; Yang, Z.; Voznyy, O.; Kim, Y.; Walters, G.; Castañeda, J.A.; Kanjanaboos, P.; Yuan, M.; Gong, X.; Fan, F.; et al. Amine-Free Synthesis of Cesium Lead Halide Perovskite Quantum Dots for Efficient Light-Emitting Diodes. *Adv. Funct. Mater.* **2016**, *26*, 8757–8763. [CrossRef]
13. Meyns, M.; Perálvarez, M.; Heuer-Jungemann, A.; Hertog, W.; Ibáñez, M.; Nafria, R.; Genç, A.; Arbiol, J.; Kovalenko, M.V.; Carreras, J.; et al. Polymer-Enhanced Stability of Inorganic Perovskite Nanocrystals and Their Application in Color Conversion LEDs. *ACS Appl. Mater. Interfaces* **2016**, *8*, 19579–19586. [CrossRef] [PubMed]
14. Yan, D.; Shi, T.; Zang, Z.; Zhou, T.; Liu, Z.; Zhang, Z.; Du, J.; Leng, Y.; Tang, X. Ultrastable CsPbBr$_3$ Perovskite Quantum Dot and Their Enhanced Amplified Spontaneous Emission by Surface Ligand Modification. *Small* **2019**, *15*, 1901173. [CrossRef] [PubMed]
15. Cai, Y.; Wang, L.; Zhou, T.; Zheng, P.; Li, Y.; Xie, R.J. Improved stability of CsPbBr$_3$ perovskite quantum dots achieved by suppressing interligand proton transfer and applying a polystyrene coating. *Nanoscale* **2018**, *10*, 21441–21450. [CrossRef] [PubMed]
16. Yoon, H.C.; Lee, S.; Song, J.K.; Yang, H.; Do, Y.R. Efficient and Stable CsPbBr$_3$ Quantum-Dot Powders Passivated and Encapsulated with a Mixed Silicon Nitride and Silicon Oxide Inorganic Polymer Matrix. *ACS Appl. Mater. Interfaces* **2018**, *10*, 11756–11767. [CrossRef] [PubMed]
17. Wang, Y.; Cao, S.; Li, J.; Li, H.; Yuan, X.; Zhao, J. Improved ultraviolet radiation stability of Mn^{2+}-doped CsPbCl$_3$ nanocrystals: Via B-site Sn doping. *CrystEngComm* **2019**, *21*, 6238–6245. [CrossRef]
18. Li, Z.J.; Hofman, E.; Li, J.; Davis, A.H.; Tung, C.H.; Wu, L.Z.; Zheng, W. Photoelectrochemically active and environmentally stable CsPbBr$_3$/TiO$_2$ core/shell nanocrystals. *Adv. Funct. Mater.* **2018**, *28*, 1704288. [CrossRef]
19. Liao, H.; Guo, S.; Cao, S.; Wang, L.; Gao, F.; Yang, Z.; Zheng, J.; Yang, W. A General Strategy for In Situ Growth of All-Inorganic CsPbX$_3$ (X = Br, I, and Cl) Perovskite Nanocrystals in Polymer Fibers toward Significantly Enhanced Water/Thermal Stabilities. *Adv. Opt. Mater.* **2018**, *6*, 1800346. [CrossRef]
20. Wang, H.-C.; Lin, S.-Y.; Tang, A.-C.; Singh, B.P.; Tong, H.-C.; Chen, C.-Y.; Lee, Y.-C.; Tsai, T.-L.; Liu, R.-S. Mesoporous Silica Particles Integrated with All-Inorganic CsPbBr$_3$ Perovskite Quantum-Dot Nanocomposites (MP-PQDs) with High Stability and Wide Color Gamut Used for Backlight Display. *Angew. Chem.* **2016**, *128*, 8056–8061. [CrossRef]
21. Leijtens, T.; Lauber, B.; Eperon, G.E.; Stranks, S.D.; Snaith, H.J. The importance of perovskite pore filling in organometal mixed halide sensitized TiO$_2$-based solar cells. *J. Phys. Chem. Lett.* **2014**, *5*, 1096–1102. [CrossRef] [PubMed]
22. Chen, X.; Li, D.; Pan, G.; Zhou, D.; Xu, W.; Zhu, J.; Wang, H.; Chen, C.; Song, H. All-inorganic perovskite quantum dot/TiO$_2$ inverse opal electrode platform: Stable and efficient photoelectrochemical sensing of dopamine under visible irradiation. *Nanoscale* **2018**, *10*, 10505–10513. [CrossRef] [PubMed]
23. Qaid, S.M.H.; Ghaithan, H.M.; Al-Asbahi, B.A.; Aldwayyan, A.S. Single-Source Thermal Evaporation Growth and the Tuning Surface Passivation Layer Thickness Effect in Enhanced Amplified Spontaneous Emission Properties of CsPb(Br$_{0.5}$Cl$_{0.5}$)$_3$ Perovskite Films. *Polymers* **2020**, *12*, 2953. [CrossRef]
24. Yang, A.; Bai, M.; Bao, X.; Wang, J.; Zhang, W. Investigation of Optical and Dielectric Constants of Organic-Inorganic CH$_3$NH$_3$PbI$_3$ Perovskite Thin Films. *J. Nanomed. Nanotechnol.* **2016**, *7*, 5–9.
25. Qaid, S.M.H.; Ghaithan, H.M.; Al-Asbahi, B.A.; Aldwayyan, A.S. Ultra-Stable Polycrystalline CsPbBr$_3$ Perovskite–Polymer Composite Thin Disk for Light-Emitting Applications. *Nanomaterials* **2020**, *10*, 2382. [CrossRef]
26. Qaid, S.M.H.; Ghaithan, H.M.; Al-Asbahi, B.A.; Aldwayyan, A.S. Investigation of the Surface Passivation Effect on the Optical Properties of CsPbBr$_3$ Perovskite Quantum Dots. *Surf. Interfaces* **2021**, *23*, 100948. [CrossRef]
27. Schaller, R.D.; Petruska, M.A.; Klimov, V.I. Tunable Near-Infrared Optical Gain and Amplified Spontaneous Emission Using PbSe Nanocrystals. *J. Phys. Chem. B* **2003**, *107*, 13765–13768. [CrossRef]
28. Yakunin, S.; Protesescu, L.; Krieg, F.; Bodnarchuk, M.I.; Nedelcu, G.; Humer, M.; De Luca, G.; Fiebig, M.; Heiss, W.; Kovalenko, M.V. Low-threshold amplified spontaneous emission and lasing from colloidal nanocrystals of caesium lead halide perovskites. *Nat. Commun.* **2015**, *6*, 8056. [CrossRef]
29. Pan, J.; Sarmah, S.P.; Murali, B.; Dursun, I.; Peng, W.; Parida, M.R.; Liu, J.; Sinatra, L.; Alyami, N.; Zhao, C.; et al. Air-Stable Surface-Passivated Perovskite Quantum Dots for Ultra-Robust, Single- and Two-Photon-Induced Amplified Spontaneous Emission. *J. Phys. Chem. Lett.* **2015**, *6*, 5027–5033. [CrossRef]
30. Qaid, S.M.H.; Ghaithan, H.M.; Al-Asbahi, B.A.; Alqasem, A.; Aldwayyan, A.S. Fabrication of Thin Films from Powdered Cesium Lead Bromide (CsPbBr$_3$) Perovskite Quantum Dots for Coherent Green Light Emission. *ACS Omega* **2020**, *5*, 30111–30122. [CrossRef]
31. Qaid, S.M.H.; Alharbi, F.H.; Bedja, I.; Nazeeruddin, M.K.; Aldwayyan, A.S. Reducing amplified spontaneous emission threshold in CsPbBr$_3$ quantum dot films by controlling TiO$_2$ compact layer. *Nanomaterials* **2020**, *10*, 1605. [CrossRef]
32. Jumali, M.H.H.; Al-Asbahi, B.A.; Yap, C.C.; Salleh, M.M.; Alsalhi, M.S. Optical Properties of Poly(9,9'-di-n-octylfluorenyl-2.7-diyl)/Amorphous SiO$_2$ Nanocomposite Thin Films. *Sains Malays.* **2013**, *42*, 1151–1157.

33. Qaid, S.M.H.; Al-Asbahi, B.A.; Ghaithan, H.M.; AlSalhi, M.S.; Al dwayyan, A.S. Optical and structural properties of CsPbBr$_3$ perovskite quantum dots/PFO polymer composite thin films. *J. Colloid Interface Sci.* **2020**, *563*, 426–434. [CrossRef]
34. Woo, J.Y.; Kim, Y.; Bae, J.; Kim, T.G.; Kim, J.W.; Lee, D.C.; Jeong, S. Highly Stable Cesium Lead Halide Perovskite Nanocrystals through in Situ Lead Halide Inorganic Passivation. *Chem. Mater.* **2017**, *29*, 7088–7092. [CrossRef]
35. Al-Asbahi, B.A.; Qaid, S.M.H.; Hezam, M.; Bedja, I.; Ghaithan, H.M.; Aldwayyan, A.S. Effect of deposition method on the structural and optical properties of CH$_3$NH$_3$PbI$_3$ perovskite thin films. *Opt. Mater.* **2020**, *103*, 109836. [CrossRef]
36. Wu, X.; Jiang, X.F.; Hu, X.; Zhang, D.F.; Li, S.; Yao, X.; Liu, W.; Yip, H.L.; Tang, Z.; Xu, Q.H. Highly stable enhanced near-infrared amplified spontaneous emission in solution-processed perovskite films by employing polymer and gold nanorods. *Nanoscale* **2019**, *11*, 1959–1967. [CrossRef] [PubMed]
37. Li, J.; Si, J.; Gan, L.; Liu, Y.; Ye, Z.; He, H. Simple Approach to Improving the Amplified Spontaneous Emission Properties of Perovskite Films. *ACS Appl. Mater. Interfaces* **2016**, *8*, 32978–32983. [CrossRef]
38. Stranks, S.D.; Wood, S.M.; Wojciechowski, K.; Deschler, F.; Saliba, M.; Khandelwal, H.; Patel, J.B.; Elston, S.J.; Herz, L.M.; Johnston, M.B.; et al. Enhanced Amplified Spontaneous Emission in Perovskites Using a Flexible Cholesteric Liquid Crystal Reflector. *Nano Lett.* **2015**, *15*, 4935–4941. [CrossRef]
39. Veldhuis, S.A.; Boix, P.P.; Yantara, N.; Li, M.; Sum, T.C.; Mathews, N.; Mhaisalkar, S.G. Perovskite Materials for Light-Emitting Diodes and Lasers. *Adv. Mater.* **2016**, *28*, 6804–6834. [CrossRef] [PubMed]
40. Olbright, G.R.; Klem, J.; Owyoung, A.; Brennan, T.M.; State, S.; Directorate, S.; Laboratories, S.N.; Binder, R.; Koch, S.W. Many-body effects in the luminescence of highly excited indirect superlattices. *J. Opt. Soc. Am. B* **1990**, *7*, 1473–1480. [CrossRef]
41. Yang, Y.; Ostrowski, D.P.; France, R.M.; Zhu, K.; Van De Lagemaat, J.; Luther, J.M.; Beard, M.C. Observation of a hot-phonon bottleneck in lead-iodide perovskites. *Nat. Photonics* **2016**, *10*, 53–59. [CrossRef]
42. Yan, W.; Mao, L.; Zhao, P.; Mertens, A.; Dottermusch, S.; Hu, H.; Jin, Z.; Richards, B.S. Determination of complex optical constants and photovoltaic device design of all-inorganic CsPbBr$_3$ perovskite thin films. *Opt. Express* **2020**, *28*, 15706–15717. [CrossRef]
43. Zhao, M.; Shi, Y.; Dai, J.; Lian, J. Ellipsometric study of the complex optical constants of a CsPbBr$_3$ perovskite thin film. *J. Mater. Chem. C* **2018**, *6*, 10450–10455. [CrossRef]
44. Eadie, G.B.; Rindza, M.I.B.; Lynn, R.I.A.F.; Osenberg, A.R.; Hirk, J.A.S.S. Refractive index measurements of poly (methyl methacrylate) (PMMA) from 0.4–1.6 μm. *Appl. Opt.* **2015**, *54*, 139–143.
45. De Giorgi, M.L.; Krieg, F.; Kovalenko, M.V.; Anni, M. Amplified Spontaneous Emission Threshold Reduction and Operational Stability Improvement in CsPbBr$_3$ Nanocrystals Films by Hydrophobic Functionalization of the Substrate. *Sci. Rep.* **2019**, *9*, 1–10. [CrossRef] [PubMed]
46. De Giorgi, M.L.; Anni, M. Amplified spontaneous emission and lasing in lead halide perovskites: State of the art and perspectives. *Appl. Sci.* **2019**, *9*, 4591. [CrossRef]
47. Qaid, S.M.H.; Khan, M.N.; Alqasem, A.; Hezam, M.; Aldwayyan, A. Restraining effect of film thickness on the behaviour of amplified spontaneous emission from methylammonium lead iodide perovskite. *IET Optoelectron.* **2018**, *13*, 2–6. [CrossRef]
48. Protesescu, L.; Yakunin, S.; Bodnarchuk, M.I.; Krieg, F.; Caputo, R.; Hendon, C.H.; Yang, R.X.; Walsh, A.; Kovalenko, M.V. Nanocrystals of Cesium Lead Halide Perovskites (CsPbX$_3$, X = Cl, Br, and I): Novel Optoelectronic Materials Showing Bright Emission with Wide Color Gamut. *Nano Lett.* **2015**, *15*, 3692–3696. [CrossRef]
49. Xiong, Q.; Huang, S.; Du, J.; Tang, X.; Zeng, F.; Liu, Z.; Zhang, Z.; Shi, T.; Yang, J.; Wu, D.; et al. Surface Ligand Engineering for CsPbBr$_3$ Quantum Dots Aiming at Aggregation Suppression and Amplified Spontaneous Emission Improvement. *Adv. Opt. Mater.* **2020**, *8*, 2000977. [CrossRef]
50. Chen, J.; Liu, D.; Al-Marri, M.J.; Nuuttila, L.; Lehtivuori, H.; Zheng, K. Photo-stability of CsPbBr$_3$ perovskite quantum dots for optoelectronic application. *Sci. China Mater.* **2016**, *59*, 719–727. [CrossRef]
51. Balena, A.; Perulli, A.; Fernandez, M.; De Giorgi, M.L.; Nedelcu, G.; Kovalenko, M.V.; Anni, M. Temperature Dependence of the Amplified Spontaneous Emission from CsPbBr3 Nanocrystal Thin Films. *J. Phys. Chem. C* **2018**, *122*, 5813–5819. [CrossRef]
52. Qaid, S.M.H.; Ghaithan, H.M.; Al-Asbahi, B.A.; Aldwayyan, A.S. Achieving Optical Gain of the CsPbBr$_3$ Perovskite Quantum Dots and Influence of the Variable Stripe Length Method. *ACS Omega* **2021**, *6*, 5297–5309. [CrossRef] [PubMed]

Article

The Effect of Alkyl Substitution of Novel Imines on Their Supramolecular Organization, towards Photovoltaic Applications

Paweł Nitschke [1], Bożena Jarząbek [1,*], Marharyta Vasylieva [1,2], Marcin Godzierz [1], Henryk Janeczek [1], Marta Musioł [1] and Adrian Domiński [1]

[1] Centre of Polymer and Carbon Materials, Polish Academy of Sciences, 34 M. Curie-Skłodowska Str., 41-819 Zabrze, Poland; pnitschke@cmpw-pan.edu.pl (P.N.); mvasylieva@cmpw-pan.edu.pl (M.V.); mgodzierz@cmpw-pan.edu.pl (M.G.); hjaneczek@cmpw-pan.edu.pl (H.J.); mmusiol@cmpw-pan.edu.pl (M.M.); adominski@cmpw-pan.edu.pl (A.D.)

[2] Faculty of Chemistry, Silesian University of Technology, 9 Strzody Str., 44-100 Gliwice, Poland

* Correspondence: bozena.jarzabek@cmpw-pan.edu.pl

Citation: Nitschke, P.; Jarząbek, B.; Vasylieva, M.; Godzierz, M.; Janeczek, H.; Musioł, M.; Domiński, A. The Effect of Alkyl Substitution of Novel Imines on Their Supramolecular Organization, towards Photovoltaic Applications. *Polymers* **2021**, *13*, 1043. https://doi.org/10.3390/polym13071043

Academic Editors: Juhyun Park and Sergi Gallego Rico

Received: 26 February 2021
Accepted: 24 March 2021
Published: 26 March 2021

Publisher's Note: MDPI stays neutral with regard to jurisdictional claims in published maps and institutional affiliations.

Copyright: © 2021 by the authors. Licensee MDPI, Basel, Switzerland. This article is an open access article distributed under the terms and conditions of the Creative Commons Attribution (CC BY) license (https://creativecommons.org/licenses/by/4.0/).

Abstract: Three novel conjugated polyazomethines have been obtained by polycondensation of diamines consisting of the diimine system, with either 2,5-bis(octyloxy)terephthalaldehyde or 9-(2-ethylhexyl)carbazole-3,6-dicarboxaldehyde. Partial replacement of bulky solubilizing substituents with the smaller side groups has allowed to investigate the effect of supramolecular organization. All obtained compounds have been subsequently identified using the NMR and FTIR spectroscopies and characterized by the thermogravimetric analysis, differential scanning calorimetry, cyclic voltammetry, UV–Vis spectroscopy, and X-ray diffraction. Investigated polymers have shown a good thermal stability and high glass transition temperatures. X-ray measurements have proven that partial replacement of octyloxy side chains with smaller methoxy groups induced a better planarization of macromolecule. Such modification has tuned the LUMO level of this molecule and caused a bathochromic shift of the lowest energy absorption band. On the contrary, imines consisting of N-ethylhexyl substituted carbazole units have not been so clearly affected by alkyl chain length modification. Photovoltaic activity of imines (acting as a donor) in bulk-heterojunction systems has been observed for almost all studied compounds, blended with the fullerene derivative (PCBM) in various weight ratios.

Keywords: azomethines; supramolecular organization; organic thin films; polymer:fullerene blends; organic photovoltaics

1. Introduction

From many years, a development of new, high-performance conjugated compounds and polymers has drawn much attention, due to their potential application in various optoelectronic systems, like organic photovoltaic cells (OPV), organic light emitting diodes (OLED), sensors, or organic field-effect transistors (OFET) [1–4]. Organic semiconductors used in these structures have revealed many advantages, compared to their inorganic counterparts, such as easy processing and low costs of production. The use of low-temperature, wet methods (like spin-coating, spray-coating or printing) of thin films deposition causes the significant reduction of production costs, compared to the chemical vapor deposition (CVD) or thermal vacuum evaporation (TVE) techniques. However, all these wet methods, require an appropriate solubility of used compounds. The possibility of obtaining large, elastics surfaces of organic materials, using wet methods, is also important from a practical point of view. Other advantage of organic compounds is the possibility of suitable modification of their chemical structure (by different substitutions or doping) to obtain desired properties and electronic structure, towards optoelectronic applications [5,6]. One of the interesting groups of organic semiconductors are azomethines and polyazomethines, known also as imines or Schiff-bases. Such materials are products of the condensation reaction between amines and aldehydes, where the formed imine bond (–C=N–) is proven to exhibit an isoelectronic character with a vinylene bond [7]. This condensation reaction is easy to conduct, in mild conditions, catalyzed by organic or inorganic acids, or even without using any [8,9], on the contrary to carbon-carbon or vinylene-coupled compounds, which usually require stringent reaction conditions and rather extensive purification processes [10]. Due to the relatively simple synthesis route and promising properties, conjugated azomethines and polyazomethines have been investigated for applications in optoelectronic systems, like photovoltaic solar cells [11–13], electroluminescent diodes [14,15], or electrochromic systems [16]. Highly aromatic polyazomethines are rather insoluble, so their thin films may be obtained via gas-phase condensation or by chemical vapor deposition (CVD) [17,18]. To ensure an appropriate solubility necessary for the thin films deposition by the spin coating method, a complexation of polyazomethine with the Lewis acid or di-m-cresol phosphate (DCP) [19], or much more often, a substitution with bulky side alkyl or alkoxy groups [20], may be used. It has been already proven that the introduction of alkoxy side groups modifies the electron structure of substituted compounds [21], while the length of alkyl chain only improves the solubility of materials [22]. However, such bulky side chains can affect the supramolecular organization as well [23], hindering the π-interactions, and subsequently decreasing a crystallinity and conductivity of compounds [24].

In this study, a partial replacement of bulky solubilizing substituents with smaller side groups has been investigated. Such an attitude is expected to ensure more favorable supramolecular organization, and as a result, a more planar arrangement of the macromolecules, which ought to exhibit more advantageous electrochemical and optical properties. To do so, three novel oligo- and polyazomethines have been obtained. Two of them have had an analogue chemical structure, which differed only in the length of part of the solubilizing side groups. The third imine has been substituted with both methoxy and octyloxy side groups and its properties have been compared with already reported polyazomethine with octyloxy side groups [11,22]. Thermal, electrochemical, and optical properties of these materials, together with the X-ray diffraction (XRD) patterns, have been thoroughly studied and discussed, in terms of the type of alkyl substituents. A final part of this paper presents the current-voltage (J-V) characteristics of bulk-heterojunction (BHJ) photovoltaic systems, utilizing these compounds, as donor materials, in a blend of polyazomethine (PAz) with fullerene acceptor ($PC_{61}BM$) in various weight ratios (1:1, 1:2, 1:3). All these reported results provide new insights into the effect of supramolecular engineering on physicochemical properties of these novel compounds, towards their photovoltaic applications.

2. Materials and Methods

2.1. Materials

2,5-bis(octyloxy)terephthalaldehyde (98%), 9-(2-Ethylhexyl)carbazole-3,6-dicarboxaldehyde (97%) and trifluoroacetic acid (TFA) (99%) have been purchased from Sigma-Aldrich and used as received. Diamines with diimine system, **DAAz1** and **DAAz2**, have been synthesized using a procedure described in [22]. The solvents such as toluene, methanol, chlorobenzene, and chloroform have been purchased from Avantor Performance Materials (Gliwice, Poland), and used as received. [6,6]-Phenyl-C61-butyric acid methyl ester (PC61BM) (>99% wt.) (M111) and PEDOT:PSS dispersion in water (M124) have been purchased in Ossila (Sheffield, UK) and used as received.

2.2. Characterization Methods

^{1}H NMR spectra of synthesized polyazomethines have been recorded on Avance II Ultrashield Plus spectrometer, operating at 600 MHz using deuterated chloroform as a solvent and Tetramethylsilane (TMS) as an internal reference. The FTIR spectra have been recorded on JASCO FTIR 6700 Fourier transform infrared spectrometer, in a transmittance mode, in the range of 4000–400 cm^{-1} at a resolution of 2 cm^{-1} and for 64 accumulated scans. Size exclusion chromatography (SEC) has been performed in chloroform, at 35 °C with a flow rate of 1 mL/min, using a Spectra-Physic 8800 gel permeation chromatograph with a PL-gel 5 mm MIXED-C ultra-high efficiency column and Shodex SE 61 differential refractive index detector with polystyrene standards for calibration. DSC measurements have been taken with a DSC Q2000 apparatus (TA Instruments, Newcastle, DE, USA), in a range of −50–380 °C under the nitrogen atmosphere (flow rate was 50 mL/min), using aluminum sample pans. The instrument has been calibrated with a high-purity indium. In this study, the glass transition temperature (T_g) has been taken as a midpoint of heat capacity change for amorphous samples obtained by quenching from melt in liquid nitrogen. Thermogravimetric analysis (TGA) has been performed with TGA/DSC1 Mettler-Toledo thermal analyses, in a range of 25 to 600 °C at a heating rate of 10°/min in a stream of nitrogen (60 mL/min). The obtained TGA data have been analyzed with the Mettler-Toledo Star System SW 9.30. The initial decomposition temperature has been taken as a temperature at the 5% weight loss ($T_{5\%}$). Electrochemical measurements have been performed on EDAQ E-corder 410 apparatus. The electrochemical cell has comprised of ITO (Indium Tin Oxide) quartz glass working electrode, an Ag | Ag$^+$ electrode as a pseudo-reference electrode, and a platinum wire as an auxiliary electrode. A layer of polymer has been coated on the surface of the ITO plate. Measurements have been conducted at room temperature at a potential rate of 0.1 V/s. Electrochemical studies have been undertaken in 0.1 M solutions of Bu$_4$NBF$_4$, 99% (Sigma Aldrich, Saint Louis, MO, USA) in Acetonitrile (ACN). All electronic spectra of investigated polyazomethines have been measured with the two-beam JASCO V-570 UV-Vis-NIR spectrophotometer. The absorption spectra of solutions have been recorded in ranges of 240–800 nm (chloroform) or 200–2500 nm (thin films on quartz substrates). Concentration of all investigated solutions was 5×10^{-4} M. Thicknesses of spin-coated thin films have been measured using atomic force microscope, Topo-Metrix Explorer, working in the contact mode in the air, in the constant force regime. X-Ray diffraction studies have been performed using the D8 Advance diffractometer (Bruker, Karlsruhe, Germany) with Cu-Kα cathode (λ = 1.54 Å). Due to the critical angle for conjugated polymers using copper radiation being ~0.17° [25,26] and the layer thickness of sample (~100 nm), for 2D-GIWAXS setup, the 0.18° incidence angle has been applied, which is just above the critical angle for polymer layer and below the critical angle for glass support material. The scan rate has been 0.6°/min with scanning step 0.02° in range of 2.0° to 60° 2Θ (dwell time 2 s). Measurements have been performed in 7 variations, using different φ (Phi) angle, which corresponds to the sample rotation. As a φ = 0°, the longer edge has been set as parallel to the X-Ray beam direction. The resulting φ rotation (15, 30, 45, 60, 75, and 90°) has been programmed with a resolution of 0.1° φ. Obtained 2D patterns (with width of 3.1° 2θ) for different φ angle have been integrated to 1D patterns. Background subtraction, occurring from air scattering, has been performed using DIFFRAC.EVA program. All WAXD (XRD) measurements acquired at the different Phi angle have been accumulated to obtain the representative pattern, then obtained profiles have been smoothed, using a 5-point, quadratic polynomial, the Savitzky–Golay smoothing filter. For the structural analysis, the unit cell parameter a is related to the short macromolecule axis (correlated to the planarity and side chains) and c corresponds to the long axis of polymer (the length of macromolecule), while b is related to the π-stacking period [25].

2.3. Synthesis and Structural Characterization

Compounds investigated in this paper have been obtained by solution polycondensation of equimolar amounts of 2,5-bis(octyloxy)terephthalaldehyde and diamine **DAAz1**, consisting of diimine system, substituted with methoxy side chains, resulting in the formation of **PAz-BOO-OMe** (Figure 1). Polycondensation of equimolar amounts of 9-(2-Ethylhexyl)carbazole-3,6-dicarboxaldehyde with diamines consisting diimine system (**DAAz**), substituted with either methoxy (**DAAz1**) or octyloxy (**DAAz2**) side chains, has allowed to obtain **PAz-Carb-OMe** and **PAz-Carb-OOct**, respectively (Figure 1). The reagents have been dissolved in 3 mL of toluene (dried before use over magnesium sulfate) and provided with a magnetic dipole. The temperature has been increased up to 115 °C, while stirring the reaction mixture. After this, nitrogen has been passed through the reaction system to remove moisture and then 0.45 µL of trifluoroacetic acid has been introduced into the solution. The system has been sealed, leaving a flow of an inert gas (nitrogen) through it and the reaction has been carried out for two days, at 120 °C. After completion of the reaction, the product has been precipitated in methanol, dried in air, and then purified by Soxhlet extraction with methanol. The chemical structure of obtained compounds has been investigated using ^{1}H-NMR, ^{13}C-NMR, and FTIR spectroscopies.

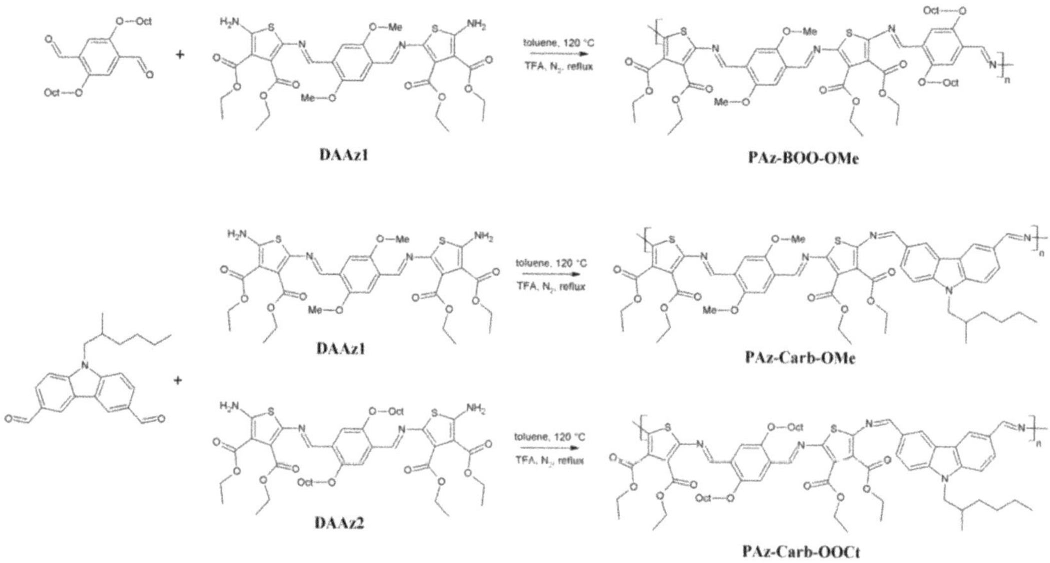

Figure 1. Synthesis procedure and chemical structures of investigated polyazomethines.

- **PAz-BOO-OMe**

    ^{1}H–NMR (600 MHz, CDCl3, ppm) δ: 9.97 (s, CHO; end group), 9.91 (s, CHO; end group) 8.11 (1H, s, CH=N), 8.07–8.03 (1H, m, CH=N), 7.89–7.87 (1H, m, Ar–H), 7.82–7.77 (4H, m, Ar–H), 7.66–7.63 (1H, m, Ar–H), 7.47–7.36 (2H, m, Ar–H), 7.24 (1H, dd, J = 4.71, 3.95 Hz, Ar–H), 7.21–7.18 (2H, m Ar–H), 7.17–7.14 (1H, m, Ar–H), 7.10–7.08 (1H, m, Ar–H), 4.43 (2H, q, J = 7.15 Hz, –O–CH$_2$–, ester), 4.26 (2H, q, J = 7.15 Hz, –O–CH$_2$– ester), 2.98 (1H, s –O–CH$_3$), 2.91 (1H, s –O–CH$_3$), 1.50–1.43 (2H, m, –CH$_2$– ether), 1.32 (3H, m,–CH$_2$ ether, –CH$_3$ ester, and ether). ^{13}C–NMR (150 MHz, CDCl3, ppm) δ: 165.57; 164.24; 160.27; 152.38; 147.02; 134.57; 130.86; 111.26; 69.13; 61.45; 60.39; 55.89; 31.78; 29.39; 25.95; 22.77; 14,29. FTIR (KBr, cm^{-1}) υ: 3314 (N–H stretching), 2924, 2852 (C–H aliphatic stretching), 1728 (C=O stretching), 1688, 1677 (C=N, imine), 1578, 1409, 1366 (vibration of the thiophene rings), 1467 (vibration of the benzene rings), 1211 (C–O ether asymmetric stretching vibrations), adnd 1063 (C–O ether symmetric stretching vibrations).

- **PAz-Carb-OMe**

    ^{1}H-NMR (600 MHz, CDCl3, ppm) δ: 10.45 (s, CHO, end group), 10.16–10.09 (m, CHO; end group), 8.68 (2H, s, CH=N), 8.61 (1H, s, CH=N), 8.50 (1H, s, CH=N), 8.42–8.34 (1H, m, Ar–H), 8.23–8.19 (1H, m, Ar–H), 8.09 (1H, d, J = 8.28, Ar–H), 7.64 (1H, s, Ar–H), 7.54 (1H, d, J = 8.66 Hz, Ar–H), 7.50–7.41 (1H, m, Ar–H), 7.35 (1H, s, Ar–H), 4.50–4.45 (1H, m, –O–CH$_2$–, ester), 4.44–4.37 (3H, m, –O–CH$_2$– ester), 4.30–4.21 (6H, m, –O–CH$_2$– ester), 3.95–3.86 (5H, m, –O–CH$_3$), 1.51–1.20 (23H, m, –CH$_2$– N-alkyl, –CH$_3$ ester), 0.96–0.91 (4H, m, –CH₃ N-alkyl), 0.87–0.82 (4H, m, –CH₃ N-alkyl). ^{13}C–NMR (150 MHz, CDCl3, ppm) δ: 165.57; 164.51; 160.00; 156.29; 153.11; 145.17; 129.80; 127.68; 124.24; 123.18; 110.20; 61.45; 60.39; 55.89; 48.20; 39.46; 30.98; 28.60; 24.36; 23.04; 14.29. FTIR (KBr, cm^{-1}) υ: 3308 (N–H stretching), 2957, 2930, 2870 (C–H aliphatic, stretching), 1728 (C=O stretching), 1689, 1628 (C=N, imine), 1588, 1432, 1385 (vibration of the thiophene rings), 1535, 1475 (vibration of the benzene rings), 1253 (C–O ether asymmetric stretching vibrations), 1030 (C–O ether symmetric stretching vibrations).

- **PAz-Carb-OOct**

^{1}**H–NMR** (600 MHz, CDCl3, ppm) δ: 10.46 (s, CHO, end group), 10.15–10.11 (m, CHO; end group), 8.68 (2H, s, CH=N), 8.63–8.59 (1H, m, CH=N), 8.47–8.37 (1H, s, CH=N), 8.09 (1H, d, J = 8.66 Hz, Ar–H), 8.06–8.00 (1H, m, Ar–H), 7.65–7.56 (1H, m, Ar–H), 7.54 (1H, d, J = 8.28 Hz, Ar–H), 7.52–7.42 (1H, m, Ar–H), 7.32 (1H, s, Ar–H), 4.50–4.33 (4H, m, –O–CH₂–, ester), 4.31–4.19 (5H, m, –O–CH₂– ester), 4.10–3.99 (4H, m, –O–CH₂– ether), 1.87–1.77 (2H, m, –CH₂–CH₂–, ether), 1.54–1.13 (37H, m, –CH₂– N-alkyl, –CH₃ ester), 0.97–0.78 (11H, m, –CH₃ N-alkyl, ether). ^{13}**C–NMR** (150 MHz, CDCl3, ppm) δ: 165.57; 164.51; 160.00; 155.76; 153.11; 152.85; 145.17; 135.36; 129.80; 127.68; 124.24; 123.18; 110.20; 68.87; 61.45; 60.39; 48.20; 39.46; 31.78; 30.98; 29.66; 28.60; 26.21; 24.36; 22.77; 14.56; 14.03; 10.85. **FTIR** (KBr, cm-1) υ: 3425, 3310 (N–H stretching), 2955, 2926, 2855 (C–H aliphatic, stretching), 1731 (C=O stretching), 1677 (C=N, imine), 1589, 1424, 1385 (vibration of the thiophene rings), 1534, 1466 (vibration of the benzene rings), 1206 (C–O ether asymmetric stretching vibrations), 1027 (C–O ether symmetric stretching vibrations).

2.4. Organic Solar Cells Preparation

Devices with the bulk-heterojunction (BHJ) structure have been prepared on ITO-coated glass substrates (6 pixels, each with an area of 4.5 mm^2, Ossila Ltd, Sheffield, UK). After cleaning the substrates with isopropanol in ultrasonic bath, a film of PEDOT:PSS has been deposited by spin coating. Solutions of the active layer have been prepared by dissolving blends of each individual polymer with the PCBM (1:1, 1:2 or 1:3 wt.) in chlorobenzene (previously dried over anhydrous magnesium sulfate). Such prepared solutions have been spin coated on the PEDOT:PSS layer and, subsequently, an aluminum counter electrode has been evaporated on the top of blend thin film. *J-V* curves of photovoltaic devices have been measured by the PV Test Solutions Solar Simulator, under the AM1.5 solar illumination and using the Keithley 2400 electrometer.

3. Results and Discussion

As shown in the literature [23], tailoring the solubilizing groups may greatly affect the molecular packing of materials through e.g., more favorable supramolecular organization. In this paper, such a modification has been achieved by partially replacing bulky solubilizing n-octyloxy groups with the shorter methoxy substituents. This modification of polyazomethine structure has been accomplished by the polymerization of diamines with diimine system, consisting of the methoxy substituted aromatic rings, with either 2,5-bis(octyloxy)terephthalaldehyde (**PAz-BOO-OMe**) or 9-(2-Ethylhexyl)carbazole-3,6-dicarboxaldehyde (**PAz-Carb-OMe** and **PAz-Carb-OOCt**). The presence of amine groups in the **DAAz** compounds has allowed to proceed with a copolymerization reaction with dialdehydes, as shown in the literature [27]. As reported in our previous research [11,22], the length of the alkyl chain in alkoxy substituent does not affect the absorption spectra and electrochemical properties of compounds, only their solubility, thermal properties [22], and photoluminescence intensity [11]. This is why the obtained polyazomethine **PAz-BOO-OMe**, consisting of both methoxy and octyloxy side chains, has been compared with an analogue polyazomethine, substituted solely with octyloxy side chains (**PAz-BOO-Oct**), previously reported in [22] (Figure 2). Apart from this, such an attempt on modification of supramolecular organization has been investigated on compounds consisting of branched N-ethylhexyl substituents, together with either methoxy (**PAz-Carb-OMe**) or octyloxy (**PAz-Carb-OOct**) side chains.

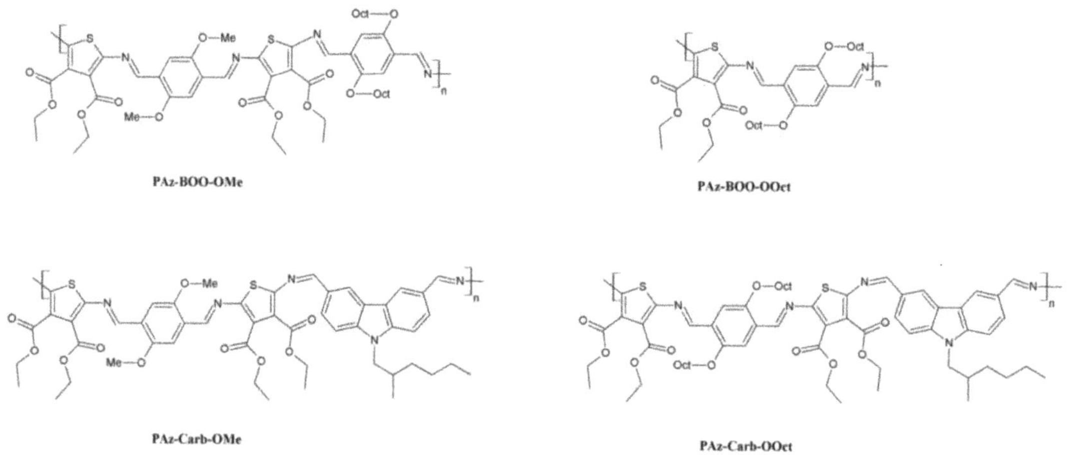

Figure 2. Chemical structures of synthesized compounds, together with polyazomethine obtained during previous studies (**PAz-BOO-Oct**) [22].

3.1. Structural and Solubility Studies

Chemical structure of obtained compounds has been confirmed using ^{1}H- and ^{13}C-NMR spectroscopies. Registered spectra have revealed signals, originating from aldehyde end groups in a range of 10.46–9.91 ppm; however, no signals of amine end groups have been observed. The presence of these signals might suggest the formation of rather oligomers than polymers, during the polycondensation reactions. Imine proton singlets have been observed in a range of 8.68–8.03 ppm, depending on the structure of imine. For **PAz-BOO-OMe**, only one group of these signals have been registered, between 8.11 and 8.03 ppm, while both **PAz-Carb** compounds spectra have revealed two groups of imine proton singlets, in ranges of 8.68–8.59 ppm and 8.50–8.37 ppm. This is due to various chemical environments of imine bonds in carbazole-consisting imines, where such bonds link together either thiophene and benzene pair or thiophene and carbazole pair. This is in contrary to **PAz-BOO-OMe**, where imine bonds link only thiophene and benzene rings, being only in one type of chemical environment. ^{1}H-NMR spectrum of **PAz-BOO-OMe** has also revealed both singlets of methoxy substituents and multiplets originating from alkyl protons, present in octyloxy side chain. Similarly, spectra of both **PAz-Carb** imines have revealed signals of alkyl protons, present in the N-alkyl chains, together with either singlet of methoxy group (**PAz-Carb-OMe**) or multiplets from octyloxy side chains (**PAz-Carb-OOct**). Carbon spectra have revealed signals of imine groups in a range of 160.27–145.17 ppm. Methoxy carbons have given signal at 55.89 ppm, while methylene groups bonded with oxygen atoms in ether and ester side groups have given signals at higher chemical shifts, in a range of 69.13–60.39 ppm. Infrared spectra have revealed signals originating from amine end groups; however, they have been of very low intensity. Bands connected with imine stretching have been found in a range of 1689–1628 cm^{-1}, while bands ascribed to stretching of carbonyl groups, present in aldehyde end groups and ester side chains, have been observed between 1731 and 1728 cm^{-1}.

The solubility of compounds has been investigated in several organic solvents, of various dielectric constants and the results are presented in Table 1, together with the solubility test results of imine **PAz-BOO-OOct**, previously reported in [22].

Table 1. Solubility of investigated compounds in solvents of various dielectric constants (ε).

Compound	NMP (ε = 33.00)	THF (ε = 7.58)	Chloroform (ε = 4.80)	n-Hexane (ε = 1.88)
PAz-BOO-OMe	-/-	-/-	+/-	-/-
PAz-BOO-OOct	+/- *	+/+ *	+/+ *	-/- *
PAz-Carb-OMe	-/-	-/-	+/-	-/-
PAz-Carb-OOct	+/+	+/+	+/+	-/-

1 mg/mL (+/+) soluble at room temperature, (+/-) soluble at boiling point, (-/-) partially soluble or insoluble at boiling point. * values from [22].

None of investigated compounds has been soluble in a non-polar n-hexane. Imines consisting only bulky substituents (**PAz-BOO-OOct**, **PAz-Carb-OOct**) have been completely or mostly soluble in the remaining solvents, both in polar NMP and also in chloroform or THF of lower dielectric constants. Partial or complete replacement of bulky side chains with methoxy groups (**PAz-BOO-OMe** and **PAz-Carb-OMe**, respectively) has drastically decreased the solubility of compounds in NMP and THF and hindered their affinity to chloroform. Observed solubility in this solvent has, nevertheless, been sufficient for the purpose of further investigations.

The molar masses and dispersity values have been calculated according to the conventional calibration with polystyrene standards. Obtained values are presented in Table 2 together with molar masses of **PAz-BOO-OOct**, reported previously in [22]. The presented values have proved that the synthesis of imine **PAz-BOO-OMe** has resulted in a formation of oligomer, while the remaining polycondensations have allowed to obtain compounds with a higher degree of polymerisation.

Table 2. Molar masses and molar mass dispersity of investigated polyazomethines.

Compound	M_n [g/mol]	M_w [g/mol]	Đ
PAz-BOO-OMe	1350	3250	2.4
PAz-BOO-OOct	2690 *	7520 *	2.8 *
PAz-Carb-OMe	4700	5300	1.2
PAz-Carb-OOct	9600	16,350	1.7

M_n—number average molar mass, M_w—mass average molar mass, Đ—molar mass dispersity, * values from [22].

Imines consisting of methoxy substituents have reached lower molar masses than their counterparts consisting solely of bulky side groups. Similarly, compounds with branched N-ethylhexyl chains have revealed higher molar masses than analogue imines with octyloxy side chains. This is probably due to the better solubility of forming macromolecules, consisting of such substituents, allowing the formation of longer polymer chains with higher molar masses. The compound **PAz-Carb-OOct**, consisting of both types of bulky substituents, has shown the highest molar mass of all investigated imines.

3.2. Thermal Properties

Glass transition temperatures from differential scanning calorimetry (DSC) measurements and weight loss data, obtained from thermal gravimetric analysis (TGA) of synthesised polymers, together with analogue values, previously reported in [22] for **PAz-BOO-OOct**, are gathered in Table 3. The DSC curves obtained during the first heating scans of investigated polyazomethines have not revealed endotherms related to melting, and during the second heating stage, after a rapid cooling, all have shown a characteristic deflection, connected to the glass-transition (Figure S1).

Table 3. Thermal properties of synthesized polymers.

Compound	T_g [°C]	$T_{5\%}$ [°C]	$T_{10\%}$ [°C]	T_{max} [°C]
PAz-BOO-OMe	318.0	n/d	n/d	n/d
PAz-BOO-OOct	30.8 *	350.0 *	364.8 *	380.0 *
PAz-Carb-OMe	264.0	341.5	368.8	391.5
PAz-Carb-OOct	291.0	331.3	365.5	393.7

T_g—glass transition temperature, $T_{5\%}$, $T_{10\%}$—temperatures of 5% and 10% weight loss, T_{max}—temperature of the maximum mass loss rate, n/d—not designated, * values from [22].

The analysis of registered parameters has revealed that utilization of diamines with diimine system has allowed to obtain compounds with much higher glass transition temperatures, in a range of 264.0–318.0 °C, compared to the imine, synthesized from the simpler monomers (**PAz-BOO-OOct**), which have shown T_g = 30.8 °C. Such a difference may be due to the large amount of long n-alkyl chains, together with a moderate oligomer chain length. All investigated compounds have shown a one-step thermal degradation process, with similar initial decomposition temperatures (Figure S2). Unfortunately, due to the low amount of **PAz-BOO-OMe**, TGA measurements have not been completed for this compound. Designated $T_{5\%}$ values have been observed to be lower for imines with branched ethylhexyl side chains (**PAz-Carb**) and have been the lowest for polyazomethine consisting solely of bulky substituents (**PAz-Carb-OOct**). The temperatures of 10% mass loss have been almost identical for all investigated compounds and have probably been connected to the thermal degradation of ester side chains, present in all of those structures.

3.3. Electrochemical Measurements

Electrochemical measurements have been conducted using the cyclic voltammetry (CV) technique, performed in the range of all noticeable oxidation and reduction peaks, within the range of electrolyte stability window. Obtained voltammograms have revealed that both oxidation and reduction processes, of all investigated polyazomethines, have been electrochemically irreversible (Figure 3a). The onsets of electrochemical processes have been designated, which has allowed to calculate the energies of HOMO (the Highest Occupied Molecular Orbital) and LUMO (the Lowest Unoccupied Molecular Orbital) levels, together with the energy gaps (E_g^{CV}). All of these parameters have been gathered in Table 4, and presented in Figure 3b, together with the values previously reported for **PAz-BOO-Oct** [11].

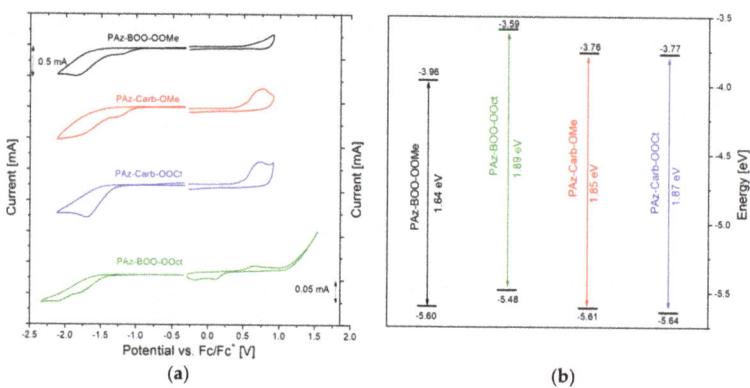

Figure 3. (a) Voltammograms, recorded during oxidation and reduction processes (v = 0.1 V/s; 0.1 M Bu$_4$NPF$_6$/ACN) of studied oligo- and polyazomethine thin films, deposited on the glass with ITO together with voltammogram obtain for **PAz-BOO-Oct**, presented previously in [11]; (b) the energy levels of HOMO and LUMO orbitals, designated based on the results of cyclic voltammetry measurements.

Table 4. Electrochemical properties of investigated compounds.

Compound	E_{ox}^{onset} [V]	E_{red}^{onset} [V]	E_{HOMO} [eV]	E_{LUMO} [eV]	E_g^{CV} [eV]
PAz-BOO-OMe	0.50	−1.14	−5.60	−3.96	1.64
PAz-BOO-OOct	0.38 *	−1.51 *	−5.48 *	−3.59 *	1.89 *
PAz-Carb-OMe	0.51	−1.34	−5.61	−3.76	1.85
PAz-Carb-OOct	0.54	−1.33	−5.64	−3.77	1.87

E_{ox}^{onset}—the onset of oxidation potential, E_{red}^{onset}—the onset of reduction potential, $E_{HOMO} = -e^-(5.1 + E_{ox}^{onset})$, $E_{LUMO} = -e^-(5.1 + E_{red}^{onset})$ according to [28], E_g^{CV}—electrochemical energy gap = E_{LUMO}-E_{HOMO}, * values from [11].

Analysis of presented values of **PAz-BOO** imines has revealed an increase of oxidation onset potential and a simultaneous decrease of the E_{red}^{onset}, upon a partial replacement of octyloxy side chains (**PAz-BOO-OOct**) with shorter methoxy substituents (**PAz-BOO-OMe**). The reduction potential, nevertheless, has been decreased more significantly than the oxidation potential has been increased, which has resulted in more lowered LUMO orbital and, subsequently, in a decrease of the band gap width by 0.25 eV of methoxy-substituted imine, compared to the analogue compound, consisting solely of bulky side chains. Since **PAz-BOO-OMe** has had lower molar mass, which has also an impact on electrochemical properties [29], this is most probably due to adopting a more planar geometry by the oligoimine **PAz-BOO-OMe**, due to the lower amount of bulky alkyl chains. Such planarity adjustment has tuned the LUMO level of compound, with the slight effect on HOMO level [30]. Similar manipulation of substituents' length has almost not affected either oxidation or reduction potentials of carbazole-consisting compounds (**PAz-Carb**). They have undergone these processes at nearly identical potentials and have subsequently revealed very similar energies of HOMO and LUMO orbitals and similar energy gap widths. This might suggest that disruption of macromolecule planarity, caused by branched N-ethyl-hexyl chains, cannot be influenced by the manipulation with alkoxy substituent lengths.

3.4. Optical Properties

UV-Vis absorption spectra have been measured for imines solutions in chloroform and for their thin films, deposited on quartz substrates, using the spin-coating technique. Electronic spectra of investigated oligo– and polyazomethines solutions (Figure 4a) have shown absorption bands in a range of 263.5–684.5 nm. The bands localized at the lowest energies have been attributed to the $\pi \to \pi^*$ electronic transitions. Absorption bands localised at higher energies are due to the electron transitions between the $\sigma \to \pi^*$ or $\pi \to \sigma^*$ levels. Position of the low energy absorption bands have been influenced by the length of alkoxy substituent and by the chemical structure of imine, while bands registered at shorter wavelengths have revealed similar positions for all compounds, although their intensity have been much higher for **PAz-Carb** imines.

The comparison of oligoazomethine **PAz-BOO-OMe** spectrum with that recorded previously for polyazomethine **PAz-BOO-OOct** [22] has revealed a distinct bathochromic shift of the absorption band connected with $\pi \to \pi^*$ transitions (from 568.0 to 684.5 nm). Due to a lower molar mass of this compound, such a shift could have not resulted from larger π-conjugation area on longer polymer chains [31], which is consistent with the electrochemical data, and most probably has been due to adopting the more planar geometry by this molecule. Similar manipulation with substituents' length in **PAz-Carb** imines has resulted in a much smaller bathochromic shift of the band (from 446.5 to 513.0 nm), after the complete replacement of octyloxy substituents (**PAz-Carb-OOct**) with methoxy side groups (**PAz-Carb-OMe**). This might also suggest an increase of the molecule planarity, upon modification of side chains length, although in a much smaller extent.

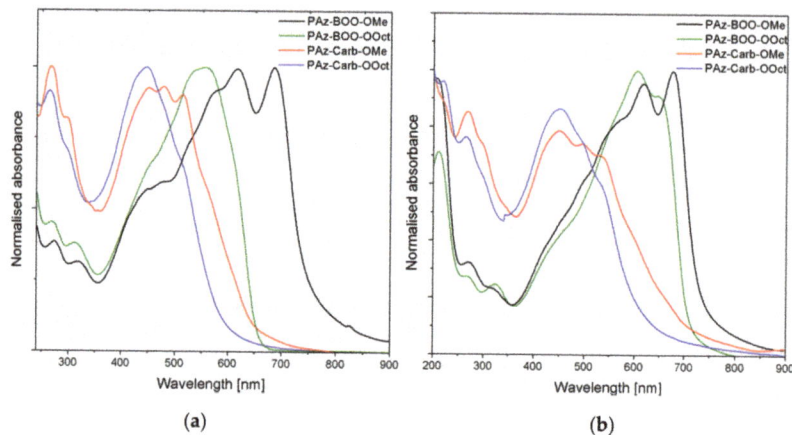

Figure 4. (a) Normalized absorbance of investigated oligo- and polyazomethines solutions in chloroform and (b) their thin films, deposited on quartz substrates.

Absorbance spectra of thin films of investigated compounds (Figure 4b) have revealed additional bands, localized at high energies (from 212.5–220.0 nm), which are connected to the $\sigma \rightarrow \sigma^*$ electron transitions and could have not been observed in solution, due to the solvent measurement window (the spectral range of solvent permeability). Almost all imines have shown a bathochromic shift of the absorption band connected with the $\pi \rightarrow \pi^*$ electron transitions, after deposition of thin film, compared to the solution (Table 5). This suggests the formation of J-type aggregates during spin-coating [32]. Unlike others, absorption band of **PAz-BOO-OMe** thin film has shifted hypsochromically, which is probably due to formation of H-type aggregates in this individual imine thin film [32]. The absorption bands, nevertheless, have been still observed at lower energies for imines consisting of methoxy side groups, than for their octyloxy-substituted analogues.

Table 5. Positions of absorption bands connected with $\pi \rightarrow \pi^*$ transitions of imine solutions and thin films, together with exciton band widths and absorption edge parameters of thin films.

Compound	λ_{max} [nm]/(E_{max} [eV])		W [meV]	Eg [eV]	E_U [meV]
	Solution	Thin Film			
PAz-BOO-OMe	684.5/(1.81)	676.0/(1.83)	4	1.59	93
PAz-BOO-OOct	556.0/(2.23) *	604.0/(2.05) *	46	1.69 *	62 *
PAz-Carb-OMe	513.0/(2.42)	535.0/(2.32)	30	1.42	349
PAz-Carb-OOct	446.5/(2.78)	450.0/(2.76)	85	1.77	233

λ_{max}—position of lowest-energy distinct absorption band, W—exciton bandwidth, Eg—energy gap designated according to Tauc model, E_U—Urbach energy, * values from [22].

Moreover, all absorption bands connected with the $\pi \rightarrow \pi^*$ electron transitions have shown a vibronic structure (Figure 5). Low energy bands of the methoxy-substituted compounds (**PAz-BOO-OMe** and **PAz-Carb-OMe**) have shown a clear vibronic structure, while individual vibronic peaks of remaining imines have been found using the second derivative method (i.e., minimum of the second derivative of absorption corresponds to the absorption maximum). Afterwards, the vibronic progression of imine thin films bands have been deconvoluted, with the modified Fourier self-deconvolution and Finite Response Operator (FIRO) methods [33]. Observed vibronic peaks have been assigned according to the Franck-Condon principle, assuming the stationary nuclear framework [34]. Energy differences between individual vibronic peaks have been in the range of 0.16–0.25 eV, which clearly suggests presence of the electron–phonon interaction, which are connected to the benzene ring stretching mode [35].

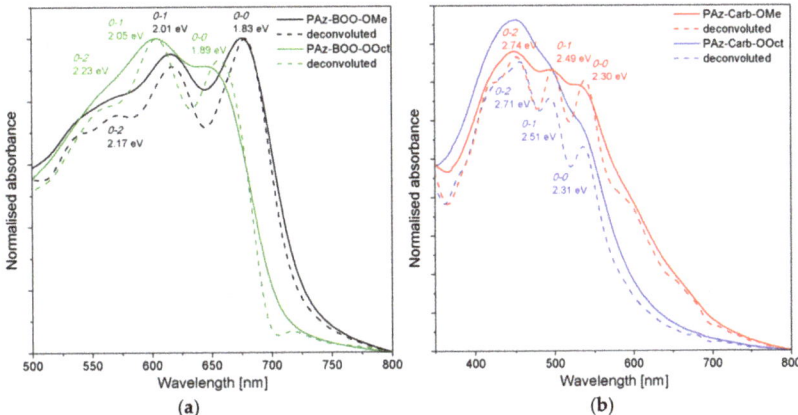

Figure 5. Normalized (solid line) and deconvoluted (dashed line) spectra of **PAz-BOO** (a) and **Paz-Carb** (b) imine thin films.

Spectra of **PAz-BOO** thin films (Figure 5a) have revealed the most pronounced peaks connected to transitions between *0-0* and *0-1* levels, while peaks connected to transitions of higher energies have been of much smaller intensity. In contrary, deconvoluted spectra of **PAz-Carb** imines (Figure 5b) have shown a gradual increase of intensity as the vibronic peak energy increased. Assuming that the main intramolecular vibration (E_p), coupled to the electronic transition, is the C=C symmetric stretch at 0.18 eV [36], a ratio of intensities of *0-0* and *0-1* vibronic peaks has allowed to calculate values of exciton bandwidths (W) for each of individual thin film, according to the Equation (1) [37].

$$\frac{A_{0-0}}{A_{0-1}} = \frac{1 - 0.24 W/E_p}{1 + 0.073 W/E_p} \quad (1)$$

All values of calculated exciton bandwidths (W) are gathered in Table 5. This parameter is connected to the effective conjugation length, where an increase of the conjugation will lead to a decrease of exciton bandwidth [38]. Such a correlation, nonetheless, is only true in systems with similar interchain order, thus calculated values have been compared only within systems with analogue chemical structure. For both groups that are **PAz-BOO** and **PAz-Carb** imines, the value of W has been much smaller for compounds consisting of methoxy side groups (**PAz-OMe**) than for their octyloxy-substituted counterparts (**PAz-OOct**). This clearly indicates that shortening of the alkyl chain length has tuned the macromolecule geometry, allowing for. adopting the more planar configuration, with an enhanced effective π-conjugated length

The edges of absorption bands, connected to the $\pi \to \pi^*$ electron transitions, of investigated imines thin films have allowed to determine absorption edge parameters, which is the Urbach energy (E_U) and the energy gap width (E_g). Values of the Urbach energy have been calculated based on the slope of exponential edges, which follow the Urbach relation (2) [39], as it is depicted in Figure 6a.

$$\alpha \propto exp\left(\frac{E}{E_U}\right) \quad (2)$$

Energy gaps have been obtained, using a linear approximation to energy axis of the $(\alpha E)^{1/2}$ dependence, according to the Tauc relation (3) [40]:

$$\alpha \propto (E - E_g)^2 \quad (3)$$

(true for the energy range $E > E_g$), typical for amorphous semiconductors, often used also for polymers thin films [17,18,21,22]. The way of determination the energy gaps for investigated films is shown in Figure 6b. All obtained absorption edge parameters have been gathered in Table 5.

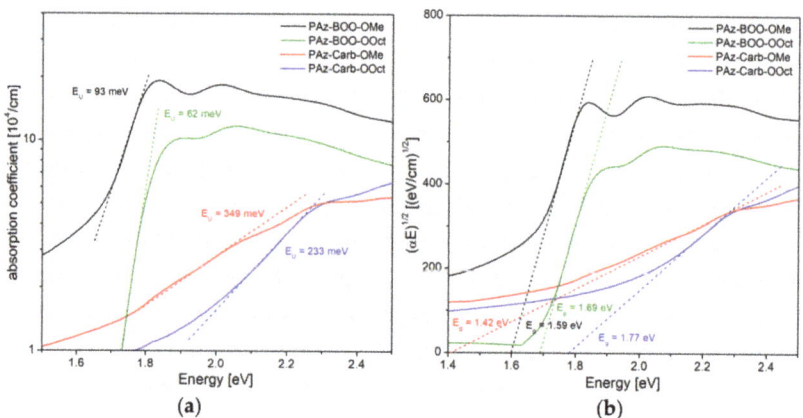

Figure 6. Absorption edge parameters of investigated imines: (**a**) Urbach energies and (**b**) energy gaps.

Generally, carbazole consisting imines have shown higher values of the Urbach energy, indicating a higher amount of structural disorder defects, which have introduced localized energy states within the energy gap. Especially the absorption edge of **PAz-Carb-OMe** thin film has shown high value of $E_U = 349$ meV, suggesting the large number of localized states within the energy gap. All energy gap values obtained using the Tauc model, have been lower than these designated using cyclic voltammetry. Compounds **PAz-BOO** have shown rather close values of energy gaps calculated by these two methods, and a partial replacement of octyloxy side chains with methoxy groups (**PAz-BOO-OMe**) has caused a decrease of E_g, compared to imine solely substituted with bulky alkyl chains (**PAz-BOO-OOct**), from 1.69 to 1.59 eV. Energy gap width of **PAz-Carb-OOct** has been also very similar (1.77 eV) to that, obtained by electrochemical measurements, although the replacement of octyloxy chains with methoxy groups (**PAz-Carb-OMe**) has caused a significant decrease of E_g, to a value much lower (1.42 eV) than the electrochemically determined. Such a low value has been most probably caused by the high amount of localized states within the energy gap, which allows energy transitions, of lower energy than the energy gap width.

3.5. Morphology of Thin Films

Morphology of investigated imines thin films has been observed using the Grazing-Incidence Wide-Angle X-ray Scattering (GIWAXS) measurements. Obtained diffractograms (Figure 7) have revealed that most of the investigated imines have shown rather low-ordered structure, showing only one broad signal, localized at similar positions, between 7.58°–8.22° 2θ, with a small broadening in a range of 4.74°–5.02° 2θ, which indicates a presence of weak signal. This peak has become a clearer signal for **PAz-BOO-Oct**. Only imine **PAz-BOO-OMe** has shown a much more ordered structure with multiple peaks.

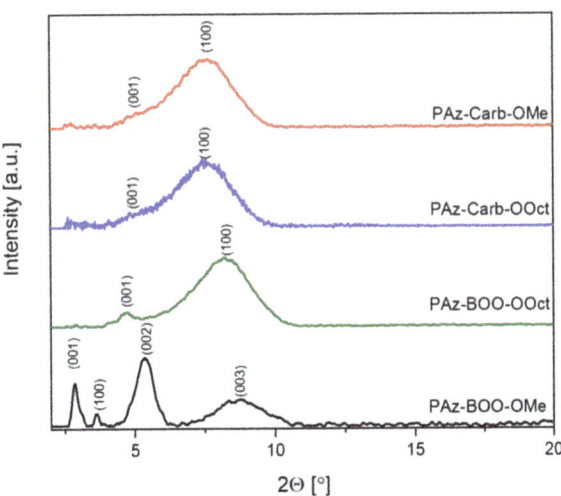

Figure 7. XRD patterns of imine thin films with possible assignment of Miller indexes.

The structure order, visible in the form of peaks in WAXD patterns, is related to the macromolecules planarity and intermolecular interactions, but is not always related to π-stacking peak. In the case of thin-film examination, a presence of π-stacking peak indicates a highly ordered structure. In ordered structures, with few peaks visible, more planar molecules might be characterized by a (100) peak position—higher *d*-spacing values show higher macromolecule planarity [25,26]. After the assignment of Miller indexes, it has been noticed that both carbazole-consisting imines (**PAz-Carb**) have shown only a peak corresponding to the *a* parameter, connected to the short polymer axis, and thus with the planarity of macromolecules. Almost identical positions of these peaks (7.6° and 7.7° 2θ for **PAz-Carb-OOct** and **PAz-Carb-OMe,** respectively) has indicated lack of any changes in the molecule geometry, regardless of alkyl chain variation. This is in agreement with the electrochemical data, registered for these compounds. Polyazomethine substituted solely with octyloxy groups (**PAz-BOO-Oct**) has revealed that the peak, corresponding to planarity of the molecule, shifted towards higher angles (8.2° 2θ), suggesting a narrower *a* axis dimensions, which may be connected to the presence of linear, n-octyloxy side chains. Apart from that, a peak ascribed to the *c* parameter, connected to the length of molecules, at 4.8° 2θ, has become more developed. The partial replacement of bulky octyloxy groups with methoxy substituents (**PAz-BOO-OMe**) has resulted in a significant increase of solid order. A peak, ascribed to the *c* parameter, has been visible at lower angles (2.8° 2θ), suggesting larger dimensions in this axis. Moreover, the diffractogram of this imine thin film has shown peaks ascribed to multiples of this parameter (5.3° and 8.8° 2θ). The signal corresponding to the planarity of macromolecule, assigned to the *a* parameter, has been observed at lower angles for **PAz-BOO-OMe** (3.6° 2θ), compared to **PAz-BOO-Oct,** (8.2° 2θ). This clearly suggests the larger dimension in this axis, which suggests the enhanced planarity of this imine in respect to its counterpart substituted solely with octyloxy side chains. According to [25,26], spin-coated thin films exhibits lower order than solid samples, especially for high molar masses of macromolecules.

3.6. Preliminary Photovoltaic Activity Tests

All of the investigated compounds (PAz) have been utilized in photovoltaic bulk-heterojunction (BHJ) systems, acting as donor, together with the fullerene derivative (PC$_{61}$BM), acting as acceptor. The conventional architecture ITO/PEDOT:PSS/PAz:PC$_{61}$BM/Al has been chosen, where various weight ratios of donor and acceptor have been studied (1:1, 1:2 and

1:3 wt.). Registered *J-V* characteristics (Figure 8) have allowed to designate parameters of prepared photovoltaic cells (Table 6).

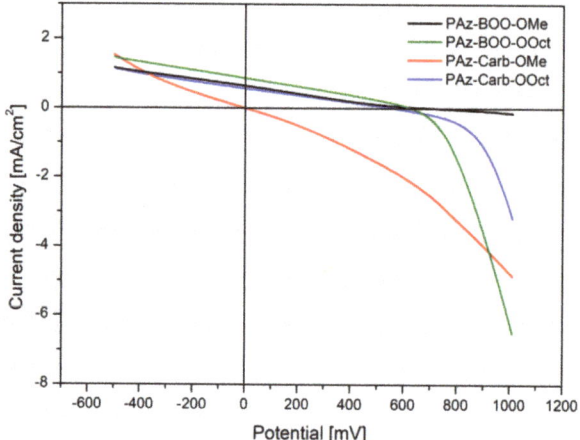

Figure 8. Current density–voltage (*J-V*) characteristics of bulk-heterojunction (BHJ) systems, consisting investigated oligo- and polyazomethines. Donor:acceptor ratios 1:2 for all, except for **PAz-BOO-OMe**, where the ratio D:A is 1:1.

Table 6. Photovoltaic parameters of the BHJ photovoltaic systems, consisting investigated imines.

System	V_{OC} [mV]	J_{SC} [mA/cm2]	FF	η [%]
PAz-BOO-OMe:PCBM (1:1)	728.2	0.65	0.20	0.09
PAz-BOO-OMe:PCBM (1:2)		less than 0.01%		
PAz-BOO-OMe:PCBM (1:3)				
PAz-BOO-Oct:PCBM (1:1)	834.8	0.72	0.16	0.16
PAz-BOO-Oct:PCBM (1:2)	630.8	0.87	0.29	0.17
PAz-BOO-Oct:PCBM (1:3)		less than 0.01%		
PAz-Carb-OMe:PCBM		less than 0.01% for each ratio		
PAz-Carb-OOct:PCBM (1:1)	210.6	0.22	0.47	0.02
PAz-Carb-OOct:PCBM (1:2)	615.1	0.58	0.26	0.09
PAz-Carb-OOct:PCBM (1:3)		less than 0.01%		

V_{OC}—open circuit voltage, J_{SC}—short circuit current density, *FF*—fill factor, η—power conversion efficiency.

Almost all of the studied compounds have shown a photovoltaic effect, while acting as a donor, with [6,6]-phenyl-C61-butyric acid methyl ester (PCBM) as an acceptor, except for **PAz-Carb-OMe**, which has shown no activity in all systems. Lack of any photo-response of this compound may be caused by a large amount of structural disorder effects, which have introduced a localized energy states within energy gap and are trapping generated charge carriers. Power conversion efficiencies of studied systems have been in the range of 0.02–0.17%. Such values are similar to others, reported for BHJ systems with phenylene or thiophene-phenylene imines, described in the literature [41,42]. The highest power conversion efficiency has been observed for systems consisting of **PAz-BOO-Oct** imine. Partial replacement of octyloxy side chains with methoxy groups (**PAz-BOO-OMe**), despite providing more favorable optical and electrochemical properties, has decreased efficiency of photovoltaic cell, where modified imine has been utilized. The main parameter, which has been responsible for such a decrease, has been the short-circuit current density (J_{SC}). A decrease of this parameter has probably been most connected with the lower molar mass of the **PAz-BOO-OMe**. For shorter macromolecules, the ratio of intermolecular charge transport in relation to the intramolecular part is higher, which causes a decrease of the overall conductivity in the material [43]. The increase of the donor:acceptor ratio to 1:2 has enhanced the efficiency of photovoltaic systems consisting **PAz-BOO-OOct** and **PAz-Carb-OOct**, while it has completely ceased any activity of **PAz-BOO-OMe**. Further increase of the acceptor quantity has ceased activity of the remaining imines.

4. Conclusions

In this paper, three novel oligo- and polyazomethines have been obtained and the influence of alkyl side chains lengths on their supramolecular organization has been observed. Such a modification has been accomplished by the condensation of diamines with diimine systems that, to the best of our knowledge, is presented for the first time in literature.

The realization of such an approach has been confirmed using ^{1}H–, ^{13}C–NMR, and FTIR spectroscopies. All of new compounds have shown good thermal stability, and high glass transition temperatures. They have been electrochemically active, and revealed narrow energy gaps, being in the range of 1.64–1.87 eV. Electronic spectra of all new imines solutions and thin films have revealed the broad absorption range. During the spin-coating process of most imines, the J type aggregates have been formed, except for the compound with both octyloxy and methoxy side groups. The partial replacement of the bulky octyloxy side chains with the shorter methoxy groups has induced the adoption of a more planar geometry by a macromolecule, tuning the LUMO orbital and subsequently decreasing the energy gap. This has provided more favorable optical properties, in terms of application in photovoltaic systems, shifting the absorption band connected with the $\pi \rightarrow \pi^*$ electron transitions. Variation of the alkoxy side chains length, nevertheless, has not influenced the electrochemical properties and only slightly affected electronic spectra, when branched N-ethylhexyl chains have been present in the polymer structure.

Almost all compounds have shown the activity in photovoltaic devices, acting as a donor in the blend with fullerene. It has been demonstrated that the chemical structure of investigated imines (both the main group, as side chains) determined the photovoltaic properties, but also the ratio polymer:fullerene in the BHJ active layers is very important in these organic solar cells. Generally, carbazole consisting imines have exhibited a worse photovoltaic properties than their thiophene-phenylene counterparts, while the presence of octyloxy side chains enhance these properties for both groups of imine compounds.

This paper has shown that through the variation of substituents length, it is possible to change the supramolecular structure, which influences the electrochemical and optical properties of materials. Such an approach provides useful information, which may be used during the designing of novel compounds with properties, desired for the application in optoelectronic systems.

Supplementary Materials: The following are available online at https://www.mdpi.com/article/10.3390/polym13071043/s1, Figure S1: DSC curves obtained during the second heating stage of investigated compounds, Figure S2: TGA (a, c) and DTG (b, d) curves of investigated compounds: PAz-Carb-OMe (red lines) and PAz-Carb-OOct (blue lines).

Author Contributions: P.N.—conceptualization, methodology, investigation, writing—original draft; B.J.—conceptualization, supervision, writing—review and editing; M.V.—methodology, investigation; M.G.—methodology, investigation; H.J.—methodology, investigation, M.M.—methodology, investigation; A.D.—methodology, investigation. All authors have read and agreed to the published version of the manuscript.

Funding: This research received no external funding.

Institutional Review Board Statement: Not applicable.

Informed Consent Statement: Not applicable.

Data Availability Statement: The data presented in this study are available on request from the corresponding author.

Conflicts of Interest: The authors declare no conflict of interest.

References

1. Brabec, C.J.; Gowrisanker, S.; Halls, J.J.M.; Laird, D.; Jia, S.; Williams, S.P. Polymer-fullerene bulk-heterojunction solar cells. *Adv. Mater.* **2010**, *22*, 3839–3856. [CrossRef]
2. Sirringhaus, H. 25th anniversary article: Organic field-effect transistors: The path beyond amorphous silicon. *Adv. Mater.* **2014**, *26*, 1319–1335. [CrossRef]
3. Yang, X.; Xu, X.; Zhou, G. Recent advances of the emitters for high performance deep-blue organic light-emitting diodes. *J. Mater. Chem. C* **2015**, *3*, 913–944. [CrossRef]
4. Drewniak, A.; Tomczyk, M.D.; Hanusek, L.; Mielanczyk, A.; Walczak, K.; Nitschke, P.; Hajduk, B.; Ledwon, P. The effect of aromatic diimide side groups on the π-Conjugated polymer properties. *Polymers* **2018**, *10*, 487. [CrossRef] [PubMed]
5. Iwan, A.; Boharewicz, B.; Tazbir, I.; Filapek, M.; Korona, K.P.; Wróbel, P.; Stefaniuk, T.; Ciesielski, A.; Wojtkiewicz, J.; Wronkowska, A.A.; et al. How do 10-camphorsulfonic acid, silver or aluminum nanoparticles influence optical, electrochemical, electrochromic and photovoltaic properties of air and thermally stable triphenylamine-based polyazomethine with carbazole moieties? *Electrochim. Acta* **2015**, *185*, 198–210. [CrossRef]
6. Bejan, A.E.; Damaceanu, M.D. Acid-responsive behavior promoted by imine units in novel triphenylamine-based oligomers functionalized with chromophoric moieties. *J. Photochem. Photobiol. A Chem.* **2019**, *378*, 24–37. [CrossRef]
7. Bolduc, A.; Al Ouahabi, A.; Mallet, C.; Skene, W.G. Insight into the isoelectronic character of azomethines and vinylenes using representative models: A spectroscopic and electrochemical study. *J. Org. Chem.* **2013**, *78*, 9258–9269. [CrossRef]
8. Nitschke, P.; Jarząbek, B.; Vasylieva, M.; Honisz, D.; Małecki, J.G.; Musioł, M.; Janeczek, H.; Chaber, P. Influence of chemical structure on thermal, optical and electrochemical properties of conjugated azomethines. *Synth. Met.* **2021**, *273*, 116689. [CrossRef]
9. Yağmur, H.K.; Kaya, İ.; Aydın, H. Synthesis, characterization, thermal and electrochemical features of poly (phenoxy-imine)s containing pyridine and pyrimidine units. *J. Polym. Res.* **2020**, *27*, 356. [CrossRef]
10. Peng, H.; Sun, X.; Weng, W.; Fang, X. 2-Synthesis and Design of Conjugated Polymers for Organic Electronics. In *Polymer Materials for Energy and Electronic Applications*; Academic Press: Cambridge, MA, USA, 2017; pp. 9–61.
11. Nitschke, P.; Jarząbek, B.; Damaceanu, M.-D.; Bejan, A.-E.; Chaber, P. Spectroscopic and electrochemical properties of thiophene-phenylene based Schiff-bases with alkoxy side groups, towards photovoltaic applications. *Spectrochim. Acta Part A Mol. Biomol. Spectrosc.* **2021**, *248*, 119242. [CrossRef] [PubMed]
12. Petrus, M.L.; Bouwer, R.K.M.; Lafont, U.; Athanasopoulos, S.; Greenham, N.C.; Dingemans, T.J. Small-molecule azomethines: Organic photovoltaics via Schiff base condensation chemistry. *J. Mater. Chem. A* **2014**, *2*, 9474–9477. [CrossRef]
13. Bogdanowicz, K.A.; Jewłoszewicz, B.; Iwan, A.; Dysz, K.; Przybyl, W.; Januszko, A.; Marzec, M.; Cichy, K.; Świerczek, K.; Kavan, L.; et al. Selected Electrochemical Properties of 4,4′-((1E,1′E)-((1,2,4-Thiadiazole-3,5-diyl)bis(azaneylylidene))bis(methaneylylidene))bis(N,N-di-p-tolylaniline) towards Perovskite Solar Cells with 14.4% Efficiency. *Materials* **2020**, *13*, 2440. [CrossRef] [PubMed]
14. Gnida, P.; Pająk, A.; Kotowicz, S.; Malecki, J.G.; Siwy, M.; Janeczek, H.; Maćkowski, S.; Schab-Balcerzak, E. Symmetrical and unsymmetrical azomethines with thiophene core: Structure—properties investigations. *J. Mater. Sci.* **2019**, *54*, 13491–13508. [CrossRef]
15. Kotowicz, S.; Siwy, M.; Golba, S.; Malecki, J.G.; Janeczek, H.; Smolarek, K.; Szalkowski, M.; Sek, D.; Libera, M.; Mackowski, S.; et al. Spectroscopic, electrochemical, thermal properties and electroluminescence ability of new symmetric azomethines with thiophene core. *J. Lumin* **2017**, *192*, 452–462. [CrossRef]
16. Barik, S.; Bishop, S.; Skene, W.G. Spectroelectrochemical and electrochemical investigation of a highly conjugated all-thiophene polyazomethine. *Mater. Chem. Phys.* **2011**, *129*, 529–533. [CrossRef]
17. Jarząbek, B.; Hajduk, B.; Domański, M.; Kaczmarczyk, B.; Nitschke, P.; Bednarski, H. Optical properties of phenylene–thiophene-based polyazomethine thin films. *High Perform. Polym.* **2018**, *30*, 1219–1228. [CrossRef]
18. Jarzabek, B.; Weszka, J.; Domański, M.; Jurusik, J.; Cisowski, J. Optical studies of aromatic polyazomethine thin films. *J. Non Cryst. Solids* **2008**, *354*, 856–862. [CrossRef]
19. Yang, C.J.; Jenekhe, S.A. Conjugated Aromatic Poly(azomethines). 1. Characterization of Structure, Electronic Spectra, and Processing of Thin Films from Soluble Complexes. *Chem. Mater.* **1991**, *3*, 878–887. [CrossRef]
20. Thomas, O.; Inganäs, O.; Andersson, M.R. Synthesis and properties of a soluble conjugated poly(azomethine) with high molecular weight. *Macromolecules* **1998**, *31*, 2676–2678. [CrossRef]
21. Jarząbek, B.; Kaczmarczyk, B.; Jurusik, J.; Siwy, M.; Weszka, J. Optical properties of thin films of polyazomethine with flexible side chains. *J. Non Cryst. Solids* **2013**, *375*, 13–18. [CrossRef]
22. Nitschke, P.; Jarząbek, B.; Wanic, A.; Domański, M.; Hajduk, B.; Janeczek, H.; Kaczmarczyk, B.; Musioł, M.; Kawalec, M. Effect of chemical structure and deposition method on optical properties of polyazomethines with alkyloxy side groups. *Synth. Met.* **2017**, *232*, 171–180. [CrossRef]
23. Himmelberger, S.; Duong, D.T.; Northrup, J.E.; Rivnay, J.; Koch, F.P.V.; Beckingham, B.S.; Stingelin, N.; Segalman, R.A.; Mannsfeld, S.C.B.; Salleo, A. Role of side-chain branching on thin-film structure and electronic properties of polythiophenes. *Adv. Funct. Mater.* **2015**, *25*, 2616–2624. [CrossRef]

24. Chen, S.; Sun, B.; Hong, W.; Aziz, H.; Meng, Y.; Li, Y. Influence of side chain length and bifurcation point on the crystalline structure and charge transport of diketopyrrolopyrrole-quaterthiophene copolymers (PDQTs). *J. Mater. Chem. C* **2014**, *2*, 2183–2190. [CrossRef]
25. Zajaczkowski, W.; Nanajunda, S.K.; Eichen, Y.; Pisula, W. Influence of alkyl substitution on the supramolecular organization of thiophene- and dioxine-based oligomers. *RSC Adv.* **2017**, *7*, 1664–1670. [CrossRef]
26. Schuettfort, T.; Thomsen, L.; McNeill, C.R. Observation of a Distinct Surface Molecular Orientation in Films of a High Mobility Conjugated Polymer. *J. Am. Chem. Soc.* **2013**, *135*, 1092–1101. [CrossRef] [PubMed]
27. Barik, S.; Skene, W.G. Selective chain-end postpolymerization reactions and property tuning of a highly conjugated and all-thiophene polyazomethine. *Macromolecules* **2010**, *43*, 10435–10441. [CrossRef]
28. Bujak, P.; Kulszewicz-Bajer, I.; Zagorska, M.; Maurel, V.; Wielgus, I.; Pron, A. Polymers for electronics and spintronics. *Chem. Soc. Rev.* **2013**, *42*, 8895–8999. [CrossRef]
29. Zhang, F.B.; Ohshita, J.; Miyazaki, M.; Tanaka, D.; Morihara, Y. Effects of substituents and molecular weight on the optical, thermal and photovoltaic properties of alternating dithienogermole—dithienylbenzothiadiazole polymers. *Polym. J.* **2014**, *46*, 628–631. [CrossRef]
30. Wang, R.; Chen, Q.; Feng, H.; Liu, B. Simple adjustments to the molecular planarity of organic sensitizers: Towards highly selective optimization of energy levels. *New J. Chem.* **2017**, *41*, 11853–11859. [CrossRef]
31. Zhou, C.; Liang, Y.; Liu, F.; Sun, C.; Huang, X.; Xie, Z.; Huang, F.; Roncali, J.; Russell, T.P.; Cao, Y. Chain length dependence of the photovoltaic properties of monodisperse donor-acceptor oligomers as model compounds of polydisperse low band gap polymers. *Adv. Funct. Mater.* **2014**, *24*, 7538–7547. [CrossRef]
32. Deng, Y.; Yuan, W.; Jia, Z.; Liu, G. H- and J-aggregation of fluorene-based chromophores. *J. Phys. Chem. B* **2014**, *118*, 14536–14545. [CrossRef]
33. Jones, R.N.; Shimokoshi, K. Some Observations on the Resolution Enhancement of Spectral Data by the Method of Self-Deconvolution. *Appl. Spectrosc.* **1983**, *37*, 59–67. [CrossRef]
34. Barford, W. *Electronic and Optical Properties of Conjugated Polymers*; Oxford University Press (OUP): Oxford, UK, 2013.
35. Jarzabek, B.; Weszka, J.; Domanski, M.; Jurusik, J.; Cisowski, J. Optical properties of amorphous polyazomethine thin films. *J. Non Cryst. Solids* **2006**, *352*, 1660–1662. [CrossRef]
36. Louarn, G.; Trznadel, M.; Buisson, J.P.; Laska, J.; Pron, A.; Lapkowski, M.; Lefrant, S. Raman Spectroscopic Studies of Regioregular Poly(3-alkylthiophenes). *J. Phys. Chem.* **1996**, *100*, 12532–12539. [CrossRef]
37. Clark, J.; Chang, J.F.; Spano, F.C.; Friend, R.H.; Silva, C. Determining exciton bandwidth and film microstructure in polythiophene films using linear absorption spectroscopy. *Appl. Phys. Lett.* **2009**, *94*, 2007–2010. [CrossRef]
38. Beljonne, D.; Cornil, J.; Silbey, R.; Millié, P.; Brédas, J.L. Interchain interactions in conjugated materials: The exciton model versus the supermolecular approach. *J. Chem. Phys.* **2000**, *112*, 4749–4758. [CrossRef]
39. Cody, G.D. Chapter 2: The Optical Absorption Edge of a-Si: H. *Semicond. Semimet.* **1984**, *21*, 11–82. [CrossRef]
40. Tauc, J.; Menth, A. States in the gap. *J. Non Cryst. Solids* **1972**, *8–10*, 569–585. [CrossRef]
41. Hindson, J.C.; Ulgut, B.; Friend, R.H.; Greenham, N.C.; Norder, B.; Kotlewski, A.; Dingemans, T.J. All-aromatic liquid crystal triphenylamine-based poly(azomethine)s as hole transport materials for opto-electronic applications. *J. Mater. Chem.* **2010**, *20*, 937–944. [CrossRef]
42. Iwan, A.; Boharewicz, B.; Tazbir, I.; Filapek, M. Enhanced power conversion efficiency in bulk heterojunction solar cell based on new polyazomethine with vinylene moieties and [6,6]-phenyl C61 butyric acid methyl ester by adding 10-camphorsulfonic acid. *Electrochim. Acta* **2015**, *159*, 81–92. [CrossRef]
43. Mihailetchi, V.D.; Xie, H.; De Boer, B.; Koster, L.J.A.; Blom, P.W.M. Charge transport and photocurrent generation in poly(3-hexylthiophene): Methanofullerene bulk-heterojunction solar cells. *Adv. Funct. Mater.* **2006**, *16*, 699–708. [CrossRef]

Article

A Novel Poly-N-Epoxy Propyl Carbazole Based Memory Device

Ahmed. N. M. Alahmadi [1,*] and Khasan S. Karimov [2]

[1] Electrical Engineering Department, Umm-Al-Qura University, Makkah 21955, Saudi Arabia
[2] Ghulam Ishaq Khan Institute of Engineering Sciences and Technology, Topi 23640, Khyber Pakhtunkhwa, Pakistan; khasan@giki.edu.pk
* Correspondence: anmahmadi@uqu.edu.sa

Abstract: Generally, polymer-based memory devices store information in a manner distinct from that of silicon-based memory devices. Conventional silicon memory devices store charges as either zero or one for digital information, whereas most polymers store charges by the switching of electrical resistance. For the first time, this study reports that the novel conducting polymer Poly-N-Epoxy-Propyl Carbazole (PEPC) can offer effective memory storage behavior. In the current research, the electrical characterization of a single layer memory device (metal/polymer/metal) using PEPC, with or without doping of charge transfer complexes 7,7,8,8-tetra-cyanoquino-dimethane (TCNQ), was investigated. From the current–voltage characteristics, it was found that PEPC shows memory switching effects in both cases (with or without the TCNQ complex). However, in the presence of TCNQ, the PEPC performs faster memory switching at relatively lower voltage and, therefore, a higher ON and OFF ratio ($I_{ON}/I_{OFF} \sim 100$) was observed. The outcome of this study may help to further understand the memory switching effects of conducting polymer.

Keywords: memory device; organic semiconductors; poly-N-epoxy-propylcarbazole; tera-cyanoquino-dimethane

Citation: Alahmadi, A..N.M.; Karimov, K.S. A Novel Poly-N-Epoxy Propyl Carbazole Based Memory Device. *Polymers* 2021, 13, 1594. https://doi.org/10.3390/polym13101594

Academic Editor: Bożena Jarząbek

Received: 10 April 2021
Accepted: 12 May 2021
Published: 15 May 2021

Publisher's Note: MDPI stays neutral with regard to jurisdictional claims in published maps and institutional affiliations.

Copyright: © 2021 by the authors. Licensee MDPI, Basel, Switzerland. This article is an open access article distributed under the terms and conditions of the Creative Commons Attribution (CC BY) license (https://creativecommons.org/licenses/by/4.0/).

1. Introduction

Polymer-based electronic devices have been a popular research area in recent decades due to their unmatched properties, such as light weight, flexibility, low cost, scale-ability, low-temperature processing, tenability, etc. [1–4]. As a result, these devices are an excellent choice for light-emitting diodes, solar cells, low-cost RFID, sensors, and many other electronic devices [5–8]. For a number of applications, such as mobile phones, smart watches, and other electronic devices, low-cost and flexible memory devices are essential [9,10]. Therefore, polymer-based memory devices, compared to Si-based memory devices, represent a new trend that not only exploits the reported advantages of polymer, but also improves the storage capacity of memory space for future disposable electronic devices [10].

For this purpose, many semiconducting polymer materials have been previously reported to show capacitive, resistive, and transistor-based memory effects; among these, carbazole-containing materials (e.g., poly-N-vinylcarbazole, PVC) are gaining considerable attention [11–13]. The observed resistive memory response of PVK thin film is accounted due to the change in the electrical resistivity as a function of the applied electric field [14–16]. The possible mechanisms for resistive switching memory in conductive polymer can be approximately categorized into three types: reduction/oxidation, electronic, and thermal. In the reduction/oxidation type of resistive memory device, the ions are migrated towards the corresponding electrodes due to a series of electrochemical reactions, and hence form a low resistive (ON) and high resistive (OFF) conducting path between electrodes under the influence of the applied electric field [14,17]. Similarly, for the electronic type of resistive memory device, the injected carriers are trapped (high resistance) and released (low resistance) during the charge transport process. Electronic charge trapping mechanisms are further classified as charge trapping inside the band gap, bulk charge trapping in the presence of the space charge, and charge trapping at metal–polymer interfaces. By comparison, in the thermal type of resistive memory device, the memory switching effects are originated by the formation and rupture of the conductive filamentary path initiated by the local Joule-heating effects [14,18].

In the trapped space charge limited current model, it is generally accepted that both the positional and energetic traps have the capability to capture free charge carriers, which are distributed throughout the polymer layer and degrade the free carrier mobility (high resistance, OFF state), particularly at a lower operating voltage. In a higher applied electric field, many free carriers overcome the trap barrier potential to form a space charge, increasing the mobility and hence the conductivity (ON state) of the polymer thin film [15,19–23]. Such resistive ON and OFF behavior can be observed as hysteresis due to the film's current–voltage characteristics. Therefore, the trapped space charge effect plays a vital role in defining the memory effect for many organic and polymer-based electronic devices [24].

Organic molecules, and particularly TCNQ-based charge transfer complexes, have a long history of use in memory and other electronic devices [25–28], and the highly stable and reliable memory response was noted for a Cu/Cu-TCNQ/Al device, as reported by numerous researchers [29,30]. Therefore, in this study, a novel Cu/PEPC-TCNQ/Ag device was fabricated and investigated. It was observed that the device shows a high rectification ratio (I_{ON}/I_{OFF} ~ 100) with extended retention time, which is highly suitable for memory devices.

2. Device Fabrication

For the resistive memory device, a simple metal–polymer–metal like diode structure was fabricated at room temperature. For the active polymer layer, PEPC and TCNQ complex materials were selected, based on the advantages discussed above. The chemicals TCNQ (CAS number: 1518-16-7; molecular weight: 204.19; empirical formula (Hill Notation): $C_{12}H_4N_4$) and tetrahydrofuran (CAS number: 109-99-9; molecular weight: 72.11; empirical formula (Hill Notation): C_4H_8O) were purchased from Sigma–Aldrich (Karachi, Pakistan), and PEPC was locally developed, for which detailed information can be found elsewhere [12,13,28]. The purchased chemicals were used without any further purification. The molecular structure of PEPC (1400 amu) and TCNQ is shown in Figure 1a,b, respectively. Generally, the doping of PEPC with TCNQ makes a charge transfer complex, where PEPC behaves as an electron donor, whereas the low-molecular-weight organic material TCNQ behaves as an electron acceptor [28]. The conductive TCNQ was obtained after successive processes of re-crystallization with acetonitrile solvent. Because both PEPC and TCNQ are soluble in tetrahydrofuran as an organic solvent, a solution was made between PEPC and TCNQ (4:1 ratio) with 8% by weight in the solvent tetrahydrofuran. Thin films of both PEPC and PEPC-TCNQ solution were deposited by the spin-coating method (1000 rpm, 30 s) over 99.99% Cu substrate, separately. The thickness of the films was in the range of 500 nm–1.2 μm, and the average surface area of the films was in the range of 1.6–2.1 cm^2. For another electrode, highly conductive silver paste was deposited onto the PEPC-TCNQ thin films. For characterization, three samples were fabricated for Cu/PEPC/Ag and Cu/PEPC-TCNQ/Ag devices, and the median response (which was very close to the average response for most cases) among the three samples was selected for the further analysis that led to the conclusions. A schematic cross-section of the sample is shown in Figure 2. Using the hot probe method, it was observed that both PEPC and PEPC-TCNQ behave as a p-type semiconductor [28]. For forward bias current–voltage characteristics, the positive and negative terminal of the battery was connected to the top electrode (Ag) and bottom (Cu) electrode, respectively, for both devices, as shown in Figure 2.

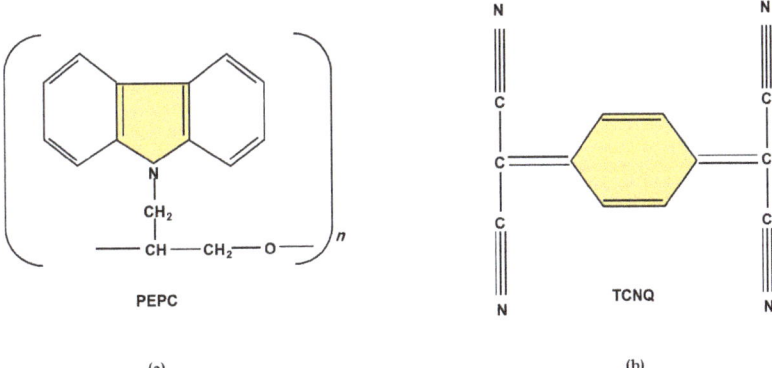

Figure 1. Molecular structure of (**a**) poly-N-epoxy-propylcarbazole (PEPC); and (**b**) 7,7,8,8-tetra-cyanoquino-dimethane (TCNQ).

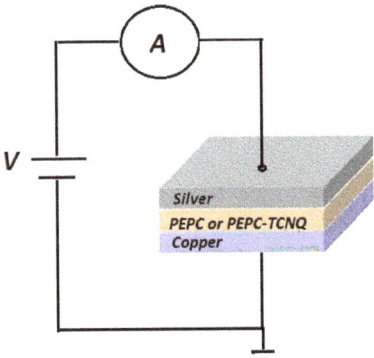

Figure 2. A cross-sectional view of a metal/polymer/metal diode fabricated using p-type PEPC, with or without TCNQ, as a Cu/PEPC/Ag- and Cu/PEPC-TCNQ/Ag-based memory device; for forward bias, the positive and negative terminals of the battery were connected to the top (Ag) and bottom (Cu) electrodes, respectively, for both devices.

3. Results and Discussion

The current–voltage characteristics of both the Cu/PEPC/Ag and Cu/PEPC-TCNQ/Ag devices were measured at room temperature, as shown in Figure 3. Because we were mainly interested in qualitatively determining and comparing the memory effects for the PEPC and PEPC-TCNQ devices, we measured the current–voltage hysteresis loop for both devices at a sweeping rate of 100 mV/sec for simplicity. The figure shows that both devices exhibited electrical switching effects in both forward and reverse cycles. In the first cycle, the applied voltage increased in the forward bias from zero to 3.25 V for the Cu/PEPC/Ag devices, and from 0 to 1.5 V for the Cu/PEPC-TCNQ/Ag device, whereas in the second cycle, the applied voltage decreased in the forward bias from 3.25 and 1.5 V to zero voltage, respectively. The sweeping rate for both directions of the cycle was maintained at 100 mV/sec without any hold time. Both devices followed different current paths for different cycles, clearly demonstrating an electrical memory effect (transition from high resistance to low resistance) for both devices. The transition from high resistance state to low transition state was observed at ~2.5 V for the Cu/PEPC/Ag device, whereas a sharp transition was observed at 1.5 V for Cu/PEPC-TCNQ/Ag. This result indicates that doping of PEPC with TCNQ improves the electrical bi-stability of the memory behavior.

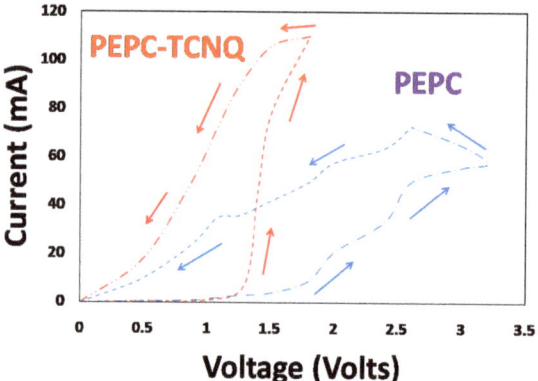

Figure 3. Current–voltage characteristics of Cu/PEPC/Ag and Cu/PEPC-TCNQ/Ag diodes measured at room temperature, showing a typical memory hysteresis curve for both devices. The sweeping path in the figure is shown by the arrow directions.

To further investigate the likely switching mechanism of the Cu/PEPC-TCNQ/Ag device, the ln(current) vs. ln(voltage) characteristics were further explored. Four well-defined charge transport regions were found as follows: (1) ohmic region; (2) trapped space charge region; (3) trapped filled voltage region (VTFL); and (4) trap-free space charge region [31–33], as shown in Figure 4. At an early stage, the devices follow ohmic response, leading to the trapped space charge limited current. This can be explained by the effects of mobility on charge concentration (holes) in the devices. With increasing voltage, the charge density increased. The sharp increase in the charge mobility with the hole density due to the trap filling was directly confirmed by [34]. By further increasing the voltage, the device reached a trapped-free region after passing through the trapped-filled voltage, as shown in Figure 3. The transformation of the trapped-space charge region to the trapped-free region is mainly accountable for the switching response of the Cu/PEPC-TCNQ/Ag device.

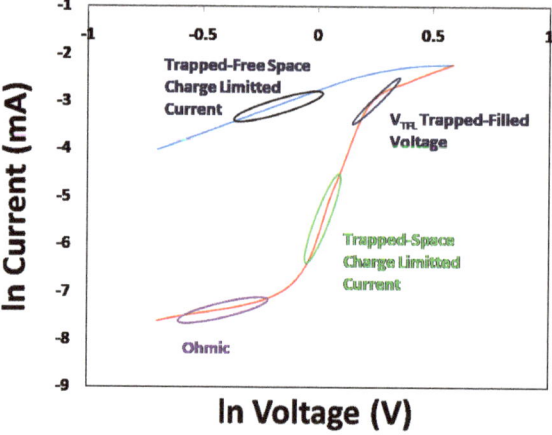

Figure 4. ln Current vs. ln Voltage characteristics of the Cu/PEPC-TCNQ/Ag diode, which clearly shows possible charge transport mechanisms as: (1) ohmic region; (2) trapped-space charge region; (3) trapped-filled voltage region; and (4) trapped-free space charge region.

The ON–OFF current ratio (I_{ON}/I_{OFF}) is an important factor to define the performance of memory devices and can be estimated from the resistive hysteresis of the current–voltage response of memory devices. Because the resistive hysteresis is originated by the random trapping and de-trapping of charged carriers inside the bulk region of the polymer, the ON–OFF ratio indicates the dynamic behavior of charged carriers. Hence, a high value of the ON–OFF current ratio indicates fast switching transition, whereas a low value of the ON–OFF current ratio reveals a slow switching transition inside the bulk region of the polymer. Figure 5 shows the ON–OFF current ratio for both Cu/PEPC-TCNQ/Ag and Cu/PEPC/Ag memory devices as a function of applied voltage. The ON–OFF current ratio increases with increasing voltage up to a certain voltage with a maximum ON–OFF ratio; then, the given ratio begins to decrease as voltage increases further for both memory devices due to the complex dynamic behavior of deep-level traps. However, the domain of the voltages and the range of the ON–OFF current ratios are different for the two memory devices, as shown in the figure. The maximum and steeper I_{ON}/I_{OFF} ratio (~100) are observed for the Cu/PEPC-TCNQ/Ag device at nearly 0.85 V, whereas the same maximum and broad I_{ON}/I_{OFF} ratio (~23) is observed for the Cu/PEPC/Ag device at 1.4 V. The figure clearly demonstrates that PEPC-TCNQ shows very fast memory switching with an excellent I_{ON}/I_{OFF} ratio compared to the PEPC-based memory device [35]. Traps for PEPC-TCNQ are simply defects or structural disorders in PEPC, which can be formed during the device fabrication process due to many known and unknown reasons, such as external or internal impurities and chemical defects [36]. It is generally accepted that the traps are energetically distributed between HOMO (highest occupied molecular orbital, valence band) and LUMO (lowest unoccupied molecular orbital, conduction band), and cause the electrical properties of polymer-based electronic devices to be degraded. Because the energy band gap of PEPC is much higher than the TCNQ energy band gap, TCNQ molecules may offer favorite sites for charge carriers hopping inside PEPC, which helps to improve the charge transport process at the applied voltage. Therefore, a visible improvement in current–voltage characteristics was observed for the Cu/PEPC-TCNQ/Ag device compared to the Cu/PEPC/Ag device.

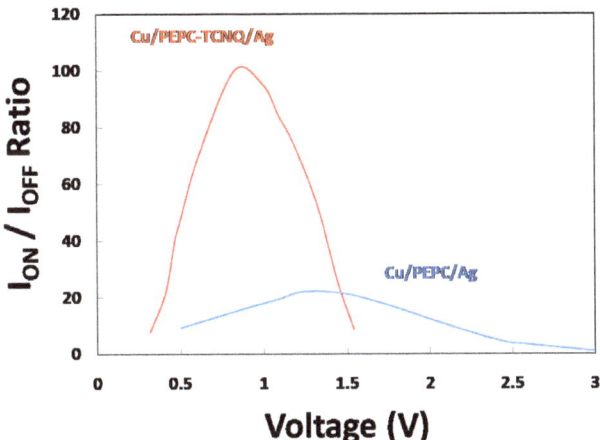

Figure 5. I_{ON} and I_{OFF} ratio (I_{ON}/I_{OFF}) for Cu/PEPC/Ag and Cu/PEPC-TCNQ/Ag devices as a function of applied voltage.

In the trapped-space charge region, when applied voltage is gradually increased from the ohmic region, the injected hole charges from the Cu electrode are caught by traps and are no longer available as free carriers, and, overall, the device offers high resistance (OFF state) to the flow of charges. By continuously increasing voltage in this region, traps are continuously filled and a stage is reached when nearly all traps are filled by injected carriers. Because filled traps do not play any further significant role in the charge transport mechanism, this stage is called the trapped filled voltage point. After this stage, both PEPC with or without TCNQ behaves like a trapped-free space charge limited current and offers very low resistance (ON state), yielding a higher I_{ON}/I_{OFF} ratio, as shown in Figure 5. However, the striking difference observed is that, for PEPC-TCNQ, the trapped-free region is achieved at a much earlier voltage (1.5 V) compared to PEPC (3.25 V), which clearly demonstrates that TCNQ improves the overall conductivity of PEPC for the ON state with fast switching for the Cu/PEPC-TCNQ/Ag memory device.

Similar to the fast switching characteristics, the stability in terms of retention of the ON and OFF states of the memory device for a longer period is another important parameter required for the performance of polymer memory devices. Therefore, the retention ON and OFF state resistance was investigated, measured from the current–voltage characteristics. For this purpose, the samples were heated up to 60 °C and held for 15 min to achieve thermal equilibrium. After thermal equilibrium, the resistance was measured for up to approximately 8 continuous hours (480 min) under the same laboratory environmental conditions. In addition, the temperature was continuously monitored to maintain it at 60 °C throughout the experiment for both devices. The output results are shown in Figure 6. Therefore, the marker points in Figure 6 correspond to the resistance of both memory devices in the ON/OFF states as a function of the same time interval. These resistance measurements were carried out in a very similar way to that discussed in Figure 3, but the resistances were calculated at 60 °C (1 volts) as a function of ageing time. The figure clearly shows that both ON state (relatively low) and OFF state (relatively high) resistance demonstrate nearly negligible degradation for both devices, illustrating the excellent stability of the Cu/PEPC-TCNQ/Ag- and Cu/PEPC/Ag-based memory devices under the given conditions.

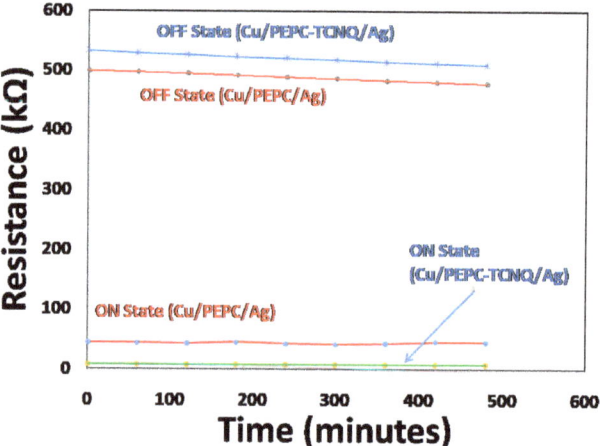

Figure 6. Degradation of ON state and OFF state resistance as a function of time for Cu/PEPC/Ag and Cu/PEPC-TCNQ/Ag devices after heating and holding the sample at 60 °C for ~1 h duration.

4. Conclusions

To investigate memory switching effects, the electrical characterization of a single layer memory device (metal–polymer–metal) containing PEPC doped with and without TCNQ charge transfer complexes was performed for Cu/PEPC/Ag and Cu/PEPC-TCNQ/Ag memory devices. From the current–voltage characteristics, it was observed that both devices showed memory switching effects. However, the doping of PEPC with TCNQ improved the quick response of the memory device, resulting in a high ON and OFF ratio ($I_{ON}/I_{OFF} \sim 100$). This memory switching effect may be due to the observed space charge limited behavior in the presence of trap distributions for both devices. Finally, both devices demonstrated a high degree of retention stability for the required memory operations.

Author Contributions: Conceptualization, A.N.M.A.; resources, A.N.M.A. and K.S.K.; validation, A.N.M.A. and K.S.K.; writing—original draft, A.N.M.A.; writing—review and editing, K.S.K. All authors have read and agreed to the published version of the manuscript.

Funding: This research received no external funding.

Institutional Review Board Statement: Not applicable.

Informed Consent Statement: Not applicable.

Data Availability Statement: The data presented in this study are available on request from the author.

Conflicts of Interest: The author declares no conflict of interest.

References

1. Ying, S.; Ma, Z.; Zhou, Z.; Tao, R.; Yan, K.; Xin, M.; Li, Y.; Pan, L.; Shi, Y. Device Based on Polymer Schottky Junctions and Their Applications: A Review. *IEEE Access* **2020**, *8*, 189646–189660. [CrossRef]
2. Zheng, Y.; Fischer, A.; Sawatzki, M.; Doan, D.H.; Liero, M.; Glitzky, A.; Reineke, S.; Mannsfeld, S.C.B. Introducing PinMOS Memory: A Novel, Nonvolatile Organic Memory Device. *Adv. Funct. Mater.* **2020**, *30*, 1907119. [CrossRef]
3. Moiz, S.A.; Alahmadi, A.N.M.; Karimov, K.S. Improved Organic Solar Cell by Incorporating Silver Nanoparticles Embedded Polyaniline as Buffer Layer. *Solid State Electron.* **2020**, *163*, 107658. [CrossRef]
4. Moiz, S.A.; Alahmadi, A.N.M.; Aljohani, A.J. Design of Silicon Nanowire Array for PEDOT:PSS-Silicon Nanowire-Based Hybrid Solar Cell. *Energies* **2020**, *13*, 3797. [CrossRef]
5. Melling, D.; Martinez, J.G.; Jager, E.W.H. Conjugated Polymer Actuators and Devices: Progress and Opportunities. *Adv. Mater.* **2019**, *31*, 1808210. [CrossRef]
6. Lewis, J. Material Challenge for Flexible Organic Devices. *Mater. Today* **2006**, *9*, 38–45. [CrossRef]
7. Kruijne, W.; Bohte, S.M.; Roelfsema, P.R.; Olivers, C.N.L. Flexible Working Memory through Selective Gating and Attentional Tagging. *Neural Comput.* **2021**, *33*, 1–40. [CrossRef]
8. Yang, Y.; Ouyang, J.; Ma, L.; Tseng, R.J.-H.; Chu, C.-W. Electrical Switching and Bistability in Organic/Polymeric Thin Films and Memory Devices. *Adv. Funct. Mater.* **2006**, *16*, 1001–1014. [CrossRef]
9. Kim, S.-J.; Lee, J.-S. Flexible Organic Transistor Memory Devices. *Nano Lett.* **2010**, *10*, 2884–2890. [CrossRef] [PubMed]
10. Li, L.; Ling, Q.-D.; Lim, S.-L.; Tan, Y.-P.; Zhu, C.; Chan, D.S.H.; Kang, E.-T.; Neoh, K.-G. A Flexible Polymer Memory Device. *Org. Electron.* **2007**, *8*, 401–406. [CrossRef]
11. Wu, H.-C.; Liu, C.-L.; Chen, W.-C. Donor–Acceptor Conjugated Polymers of Arylene Vinylene with Pendent Phenanthro[9,10-d] Imidazole for High-Performance Flexible Resistor-Type Memory Applications. *Polym. Chem.* **2013**, *4*, 5261–5269. [CrossRef]
12. Moiz, S.A.; Karimov, K.S.; Ahmed, M.M. Effect of Gravity Condition on Charge Transport Properties of Polymer Thin Film Deposited by Centrifugal Method. *Optoelectron. Adv. Mater. Rapid Commun.* **2011**, *5*, 577–580.
13. Moiz, S.A.; Ahmed, M.M.; Karimov, K.H.S.; Mehmood, M. Temperature-Dependent Current–Voltage Characteristics of Poly-N-Epoxypropylcarbazole Complex. *Thin Solid Film.* **2007**, *516*, 72–77. [CrossRef]
14. Chen, W.-C. *Electrical Memory Materials and Devices*; Polymer Chemistry Series; The Royal Society of Chemistry: London, UK, 2016. [CrossRef]
15. Saitov, S.R.; Amasev, D.V.; Tameev, A.R.; Kazanskii, A.G. A Simple Approach for Determination of Density of States Distribution in an Organic Photoconductor. *Org. Electron.* **2020**, *86*, 105889. [CrossRef]
16. Sun, Y.; Wen, D.; Sun, F. Influence of Blending Ratio on Resistive Switching Effect in Donor-Acceptor Type Composite of PCBM and PVK-Based Memory Devices. *Org. Electron.* **2019**, *65*, 141–149. [CrossRef]
17. Jeong, D.S.; Thomas, R.; Katiyar, R.S.; Scott, J.F.; Kohlstedt, H.; Petraru, A.; Hwang, C.S. Emerging Memories: Resistive Switching Mechanisms and Current Status. *Rep. Prog. Phys.* **2012**, *75*, 076502. [CrossRef] [PubMed]
18. Ling, Q.-D.; Liaw, D.-J.; Zhu, C.; Chan, D.S.-H.; Kang, E.-T.; Neoh, K.-G. Polymer Electronic Memories: Materials, Devices and Mechanisms. *Prog. Polym. Sci.* **2008**, *33*, 917–978. [CrossRef]
19. Majumdar, H.S.; Bandyopadhyay, A.; Bolognesi, A.; Pal, A.J. Memory Device Applications of a Conjugated Polymer: Role of Space Charges. *J. Appl. Phys.* **2002**, *91*, 2433–2437. [CrossRef]
20. Murari, N.M.; Hwang, Y.-J.; Kim, F.S.; Jenekhe, S.A. Organic Nonvolatile Memory Devices Utilizing Intrinsic Charge-Trapping Phenomena in an n-Type Polymer Semiconductor. *Org. Electron.* **2016**, *31*, 104–110. [CrossRef]
21. Bozano, L.D.; Kean, B.W.; Beinhoff, M.; Carter, K.R.; Rice, P.M.; Scott, J.C. Organic Materials and Thin-Film Structures for Cross-Point Memory Cells Based on Trapping in Metallic Nanoparticles. *Adv. Funct. Mater.* **2005**, *15*, 1933–1939. [CrossRef]
22. Moiz, S.A.; Ahmed, M.M.; Karimov, K.S. Estimation of Electrical Parameters of OD Organic Semiconductor Diode from Measured I-V Characteristics. *ETRI J.* **2005**, *27*, 319–325. [CrossRef]
23. Karimov, K.S.; Ahmed, M.M.; Moiz, S.A.; Babadzhanov, P.; Marupov, R.; Turaeva, M.A. Electrical Properties of Organic Semiconductor Orange Nitrogen Dye Thin Films Deposited from Solution at High Gravity. *Eurasian Chem. Technol. J.* **2007**, *5*, 109–113. [CrossRef]

24. Xu, X.; Li, L.; Liu, B.; Zou, Y. Organic Semiconductor Memory Devices Based on a Low-Band Gap Polyfluorene Derivative with Isoindigo as Electron-Trapping Moieties. *Appl. Phys. Lett.* **2011**, *98*, 063303. [CrossRef]
25. Potember, R.S.; Poehler, T.O.; Cowan, D.O. Electrical Switching and Memory Phenomena in Cu-TCNQ Thin Films. *Appl. Phys. Lett.* **1979**, *34*, 405–407. [CrossRef]
26. Karimov, K.S. Transversal Tensity Resistive Effect in TEA (TCNQ)2 Crystals. *Synth. Met.* **1991**, *44*, 103–106. [CrossRef]
27. Karimov, K.S. Electrical Conductivity of TEA(TCNQ)2 Crystals under Uniaxial Tension and Compression. *Solid State Commun.* **1994**, *89*, 1029–1031. [CrossRef]
28. Ahmed, M.M.; Karimov, K.S.; Moiz, S.A. Temperature-Dependent I-V Characteristics of Organic-Inorganic Heterojunction Diodes. *IEEE Trans. Electron. Devices* **2004**, *51*, 121–126. [CrossRef]
29. Erlbacher, T.; Jank, M.P.M.; Ryssel, H.; Frey, L.; Engl, R.; Walter, A.; Sezi, R.; Dehm, C. Self-Aligned Growth of Organometallic Layers for Nonvolatile Memories: Comparison of Liquid-Phase and Vapor-Phase Deposition. *J. Electrochem. Soc.* **2008**, *155*, H693. [CrossRef]
30. Zhang, Q.; Kong, L.; Zhang, Q.; Wang, W.; Hua, Z. The Effect of Heat Treatment on Bistable Ag-TCNQ Thin Films. *Solid State Commun.* **2004**, *130*, 799–802. [CrossRef]
31. Rose, A. Space-Charge-Limited Currents in Solids. *Phys. Rev.* **1955**, *97*, 1538–1544. [CrossRef]
32. Campbell, A.J.; Bradley, D.D.C.; Lidzey, D.G. Space-Charge Limited Conduction with Traps in Poly(Phenylene Vinylene) Light Emitting Diodes. *J. Appl. Phys.* **1997**, *82*, 6326–6342. [CrossRef]
33. Moiz, S.A.; Younis, W.A.; Yilmaz, K.S.K.E.-F. Space Charge–Limited Current Model for Polymers. In *Conducting Polymers*; Khan, I.A., Ed.; IntechOpen: Rijeka, Croatia, 2016; p. 5. [CrossRef]
34. Toman, P.; Menšík, M.; Bartkowiak, W.; Pfleger, J. Modelling of the Charge Carrier Mobility in Disordered Linear Polymer Materials. *Phys. Chem. Chem. Phys.* **2017**, *19*, 7760–7771. [CrossRef] [PubMed]
35. Shah, M.; Karimov, K.S.; Ahmad, Z.; Sayyad, M.H. Electrical Characteristics of Al/CNT/NiPc/PEPC/Ag Surface-Type Cell. *Chin. Phys. Lett.* **2010**, *27*, 106102. [CrossRef]
36. Kadashchuk, A.; Weiss, D.S.; Borsenberger, P.M.; Ostapenko, N.; Zaika, V.; Skryshevski, Y. Effect of Extrinsic Traps on Thermally Stimulated Luminescence in Molecularly Doped Polymers. *Synth. Met.* **2000**, *109*, 177–180. [CrossRef]

www.ingramcontent.com/pod-product-compliance
Lightning Source LLC
LaVergne TN
LVHW070158100526
838202LV00015B/1964